CLASSICAL THEORY

The Theory of Interacting Systems | Volume 2

The Theory of Interacting Systems

CLASSICAL THEORY

The Theory of Interacting Systems | Volume 2 | Paul McEvoy

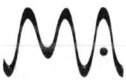

MICROANALYTIX
www.microanalytix.com   2002   San Francisco

This book is dedicated to my family
Ruth, Heather and Kyle.

ISBN 1-930832-02-8

First Edition, revision 1.3

Cover and title page designs and
production by Julie R. Wheeler.

This book was typeset with $\mathcal{AMS}$-TeX,
the TeX macro system of the
American Mathematical Society.

Illustrations were done with PiCTeX,
the TeX macro package of Michael Wichura.

Typesetting, line illustrations, and production
of this book were done by the author.

Translations are by the author,
unless otherwise noted.

CIS Preface

This is the second volume in a six volume series. The series as a whole is titled the *Theory of Interacting Systems* and designated by TIS. It is concerned with the relation of the microscopic world of particles to the macroscopic world of our experience. The first volume has already appeared. Three other volumes have been written and the sixth is in progress. This volume is devoted to the development of a general time-dependent thermodynamics based on the classical statistical mechanics of interacting systems of particles. The resulting theory is a local, nonequilibrium thermodynamics that yields equilibrium thermodynamics when the underlying particle system is described by the equilibrium particle distribution.

The work in this series stemmed from a statement made by Niels Bohr in his 1930 lecture to the Faraday Society in England. He felt that new classical ideas, similar to the concept of complementarity in quantum mechanics, would play a role in the proper understanding of the relation of thermodynamics to statistical mechanics. When I turned to the literature for an explanation of how thermodynamics is related to the interacting particles described by statistical mechanics, I found that no general theory of the connection between the quantities defined in thermodynamics and the underlying particles existed. The existing theory of this relationship at equilibrium cannot deal with change and no generalization into non-equilibrium thermodynamics had been widely considered as successful. The work of Maxwell and then Boltzmann came the closest to providing a nonequilibrium theory, but their work was marred by persistent paradoxes associated with the conflict between the irreversibility of thermodynamics and the reversibility and recurrence properties of the equations of particle motion. Examining these issues, I found that the standard resolutions offered for the paradoxes surrounding irreversibility, recurrence, the second law of thermodynamics and entropy, were neither satisfying physically nor mathematically cogent.

Because Bohr's analysis of the epistemological issues in physics was the impetus for this investigation, his understanding of the fundamental issues has played an important role. For this reason, the first volume in the series was concerned with the Bohr's analysis of the connections between knowledge, theory, and experiment, in light of the changes in physics brought by quantum mechanics.

In the face of interpretational problems associated with the uncertainty relations and the wave-particle duality, Bohr was concerned with the stating conditions sufficient for obtaining unambiguous answers in experiments. Bohr addressed two important issues in his analysis of this issue that have significant implications for thermodynamics and statistical mechanics. The first is the requirement that the variables associated with distinct interacting macroscopic systems of particles are independent. This is in accord with the standard practice in physics of singling out a particle system for study and representing the forces exerted on it by its environment using arbitrary potentials and boundary conditions. In quantum theory, the requirement of independent variables itself turned out to be important in determining the representation of the particle system in the quantum formalism and the interpretation of experiments. This independence requirement also plays a central role in the Theory of Interacting Systems as a condition on the mathematical representation of interacting systems.

The second issue that Bohr addressed with significance to thermodynamics is concerned with the need to distinguish the proper domains of application for our concepts. Pursuing this in the TIS investigation of thermodynamics led to a distinction between concepts that are macroscopic in origin, such as the volume occupied by a system of particles, and those that are microscopic, such as the particle momentum. This distinction plays an important role in conjunction with the independence requirement in understanding the entropy and the resolution of the paradoxes of thermodynamics.

In a review of the literature, undertaken as part of the examination of the foundations of thermodynamics in TIS, other issues came readily to hand. As I looked more deeply into the formulas of standard equilibrium thermodynamics, for example, I could not see clearly how they emerged from the underlying particle dynamics. As one example of this problem of interpretation, I recall that I was often confronted in the literature by formulas of the form

$$\mathrm{P} = - \int d\Gamma \, F(\Gamma, t) \frac{\partial \mathcal{H}(\Gamma, t)}{\partial V}.$$

This formula expresses the pressure P of a system of particles as minus the average over the particle probability distribution density F of the volume derivative of the system Hamiltonian energy $\mathcal{H}$. For a system with $\mathcal{N}$ particles, the quantity $d\Gamma$ represents a differential for the particle variables taking values in a $6\mathcal{N}$-dimensional particle phase space. The quantity $F(\Gamma, t)$ represents the system probability distribution density as a function of these phase space variables and the time. $\mathcal{H}(\Gamma, t)$ is the system Hamiltonian and $\partial \mathcal{H}/\partial V$ is the derivative of this Hamiltonian with respect to the volume.

There is a major problem with this formula and others like it because the particle system Hamiltonian is not a function of the volume. The volume we attribute to a collection of particles is based on a macroscopic choice of where to place the boundaries of the macroscopic object to which they belong. There is no intrinsic connection between this choice and the properties of the underlying microscopic mechanics. In other words, the volume of a collection of particles is not part of the Hamiltonian mechanics of the particle system. There were attempts to get around this problem by Clausius, Boltzmann, and Gibbs, by including a set of "constraint parameters" in the system Hamiltonian that in most cases used a large potential at the boundary to confine the particles. As discussed in detail in the book, this solution is not acceptable in TIS because it introduces macroscopic parameters into the microscopic quantities that determine the particle motion. In short, I could attach no physical meaning to the symbol $\partial H/\partial V$ in the standard theory.

I set out in 1975 to resolve these problems with the standard theory. This turned out to be a task far greater than I had imagined. During the development of the ideas that became TIS, the need to root thermodynamics closely in particle mechanics and the development of an associated mathematical formalism for nonequilibrium thermodynamics slowly emerged. Understanding the proper approach to thermodynamics and its relation to statistical mechanics required that I shake off one by one many of the viewpoints that were part of the body of accumulated knowledge established during the long history of equilibrium thermodynamics. I was very reluctant to do this because some of these viewpoints had been considered for generations to be fundamental and well-established principles at the root of thermodynamics. I realized that the more differences there were between the previous work and the current theory, the harder it would be to gain acceptance for the new ideas.

A number of concerns eventually came to play a role in shaping the emerging theory. While these can be expressed in a logical order, the actual evolution of the ideas underlying the theory did not follow a logical progression. In response to the problem illustrated above concerning formulas used in thermodynamics that had no meaning, it became a requirement that every physical quantity in thermodynamics must have a microscopic analog that is a function of the particle phase space coordinates. Using a general time-dependent particle distribution density, a thermodynamic quantity was defined as the time-dependent phase average of its analog over the distribution. The next step was to compute the evolution of these thermodynamic quantities in time as a consequence of the evolution of the underlying particle distribution. This led to an analysis of the boundaries of systems, their motions, the flows of physical quantities across these boundaries, and the development of a formalism to account for all of this. The formalism supports localization of those quantities that are properties of the particles, so the result was a local nonequilibrium thermodynamics in spacetime that can track both the forces between macroscopic bodies and the flows of conserved quantities between them.

The next concern was the problem of how to define the derivatives of thermodynamic quantities with respect to the macroscopic volume and temperature. Because the volume and temperature are not part of the Hamiltonian description of the system, it was necessary to find operators, acting at the particle level, which would map the analog of a physical quantity to an analog corresponding to its derivative with respect to volume or temperature. The phase average of the resulting analog is then identified as the thermodynamic quantity corresponding to the derivative of the original thermodynamic quantity.

Carrying this program through in the case of general particle distributions evolving in time eventually yielded a complete general nonequilibrium thermodynamics. The basic components of the resulting thermodynamics differ in some respects from those of standard equilibrium thermodynamics, but much of that structure is retained. Moreover, the quantities and relations defined in equilibrium thermodynamics are obtained from the general formalism when the underlying particle system is described by the equilibrium distribution.

The basic work on this book was completed in 1988. Although the issues concerning the structure of the basic theory that had come up were resolved, the next set of questions was concerned with whether the formalism could account for the full range of thermodynamic issues. These issues included the theories of phase boundaries and thermodynamic surfaces, and whether particular explicit force laws between particles and an approximation procedure could be used to obtain quantitative results. There were also more distant questions of whether this work could be successfully generalized to quantum mechanics.

After a number of fruitless attempts to publish this work as a standalone volume, I gave up and turned to the equilibrium theory of phase boundaries, thermodynamic surfaces, and what I called "computable thermodynamics". This work took 8 years and evolved into Volume 3, Equilibrium Theory. That volume was completed in 1996. The volumes on Bohr, quantum mechanics, and quantum thermodynamics followed in the next 4 years.

Throughout this investigation into thermodynamics and statistical mechanics, the work of original authors has proved crucial to sorting out the fundamental issues and to penetrating the glosses of later writers over the fundamental problems and paradoxes that were at its root. The task of making conceptual sense of the progression of ideas in this area has been assisted by the work of a number of excellent historians of physics. They have provided useful accounts of some of the emerging physical ideas, and the work of those that criticized them, and sometimes pointed the way to wonderful treasures in a forgotten paper of long ago.

Table of Contents and Index.

The analytical Table of Contents presents the major headings within each chapter. This is the place to locate the treatments of various concepts important to the analysis of Bohr's work. The Index is a search tool that lists exhaustively the locations where various ideas are discussed or mentioned.

Bibliography.

Bibliographical references take the form Boltzmann [**1872**]. The authors are listed in the Bibliography in alphabetical order with the year of the publication in the left margin.

Translation.

The French quotation by Henri Poincaré that appears at the beginning of Part IV is

La théorie cinétique des gaz laisse encore subsister bien des points embarrassants pour ceux qui sont accoutumés à la riguer mathématique; bien des résultats, insuffisamment précisés, se présentent sous une forme paradoxale et semblent engendrer des contradictions quis ne sont d'ailleurs qu'apparentes.

A translation is

The kinetic theory of gases still contains many aspects troublesome to those accustomed to mathematical rigor; many of its results, insufficiently specified, are presented in a paradoxical form and appear to give rise to contradictions that are really only apparent.

Acknowledgments.

A number of people have contributed directly and indirectly to the writing of this book. To them I express my deep appreciation. In particular, I would like to especially thank my partner Ruth Roberts for her ongoing support during the development of the whole theory and for comments on an earlier version of this book. I also express appreciation to my children Heather and Kyle for their support during the long years it has taken to complete this work.

I would also like to thank Joan Aufiero, Dr. Jean Chambers, Dr. Jack Martinelli, Mary Merryman, M. F. Suzanne Pinette, Dr. David Vampola, and Charles Reynolds for valuable bibliographic assistance. In addition, I wish to express my appreciation to Clifford Truesdell IV for his support for this project, some bibliographic assistance, and other help. I express my appreciation also to Professor Clifford Truesdell III for some encouragement concerning an earlier version of the theory.

Discussions with Dr. Jane Schulman on statistics and Mr. M. Shanahan Lavine on the work of Prigogine were helpful as well. Dr. Larry W. Lamoreux and Mr. Julian Allen provided computer assistance in the production of an earlier version of this manuscript. Poynter McEvoy provided some equipment used in the production of an early version of this manuscript. I would also like to acknowledge many helpful conversations with Dr. Helen Desfosses concerning the attempt to publish this manuscript in 1988.

Paul McEvoy
MicroAnalytix, 2002

Contents

Part II: Classical and Equilibrium Thermodynamics.

Part III: The Maxwell-Boltzmann Approach.

Part I

The Theory of
Interacting Systems

Something there is that doesn't love a wall.

…

Before I'd built a wall I'd ask to know
What I was walling in or walling out.

Robert Frost, *Mending Wall,* 1914

CHAPTER 1

Introduction to the Theory
of Interacting Systems

1.1 Prologue

We live in a changing world. Mountains rise and are eventually washed away. Oceans recede, sand dunes become forests, and jungles turn into deserts. In this seeming chaos, some have blamed the butterflies in Brazil for answering the call of a Lorentz attractor and causing a gale in the Bahamas by flapping their wings. Others have found a rising tide of entropy in the deterministic march of atoms to the heat death of the universe. It is this world that is the subject of thermodynamics.

In spite of the change that is part of our world, thermodynamics has been unable to encompass it. It describes a world reminiscent of Victorian steam engines that move infinitely slowly when they move at all. Attempts to add vitality to this lifeless world have resulted in paradox and contradiction.

1.2 Introduction

Thermodynamics is the macroscopic theory of matter. It describes the exchange of physical quantities between macroscopic bodies and tracks the relationships between these quantities in a body as the temperature, volume, or pressure of the body changes. While this definition includes domains that have traditionally been part of hydrodynamics and macroscopic mechanics, it fits the natural range of elements that belong to a thermodynamics that is closely associated with the mechanics of matter.

Thermodynamics has long maintained both a purely macroscopic approach, based on empirical or phenomenological laws, and a hybrid approach including both macroscopic and microscopic components, based on obtaining thermodynamic laws from the description of the underlying matter by particle statistical mechanics or continuum mechanics. The Euler-Cauchy and Navier-Stokes equations are two examples of sets of equations based on a macroscopic phenomenological analysis. The second law of thermodynamics is often considered a phenomenological law as well.

Until almost the middle of the twentieth century, the phenomenological approach was viewed as better established and secure than those approaches that tried to obtain thermodynamic laws from the behavior of matter at a microscopic level. A major example of this practice appeared in the groundbreaking work of John von Neumann [**1932c**] on quantum mechanics. Von Neumann employed equations of phenomenological thermodynamics to guide the analysis and determine in part the shape of the quantum thermodynamics under development. This practice survives today in papers that use entropy to explain why particles in certain circumstances line up in particular ways.

Other workers who tried to explain the laws concerning the behavior of macroscopic bodies in terms of the actions of microscopic components that compose them focused on the microscopic components themselves. While theories based on postulated microscopic particles had appeared in ancient India and Greece, scientific inquiries into the behavior of particles in the modern sense began in the sixteenth and seventeenth centuries. These explanations divided into two main camps—with some supporting the view that matter is a collection of interacting microscopic particles and others the view that it is a continuous medium. Over the centuries, microscopic particle explanations have generally been favored over continuum explanations, but continuum theories never died out completely.

A group calling themselves energeticists initiated a continuum approach in the nineteenth century. They felt that explanations of macroscopic phenomena in terms of unobservable microscopic particles were philosophically and scientifically unacceptable. They obtained equations of thermodynamics from rational continuum mechanics and the continuum theories of elasticity and electricity.[1] More recent theories based on rational continuum mechanics, such as the one by Bernard D. Coleman and Walter Noll [**1963**], have found adherents who do not deny the existence of particles at a lower level. Each version of continuum thermodynamics postulates a different set of assumed axioms and constitutive relations that describe the behavior of matter in bulk. Each version also has some method for defining a temperature scale and the entropy. However, the absence of some definitive and independent way of establishing the properties of these continua meant that these theories were never very popular. Continuum theories will play a very small role in this series of volumes.

Particle based thermodynamics has a long history that spans at this point well over 150 years and its roots are more than 200 years older. Most authors have viewed the issues surrounding its basic principles as long settled. In spite of this belief, a number of important problems in the conceptual foundations of the theory have remained unresolved during the past 150 years. The central problem is that a version of thermodynamics compatible with the time-dependent mechanics of the underlying particles has never been found.

Physicists have been aware of the need to reconcile thermodynamics and mechanics. Several prominent authors in the middle to late nineteenth century attempted to address mathematically the relation of thermodynamics to mechanics. These authors tried to show that the second law of thermodynamics is in some way a direct consequence of the evolution of a particle system in accord with Hamiltonian mechanics. However, the proofs they provided all required specialized and implausible assumptions, so later authors rejected them as failures.

Writers working on these issues in the late nineteenth century expressed the central problems in theories relating thermodynamics to mechanics in the form of several paradoxes. Two of these were the reversibility and recurrence paradoxes. The reversibility paradox refers to the incompatibility of the thermodynamic principle that the entropy in a system will always increase, unless the system is at equilibrium, with the mechanical reversibility of the laws of motion of the underlying particles. The second paradox is concerned with the implications of Poincaré's recurrence theorem, which states that the mechanical state of a closed system in a fixed volume will return to the neighborhood of an initial mechanical state in a finite time. The attempts by Ludwig Boltzmann, James Clerk Maxwell, Josiah-Willard Gibbs, and a long list of later authors and textbook writers, to resolve these and other paradoxes, such as Gibbs' paradox concerning the entropy, have resulted in the employment of a number of physical assumptions and mathematical mechanisms that are not very plausible when scrutinized carefully.[2]

---

[1]See E. Mach [**1896**] for a discussion of energetics.

[2]For an accessible treatment from the standard perspective, see C. J. Thompson [**1972**]. For a succession of treatments of statistical mechanics and thermodynamics in

Physics abhors a paradox. Galileo criticized the Aristotelian theory that heavier bodies fall faster than light bodies by a thought experiment that showed the paradox implicit in it. Albert Einstein discovered special relativity while reconciling the transformation properties of electrodynamics and mechanics. Niels Bohr struggled with the paradoxes of the wave-particle duality in quantum physics.

Paradoxes describe conceptual mismatches—often between two deeply held views of the nature of things. Their resolution tells us something important about our relation to the world. Galileo was able to resolve the paradoxes associated with Aristotle's theory by presenting an alternative theory, Einstein's special relativity successfully joined mechanics and electrodynamics, and Bohr's principle of complementarity kept the quantum contradictions at bay. However, thermodynamics has not been so fortunate. The answers to the paradoxes of thermodynamics provided by Maxwell, Boltzmann, Gibbs, and others after them, have not won the hearts and minds of their successors. Each new textbook makes obligatory reference to these paradoxes and makes proposals for their resolution, sometimes with a new twist, but these solutions have never seemed fully convincing.

An examination of standard equilibrium theory shows that there are a number of components of the theory that cannot be attached to the underlying particles and their behavior. A case in point is the free energy. It is demonstrated in Chapter 19 that, by contrast with the energy and momentum, there is no particle analog corresponding to the free energy. Because the free energy is central to the theory of phase change in standard thermodynamics, this result implies that an alternative theory of phase change is needed if equilibrium thermodynamics is to be understood from the particle perspective. A similar problem is the lack of an interpretation of the thermodynamic derivatives of physical quantities with respect to volume and temperature in terms of the underlying particle description. The difficulty, in brief, is that the formalism of standard equilibrium thermodynamics has never been able to make a complete and proper connection with the mechanics of the underlying particles. This has led to the fragmentation of work in thermodynamics into separate domains founded on unrelated concepts and incompatible formalisms based on different perspectives on the relation of thermodynamics to statistical mechanics.

The most fundamental difficulty with standard thermodynamics stems from the fact that while we live in a changing world, the majority of thermodynamic theories, and the best-developed ones, are equilibrium theories. To account for change in these theories, the concept of infinitely slow quasistatic processes was introduced in which the equilibrium description of the system remains valid during the evolution of the system in time. However, almost no

---

the twentieth century, see R. C. Tolman [1938], J. Kirkwood [1949], T. L. Hill [1958], and M. Toda, R. Kubo and N. Saito [1983].

real processes come close to fitting the assumptions underlying the employment of quasistatic processes, so equilibrium thermodynamics is essentially useless in accounting for change.

Other attempts to extend equilibrium thermodynamics into a time dependent nonequilibrium theory have modified microscopic particle mechanics in order to obtain the irreversible mechanical behavior thought to be essential for the explanation of thermodynamic irreversibility. In other cases, the behaviors of some macroscopic thermodynamic functions are simply postulated in conjunction with a series of unprovable assumptions about how these functions are related to the underlying particle distribution and its evolution in time. None of these theories has provided an adequate demonstration that its predictions are a consequence of the mechanics of the underlying particles—or even that they are consistent with this mechanics.

A consequence of the lack of an alternative to equilibrium thermodynamics has been the adoption of a number of beliefs and practices in thermodynamics that are not warranted by the underlying particle mechanics and often contradict it. These beliefs are sometimes accorded the status of "theorems" or empirically derived "phenomenological axioms". Typical examples are: 'the thermodynamic limit is necessary to obtain thermodynamics from statistical mechanics' or 'entropy will always increase in an isolated system that is not at equilibrium'. Other examples include the use of unsupported assumptions in many thermodynamic proofs and derivations. Examples are the assumptions that the system is at 'local thermodynamic equilibrium' or that certain kinds of particle correlations decay more rapidly than others do. These examples, and the others given above, will be discussed in detail in succeeding chapters.

In a situation like this, Bohr emphasized that it is necessary to examine the roots of our conceptions and our understanding of experiments in order to extract the paradigmatic content from them. He had a great interest in thermodynamics and a special fondness for the work of Gibbs. Bohr [1930] felt that in the relation of thermodynamics to statistical mechanics an aspect of complementarity might be found that is similar to that in quantum theory. It was Bohr's suggestion that prompted this reexamination of thermodynamics and statistical mechanics.

In order to reunite disparate viewpoints, compare results, and resolve questions, it is necessary to examine the original work in each of these areas within a common conceptual framework and formalism. This has led to a reexamination of much of the history of thermodynamics with an eye to fitting the concepts that were introduced over many years into a common mathematical formalism. The ultimate goal of the Theory of Interacting Systems is to provide a framework in which thermodynamics and mechanics may be united.

Changing the 150 year old structure of equilibrium thermodynamics to accommodate a changing world requires deconstructing it and building it anew on the back of the mechanical elephant that carries the world. Such changes, however, come at a cost that may be for some to high a price to pay. The reconciliation of thermodynamics and mechanics is required to resolve the old paradoxes. This means coming to terms with a different view of entropy and equilibrium and a necessary intrusion of epistemology into classical physics. The outcome is a natural thermodynamics in space and time that does not include arbitrary restraints and assumptions.

Epistemological issues intrude when we divide the world into pieces for separate study. This division is connected with what Bohr has called the subject/object distinction, which is used in this context to disentangle one macroscopic system from others that are interacting with it. The resulting separation is fundamental to showing how the irreversibility of thermodynamics can coexist with the reversibility of the underlying mechanics. It turns out, unsurprisingly, that these issues are bound up with how we understand entropy and the role it plays in our theories.

This series of volumes devoted to the Theory of Interacting Systems (TIS) shows how a collection of particles can be subdivided into interacting macroscopic bodies and how these bodies can be described in space and time by a macroscopic formalism that is explicitly consistent with the mechanical formalism chosen to describe the underlying particle system. The analysis of Bohr's work in the first volume is referred to as RSO, the classical version of the thermodynamics and statistical mechanics of interacting systems presented in this book is designated by CIS, the equilibrium version of the theory in the third volume is indicated by EIS, the quantum version in the fourth volume is designated by QIS, and the fifth volume on the quantum thermodynamics of interacting systems is referred to as QTS. The resulting general formalism developed in this series, where the term 'general' refers to elements of the theory that can either be time-dependent or at equilibrium, is free of the defects of current theories and not subject to the paradoxes.

1.3 The Development of Equilibrium Thermodynamics

Thermodynamics originated in the seventeenth century in the work of Daniel Bernoulli on the kinetic theory of pressure and prior studies of equations of state of matter. One such study resulted in the Robert Boyle-Richard Towneley law, which showed a quantitative relation between the pressure in a system and the system volume. An important advance from the thermodynamic perspective was the separation of the concepts of heat and temperature at the end of the eighteenth century. This led to the discovery of latent and specific heats and the development of classical thermodynamics. Engineering studies concerned with steam engines used these results to describe the flows of heat and work between macroscopic bodies.

Classical thermodynamics emerged at the beginning of the nineteenth century with the theory of calorimetry. The theory of calorimetry was concerned with heat and the flow of heat in relation to the temperature, volume, and pressure of a body and their changes. The 1824 work of Carnot on cyclic heat engines established a quantitative relation between heat and work in the operation of an engine. A quantitative relation was introduced into the study of phase phenomena in 1834 by Émile Clapeyron.

One of the first definite particle theories used in thermodynamics was the particulate theory of Pierre Simon Laplace in which heat was represented by caloric particles. Other particle theories employed vortex atoms to provide explanations for macroscopic physical phenomena.

By the middle of the nineteenth century, macroscopic equilibrium thermodynamics succeeded classical thermodynamics. Equilibrium thermodynamics included the newly discovered law of the conservation of energy and a phenomenological law concerned with the direction of thermodynamic processes involving the flow of heat. These were soon called the first and second laws of thermodynamics. About the same time, kinetic theory took shape in the work of Auguste Krönig and Rudolph Clausius. It was based on the postulate that matter is composed of impenetrable microscopic particles in incessant motion. Krönig and Clausius introduced simple particle distributions and mean free path calculations to compute values for diffusion, viscosity, and other quantities in systems at equilibrium.

In the last half of the nineteenth century, thermodynamics split into theories based on a particulate perspective and those based on a purely macroscopic or phenomenological viewpoint. Maxwell's [**1860**], [**1866**], introduction of statistical particle distributions into theoretical calculations of the transport of physical quantities gave particulate theories a strong conceptual boost. This was soon followed by the use of probability distributions and the development of a dynamical theory in the work of Boltzmann [**1872**]. Further developments included an association of the entropy with probability in the context of discrete particle distributions by Boltzmann [**1877b**] and later calculations of viscosity and diffusion using approximations of Boltzmann's equation by others.[3]

On the macroscopic side, Clausius [**1865**] defined a quantity S he called the entropy and stated an inequality $dS \geq dQ/T$ that governs its change. This development provided a physical interpretation of the macroscopic second law and a formalism for using it. Gibbs [**1875**] introduced and codified a number of results concerning the behavior of macroscopic thermodynamic

---

[3]For more historical information on the introduction of statistical ideas into kinetic theory and thermodynamics, see Stephen Brush [**1970**], pp. 583–654. For more on the introduction of probability into physics, and the implications that follow, see the discussions of different aspects in RSO, EIS, and QTS.

systems. His work included the use of thermodynamic potentials, a discussion of stability principles in thermodynamic systems, and a formalism for predicting the formation of thermodynamic phase boundaries. Concern with the fact that particles are unobservable entities led a number of researchers during this period, including William John Macquorn Rankine in the middle of the century, followed by Ernst Mach, and then Wilhelm Ostwald at the end of the century, to work out a purely macroscopic approach to thermodynamics mentioned above as *energetics*. Energeticists rejected particles altogether as unobservable and unnecessary to thermodynamics. The controversy between those who worked with particle based theories and the energeticists was not resolved until experiments demonstrating the existence of particles were performed around the beginning of the twentieth century.

The microscopic approach was stimulated during the nineteenth century by the development of a sophisticated mathematical apparatus for Newtonian mechanics. These innovations appeared first in the mechanics of Siméon-Denis Poisson, Pierre Simon Laplace, and Joseph Louis Lagrange, and then in the mechanics of William Rowan Hamilton. Hamilton's work subsequently evolved into Hamiltonian statistical mechanics. These developments allowed mechanics to be applied to the explanation of Brownian motion, viewed as the effects of individual atoms on larger observable particles, and to the explanation of other phenomena in terms of the behavior of microscopic particles.

The work of Gibbs [**1902**] began with a general approach to thermodynamics based in statistical mechanics and then quickly focused on equilibrium theory. Gibbs was able to reproduce many of the formulas of macroscopic equilibrium thermodynamics by showing how thermodynamic quantities, defined as phase averages over analog functions, led to these relationships when the particles were represented by the equilibrium distribution. Gibbs' work added to and supported the body of results associated with equilibrium theory. However, Gibbs' thermodynamics was not a complete solution. One element missing from Gibbs' equilibrium theory, for example, was an effective method of defining the volume derivative of a thermodynamic quantity as a thermodynamic quantity with its own analog.

From an experimental perspective, systems at equilibrium have the useful property that they are in a state that is reproducible by macroscopic means and the quantities observed are less subject to the transient effects of changing flows of energy, momentum, and matter. This means that quantities such as the specific heat of a substance need to be measured at equilibrium in order to obtain reproducible results. For similar reasons, it is necessary to use equilibrium theory with a formalism for predicting thermodynamic phases and for the study of the influence of the microscopic parameters that determine the strength and range of interparticle forces on macroscopic physical quantities.

1.4 Nonequilibrium Thermodynamics and Statistical Mechanics

In spite of the dominance of theories of equilibrium thermodynamics, a number of attempts have been made to create a nonequilibrium thermodynamics based on a nonequilibrium statistical mechanics. The initial particle theories, in the form of the transport equations of Maxwell [**1866**] and the dynamic theory of Boltzmann [**1872**], are both nonequilibrium theories. Since that time, a number of other nonequilibrium theories have been proposed. The main difficulties for these theories are the complexity of the laws of motion for a collection of 10^{23} particles and the difficulty of extracting significant regularities from this situation. An explicit solution of the system evolution equation for the particle distribution is available only in the equilibrium case. This implies that the nonequilibrium theories must employ some significant approximations or find another way to establish relationships between thermodynamic quantities if they are to lead to definite and usable results.

Early nonequilibrium results in particle based theories were obtained by using approximations of the Boltzmann evolution equation or of Hamilton's evolution equations. These theories include the work of Sydney Chapman and David Enskog approximating the Boltzmann equation, the Fokker-Planck equation, Wolfgang Pauli's master equation, and theories of "near-equilibrium states". Other theories of this type attempted generalizations of Gibbs canonical formalism or used specially constructed particle distributions that depend on macroscopic thermodynamic parameters characterizing the system. Still other theories were based on special assumptions concerning how states can be decomposed, used asymptotic results with special limits, or made specific modifications of the particle dynamics. Each of these theories has an evolution equation that follows from its particular assumptions.

There have been attempts to create purely macroscopic nonequilibrium theories without particle assumptions by selecting an appropriate set of axioms. Some of the early statements of the second law took the form of the impossibility of constructing perpetual motion machines of particular kinds. Other versions spoke of the direction of heat flow in relation to temperature gradients. Still other versions are expressed in terms of inequalities involving the entropy. Recent versions of these theories are often based on the principles of continuum mechanics and use macroscopic axioms that are either pure assumptions or are extracted from empirical observations. [4]

To illustrate the issues involved in creating a particle-based, nonequilibrium theory, two related nonequilibrium theories from the middle of the twentieth century will be mentioned briefly here. These theories originated in the work of N. N. Bogoliubov [**1946**] and Illya Prigogine [**1962**]. Both of these theories involve special assumptions on the system distribution. Some of the

---

[4]For a recent example of the use of macroscopic axioms based on empirical observations in proving Clausius' inequality and other results in nonequilibrium thermodynamics, see W. Muschik [**1990**] and the work in the references cited there.

other theories, such as the relatively recent theory called Irreversible Thermo-dynamics, use states based on both microscopic and macroscopic quantities associated with a particle system. These theories will be discussed in Chapter 16.

The work of Bogoliubov and Prigogine was explained and compared by J. Stecki and H. S. Taylor [**1965**]. They showed how Bogoliubov's theory is developed and the relations between it and the work of Prigogine and his collaborators. Stecki and Taylor observed that an advantage of the approach initiated by Bogoliubov is that it is a pure statistical mechanical theory that "does not make recourse to thermodynamic arguments and concepts". They meant that Bogoliubov did not use the macroscopic second law of thermody-namics to arbitrate between microscopic approaches to a problem.

The theory of Bogoliubov is based on the Hamiltonian mechanics of a par-ticle system. In a notation that will be introduced formally below, the particle system k contains $\mathcal{N}_k$ particles, is represented by an $\mathcal{N}_k$-particle probability distribution density, and occupies a volume δ_k is assigned to the system. The evolution equation of the probability distribution density $F_k^{(n)}$ representing this system is $dF_k^{(n)}/dt = 0$, which is usually called the Liouville equation. Bogoliubov's theory is developed in terms of what are called "reduced dis-tribution functions" that are various marginalizations of the complete $\mathcal{N}_k$ particle distribution $F_k^{(n)}$ to the set $\{ F_s^{(n)k} \}$ of normalized distributions for s-particles, where $1 \le s \le \mathcal{N}_k$. This set of reduced distributions is used in the derivation of a set of coupled equations from the Liouville equation called the BBGKY (Bogoliubov, Born, Green, Kirkwood, Yvon) hierarchy. To achieve closure for this set of equations, Bogoliubov assumed that the functions $F_s^{(n)k}$ for $s > 2$ are all functions of $F_1^{(n)k}$, the one particle function.

Prigogine's version of the theory uses a Fourier analysis to separate each $F_s^{(n)k}$ into a collection of various components that are assumed to decay according to different time scales. A closed theory results if the faster decaying components are neglected.

As Stecki and Taylor, p. 764, point out, Prigogine avoids the Poincaré recurrence theorem and evades the reversibility of the underlying mechanics by viewing irreversibility as an asymptotic property of a system in the ther-modynamic limit. In this version of the thermodynamic limit, the number of particles and the volume are allowed to go to infinity, while requiring that the particle number density $\mathcal{N}_k/\delta_k$ remains finite and constant. Strictly speaking, it follows that the equations of Prigogine and his coworkers are valid only for an infinite system.

A formal solution of the Liouville equation is obtained in these circum-stances by making a Fourier transform of both the system distribution and the Liouville equation. The separation of the transformed solution into a rapidly decaying component, in which particle correlations are assumed to quickly die out, and a slowly decaying component, in which the correlations are assumed to persist, leads to a set of kinetic equations. After a sufficient time, the system is assumed to be in a Markovian regime and the resulting evolution equation for the persistent component is analyzed and solved as a Markovian differential equation.[5]

While this brief summary hardly does these theories justice, several important aspects are already clear in this description. This type of theory, including similar ones by other authors, is developed as a sequence of approximations culminating in a hierarchy of partial solutions. In some versions of theories of this type, the thermodynamic limit approximation is used. In other cases, such as the Fokker-Planck and Master equations, a discrete state distribution replaces the system distribution and transition probabilities represent the transitions between system states. The resulting sets of approximate evolution equations are far from the exact evolution equation of Hamiltonian mechanics from which they originated. In particular, it will be shown below that the predictions concerning the behavior of the entropy will be quite different for the exact and approximate theories.

A technique for making calculations easier was the discretization of the system distribution by Boltzmann [**1872**], [**1877b**]. It was later called 'coarse-graining the system distribution' and employed by Gibbs [**1902**] to make the entropy increase as a function of time. Coarse-graining is a significant and popular approximation that will be analyzed critically in succeeding chapters.

Several other significant problems with the approach used in the development of the evolution equation in these theories will be investigated after the introduction of the formalism of the Theory of Interacting Systems. To justify the assumptions made by Bogoliubov, for example, requires a physical model, based on particles and their interactions, that shows how the assumption that the higher order reduced distributions depend on the first order reduced particle distributions can work. There is a similar issue concerning the replacement of the two-particle distribution f_{ij} by the product $f_i f_j$ of one-particle distributions in the collision term used in Boltzmann's equation that is discussed in Part III of this book.

Some of the more recent work on statistical mechanics makes use of field theoretic methods. At the root, these methods are still based on the same fundamental assumptions and approximations as the work of Bogoliubov or

---

[5]From a mathematical point of view, Stephen Brush [**1961**] has shown a relation between the functional integrals of theories like this and those that appear in the 'sum over histories' work of Feynman. Clifford Truesdell III [**1969**] has raised serious objections to the assumptions concerned with the decay of correlations because they are not founded on any objective physical aspects of the situation that is based on properties of the particles that the theory is describing.

Prigogine. These more sophisticated techniques for approximating solutions do not resolve the objections raised above.[6]

The failure of all attempts to prove that the second law of thermodynamics is a consequence of mechanics indicates that there is a disjuncture between the mechanics of the microscopic particles and the view that the second law and its thermodynamic consequences are inherited from this mechanics. The approximations proposed by Bogoliubov and Prigogine, which involve defining the marginalized distribution F_s^k for $s > 2$ as the product of the one-particle component distributions F_1^k, represent attempts to resolve this disjuncture. These approximations do lead to an increasing entropy function. This is a consequence of making the replacement $F_s^k \to [F_1^k]^s$ because the information in F_s^k on the correlations between the particles is discarded. It will be shown below that the loss of information in moving from F_s^k to the approximation $[F_1^k]^s$ increases the system entropy all by itself.

Up to now, nonequilibrium theories have played only a minor role in thermodynamics because the assumptions described above, which are needed to make them work, do not have wide appeal. Apart from maintaining the subject/object distinction with regard to separate systems as an epistemological requirement, the Theory of Interacting Systems will not require the use of special assumptions concerning particles and their representation.

1.5 Specific Problems in the Foundations of Thermodynamics

A careful examination of the catalog of results accepted as established in thermodynamics shows that many of them were demonstrated using questionable assumptions. Because these results are often quoted or applied without stating the assumptions used to obtain them, it is often not clear whether a particular use is justified in a particular setting. Some typical examples illustrating this problem will be reviewed next.

In the standard discussions of the relation of thermodynamics to statistical mechanics, the system k under consideration is usually assumed to be isolated. If it is surrounded by other systems that are interacting with it, these other systems are either ignored or are represented by arbitrary potentials and boundary conditions in the equations describing the k system.[7] An evolution equation and a particle distribution are then attributed to the k system and results are computed. This procedure ignores the fact that the other systems sharing a boundary with the k system are also made of particles and are evolving, so the potentials associated with them should reflect this fact. The

---

[6]For treatments of nonequilibrium thermodynamics and statistical mechanics from the standard point of view, see, for example, David Ruelle [**1969**] and Göran Lindblad [**1983**].

[7]The work of Pierre Duhem [**1911**] is an exception. He considered an isolated system divided into interacting systems with boundaries that could move and discussed the relations between these subsystems as part of his study of the approach to thermodynamics from the standpoint of energetics.

use of arbitrary external potentials and boundary conditions to describe the k system means that we cannot tell if the predicted results are consistent with what would be expected in nature. In addition, the motion of the k system boundaries and the conserved quantities exchanged through these boundaries are usually ignored as well.

The justification usually given for neglecting the exchange of physical quantities between systems is that the systems are "weakly interacting", so that the exchange of energy and momentum between them, while essential to establishing equilibrium, is otherwise unimportant to the evolution of the systems in time. This statement is generally untrue and in any case does not resolve the logical problem of how to deal with the interaction between systems in a way that is consistent with the fact that they are both parts of a larger system and governed by the mechanics of the larger system. This gap in the application of the standard theory affects the definition of the thermodynamic quantities involved and has contributed the budget of paradoxes associated with entropy, irreversibility, and recurrence.

Much analytical effort has been expended in attempting to answer these paradoxes and show how macroscopic irreversibility arises from microscopic equations of motion that are reversible in time.[8] Various explanations of how this happens have been given in terms of the "complexity" of macroscopic systems, the "great disparity between the macroscopic and microscopic scales", the "initial very low entropy condition of the universe", or the use of statistical methods and probability. These arguments employ statements such as "S_B [Boltzmann's entropy] will *typically* increase in a way which *explains* qualitatively the evolution towards equilibrium of macroscopic systems", or refer to "systems with good ergodic properties", or draw conclusions of the form "It is therefore reasonable that", and so on. In many cases, the author puts crucial words within quotation marks that indicate some doubt about their use, meaning, or validity, in that context. Statements such as those in the quotations above are all parts of a plausibility argument and not part of a logical or mathematical argument.[9]

Many of the problems in thermodynamic reasoning are connected with the entropy and the second law of thermodynamics. It is usually treated as axiomatic, for example, that the net entropy of the universe increases in the normal evolution of a particle system. To judge the correctness of such a statement, it is necessary to have a clear idea of the analog for the entropy, how its phase average representing the macroscopic entropy changes in time as the underlying particle system evolves, and what role either the entropy

---

[8]The examples mentioned in this paragraph are taken from the recent review article by Joel L. Lebowitz [**1999**] on statistical mechanics, but are representative of very many standard treatments of this issue.

[9]Some of the arguments of this type raised by Boltzmann will be discussed in Part III of this book. Other plausibility arguments are discussed in connection with ergodicity in EIS and QTS.

analog or the entropy actually plays, if any, in determining the course of history for a physical system. In the standard approach, the problem is approached piecemeal and not within the context of a guiding theory concerning the relation between the entropies of systems in interaction.

The thermodynamic entropy is the phase average of a particle-based analog in the Theory of Interacting Systems. This implies immediately that some of the standard assumptions concerning the entropy are not valid in the Theory of Interacting Systems. A simple example is the requirement that the entropy is an extensive quantity proportional to the number of particles in a system. However, the entropy analog selected in TIS is proportional to the logarithm of the $\mathcal{N}_k$ particle system distribution—just as it was for Gibbs. It will be shown below that the entropy, obtained in the general theory as the phase average of this analog, is an extensive quantity only when the system distribution factors into a product of one-particle distributions. Similarly, the entropy associated with this analog is not a homogeneous function of the macroscopic parameters describing the system and does not meet other thermodynamic requirements that are sometimes imposed. Moreover, in the general TIS theory the entropy can be time dependent, there are no sources of entropy internal to a system, and there are no entropy currents. There is also no "entropy balance equation". On the other hand, the entropy defined in the TIS version of Boltzmann's equation presented in Part III is an extensive quantity and an entropy current is defined as well.

For a purely macroscopic or phenomenological thermodynamics, it is necessary to use some assumptions about the nature of thermodynamic processes in order to prove the second law. As an example, the proof of the second law in the survey of thermodynamics by Martin Bailyn [**1994**] will be reviewed.[10] As part of the proof, Bailyn uses the concept of *local thermodynamic equilibrium*, LTE, which he, p. 133, defined as "situations where macroscopic changes are much slower than microscopic changes." Bailyn, [**1994**], p. 135, presented this as a general principle of thermodynamics: "We take the point of view that thermodynamic processes involve LTE states generally . . . ." He went on to use this idea in his proof of "the entropy inequality form of the second law," i.e., the form of the second law that states that $\Delta S \geq \int \, dQ/T$. As part of the proof, he required that the system process pass through two LTE states that can be connected by a reversible process. If the process does not contain any LTE states, he, p. 136, acknowledged that the proof fails, but stated that in this case "the process is hardly thermodynamic."

The problem with this argument is that there is no independent way of knowing when, if ever, a system state meets the criteria of being an LTE

---

[10]Bailyn [**1994**] has presented an authoritative and detailed survey of the standard theories of thermodynamics in their historical context that will be used as a reference on a number of occasions. The remarks below should not be construed as a criticism of Bailyn's excellent work in presenting thermodynamics. His statements reflect faithfully the thermodynamic reasoning typical in the field.

state and, if so, whether these states can be connected by a reversible process. Because there is no independent way of knowing if there are any LTE thermodynamic processes and when they occur, this proof is not sufficient in a practical sense to demonstrate that the theorem is valid for real systems. Rejecting an ordinary process as "hardly thermodynamic" because it does not match these assumptions simply shows the conceptual gaps in the theory. Moreover, leaving such processes out of consideration opens the door to the possibility that these processes may violate the entropy inequality being proved.

As a second example, consider the entropy change in an ideal gas undergoing free expansion from an initial volume V_i to a final volume V_f. Bailyn [1994], pp. 127–128, obtained the standard result $\Delta S = k_B \mathcal{N}_k \ln(V_f/V_i)$. Similarly, he, pp. 141–143, obtained $\Delta S = C_{k,\mathrm{P}} \ln[T_2/T_1]$ for a reversible heat flow that raises the temperature T_1 of a system to T_2 under the assumption that the heat capacity at constant pressure, $C_{k,\mathrm{P}}$, is constant. To obtain these results requires the integration of macroscopic quantities between finite limits. The problem with this phenomenological procedure is that there is no way of knowing whether or not such an integration, which in this case is the mathematical representation of a thermodynamic process in a system moving between two volumes or temperatures, is warranted by the behavior of the underlying particle system.

Careful treatments of thermodynamic derivations state the limitations or basic assumptions on which the derivation is based. This has the consequence that qualifications such as (Bailyn, p. 169) "provided again that the motions are sufficiently slow and smooth enough for dQ/T to be identified with dS" are often found in the proofs of thermodynamic theorems. Leaving aside the problems of determining how slow these motions must be and how we would know if the condition is fulfilled, qualifications such as this one should appear in every theorem or formula associated with the standard approach to thermodynamics in which there is change. Strictly speaking, of course, no change can really be justified because the standard equilibrium formalism does not support it.

Another significant problem with the standard formalism is called the *Gibbs' paradox*. Gibbs computed the entropy change due to the mixing of dissimilar particles (entropy of mixing) when two systems of different types of particles are combined into one system. When he applied this formula to a situation in which two systems containing identical particles are combined, he found an entropy change of the same magnitude as expected for the entropy of mixing when the two systems contain different particles. This contradicted his feeling that the entropy should not change in the latter case, because the total entropy of the system made by combining two portions of a gas of the same particles should be the same as the sum of the entropies of the portions of gas.

In order to evade this paradox and obtain the proper behavior for the entropy, Gibbs made a distinction between what he called *generic* and *specific* phases. A specific phase function is a function that takes a value for a particular particle arrangement. A generic phase function is an average of a specific phase function over each member of the set of $\mathcal{N}_k!$ possible permutations of the particles. The computation of the entropy as the expectation value of the entropy analog was deemed by Gibbs to be a generic phase computation, because the entropy met his requirement when it was divided by $\mathcal{N}_k!$. This is equivalent to dividing Gibbs' entropy analog by $\mathcal{N}_k!$ before the phase averaging and calling this new function the entropy analog. While dividing either the thermodynamic entropy or the entropy analog by $\mathcal{N}_k!$ does evade Gibbs' paradox, it is an arbitrary modification of either the phase averaging procedure or of the entropy analog without any other justification for doing it. Later writers have tried to justify the division by $\mathcal{N}_k!$ in terms of quantum particle symmetrization, but this is inappropriate for the classical theory.[11] In addition, these writers and others have often added other superfluous quantum ideas to the classical theory, such as a minimum size for cells in phase space, and used this to insert factors proportional to $h^{\mathcal{N}_k}$ into the classical formalism without justification.

The use of such arbitrary assumptions and qualifications indicates that there is a basic problem at the root of the standard approach to thermodynamics. This problem has the consequence that the existing macroscopic formalisms and proofs, based on these assumed microscopic underpinnings, are often not compatible with the mechanics of the underlying particle system. In the TIS version of thermodynamics presented in this book, the axioms and theorems of thermodynamics are direct mathematical consequences of the classical mechanics of the underlying particles in conjunction with the phase averaging procedure. Only those relations that follow from this procedure have general validity and only these are accepted in TIS. For this reason, the qualifications and limitations mentioned above are not needed.

1.6 The Reduction of Thermodynamics to Statistical Mechanics

The previous discussion has indicated that a number of issues need to be considered in the reduction of the thermodynamics of interacting macroscopic systems to statistical mechanics. A review of these issues indicates that an adequate reduction must be concerned with

- establishing the relation between the microscopic and macroscopic levels of description;
- finding the proper statistical mechanical analogs for thermodynamic quantities, especially the entropy, temperature, and pressure;

---

[11]The controversy over dividing the quantum entropy by $\mathcal{N}_k!$ is discussed in detail in QTS.

- reconciling the reversibility and recurrence properties of the underlying particle mechanics with the irreversibility of thermodynamics;
- proving the existence of reversible processes;
- defining equilibrium and stationary particle distributions and corresponding macroscopic equilibrium and stationary states;
- defining various other special system distributions, such as the absolute zero and microcanonical distributions;
- working out the correct employment and interpretation of the components of the theory, including the thermodynamic derivatives with respect to time, temperature, and volume;
- showing that the equilibrium theory follows from the general theory when the underlying system distribution is the equilibrium distribution;
- placing Boltzmann's equation within the framework of the general theory; and
- establishing the theory on a mathematically rigorous foundation.

Each of these issues will be addressed in the Theory of Interacting Systems.

The Theory of Interacting Systems avoids the problems of the standard theory by employing several innovations. The first is an explicit concern with the proper use and interpretation of the statistical elements of the theory from the outset. Second, each of the interacting systems is treated on an equal footing in the formalism—although one may be singled out for study. Third, there is an explicit treatment of the boundary and the quantities transmitted through it. A major advantage of this symmetric treatment of the collection of systems and their boundaries is that it allows an adequate understanding of how the entropy of the collection changes in time.

An important goal of this approach is to show explicitly the roots of all functions and formulas of thermodynamics in the underlying microscopic particle dynamics. This is the reason that each thermodynamic function is required in TIS to be the phase average of a specific analog phase function. The phase analog functions themselves use only the particle coordinate and momentum phase variables and the particle interaction potential energy functions. The interaction potential energy functions, in turn, depend only on the distances between pairs of the particles. Because each quantity used in TIS thermodynamics is a phase average over a microscopically defined analog function, its time evolution is guaranteed to be consistent with the time evolution of the underlying particle system. In parallel with this, the fourth innovation is the use of specific operators to map thermodynamic functions to other thermodynamic functions representing their time, volume, and temperature derivatives. This allows the association of thermodynamic relations involving derivatives of the time, volume, or temperature with the underlying particle system.

As the fifth innovation, the structure of the theory is set up so that an explicit Galilean transformation of the frame of reference in phase space leads to an explicit transformation of the system thermodynamic quantities. The transformations of the thermodynamic entropy, pressure, and temperature, follow easily from the transformation properties of their analogs. For both physical and mathematical reasons, classical systems are limited to a finite number of particles and to bounded domains in phase space. This implies that all particle energies, momenta, and coordinates, have finite values at all times. The fact that these quantities are finite means that the relation between the energy of an isolated system and the energies of its component subsystems is always well defined. There is no loss of generality in this approach because all actual physical systems are bounded in this way. A mathematical formalism in which all physical quantities are bounded is more realistic than the usual one that allows integrations over unbounded energies and momenta. Finally, this approach leads to a number of mathematical relationships that are not well defined in the unbounded momentum and energy case.

In the development of the theory, there is a clear distinction between concepts that are macroscopic in origin, such as the volume of a system (with the associated concepts of its boundary and the motion of the boundary), and those with roots in the underlying microscopic particle dynamics. This distinction is important, for example, when considering the division of the microscopic energy flow through a boundary into the macroscopic concepts of heat flow and work flow. Special attention has also been paid to understanding the distinction between concepts that are mechanical in origin, such as particle number, momentum, angular momentum, and energy, and those that originate in thermodynamics: entropy, temperature, and pressure. The analog phase functions for momentum and energy, for example, depend on the momentum and energy of the microscopic elements. The analog phase functions for the entropy and temperature, on the other hand, reflect their non-mechanical origin in that each involves the system distribution in its definition.

The approach to the issues of irreversibility and recurrence taken here centers on the analysis of the boundary and how it modifies the evolution equation of the system. Thus, when the boundary is described stochastically, the system evolves irreversibly. On the other hand, when the particle interactions at the boundary are described exactly, the evolution of the system is reversible. The boundary formalism allows irreversible behavior for subsystems of an isolated system within the interacting systems approximation while not contradicting the reversible behavior of the isolated system as a whole. This point is reflected in the proof of Clausius's inequality in Chapter 9. The existence of reversible and equilibrium distributions and their relation to each other will also be shown within this formalism. Other specific system distributions and their properties are presented and discussed as well. Finally, as part of the examination of the mathematical foundations of the theory in Part IV, it is shown that the system equilibrium distribution is approached asymptotically from almost any initial distribution when the boundary conditions are constant.

1.7 The Role of Probability and Statistics

Probability plays an essential role in the TIS approach, as it does in all other theories of the relation of thermodynamics to statistical mechanics, but in a somewhat modified form. A distinction is made in TIS between the description of a system by a normalized *probability distribution density* and an unnormalized *system distribution density*. The system distribution density is a solution of the system evolution equation. The probability distribution density is obtained from the system distribution density by dividing it by its norm.

In ordinary statistical mechanics, a probability distribution density describes a microscopic particle system. In TIS, on the other hand, the system distribution density describes a system of particles. The standard equilibrium distribution density in statistical mechanics is $F_k^\epsilon = e^{-\beta_k \mathcal{H}_k^\epsilon}$ and the normalized equilibrium system probability distribution density is $F_k^{(n)\epsilon} = [Z_k^\epsilon]^{-1} e^{-\beta_k \mathcal{H}_k^\epsilon}$. The quantity $\mathcal{H}_k^\epsilon$ is the equilibrium Hamiltonian energy function and Z_k^ϵ is the partition function associated with the system distribution density. In the literature, the term 'system distribution density' is often used interchangeably with 'system probability distribution density', so the terminology used here is not standard in that respect. The reason for basing the TIS theory on the system distribution density rather than the system probability distribution density will be taken up in Chapter 5.

The system distribution changes in time due to (1) motion of the system particles, (2) their interactions with each other, (3) the interaction of the system particles with those of other systems at a shared boundary, and (4) the interactions of the system particles with an external potential contributed by other systems. The change in the system distribution density due to the motion of the system particles is described as a Hamiltonian flow. In Boltzmann's theory, the change in the system distribution due to particle interactions is described by a separate collision term. In Gibbs' theory, particle interactions are described by the particle interaction potentials and are an aspect of the Hamiltonian flow.

As outlined above, the TIS theory adds a boundary formalism that describes the changes in a given system distribution density due to the transmission of physical quantities through the system boundary. For simplicity, the classical version of TIS assumes closed systems, so no particles are exchanged between systems. From the microscopic perspective, physical quantities transported into a boundary by microscopic particles are either reflected back into the system, carried by the particles that brought them, or transmitted through the boundary to particles in other systems. The boundary formalism is constructed so that the conserved quantities of all the systems taken together are properly accounted for.

From the statistical point of view, the functions of thermodynamics are statistical measures on the microscopic system. That this is the essential point of thermodynamics was recognized by Hendrik Kramers [**1949**] when he referred to thermodynamics as "thermostatistics".[12] The TIS rule that each observable thermodynamic quantity must have an associated microscopic analog function means that the macroscopic phase average representing this quantity is a statistical measure on the particle phase space conditioned by the system distribution. The primary statistic used in TIS is the *mean* or *expectation value,* which is usually called the *phase average* in statistical mechanics, but the variances of physical quantities and correlations between them will be used when needed. The *mode* of the system distribution density is another statistic that was used by Boltzmann in many of his calculations.

Gibbs represented a system k containing $\mathcal{N}_k$ particles by a general time-dependent $\mathcal{N}_k$ particle probability density distribution F_k. He presented the probability issues in terms of the visualizable metaphor of an ensemble of systems. An *ensemble* according to Gibbs is a collection of identical systems with the same boundary conditions. The state of the system is different in each member of the ensemble and a member of the ensemble represents every possible mechanical state of the system.[13] The idea is that the likelihood that the system mechanical state is in the neighborhood of a given mechanical state is equal to the relative number of members of the ensemble with states in that neighborhood. The ensemble metaphor therefore represents probabilities as relative frequencies and is equivalent from the point of view of probability theory to viewing the system mechanical state as a random variable in $6\mathcal{N}_k$-dimensional phase space.

While Gibbs' ensemble metaphor is used almost universally in discussions of thermodynamics in relation to statistical mechanics, it is ultimately misleading. Gibbs [**1902**] himself stated that his ensembles are "creations of the imagination" that have "no simultaneous objective existence".[14] The ensemble of a system is often manipulated mathematically as if it were a finite set of representatives for the system and a number $\mathcal{N}_{\mathcal{E}}$ is used to represent the number of members of the ensemble $\mathcal{E}$. This idea was applied to

---

[12]Other authors who have emphasized this are Lazlo Tisza [**1961**], Tisza and Paul M. Quay [**1963**], and Benoit Mandelbrot [**1964**].

[13]P. and T. Ehrenfest [**1912**], note 84, pointed out that Boltzmann [**1871a**] and Maxwell [**1879b**] had already used an equivalent idea.

[14]See Bailyn [**1994**], Chapter 11, for a discussion of these quotations and for examples of ensemble calculations based on Gibbs work. See also the influential discussion in Tolman [**1938**], Chapter III. Richard von Mises took up ensemble theory in 1928 as the basis for his approach to the theory of probability based on relative frequencies.

quantum mechanics in particular so that ensembles, viewed as collections of distinct quantum states of a system, could be averaged over. In spite of its widespread use in thermodynamics and statistical mechanics, the ensemble metaphor adds nothing to the mathematical representation of the system and is not used in TIS.

The crux of the conceptual issue concerning the relation of thermodynamics to classical statistical mechanics is that the values of the physical quantities of a system are determined at a given time by the mechanical state represented by the point in $6\mathcal{N}_k$-dimensional phase space that the system occupies at that time. This fact made the calculation of thermodynamic quantities as phase averages over a range of system mechanical states in the system distribution puzzling. Both Maxwell and Boltzmann felt the need to justify this procedure and proposed what came to be called the *ergodic hypothesis* to explain it. The ergodic hypothesis maintains that a system of particles will visit each possible system state in phase space if left to evolve for a sufficient time. If this is true, it is possible to show in some cases that the infinite time average of a physical quantity is equivalent to the phase average of that quantity at some given time. Gibbs felt that this hypothesis was the justification for his use of the microcanonical ensemble. P. and T. Ehrenfest [**1912**], pp. 22–23, agreed with this view and claimed that the only justification for the use of Gibbs' canonical and microcanonical ensembles is the ergodic hypothesis. As will be discussed below, and in more detail in EIS, ergodic theory is neither needed nor used in TIS.

The TIS formalism is based mathematically on modern probability theory in topological vector spaces. Asymptotic results of probability theory in particular will be useful in the development of an asymptotic form for thermodynamics and for constructing approximate equations of state. Equations of state and the approximation of thermodynamic formulas will be taken up primarily in EIS.

1.8 The Framework of the Theory of Interacting Systems

The requirement that each thermodynamic object that represents a physical quantity have an analog function, and that the value of that quantity at any given time is the phase average of its analog function weighted by the particle distribution, makes definite in TIS the explicit connection between thermodynamic quantities and the underlying mechanical particle model. The spirit of most of Gibbs' work is preserved in TIS in that no approximations, maximizing principles, special nonequilibrium states, or applications of the thermodynamic limit, are used to define the computed thermodynamic quantities or their evolution.

The concept of a system boundary and its formal representation facilitates the inclusion of the dynamical effects of the boundary motion on thermodynamic quantities and their time derivatives as a natural consequence of the formalism. The TIS principle of treating each system on an equal footing insures that the boundary conditions for each system, and for the collection of systems taken together, are compatible with each other and with the underlying particle mechanics.

Theoretically, a thermodynamic body is described in TIS as an abstract dynamical system with a boundary that evolves in time in accord with an evolution equation and its boundary conditions. The connection between the macroscopic and microscopic levels of description provided by the phase average map means that the evolution of the microscopic particle system will result in the evolution of macroscopic quantities that depend on it. There is also a parallel between the mapping of thermodynamic functions to other thermodynamic functions by the thermodynamic derivative operators and by Galilean transformations. The central role of evolving statistical measures in TIS means that TIS is in essence a branch of *dynamical statistics*.

On the most abstract level, the TIS formalism is a metatheory that is made concrete by specifying the particular underlying particle mechanics, with its associated particle interaction potentials, and the particular phase average map to be used. While the values computed for the physical quantities of each version of the theory may differ, the macroscopic formalism itself is relatively insensitive to the choice of an underlying mechanics and phase average map. In this book, the theory is based on classical Hamiltonian statistical mechanics and the phase average map is defined as a standard expectation computation of a particle analog over the particle probability distribution density.

This metatheoretic schema is made concrete in the classical case by (1) choosing classical Hamiltonian mechanics with action at a distance potentials for the description of the microscopic elements of a system and their interactions, (2) using a particular stochastic description of the boundaries between systems to be specified below, and (3) choosing a normed measure on this space. Since the concern is with systems with boundaries, a version of Hamiltonian mechanics appropriate for this situation is required.

The primary theoretical entity that will be used in the work that follows is the system distribution density defined on the system phase space. Because all system distribution densities are required to be normed measures, they are normalizable and a normalization factor will always be included when expectation values are being computed to give the proper results. The use of the unnormalized version of a system distribution density is shown below to be required for the proper treatment of the underlying particle mechanics and the entropy.

The next step is to place some conditions on how the particles interact. Mathematically, this is expressed as a set of conditions on the collection of functions $\{\phi_{ij}(q_i - q_j)\}$, which represent for each pair of particles, i and j, the potential energy of the interaction between them as a function of the distance between them. These functions are the system *action at a distance potentials*. There is actually a hierarchy of possible choices for the form of the particle-particle interactions in the theory, and each level leads to a distinct formulation of thermodynamics. On the first level, there is a *local* theory in which (i) the particle-particle action at a distance interactions are absent, as in the ideal gas case, (ii) the interparticle interactions are treated as local hard-sphere or point particle collisions, or (iii) the interparticle interactions are mediated by explicit fields with local particle-field interactions. There are no action at a distance potentials in a local theory and the concept of force does not play a direct role. The ideal gas and hard-sphere models will be used as examples of local theories and the particle-field theories will be ignored in this book.

The second level is that of a *quasilocal* theory in which the interactions between particles are described by action at a distance potentials (i.e., forces) of finite range. Quasilocal forces that have a range that is small compared to macroscopic dimensions lead to the most interesting version of the theory from the standpoint of thermodynamics. This version of the theory will occupy most of our attention.

The third level is that of a *global* theory in which the interactions between the particles are described by action at a distance potentials of arbitrarily long range. While this level is important for electromagnetic and gravitational forces, it will be given only a brief treatment.

Bohr has emphasized the importance of selecting a definite and explicit conceptual structure within which a theory operates. In this book, the conceptual structure of classical mechanics is strictly adhered to. This helps avoid the pitfalls created by the premature introduction of quantum mechanical considerations into the theory.

Within this classical framework, there is no *in principle* physical distinction between the macroscopic and the microscopic domains. Nor does the theory depend on assumptions concerning the number of microscopic particles, the "complexity" of their interactions, or the "number of degrees of freedom" in a system. For this reason, the formulas in this theory are valid for any finite number of particles. The thermodynamic limit is neither required nor used in the development of the theory.

Mathematical concerns have played an essential role throughout the theory. The intention has been to develop a mathematically consistent and tractable formalism that will not be subject to the problems of other formulations. Of necessity, the mathematical framework supporting this theory is wider than the usual one. For example, certain singular distributions are needed to characterize a physical system under certain circumstances. For this reason, the theory finds its natural expression in the framework of topological vector spaces. Various aspects of the theory of topological vector spaces will be introduced as needed, but the bulk of the purely mathematical considerations will be reserved for Part IV of the book.

The standard thermodynamic convention regarding the labeling of partial derivatives of thermodynamic quantities is not used here. This reflects the different perspective and mathematical approach of TIS. Standard mathematical practice, rather than the standard thermodynamic practice, will be used for representing thermodynamic formulas and their derivatives.

While the focus is exclusively on classical physics in this volume and the next one, the theory is constructed in such a way that the changes required to extend it into the quantum and relativistic domains in later volumes are minimized.

1.9 The Theory of Interacting Systems

Work on the Theory of Interacting Systems was begun in response to the conjecture by Bohr, mentioned above, that the relationship between statistical mechanics and equilibrium thermodynamics might exhibit some sort of complementarity similar to that in atomic physics. Bohr's conjecture offered the prospect of testing his epistemological work in a physics setting distinct from quantum mechanics. To pursue this prospect, a study of thermodynamics and statistical mechanics was initiated. The volumes in this series are the result of that study.

This is the second volume in a series of six concerned with the relationship between the macroscopic description of matter as we ordinarily experience it and the microscopic description of this matter in terms of the particles composing it and their interactions. The overall title of the series is *The Theory of Interacting Systems*. The purpose of the theory is to address and resolve the problems outlined above concerning the extension of thermodynamics into the spacetime arena.

This volume is referred to as CIS and presents the classical version of TIS. The preceding volume, *The Theory of Interacting Systems, Volume 1, Niels Bohr: Reflections on Subject and Object,* was called RSO above. It was concerned with a number of philosophical and scientific issues fundamental to the macroscopic/microscopic relationship and their implications for physics. These issues were approached through an analysis of the epistemology presented by Niels Bohr, who was the first to realize how deep a conceptual break with the classical worldview of physics was required by the new quantum physics. The work in this book extends his concern with the subject/object distinction to classical physics.

Equilibrium theory is the focus of the third volume, *The Theory of Interacting Systems, Volume 3, Equilibrium Theory,* which was referred to as EIS above. In EIS, the basic TIS formalism is given a systematic introduction. This is followed by an investigation of the thermodynamic properties of the electromagnetic field and thermoelectricity. An inquiry into the issues connected with the approach of a system to equilibrium leads into a treatment of TIS equilibrium theory, which includes the calculation using the TIS formalism of a number of standard equilibrium thermodynamic coefficients. Recent work on the subject of thermodynamic surfaces is examined next and used in conjunction with a new formalism for determining thermodynamic phase boundaries. Finally, an approximation procedure for equilibrium systems is used to obtain a formalism for a computable thermodynamics. This approximation formalism shows how the microscopic parameters describing particle interactions are translated into the quantitative values for macroscopic physical quantities associated with collections of these particles.

Quantum mechanics from the TIS perspective is the concern in *The Theory of Interacting Systems, Volume 4, Quantum Theory*, called QIS above. The version of quantum mechanics presented in the fourth volume adheres closely to Bohr's perspective on the proper epistemological foundation for the theory. The basic TIS formalism is moved into the quantum setting and a quantum theory of macroscopic bodies is pursued as a foundation for quantum thermodynamics.

The fifth volume, *The Theory of Interacting Systems, Volume 5, Quantum Thermodynamics*, was called QTS above. This volume uses the formalism set up in QIS to extend thermodynamics into the quantum domain. As part of this work, the history of the analogs representing the entropy in the nineteenth and early twentieth centuries and the issues surrounding the entropy are explored carefully in an analysis of the work of Boltzmann, Planck, Lorentz, Einstein, P. and T. Ehrenfest, Sakur, Tetrode, Fowler, and others. This analysis is used as the basis for sorting out the essential from the accidental in relation to the entropy. The resulting quantum thermodynamics differs in a number of significant respects from that proposed by von Neumann.

The sixth volume, *The Theory of Interacting Systems, Volume 6, Relativity Theory*, is under construction and will be referred to as RIS. It is concerned with the extension of the spacetime thermodynamics of the previous volumes to the realms of special and general relativity theory.

The theory presented in this book is a synthesis and extension of the work on the relation between thermodynamics and statistical mechanics begun in the middle of the eighteenth century. Many theorists have obtained individual results and made definitions in mechanics, equilibrium thermodynamics, and theories of gas, liquid and solid states without a suitable framework to incorporate them all. The Theory of Interacting Systems provides that framework.

1.10 Philosophical, Historical, and Mathematical Background

Philosophically, the approach followed in this book has been influenced primarily by the ideas expressed by Maxwell, Boltzmann, Clausius, Gibbs, James Jeans [1925], Aleksandr I. Khinchin [1943], John Blatt [1959], Truesdell [1969], Walter Noll [1955], Edwin Jaynes [1957], Henry Bent [1965], Dmitrii N. Zubarev [1971], and the epistemological work of Bohr. The approaches of Maxwell [1866] and Gibbs [1902] are closest in spirit to that of the present work. The work of Gibbs is the basis for the focus on thermodynamic analog functions and that of Noll contained the germ of the mathematical treatment of them. Discussions of specific concepts, such as the critique of the concept of entropy by Blatt and Bent's discussion of the second law, in particular, were very useful. The works of Maxwell and Boltzmann have been cited explicitly on many occasions and have played a central role in the development of the theory. Truesdell's treatment of rational thermodynamics was very influential as were his historical accounts. Jeans' work was helpful in illuminating a number of issues with experimental relevance. The work of Bohr concerning theoretical descriptions, the subject/object distinction, and epistemological requirements on theories is used here in specific ways cited below. Other works that were important in the development of specific components of the theory are referred to in the appropriate chapters.

The works of many other founders of statistical mechanics and thermodynamics have been used extensively. These discussions often provided insights into the conceptual tensions and uncertainties that underlay their theories, which subsequent textbook accounts have glossed over. In addition to the original works, several good analytical and historical accounts of thermodynamics, kinetic theory, and statistical mechanics, have proved very useful: Paul and Tatiana Ehrenfest [1912], Stephen Brush [1976], [1985], Truesdell [1980], Truesdell and R. G. Muncaster [1980], James R. Partington [1949], and Bailyn [1994].

Certain aspects of probability theory in topological vector spaces have been important to the development of the theory. In particular, use has been made of the concept of dual spaces to define a function space for system distributions, which guarantees that all thermodynamic phase averages will be finite. In addition, the differential equations for the system evolution are analyzed in this function space. A number of important calculations, and those involving the entropy in particular, require that the solution of the evolution equation belongs to a certain Sobolev subspace of the original function space. For further information on the mathematical basis of this theory, see François Treves [1967] on topological vector spaces, Robert W. Carroll [1969] for more on abstract differential equations in function spaces, Robert Adams [1975] for information on differential equations in Sobolev spaces, and R. G. Laha and Vijay K. Rohatgi [1979] on probability theory.

There are a number of issues concerning the nature of physical models, experiments, theoretical descriptions, and many other topics, that were taken up in the first volume, RSO. In particular, the distinction between microscopic and macroscopic concepts was introduced there. These discussions form a valuable background to the work done here, but the particular definitions required in this volume will be introduced as needed so the current enterprise is independent of the first volume.

1.11 The Presentation of the Theory

Thermodynamics is presented in chapters 2 and 3, as a set of primitive terms, definitions, and axioms. The axioms fall into two major camps: the mechanical axioms deal with the rate of change of the local matter, momentum, angular momentum, and energy densities; the thermodynamic axioms deal with entropy, pressure, and temperature. It is the task of the rest of Part I to introduce the appropriate analog functions and show that these axioms and relations are consequences of the underlying particle mechanics and the phase averaging procedure.

The next step is the introduction of the proper version of statistical mechanics. This takes the form in Chapter 4 of the classical Hamiltonian statistical mechanics of closed systems with boundaries. In Chapter 5, the system evolution equation is established with the system distribution as its solution. The phase averaging formalism is introduced and the formalism for computing the time derivatives of phase averages is presented. In Chapter 6, the consequences of dividing the total system into subsystems are discussed. This includes the development of a wall formalism that is consistent with the requirement that all systems are treated on an equal footing in the formalism. Based on these ideas, the mechanical axioms of thermodynamics are shown in chapter 7 to be theorems or identities of the Hamiltonian formalism in conjunction with the phase averaging procedure.

The treatment of the specifically thermodynamic aspects of the theory begins with a discussion of the conceptual issues surrounding the second law of thermodynamics and the entropy in Chapter 8. An entropy analog based on the work of Boltzmann, Planck, and Gibbs is introduced in Chapter 9. This definition and the work on the evolution equation for thermodynamic quantities is used to compute a bound on the rate of change of the entropy in time. This leads in turn to a general proof of Clausius' inequality that is valid in both the nonequilibrium and equilibrium domains.

The concept of volume, and the associated volume derivative operator, are presented in Chapter 10. These definitions are used in the development of the formalism for the thermodynamic pressure as the volume derivative of the Hamiltonian energy in Chapter 11. A phase analog for the temperature, its associated thermodynamic formula, and the temperature derivative operator formalism are presented in Chapter 12. Reversible and equilibrium system states are examined in detail in Chapter 13. This is followed by a treatment of the absolute zero state and "Nernst's theorem" in Chapter 14. Chapter 15 is devoted to the behavior under Galilean coordinate transformations of the particle distributions and the various special states and formulas that were developed in previous chapters. The basic theory is completed in Chapter 16 with a discussion and critique of the relationship between the concept of entropy and that of information.

The discussion then moves in Part II to the second major area of concern: classical and equilibrium thermodynamics. It is shown in chapters 17 and 18 that the familiar formulas of both classical and equilibrium thermodynamics can be obtained as consequences of the framework of the Theory of Interacting Systems when attention is paid to the difference between quasistatic paths in equilibrium space and the system trajectory in phase space. Problems in the definitions of the free energy, free enthalpy, and chemical potential from the TIS perspective are connected to the tension between the spacetime approach and the equilibrium space approach to thermodynamics in Chapter 19.

While the basic theory is constructed in Parts I and II from the perspective of Gibbs' conceptualization of the relation between statistical mechanics and thermodynamics, the theories developed by Maxwell and Boltzmann are the concerns of Part III. The Boltzmann equation is the centerpiece of this discussion. In the TIS version of the Boltzmann equation, the set $\{\, f_i(q_i, p_i, t) \,\}$ of one-particle distributions, with one member for each particle in the system, replaces the single one-particle distribution $f(q, p, t)$ that was used by Boltzmann. The collision term that results from this approach is the sum of a set of coupled linear equations in the individual particle distributions rather than Boltzmann's single nonlinear quantity based on his generic one-particle distribution. The kernel of the integral equation representing the TIS version of the collision term in the Boltzmann equation is shown to be a tempered distribution. Boltzmann's H theorem and entropy are discussed next, followed by an analysis of the way in which the system state concept has been used by Boltzmann and others.

The mathematical structure of the theory and questions of rigor appear in Part IV. The components of the theory are shown to be well defined within the mathematical structure that has been constructed. In particular, the integrals and differential equations used in the preceding work are all shown to be well defined in the function spaces that were chosen. In addition, existence proofs are presented for the solutions of the evolution equations of the basic theory and for Boltzmann's equation. Finally, the asymptotic equilibrium state is shown to be the final steady state resulting when the boundary conditions are fixed properly.

CHAPTER 2

The Thermodynamics of Closed
Systems: Basic Formalism

In this chapter, the formal groundwork for TIS thermodynamics is prepared by providing a set of definitions that are adequate for the description of local time-dependent mechanical and thermodynamic quantities.

2.1 Introduction

The components of TIS thermodynamics are introduced and relations between them are presented as axioms. These axioms parallel those that are sometimes found in works on equilibrium thermodynamics. The relationships expressed in the general TIS thermodynamic axioms are valid for both nonequilibrium and equilibrium cases. They are actually theorems of Hamiltonian statistical mechanics in conjunction with the phase average map, as will be demonstrated in subsequent chapters using the TIS formalism. Moreover, this set of axioms is extensible in that new analog functions for physical quantities may be introduced. It is not difficult to use the TIS formalism with these new analogs and chart the relationships of the new quantities to other quantities of thermodynamics and to compute their derivatives with respect to time, volume, and temperature. With this information, the new quantities can be added to the formalism of TIS thermodynamics.

Microscopic particles are not observables in thermodynamics. It follows that in the derivation of thermodynamic formulas from statistical mechanics, all explicit references to particles must disappear. The particles make their presence felt at this level in the phrases "quantity of matter", "matter density", and "matter flow" that are used in this chapter and the next one. These locutions may be understood as macroscopic references to the "number of particles", "particle number density", and "particle flow", respectively. The thermodynamic notion of a 'body' is translated here as 'system' and a 'thermodynamic process' is defined as a set of macroscopic physical quantities that evolve in time in accord with the evolution of the underlying particle system.

Certain aspects of the physical situation are usually neglected or assumed to be trivial in the standard approaches to thermodynamics. These are concerned with the description of the macroscopic arena in which the thermodynamic processes take place. The macroscopic arena is chosen by the experimenter and described in terms of the volume occupied by a system and its boundary. Because moving boundaries are allowed in the TIS formalism, it is necessary to develop a formalism suitable for describing the time-dependent volume of a system, its moving boundary, and the fluxes of physical quantities passing through this moving boundary. For this reason, a volume descriptor formalism will be introduced first. This is followed by the introduction of the basic physical concepts used in the theory and a statement of their behavior under Galilean transformations of the coordinate frame. In the next chapter, a connection will be drawn between the thermodynamic concepts introduced here and some of their counterparts in the formulas of the standard approaches to thermodynamics.

All of the formulas in this book are assumed to be valid for each value of the time in the half-open interval $[0, \alpha)$, for $\alpha > 0$, even if this is not explicitly mentioned. When certain analytical conditions on the evolution equations are met, α may take arbitrarily large values. These conditions will be investigated in Part IV. From now on, it will be assumed that $t \in [0, \alpha)$ unless otherwise stated.

2.2 The System Volume Descriptor Formalism

The concepts of the volume of a quantity of matter and its boundary are not intrinsic to the microscopic definition of the collection of matter itself and do not appear in its Hamiltonian description. This implies that the volume and its boundary are macroscopic concepts. The development of TIS thermodynamics begins with a quantity of matter located in a bounded coordinate volume that is isolated from the rest of the universe. This coordinate volume is partitioned exhaustively into subvolumes that become the volumes of subsystems. The boundaries of the volumes of these subsystems are allowed to move, but certain consistency conditions will be imposed on this motion to maintain the integrity of the subsystems.

To make these ideas precise, consider a quantity of matter located in a bounded, macroscopically defined, time-dependent 3-dimensional volume. This volume is connected and surrounded by a 2-dimensional boundary that is piecewise smooth, does not intersect itself, and has a bounded area. The volume and the matter in it are called the *total system* and labeled by the letter t. The context will indicate whether t refers to the t system or to the time. The t system is *isolated*, which means, as will be expressed in formal terms later, that no external forces act on it and no moments of momentum cross its boundary. It is also assumed that t is a *closed system*, which means that no matter crosses its volume boundary either. The matter in the t system is assumed to have a finite kinetic energy and therefore a finite momentum and angular momentum. It is further assumed that the matter in t retains its form. This excludes modifications in it due to chemical reactions or to the formation of excited states. Finally, electromagnetic radiation is temporarily excluded from the theory. It will appear in the thermodynamics presented in EIS.

Let $\mathbf{R}$ be the set of real numbers and let $\mathbf{R}^3$ be the 3-dimensional direct product, also called the Cartesian product, of $\mathbf{R}$. Choose a rectangular coordinate frame in $\mathbf{R}^3$. This frame is assumed initially to be an inertial frame, so that only Newtonian forces appear in the theoretical description of matter in the frame, but, as will be discussed in Chapter 15, this is not a requirement. The frame is otherwise arbitrary.

Because the boundaries of the t system can move, the set of points assigned to the t system can change in time. This is accounted for by defining in the t frame the set $d_t(t) \subset \mathbf{R}^3$ as an open, connected, bounded, time-dependent Borel set that contains all the points at which the matter is located in the t system at time t. The quantity $d_t(t)$ is called the *volume set* of the

t system. The volume set $d_t(t)$ is endowed with a closed piecewise smooth boundary, $\partial d_t(t)$, that never intersects itself.

A volume in TIS thermodynamics is a connected open set of points that is required to have a well-behaved boundary. Because the system t contains a finite set of particles that are located at the set of points $\{q_i\}$, there is an element of arbitrariness concerning what other points are included in $d_t(t)$ as part of the volume set. The volume set is usually chosen, when possible, so that the surface of a system matches the average macroscopic surface of the material in the system. For enclosed gases, the volume set is the set of points enclosed within the container. In other cases, this choice is more arbitrary.

The arbitrariness in the definition of the volume set of a system of particles is the first indication of the degree of latitude and the need for conventions that are involved in moving from the concepts appropriate to the microscopic description of matter to the macroscopic description. It also illustrates the significance of the distinction made in RSO between concepts that are macroscopic in origin and those that refer to the elements of microscopic systems. That distinction will play a large role in this work.

The set $\partial d_t(t)$ is called the *boundary set* of the t system. The boundary set is allowed to have disjoint components in those cases in which the volume is contained between two or more disconnected surfaces. As stated above, this set does not intersect itself. The boundary does not belong to the volume set but does belong to its closure $\overline{d_k}(t)$. It can be defined in these terms as $\partial d_t(t) = \overline{d_k}(t) - d_k(t)$, which is the set of points in the closure of $d_k(t)$ that do not belong to the set $d_k(t)$.

Consider next an arbitrary finite partition of the t system into a collection of subsystems. The italic letters k, l and s are used as labels for these subsystems. Let $[t] = \{k, l, \dots\}$ be the set of labels for the subsystems in this decomposition. The decomposition is chosen so that the volume sets, $d_k(t)$, $k \in [t]$, and their boundary sets, $\partial d_k(t)$, together exhaust $\overline{d_t}(t)$. It is required that the volume sets, $d_k(t)$, for $k \in [t]$, be open Borel sets and meet the same conditions as $d_t(t)$. It is possible at any time, to repartition the volume set $d_t(t)$ into a new set of systems. This is purely a matter of macroscopic choice because there is no intrinsic requirement that the system volume sets and their boundaries be chosen to occupy any particular locations as long as the coordinates of the particles assigned to system k fall within $d_k(t)$ for $k \in [t]$ and the partition is exhaustive.

Because the decomposition of the coordinate volume of the t system is exhaustive, conditions will be imposed on the motion of the system boundaries so that gaps, consisting of points that do not belong to any system, will not appear between systems. The boundary between the k system and the l system are shared points in $\partial d_k(t)$ and $\partial d_l(t)$ that do not belong to the volume sets of either system. For any set of volumes of an acceptable decomposition, these points imply

(2.1a) $$\partial d_k(t) = \overline{d_k}(t) - d_k(t)$$

for $k = t$, $k = [t]$, and $k \in [t]$, and

(2.1b) $$\partial d_k(t) = \overline{d_k}(t) \cap \partial d_t(t) + \sum_{l \in [t] - k} \overline{d_k}(t) \cap \overline{d_l}(t),$$

for $k \in [t]$, where the sum over $l \in [t] - k$ represents the sum of each system in $[t]$ except k. In addition, the (possibly empty) kl *interface* between systems k and l is the set of points defined by

(2.1c) $$w_{kl}(t) = \partial d_k(t) \cap \partial d_l(t)$$

In accord with these definitions, the boundary set is two-dimensional surface that is a set of measure zero in $\mathbf{R}^3$. Moreover, the intersection of the k and l shared boundary with the k and s shared boundary, written $w_{kl}(t) \cap w_{ks}$ is at most a line, which is a set of measure zero in $\mathbf{R}^2$, because the volume sets are disjoint.

The k, t interface, $w_{kt}(t)$, is the part of the k system boundary that is shared with the t system boundary. This interface is a null set unless the k system intersects the boundary of the t system. The environment external to t is assumed to have no effect on k because t is isolated. The above definitions lead to the decompositions

(2.2a) $$\overline{d_t}(t) = \cup_{k \in [t]} [d_k(t) \cup (\cup_{l \in [t] - k} w_{kl}(t)) \cup w_{kt}],$$

(2.2b) $$\partial d_t(t) = \cup_{k \in [t]} w_{kt},$$

where $\cup_{k \in [t]}$ is the union over all the elements in $[t]$ and $\cup_{l \in [t] - k}$ is the union over all the elements in $[t]$ except k.

In addition to using $[t]$ to represent the collection of labels of the systems in the decomposition, the label $[t]$ will be used to represent a system called the $[t]$ *system*. The context will indicate clearly which meaning for $[t]$ is intended. The $[t]$ system refers to the system formed by the collection of all the subsystems taken together. The volume set of the $[t]$ system is disconnected in general and is therefore an exception to the general rule concerning volume sets. The volume set and boundary of the $[t]$ system are defined by

(2.3a) $$d_{[t]}(t) = \cup_{k \in [t]} d_k(t),$$

(2.3b) $$\partial d_{[t]}(t) = \cup_{k \in [t]} \partial d_k(t).$$

The $[t]$ system will play an important role in the approximation of the thermodynamics of the t system by the thermodynamics of the collection of subsystems of t.

The notation ∂k will often be used to refer to the k system boundary. At a boundary point, $q \in \partial d_k(t)$, the k *local boundary normal unit vector* is designated by $\hat{n}_{\partial k}(q, t)$, where $\hat{n}_{\partial k}$ points *into* the $d_k(t)$ volume. The k *local boundary velocity vector* is designated by $v_{\partial k}(q,t)$. At the kl interface, where $q \in w_{kl}(t)$, two consistency requirements are imposed on these vectors,

$$(2.4) \qquad \hat{n}_{\partial l}^{\mu}(q,t) = -\hat{n}_{\partial k}^{\mu}(q,t),$$

$$(2.5) \qquad \hat{n}_{\partial l}(q,t) \cdot v_{\partial l}(q,t) = -\hat{n}_{\partial k}(q,t) \cdot v_{\partial k}(q,t),$$

where $\cdot$ is the vector inner product symbol, which in this case is 3-dimensional. These two requirements insure that the k and l boundaries will not separate at the places where they coincide. If, in addition, the surfaces are not allowed to slip along each other, (2.5) is replaced for $q \in w_{kl}$ by the stronger requirement

$$(2.6) \qquad v_{\partial l}^{\mu}(q,t) = v_{\partial k}^{\mu}(q,t).$$

The vectors $\hat{n}_{\partial k}$ and $v_{\partial k}$ are assumed for mathematical reasons to be the restrictions to $\partial d_k(t)$ of the vector functions $\hat{n}_{Bk}(q,t)$ and $v_{Bk}(q,t)$. The functions $n_{Bk}(q,t)$ and $v_{Bk}(q,t)$ are at least once continuously differentiable functions of q and t in a 3-dimensional neighborhood of $\partial d_k(t)$. This assumption will be discussed further in Chapter 10 below.

2.3 Mathematical Preliminaries

Let us add next some necessary notation:
1. The *Heavyside step function* is defined for $x \in \mathbf{R}$ by

$$(2.7a) \qquad \theta(x) = \begin{cases} 1, & \text{for } x > 0, \\ 0, & \text{for } x \leq 0. \end{cases}$$

The derivative of $\theta(f(x))$ with respect to x is

$$(2.7b) \qquad \frac{d\theta(f(x))}{dx} = \delta(f(x))\frac{df(x)}{dx}.$$

2. When x and y are real numbers, the representation

$$(2.8) \qquad \delta(x-y) = \frac{1}{(2\pi)^{\frac{1}{2}}} \int d\lambda\, e^{-i\lambda(x-y)}$$

of the 1-dimensional Dirac delta measure is used in discussions of its properties. The integration is from $-\infty$ to ∞. Higher dimensional Dirac measures are products of the 1-dimensional ones. The dimension of a Dirac measure is indicated by its argument.

3. The *indicator function* of the set $d_k(t)$ is defined by

$$(2.9) \qquad \chi_{d_k(t)}(q) = \begin{cases} 1, & \text{for } q \in d_k(t), \\ 0, & \text{for } q \notin d_k(t). \end{cases}$$

Other indicator functions are defined similarly. The θ and the χ functions are projection operators. These operators are used extensively in integrals to limit their integrands to specific sets in coordinate space, momentum space, or phase space, instead of the usual procedure of assigning limits to the integrals themselves. This makes integration by parts, computing derivatives under the integral sign, and similar operations more transparent both physically and mathematically.

4. Integrals without designated domains of integration are to be taken over all of the indicated space, e.g., over all coordinate, momentum, phase space, or parameter space. Integrals over $q \in d_k(t)$ will be written in one of the equivalent forms $\int_k d^3q \cdots$ or $\int d^3q \, \chi_{d_k(t)}(q) \cdots$. Similarly, integrals over $q \in \partial d_k(t)$ are written either as $\int_{\partial k} d^2q \cdots$ or $\int d^2q \, \chi_{\partial d_k(t)}(q) \cdots$. The explicit use of an indicator function is required if partial derivatives of the integral are to be computed.

5. The measure $\mu(A)$ of a set A is defined as the Lebesgue measure of the indicator function of A. The variables of integration are those specified in the definition of A. The k system *volume* $\delta_k(t)$ is defined in these terms as the Lebesgue measure of the indicator function of the volume set:

$$(2.10) \qquad \delta_k(t) = \mu(d_k(t)) = \int d^3q \, \chi_{d_k(t)}(q).$$

The dimension of the set A determines which Lebesgue measure is being used.

6. The Greek letters μ, ν, ξ, η, τ and σ will be used for vector or tensor indices; they will range from 1 to 3; the usual summation convention will be used when a contravariant and a covariant index are the same.

7. The tensor $\delta^{\mu\nu}$ is the *diagonal unit tensor* (Kronecker delta); it will sometimes be used in the Euclidean coordinate and momentum spaces for raising or lowering tensor indices.

8. The tensor $\epsilon^{\mu\nu\xi}$ is the *totally antisymmetric unit tensor density* (Levi-Civita density), which takes the value $+1$ for even permutations of its indices, -1 for odd permutations, and 0 when an index is repeated.

9. The (symmetric) *mixing* and (antisymmetric) *alternation* of two vector or tensor indices are defined by:

$$(2.11a) \qquad a^{(\mu}b^{\nu)} = \tfrac{1}{2}[a^\mu b^\nu + a^\nu b^\mu].$$

$$(2.11b) \qquad a^{[\mu}b^{\nu]} = \tfrac{1}{2}[a^\mu b^\nu - a^\nu b^\mu].$$

When one or more indices are excluded from a mixing or alternation, they are enclosed between | symbols. The symbol $a^{(\mu}b^{|\nu|\lambda)}$ is therefore interpreted as

$$(2.11c) \qquad a^{(\mu}b^{|\nu|\lambda)} = \tfrac{1}{2}[a^\mu b^{\nu\lambda} + a^\lambda b^{\nu\mu}].$$

10. A symbol overloading convention was adopted above in which k is allowed to both designate a system of particles and represent the set of indices of these particles. The sum $\sum_{i \in k}$ and product $\prod_{i \in k}$ are two examples of operators acting over the indices of all the particles

belonging to the system k. In this notation, for example, an expression such as $i \in k - j - s$ refers to any particle in the set k except the particles j and s.

The Total Time Derivative

Some of the primitive thermodynamic quantities of the theory introduced below can be expressed as *densities*, such as the k local energy density $\mathcal{E}_k(q, t)$, which are functions of the macroscopic spacetime location (q, t). As discussed in Chapter 4, only quantities that can be represented microscopically as particle sum functions are allowed to have densities. It is assumed for now that each density is at least once continuously differentiable in q and t. In Part IV, it will be shown that this differentiability is a consequence of the analytical assumptions satisfied by the distributions describing the microscopic components of the theory.

For each quantity that can be represented as a density, there is a corresponding current density. The local energy current density $\mathcal{E}_k^\mu(q, t)$ corresponds to the flow of the local energy density $\mathcal{E}_k(q, t)$. This correspondence is important in the definition of the total time derivative of any mechanical density. The total time derivative of the thermodynamic density $G_k(q, t)$, where G_k may depend on vector or tensor indices as well, is expressed using its associated current density $G_k^\mu(q, t)$ by[1]

$$(2.12) \qquad \frac{dG_k(q, t)}{dt} = \frac{\partial G_k(q, t)}{\partial t} + \frac{\partial G_k^\mu(q, t)}{\partial q^\mu}.$$

For quantities with tensor indices, such as $G_k^{\nu\sigma\cdots}$, the convention is used that the current index is always the last contravariant index. The divergence of the current density in the total time derivative is therefore always with respect to the last contravariant index. In the discussion of the general properties of phase functions in this chapter, possible tensor indices will be ignored.

The total time derivative operator is the thermodynamic analog of the total time derivative in statistical mechanics introduced in Chapter 4 below. The total time derivative is appropriate for computing the rate of change of a quantity in a reference frame moving with the thermodynamic system. The partial time derivative $\partial/\partial t$ operator is used for the time derivative in a frame of reference fixed in the laboratory.

---

[1] An equation of this form first appears in Maxwell [**1866**]. See Brush [**1965**], Vol. 2, p. 62, equation (75). The difference between equation (2.12) and the standard formula for the total time derivative of a density, such as the mass density, will be discussed below.

Global Quantities and Bounds

For any local thermodynamic density, $G_k(q,t)$, a corresponding global quantity is obtained by integrating $G_k(q,t)$ over q:

$$(2.13) \qquad G_k(t) = \int d^3q\, G_k(q,t).$$

For macroscopic quantities that are localized to a given system k, it follows from the formalism that $G_k(q,t) = 0$ for $q \notin d_k(t)$, so the definition of $G_k(t)$ is limited to system k as well.

Among the primitive thermodynamic quantities needed are local *surface densities,* such as the local surface heating density $\mathcal{Q}_{\partial k}(q,t)$, which represents for $q \in \partial d_k(t)$ the rate at which heat is changing at the point (q,t). Surface densities are used to represent the changes of certain physical quantities at the boundary. They are singular in $\mathbf{R}^3$ in such a way that a volume integration of a system surface density over a volume containing the system boundary is the same as a surface integration over the surface of that boundary.[2] As an example of this formalism in action, a 3-dimensional coordinate volume integration over a singular local surface density $G_{\partial k}(q,t)$ becomes a 2-dimensional surface integration over the surface of that volume and yields the non-singular global surface quantity $G_{\partial k}(t)$:

$$(2.14) \qquad G_{\partial k}(t) = \int d^3q\, G_{\partial k}(q,t).$$

It is shown in Part IV that each global quantity, whether a surface quantity or not, is continuous and at least once continuously differentiable in t. For local quantities that are not densities, such as the temperature, pressure, and entropy, other global averaging procedures will be introduced below.

Consider next bounds on thermodynamic quantities. Let $G_k(q,t)$ be a thermodynamic density, $G_k^{\mu}(q,t)$ its associated current density, and $\nu_k(q,t)$ the local matter density for the k system. Then there exists a finite positive quantity, $G_k^{\max} > 0$, and a maximum speed in the system, $v_k^{\max} > 0$, such that

$$(2.15a) \qquad 0 \le |G_k(q,t)| \le G_k^{\max}\nu_k(q,t),$$
$$(2.15b) \qquad 0 \le |G_k^{\mu}(q,t)| \le v_k^{\max}G_k^{\max}\nu_k(q,t).$$

In (2.15a), the structure $|\cdot|$ is the absolute value operator and in (2.15b) it is the Euclidean norm of the vector. The quantity $-G_k^{\max}$ is a lower bound for a function that takes on negative values. These statements are proved in a chapter below using the assumptions made on the underlying particle model.

---

[2] See Chapter 5 and for further details on this formalism and the definitions involved.

2.4 The Primitive Thermodynamic Quantities

Each system is described macroscopically by a set of primitive thermodynamic concepts. The number of its tensor indices indicates the vector or tensor nature of a quantity. The list below covers the concepts of traditional concern to thermodynamics but does not exhaust the set of concepts that are, or may become, of significance to thermodynamics. As mentioned above, the theory can be easily extended by adding new analog functions to the set of observables for the system and working out the relationships between the new quantities and the other thermodynamic quantities.

The quantity β_k, used in equilibrium theory in the factor $e^{-\beta_k \mathcal{H}_k}$, has the dimensions of inverse energy. It will be called the *thermature* in TIS. The name thermature is appropriate because it is the measure of the inverse average energy of a system.[3] For reasons that will be discussed in Chapter 12, the thermature β_k rather than the temperature is adopted in TIS as the fundamental quantity and the absolute temperature T_k will be obtained from it.

The set of local primitive thermodynamic concepts that is sufficient for the thermodynamics developed here is:

I. Volume Descriptors

volume set,	$d_k(t)$;
boundary set,	$\partial d_k(t)$;
local boundary inward normal unit vector,	$\hat{n}^{\mu}_{\partial k}(q,t)$;
local boundary velocity,	$v^{\mu}_{\partial k}(q,t)$;

II. Mechanical Quantities

Matter Density

local matter density,	$\nu_k(q,t)$;
local mass density,	$\mu_k(q,t)$;

Matter Flow

local matter current density,	$\nu^{\mu}_k(q,t)$;

Momentum

local momentum density,	$\mathcal{P}^{\mu}_k(q,t)$;
local momentum current density,	$\mathcal{P}^{\mu\nu}_k(q,t)$;

Angular Momentum

local angular momentum density,	$\mathcal{J}^{\mu}_k(q,t)$;
local angular momentum current density,	$\mathcal{J}^{\mu\nu}_k(q,t)$;

---

[3] Truesdell [**1960**], p. 29, had suggested the term "coldness."

Force

local internal body force density,	$\mathcal{F}_{k,n}^{\mu}(q,t)$;
local external body force density,	$\mathcal{F}_{k,x}^{\mu}(q,t)$;
local surface force density,	$\mathcal{F}_{\partial k}^{\mu}(q,t)$;

Energy

local Hamiltonian energy density,	$\mathcal{H}_k(q,t)$;
local energy density,	$\mathcal{E}_k(q,t)$;
local kinetic energy density,	$\mathcal{K}_k(q,t)$;
local internal potential energy density,	$\Phi_{k,n}(q,t)$;
local external potential energy density,	$\Phi_{k,x}(q,t)$;

Energy Flow

local energy current density,	$\mathcal{E}_k^{\mu}(q,t)$;
local surface heating density,	$\mathcal{Q}_{\partial k}(q,t)$;
local body working density,	$\mathcal{W}_{k,b}(q,t)$;

III. Thermodynamic Quantities

local pressure density tensor,	$\mathrm{P}_k^{\mu\nu}(q,t)$;
local internal virial density tensor,	$\Xi_{k,n}^{\mu\nu}(q,t)$;
local external virial density tensor,	$\Xi_{k,x}^{\mu\nu}(q,t)$;
entropy,	$S_k(t)$;
local thermature tensor,	$\beta_k^{\mu\nu}(q,t)$.

Note that the mass density current vector is the same as the momentum density vector and is not included separately. The flows of kinetic and potential energy have also not been included separately. The surface heating and surface working densities are rates of energy change at the boundary, associated with changes in the heat and work, and should not be confused with the "amount of heat" and "amount of work" in the system. The pressure P_k is expressed in the Roman typeface to distinguish it from both the total k system momentum $\mathcal{P}_k$ and the $3\mathcal{N}_k$-dimensional momentum vector P_k in the particle theory, which are expressed in the calligraphic and mathematical italic typefaces, respectively.

The integral of each singular surface density $G_{\partial k}(q,t)$ over $q \in d_k(t)$ is assumed to be finite. Those densities in this list that are not surface densities are assumed to be at least once continuously differentiable in q and t for $t \in [0,\alpha)$ and $q \in d_k(t)$ except, possibly, for $q \in d_{\partial k}(t)$.

Temperature

The thermature has an associated microscopic analog function that gives meaning to the macroscopic local thermature function $\beta_k(q,t)$ and a formula for calculating it. In Chapter 12, the generalization of the local thermature function $\beta_k(q,t)$ to a local thermature tensor $\beta_k^{\mu\nu}(q,t)$ is discussed. The thermature is one third of the trace of the thermature tensor

$$(2.16a) \qquad \beta_k(q,t) = \tfrac{1}{3}\delta_{\mu\nu}\beta_k^{\mu\nu}(q,t).$$

The local absolute temperature $T_k(q,t)$ is defined in terms of the thermature $\beta_k(q,t)$ by the standard formula

$$(2.16b) \qquad T_k(q,t) = [k_B \beta_k(q,t)]^{-1},$$

where k_B is Boltzmann's constant.

Additional Definitions

Some additional functions required below are defined next. The *local body force density* is defined in terms of the internal and external local body force densities by

$$(2.17) \qquad \mathcal{F}_{k,b}^{\mu}(q,t) = \mathcal{F}_{k,n}^{\mu}(q,t) + \mathcal{F}_{k,x}^{\mu}(q,t).$$

The *local potential energy density* is expressed as the sum of the local internal and external potential energy densities by

$$(2.18) \qquad \Phi_k(q,t) = \Phi_{k,n}(q,t) + \Phi_{k,x}(q,t).$$

The singular *local surface working density* is given in terms of the surface force density and the boundary velocity by

$$(2.19) \qquad \mathcal{W}_{\partial k}(q,t) = \mathcal{F}_{\partial k}(q,t) \cdot v_{\partial k}(q,t).$$

Finally, the *local Hamiltonian density* is defined by

$$(2.20) \qquad \mathcal{H}_k(q,t) = \mathcal{K}_k(q,t) + \Phi_k(q,t).$$

With this formalism, it is possible to define an isolated thermodynamic system precisely. An *isolated system t* is a closed system for which all external body forces, external virials, and external potentials vanish. This means, for each $q \in d_k(t)$ and $t \in [0,\alpha)$, that

$$(2.21a) \qquad \mathcal{F}_{t,x}^{\mu}(q,t) = 0,$$
$$(2.21b) \qquad \Xi_{t,x}^{\mu\nu}(q,t) = 0,$$
$$(2.21c) \qquad \Phi_{t,x}(q,t) = 0.$$

At the boundary of the t system, that is, for $q \in \partial d_t(t)$ and $t \in [0,\alpha)$, the flows through the boundary of particles, momentum, angular momentum, and energy vanish, as well:

(2.21d) $\qquad\qquad \nu_t^\mu(q,t)\,\hat{n}_{\partial t,\mu}(q,t) = 0,$

(2.21e) $\qquad\qquad \mathcal{P}_t^{\mu\nu}(q,t)\,\hat{n}_{\partial t,\nu}(q,t) = 0,$

(2.21f) $\qquad\qquad \mathcal{J}_t^{\mu\nu}(q,t)\,\hat{n}_{\partial t,\nu}(q,t) = 0,$

(2.21g) $\qquad\qquad \mathcal{E}_{\partial t}^\mu(q,t)\,\hat{n}_{\partial t,\mu}(q,t) = 0.$

2.5 Galilean Transformations of Thermodynamic Quantities

Determining the transformation properties of thermodynamic quantities under a change of reference frames has led to controversies in the literature—particularly in discussions of the transformations of thermodynamic quantities in special relativity. While the transformation properties of the mechanical quantities are, for the most part, uncontroversial, this is not true of the temperature, entropy, and pressure. In TIS, the requirement that every macroscopic observable quantity of thermodynamics has a microscopic analog allows the determination of transformation properties of macroscopic quantities in terms of the transformation properties of their analogs.

Galilean coordinate transformations preserve the Newtonian structure of classical Hamiltonian mechanics, so they are the frame transformations that are appropriate for a thermodynamics based on an underlying classical particle theory. A general Galilean transformation consists of a coordinate translation, a velocity translation and an orthogonal rotation.[4] To define a coordinate transformation from the $\flat$ frame to the $\sharp$ frame, let $q_a^{(b)}$ be a constant coordinate vector and $v_a^{(b)}$ be a constant velocity vector defined in the $\flat$ frame. In addition, let a_r be a matrix representing an orthonormal, i.e., distance preserving, orthogonal rotation. These elements define a specific Galilean transformation $T_a = (a_r, q_a^{(b)}, v_a^{(b)})$, consisting of a coordinate translation $q_a^{(b)}$, a velocity translation $v_a^{(b)}$, and an orthogonal rotation a_r, from frame $\flat$ to frame $\sharp$. This transformation consists of the following operations performed in order: (i) at time $t = 0$, the origin of the $\sharp$ frame is set at the point $q_a^{(b)}$ in the $\flat$ frame; (ii) the origin of the $\sharp$ frame is allowed move with velocity $v_a^{(b)}$ with respect to the origin of the $\flat$ frame; and (iii) a rotation a_r of the resulting coordinate, momentum, and other vectors, is performed. This rotation acts on the vector and tensor elements defined in the $\flat$ frame. The rotation matrix a_r is defined as the inverse of the rotation a_r^{-1} that takes the $\sharp$ axes into the $\flat$ axes.

The orthonormal rotation a_r is an invertible matrix $a_{r,\nu}^\mu$ that satisfies the orthonormality condition

(2.22a) $\qquad\qquad a_{r,\lambda}^\mu a_{r,\nu}^\lambda = \delta_\nu^\mu$

---

[4]See Goldstein [**1950**], Chapter 1 and pp. 97–101.

and has the determinant

(2.22b) $|a_r| = 1.$

Using (2.22), it is easy to show that $\delta^{\mu\nu}$ and $\epsilon^{\mu\nu\xi}$ transform as

(2.23a) $^{(\sharp)}\delta^{\mu\nu} = \ ^{(\flat)}\delta^{\mu\nu},$

(2.23b) $^{(\sharp)}\epsilon^{\mu\nu\xi} = |a_r| \ ^{(\flat)}\epsilon^{\mu\nu\xi} = \ ^{(\flat)}\epsilon^{\mu\nu\xi}.$

The Galilean transformation T_a is invertible. It is used to express the components of the coordinate and velocity vectors in the $\sharp$ frame in terms of the components of the coordinate and velocity vectors in the $\flat$ frame in the form

(2.24a) $q^{(\sharp)\mu} = [T_{aq}q^{(\flat)}]^\mu = a_{r,\nu}^\mu[q^{(\flat)\nu} - v_a^{(\flat)\nu}t - q_a^{(\flat)\nu}],$

(2.24b) $v^{(\sharp)\mu} = [T_{av}v^{(\flat)}]^\mu = a_{r,\nu}^\mu[v^{(\flat)\nu} - v_a^{(\flat)\nu}],$

(2.24c) $t^{(\sharp)} = t^{(\flat)}.$

As indicated, the time coordinate is unaffected by this transformation. Since the Jacobian of this transformation is 1, it is also true that the volume measure is preserved under this transformation. The coordinate vectors $q^{(\sharp)}$ and $q^{(\flat)}$ vectors connected by T_a refer to the same physical point in the two different coordinate systems. The functional forms on the far right in (2.24a), (2.24b), and (2.24c), are said to be *induced* by the transformation. Finally, the transformation T_a preserves the distance $|q_i - q_j|$ between two coordinate vectors and the difference $|v_i - v_j|$ between two velocity vectors.

Quantities defined in a system inertial frame will usually not have any special designation. The frame of reference is indicated by a superscript symbol in formulas only when a special reference frame is used, such as the system rest frame, or transformations are under consideration.

The limit quantities (2.15), G_k^{max}, v_k^{max}, which were defined in the calculations of bounds, are generally frame dependent under Galilean transformations. Roughly speaking, the bounds transform in accord with their physical dimension. This can be verified by expressing a quantity in one frame in terms of quantities defined in another frame and computing the upper bound on each side of the equation.

Consider next a macroscopic mechanical quantity written in the k system as a local function $G_k(q, t)$, where G_k represents a physical quantity such as the matter, momentum, angular momentum, energy, or local force density, etc. Consider also two frames $\sharp$ and $\flat$, an invertible coordinate transformation T_{aq}, and two coordinate vectors $q^{(\flat)}$ and $q^{(\sharp)}$ such that $q^{(\flat)} = T_a^{-1}q^{(\sharp)}$. Because interacting systems can share moving boundaries, a transformation of the coordinates of one system requires a simultaneous transformation of the coordinates of all the other systems, including the t system, in the same way. This is a reflection of the passive role of coordinate transformations and the requirement of maintaining a consistent description of the t system and all of its subsystems. Because of this convention, it will not be hard to show that the values of the external forces and potentials at the locations of the particles in k are invariant under a Galilean frame transformation.

The System Rest Frame.

The introduction of Galilean transformations into thermodynamics complicates some issues and raises problems with others. As an example, consider the problem of defining the temperature of a system of particles in mechanical terms. Boltzmann introduced in 1871 a mechanical definition of the temperature as a quantity proportional to the average kinetic energy of the particles in the system. With this definition, a system in which all the particles are at rest is deemed to be at absolute zero. However, in a reference frame that is moving at a constant velocity with respect to the first, the particles are all moving and the system would be assigned a non-zero temperature under Boltzmann's definition. Because a frame of reference may be changed without any physical interaction with a system, this situation contradicts our intuition that the temperature of a system of particles should be independent of the frame from which it is viewed.

The problem stems from the fact that although the zero of the kinetic energy scale does have an absolute significance in Newtonian mechanics as the absence of motion in a given frame, it is not invariant under Galilean frame transformations. This means that definitions of the temperature of a system based on the average kinetic energy of its particles are not invariant with respect to Galilean transformations. Similar problems are also faced when attempting to provide a mechanical definition of the pressure based in part on the average kinetic energy of a system's particles. From the standpoint of measurements on a system, the instruments for measuring the system temperature or pressure are generally calibrated in the rest frame of the detector, so measurements made using a moving detector must either be corrected for the motion or the theoretical calculations of the quantity it measures must be made in the rest frame of the detector.

In order to make definite and fix the zero of the kinetic energy scale, a special frame is used to define quantities that depend in some way on the average kinetic energy. This special frame is called the *system rest frame* and it is defined macroscopically in terms of the motion of the boundaries of the system. The k *average boundary velocity* vector is defined by

$$(2.25) \qquad v_{\partial k}^{\mu}(t) = \int_{\partial k} d^2 r \, v_{\partial k}^{\mu}(r, t).$$

The instantaneous k *system rest frame* is the frame in which $v_{\partial k}^{\mu}(t) = 0$. This quantity is important for the calculation of the global pressure and thermature of a system because these quantities are defined in Chapters 11 and 12 in the system rest frame. The global versions of these definitions are most meaningful when $v_{\partial k}^{\mu}(t) \approx$ const. and a transformation into the frame in which $v_{\partial k}^{\mu} \approx 0$ is made. A system at equilibrium is required to be in a special rest frame, called the *equilibrium rest frame*. The equilibrium rest frame is defined as the coincidence of the system rest frame, in which the system walls are all at rest, with the *particle rest frame* introduced in Chapter 4, in which the macroscopic total momentum and angular momentum of the particle system vanishes. The equilibrium rest frame is the Galilean frame in which the equilibrium and absolute zero particle distributions will be defined.

To avoid needless qualifications and formalism, it is assumed as a temporary convention that the system rest frame of the t system is an inertial frame. It should be noted that this is not a trivial assumption because there is no guarantee that the rest frame of a collection of matter in motion is an inertial frame. A system at rest in a laboratory on earth, for example, is certainly not in an inertial frame. In the discussions of the definitions of thermodynamic concepts in later chapters, the rest frames of individual subsystems will be assumed to be inertial frames as well. This point will be reviewed and thermodynamics in non-inertial frames will be discussed in Chapter 15.

The Transformation of Primitive Thermodynamic Quantities.

The definitions $p_{a,i}^{(b)\mu} = m_i v_a^{(b)\mu}$ and $P_{a,k}^{(b)\mu} = \times_{i \in k} p_{a,i}^{(b)\mu}$ will be used in the formulas below. For quantities below defined at the boundary, it is assumed that $q^{(\sharp)} \in \partial d_k^{(\sharp)}(t)$, $q^{(b)} \in \partial d_k^{(b)}(t)$, and $q^{(\sharp)} = T_{aq}q^{(b)}$. It will be demonstrated in Chapter 15 that the induced transformations of the thermodynamic primitive quantities from the b frame into the $\sharp$ frame take the form of those in the following list:

I. Transformations of Volume Descriptors

(2.26a) $$d_k^{(\sharp)}(t) = \{\, q^{(\sharp)} \mid T_a^{-1}q^{(\sharp)} \in d_k^{(b)}(t) \,\};$$

(2.26b) $$\partial d_k^{(\sharp)}(t) = \{\, q^{(\sharp)} \mid T_a^{-1}q^{(\sharp)} \in \partial d_k^{(b)}(t) \};$$

(2.26c) $$\hat{n}_{\partial k}^{(\sharp)\mu}(q^{(\sharp)}, t) = a_{r,\nu}^{\mu} \hat{n}_{\partial k}^{(b)\nu}(q^{(b)}, t);$$

(2.26d) $$v_{\partial k}^{(\sharp)\mu}(q^{(\sharp)}, t) = a_{r,\nu}^{\mu} \left[v_{\partial k}^{(b)\nu}(q^{(b)}, t) - v_a^{(b)\nu} \right];$$

II. Transformations of Mechanical Quantities

Matter Density

(2.26e) $$\nu_k^{(\sharp)}(q^{(\sharp)}, t) = \nu_k^{(b)}(q^{(b)}, t);$$

(2.26f) $$\mu_k^{(\sharp)}(q^{(\sharp)}, t) = \mu_k^{(b)}(q^{(b)}, t);$$

Matter Flow

$$(2.26\text{g}) \qquad \nu_k^{(\sharp)\mu}(q^{(\sharp)},t) = a_{r,\nu}^\mu \left[\nu_k^{(\flat)\nu}(q^{(\flat)},t) - v_a^{(\flat)\nu}\nu_k^{(\flat)}(q^{(\flat)},t) \right];$$

Momentum

(2.26h)
$$\mathcal{P}_k^{(\sharp)\mu}(q^{(\sharp)},t) = a_{r,\nu}^\mu \left[\mathcal{P}_k^{(\flat)\nu}(q^{(\flat)},t) - v_a^{(\flat)\nu}\mu_k^{(\flat)}(q^{(\flat)},t) \right];$$

$$\mathcal{P}_k^{(\sharp)\mu\nu}(q^{(\sharp)},t) = a_{r,\xi}^\mu a_{r,\eta}^\nu [\mathcal{P}_k^{(\flat)\xi\eta}(q^{(\flat)},t) - 2v_a^{(\flat)(\xi}\mathcal{P}_k^{|(\flat)|\eta)}(q^{(\flat)},t)$$

$$(2.26\text{i}) \qquad\qquad + v_a^{(\flat)\xi}v_a^{(\flat)\eta}\mu_k^{(\flat)}(q^{(\flat)},t)];$$

Angular Momentum

$$\mathcal{J}_k^{(\sharp)\mu}(q^{(\sharp)},t) = |a_r| a_{r,\nu}^\mu [\mathcal{J}_k^{(\flat)\nu}(q^{(\flat)},t) - \epsilon^{\nu\eta\xi}\{(v_{a,\eta}^{(\flat)}t + q_{a,\eta})\mathcal{P}_{k,\xi}^{(\flat)}(q^{(\flat)},t)$$

$$(2.26\text{j}) \qquad\qquad + (q_\eta^{(\flat)} - v_{a,\eta}^{(\flat)}t - q_{a,\eta}^{(\flat)})v_{a,\xi}^{(\flat)}\mu_k^{(\flat)}(q^{(\flat)},t)\}];$$

$$\mathcal{J}_k^{(\sharp)\mu\nu}(q^{(\sharp)},t) = |a_r| a_{r,\xi}^\mu a_{r,\eta}^\nu [\mathcal{J}_k^{(\flat)\xi\eta}(q^{(\flat)},t) - \mathcal{J}_k^{(\flat)\xi}(q^{(\flat)},t)v_a^{(\flat)\eta}$$

$$- \epsilon^{\xi\sigma\lambda}\{(v_{a,\sigma}^{(\flat)}t + q_{a,\sigma}^{(\flat)})[\mathcal{P}_{k,\lambda}^{(\flat)\eta}(q^{(\flat)},t) - v_{a,\lambda}^{(\flat)}\mathcal{P}_k^\eta(q^{(\flat)},t)$$

$$- \mathcal{P}_{k,\lambda}^{(\flat)}(q^{(\flat)},t)v_a^{(\flat)\eta} + v_{a,\lambda}^{(\flat)}v_a^{(\flat)\eta}\mu_k^{(\flat)}(q^{(\flat)},t)]$$

$$(2.26\text{k}) \qquad\qquad + q_\sigma^{(\flat)}v_{a,\lambda}^{(\flat)}(\mathcal{P}_k^{(\flat)\eta}(q^{(\flat)},t) - v_a^{(\flat)\eta}\mu_k^{(\flat)}(q^{(\flat)},t))\}];$$

Force

$$(2.26\text{l}) \qquad \mathcal{F}_{k,n}^{(\sharp)\mu}(q^{(\sharp)},t) = a_{r,\nu}^\mu \mathcal{F}_{k,n}^{(\flat)\nu}(q^{(\flat)},t);$$

$$(2.26\text{m}) \qquad \mathcal{F}_{k,x}^{(\sharp)\mu}(q^{(\sharp)},t) = a_{r,\nu}^\mu \mathcal{F}_{k,x}^{(\flat)\nu}(q^{(\flat)},t);$$

$$(2.26\text{n}) \qquad \mathcal{F}_{\partial k}^{(\sharp)\mu}(q^{(\sharp)},t) = a_{r,\nu}^\mu \mathcal{F}_{\partial k}^{(\flat)\nu}(q^{(\flat)},t);$$

Energy

(2.26o)
$$\mathcal{K}_k^{(\sharp)}(q^{(\sharp)},t) = \mathcal{K}_k^{(\flat)}(q^{(\flat)},t) - v_a^{(\flat)} \cdot \mathcal{P}_k^{(\flat)}(q^{(\flat)},t) + \tfrac{1}{2}[v_a^{(\flat)}]^2 \mu_k^{(\flat)}(q^{(\flat)},t);$$

(2.26p)
$$\Phi_{k,n}^{(\sharp)}(q^{(\sharp)},t) = \Phi_{k,n}^{(\flat)}(q^{(\flat)},t);$$

(2.26q)
$$\Phi_{k,x}^{(\sharp)}(q^{(\sharp)},t) = \Phi_{k,x}^{(\flat)}(q^{(\flat)},t);$$

(2.26r)
$$\mathcal{E}_k^{(\sharp)}(q^{(\sharp)},t) = \mathcal{E}_k^{(\flat)}(q^{(\flat)},t) - v_a^{(\flat)} \cdot \mathcal{P}_k^{(\flat)}(q^{(\flat)},t) + \tfrac{1}{2}[v_a^{(\flat)}]^2 \mu_k^{(\flat)}(q^{(\flat)},t);$$

Energy Flow

$$\mathcal{E}_k^{(\sharp)\mu}(q^{(\sharp)}, t) = a_{r,\nu}^\mu [\mathcal{E}_k^{(b)\nu}(q^{(b)}, t) - \mathcal{E}_k^{(b)}(q^{(b)}, t)v_a^{(b)\nu}$$
$$- \mathcal{P}_k^{(b)\rho\nu}(q^{(b)}, t)v_{a,\rho}^{(b)} + \{\mathcal{P}_k^{(b)}(q^{(b)}, t) \cdot v_a^{(b)}\}v_a^{(b)\nu}$$
$$+ \tfrac{1}{2}[v_a^{(b)}]^2 \{\mathcal{P}_k^{(b)\nu}(q^{(b)}, t) - v_a^{(b)\nu}\mu_k^{(b)}(q^{(b)}, t)\}];$$

(2.26s)

(2.26t)
$$\mathcal{Q}_{\partial k}^{(\sharp)}(q^{(\sharp)}, t) = \mathcal{Q}_{\partial k}^{(b)}(q^{(b)}, t);$$

(2.26u)
$$\mathcal{W}_{k,b}^{(\sharp)}(q^{(\sharp)}, t) = \mathcal{W}_{k,b}^{(b)}(q^{(b)}, t) - v_a^{(b)} \cdot \mathcal{F}_{k,b}^{(b)}(q^{(b)}, t);$$

(2.26v)
$$\mathcal{W}_{\partial k}^{(\sharp)}(q^{(\sharp)}, t) = \mathcal{W}_{\partial k}^{(b)}(q^{(b)}, t) - v_a^{(b)} \cdot \mathcal{F}_{\partial k}^{(b)}(q^{(b)}, t);$$

III. Transformations of Thermodynamic Quantities

Pressure

(2.26w)
$$P_k^{(\sharp)\mu\nu}(q^{(\sharp)}, t) = a_{r,\xi}^\mu a_{r,\eta}^\nu P_k^{(b)\xi\eta}(q^{(b)}, t);$$

(2.26x)
$$\Xi_{k,n}^{(\sharp)\mu\nu}(q^{(\sharp)}, t) = a_{r,\xi}^\mu a_{r,\eta}^\nu \Xi_{k,n}^{(b)\xi\eta}(q^{(b)}, t);$$

(2.26y)
$$\Xi_{k,x}^{(\sharp)\mu\nu}(q^{(\sharp)}, t) = a_{r,\xi}^\mu a_{r,\eta}^\nu \Xi_{k,x}^{(b)\xi\eta}(q^{(b)}, t);$$

(2.26z)
$$P_k^{(\sharp)}(q^{(\sharp)}, t) = P_k^{(b)}(q^{(b)}, t);$$

Entropy

(2.26α)
$$S_k^{(\sharp)}(t) = S_k^{(b)}(t);$$

Thermature

(2.26β)
$$\beta_k^{(\sharp)\mu\nu}(q^{(\sharp)}, t) = a_{r,\xi}^\mu a_{r,\eta}^\nu \beta_k^{(b)\xi\eta}(q^{(b)}, t);$$

(2.26γ)
$$\beta_k^{(\sharp)}(q^{(\sharp)}, t) = \beta_k^{(b)}(q^{(b)}, t).$$

Notice that the angular momentum vectors are axial vectors so that the determinant $|a_r|$ appears in their transformation. The invariance of the pressure and thermature under a Galilean transformation is a consequence of the fact that they are required to be defined in the rest frame of the system.

2.6 Initial Conditions and Subsystem Bounds

The initial relationship between the t system and the $[t]$ system approximating it is established by requiring that the initial conditions on the global

$[t]$ system momentum, angular momentum and energy match those on the t system at time $t = 0$:

$$(2.27a) \qquad\qquad \nu_{[t]}(0) = \nu_t(0) \ = \mathcal{N}_t,$$

$$(2.27b) \qquad\qquad \mathcal{P}^\mu_{[t]}(0) = \mathcal{P}^\mu_t(0) = I^\mu_{P,t},$$

$$(2.27c) \qquad\qquad \mathcal{J}^\mu_{[t]}(0) = \mathcal{J}^\mu_t(0) = I^\mu_{J,t},$$

$$(2.27d) \qquad\qquad E_{[t]}(0) = E_t(0) \ = I_{H,t}.$$

This collection of initial conditions is written as the ordered set

$$(2.28) \qquad\qquad \mathcal{I}_t = (\mathcal{N}_t, I^\mu_{P,t}, I^\mu_{J,t}, I_{H,t}).$$

Because these quantities are conserved for an isolated system by the equations of motion, it follows that I_t is constant so $(\nu_t(t), \mathcal{P}^\mu_t(t), \mathcal{J}^\mu_t(t), \mathcal{H}_t(t)) = (\nu_t(0), \mathcal{P}^\mu_t(0), \mathcal{J}^\mu_t(0), \mathcal{H}_t(0)) = \mathcal{I}_t$ is valid for any time t. This ordered set therefore consists of constant scalar and vector quantities that represent the t system invariants: the total number of particles, total momentum, total angular momentum, and total energy of both systems t and $[t]$ at time $t = 0$.

The total momentum can be set to zero by an appropriate Galilean velocity transformation. The angular momentum in the new frame may be set to zero by transforming into a rotating coordinate system. This last step is not a Galilean transformation. This choice of a coordinate and momentum frame leads to

$$(2.29) \qquad\qquad \mathcal{I}_t = (\mathcal{N}_t, 0, 0, \mathcal{H}_t(0)).$$

If the momentum, angular momentum, and Hamiltonian energy, exhaust the symmetries of the Hamiltonian function in the theory, these quantities will provide a complete invariant for the t system in the sense of Truesdell [**1960**]. If this is not the case, additional conserved quantities can be added to I_t to make a complete invariant.

Although coordinate and momentum frames have been chosen so that the total momentum and angular momentum of the t and $[t]$ systems are zero, the range of these quantities in each subsystem depends on the range of momentum, angular momentum, and energy differences that are allowed in each of them. In upper the extreme, it is possible that a single system k may absorb all the available energy from each of its neighbors and have a very high energy at the expense of their low energies. On the other hand, it may give up all the energy it can to its neighbors. Let us define the lower and upper bounds on the components of the k momentum by $\mathcal{P}^{l,\mu}_k$ and $\mathcal{P}^{u,\mu}_k$, angular momentum by $\mathcal{J}^{l,\mu}_k$ and $\mathcal{J}^{u,\mu}_k$, and energy by $\mathcal{H}^l_k$ and $\mathcal{H}^u_k$. Information on the particles and their interactions will be used to compute these bounds explicitly below. These bounds will then define the ranges of momentum, angular momentum, and energy, available in system k by the set of constants

(2.30a)
$$C_{P,k}^{\mu} = \mathcal{P}_k^{u,\mu} - \mathcal{P}_k^{l,\mu},$$

(2.30b)
$$C_{J,k}^{\mu} = \mathcal{J}_k^{u,\mu} - \mathcal{J}_k^{l,\mu},$$

(2.30c)
$$C_{H,k} = \mathcal{H}_k^{u} - \mathcal{H}_k^{l}.$$

These ranges determine the range of possibilities to integrate over when computing the phase averages of thermodynamic quantities. Only the energy range $C_{H,k}$ will actually be used in most calculations. The other ranges can easily be added to calculations as needed.

CHAPTER 3

The Thermodynamics of
Closed Systems: Axioms

The macroscopic axioms of standard versions of thermodynamics express relationships between fundamental quantities that are not further analyzable within thermodynamics itself. For purposes of comparison, these relationships are expressed in TIS also as a set of "axioms" that is sufficient for charting the evolution of these quantities in time and their exchanges between regions in contact with each other. The TIS axioms will later be shown to be consequences of the underlying particle mechanics.

3.1 Macroscopic Thermodynamics

In discussions of macroscopic thermodynamics that are not based on an underlying particle model, these axioms are viewed either as postulates of a rational thermodynamics based on continuum mechanics or they are considered to be phenomenological laws rooted in experience. Stating the laws of thermodynamics in this way as a collection of plausible axioms, perhaps suggested by a phenomenological theory, allows a great deal to be accomplished, but the suggestions provided by experiments and requirements of consistency are not sufficient for a more precise formulation. In addition, there are macroscopic experiments mentioned in Chapter 1 in which heat flows from a cold region to a hot one and particles flow from a low pressure region to a high pressure one. This behavior contradicts phenomenological laws adopted, in some form, by every theory of this type. This means that the phenomenological approach provides no guarantee that the behavior of matter will respect these laws.

In 1865 Clausius introduced the concept of *entropy* as a consequence of his investigation of Carnot's analysis of reversible versus irreversible cyclic thermodynamic processes. His expression for the change in the entropy as the equivalence $dS = dQ/T$ in *reversible processes* and the inequality $dS > dQ/T$ in *irreversible processes* (TIS sign convention) offered a way to calculate bounds on the change of the entropy in these processes. The concept of entropy seemed to make precise the concept underlying the phenomenological version of the second law of thermodynamics as it was introduced by William Thomson in 1851 and discussed subsequently by many writers.

The concept of entropy was an important conceptual advance, but the association of the macroscopic second law with the concept of entropy did not clear up a number of conceptual problems. In the absence of other guiding concepts, for example, there has been a proliferation of different types of entropy in the literature. These include "configurational entropy", "entropy currents", "entropy sources", "entropy transfers", "entropic forces", and equations showing how and when these entropies are created. In spite of the fact that no means other than calculations based on interpretations of Clausius' formula have been proposed for observing these entropies or their consequences, the claims that they exist and that they conform to some particular laws has attained in some quarters the status of established fact. As mentioned in Chapter 1, the macroscopic second law was often used in the first half of the twentieth century to predict the evolution of quantities in mechanical particle theories because thermodynamics was felt to be an established and reliable guide to the behavior of matter in bulk. Definitions of various versions of the entropy, such as those stated above, are rarely questioned in the literature—probably because there is no accepted standard to judge them against.

Classical thermodynamics introduced the concepts of a (macroscopic) body and its state. Changes in this state were referred to as thermodynamic processes. The calorimetry equation was the quantitative measure in classical thermodynamics of the dependence of the change in the heat in a body on changes in its temperature or volume. It had been found by experiment that, for many bodies, setting the boundary conditions for a body by fixing any two quantities in the set $\{\beta_k, \delta_k, P_k\}$ is sufficient to determine the equilibrium thermodynamic state of the body. This state is designated macroscopically by the two quantities chosen. The change of the state of a body in classical thermodynamics is then sufficient to determine the change in its heat content by the calorimetry equation if the specific heat and latent heat as functions of the state are known.

In the step from classical to equilibrium thermodynamics, the concepts of a body, its state, and a thermodynamic process were retained, but the notion of what controls the evolution of the state of a body in a process changed. A process was redefined in equilibrium theory as the evolution over time of the values of the macroscopic physical quantities determining its equilibrium state. These processes came to be viewed as governed by the first and second axioms or laws of thermodynamics, but are otherwise unconstrained. It is seldom stated but is implicit in this statement, however, that other physical aspects of a macroscopic collection of matter, in addition to its total energy and the implications of the second law, may also take part in determining the evolution of the system.

The *bodies* of the TIS theory are macroscopic systems that occupy particular volumes with definite boundaries. These bodies can exchange energy, momentum, and other physical quantities, with each other. This definition of a body is purposely general, so that a system consisting of an empty volume of space can be a body, because even a volume devoid of material particles can transmit radiant or potential energy through its boundaries from one material body to another. This definition allows the TIS partition of the volume of the total system exhaustively into systems with boundaries that may or may not contain matter.

Because a general local time-dependent thermodynamics is the concern in TIS, the primitive quantities presented in the last chapter differ from those of standard equilibrium thermodynamics. The relations presented as axioms in this chapter also differ from those of equilibrium thermodynamics. The TIS axioms are primarily concerned with the rates of change of the primitive thermodynamic quantities with respect to the time. In spite of these differences, it will be shown in Chapter 18 that standard equilibrium thermodynamics emerges from this formalism when the underlying particle distribution is the equilibrium distribution.

The thermodynamic axioms presented in this theory depend on the properties of the microscopic particles, their interactions, and Hamiltonian mechanics. The form, for example, that the angular momentum axiom, Axiom 3, takes in this version of the theory is a consequence of the assumption that particles interact exclusively by central forces. In keeping with the fact that the primary focus of this exposition is on the quasilocal case, in which forces have finite ranges that are small compared to macroscopic dimensions, the axioms appropriate for the quasilocal case will be presented first. The changes needed for the local and the global cases will be commented on afterwards.

All thermodynamic processes are macroscopic. This means that no references to individual particles may appear in the theory. This also implies that only macroscopic means may be used to influence their progress—except in certain elliptical cases in which individual atoms are being manipulated by macroscopic instruments. A particular process in a system can therefore be defined or modified only by fixing or manipulating the macroscopic boundary conditions of a system in some specific way. This requirement that processes be macroscopic reflects the macroscopic nature of thermodynamics and the macroscopic tools available within thermodynamics to influence the direction processes will take.

3.2 The Axioms of Thermodynamics

The axioms of thermodynamics express relationships between thermodynamic quantities that are valid for any system. System specific information, which depends on the kind of matter making up the system, is contained in the constitutive relations for the type of matter in the system. In keeping with the resources available in Hamiltonian mechanics on the particle level to determine particle behavior, these constitutive relations must be obtained only from the properties of the particles and their interaction potentials.

The assumption in TIS of the boundedness of all mechanical quantities was shown in relation (2.15) to imply that all local densities of physical quantities are bounded by a quantity proportional to the local matter density and the maximum velocity. The definition of the total time derivative operator acting on macroscopic quantities in the axioms below is given in equation (2.12).

The Mechanical Axioms

Axiom 1: Matter Density.
1.1 The local matter density is non-negative:

(3.1a) $$\nu_k(q,t) \geq 0;$$

1.2 For any process, the local matter density is conserved:

(3.1b) $$\frac{d\nu_k(q,t)}{dt} = 0;$$

1.3 For any process, the global matter number is constant:

$$(3.1c) \qquad \frac{d\nu_k(t)}{dt} = 0;$$

1.4 The $[t]$ global matter number is a system sum function:

$$(3.1d) \qquad \nu_{[t]}(t) = \sum_{k \in [t]} \nu_k(t).$$

The local mass density, $\mu_k(q,t)$, is not included in this set of axioms because $\mu_i(q,t) = m_i \nu_i(q,t)$ on the particle level, so it obeys the same equations that the matter density does. Using $\mu_k(q,t)$ in place of $\nu_k(q,t)$ in equation (3.1b) yields an equation similar to the continuity equation of continuum mechanics. Next, the k *particle number* $\mathcal{N}_k$ and k *mass* $\mathcal{M}_k$ are defined in terms of the matter number $\nu_k(t)$ and mass number $\mu_k(t)$ by

$$(3.2a) \qquad \mathcal{N}_k = \nu_k(t) = \text{const.,}$$
$$(3.2b) \qquad \mathcal{M}_k = \mu_k(t) = \text{const.}$$

With this notation, Axiom 1.3 represents the conservation of particles and mass. The relations (3.1d) and (3.2a) are used with (3.1c) and (2.27a) to show

$$(3.2c) \qquad \nu_{[t]}(t) = \nu_t(t) = \mathcal{N}_t,$$
$$(3.2d) \qquad \mathcal{N}_t = \sum_{k \in [t]} \mathcal{N}_k.$$

Axiom 2: Momentum.

2.1 The local momentum current density tensor is symmetric:

$$(3.3a) \qquad \mathcal{P}_k^{\mu\nu}(q,t) = \mathcal{P}_k^{\nu\mu}(q,t);$$

2.2 For any process, the total time derivative of the local momentum density is the sum of the local body force density and the local surface force density:

$$(3.3b) \qquad \frac{d\mathcal{P}_k^\mu(q,t)}{dt} = \mathcal{F}_{k,b}^\mu(q,t) + \mathcal{F}_{\partial k}^\mu(q,t);$$

2.3 The $[t]$ global momentum is a system sum function:

$$(3.3c) \qquad \mathcal{P}_{[t]}^\mu(t) = \sum_{k \in [t]} \mathcal{P}_k^\mu(t);$$

2.4 The total global momentum is conserved in any process:

$$(3.3d) \qquad \frac{d\mathcal{P}_{[t]}^\mu(t)}{dt} = \frac{d\mathcal{P}_t^\mu(t)}{dt} = 0;$$

It follows from the relations (3.3c), (3.3d), and (2.27b) that

(3.4) $$\mathcal{P}^{\mu}_{[t]}(t) = \mathcal{P}^{\mu}_{t}(t) = \mathcal{I}^{\mu}_{P} = \text{const.}$$

Axiom 2.2 is the macroscopic version of Newton's second law. In addition, the definition (2.12) implies that equation (3.3b) can be written in the laboratory frame in terms of the negative divergence of the momentum current density tensor, the body force density, and the surface force density, as

(3.5) $$\frac{\partial \mathcal{P}^{\mu}_{k}(q,t)}{\partial t} = -\frac{\partial \mathcal{P}^{\mu\nu}_{k}(q,t)}{\partial q^{\nu}} + \mathcal{F}^{\mu}_{k,b}(q,t) + \mathcal{F}^{\mu}_{\partial k}(q,t).$$

Axiom 3: Angular Momentum.

3.1 The local angular momentum density is the vector cross product of the position and the local momentum density:

(3.6a) $$\mathcal{J}^{\mu}_{k}(q,t) = \epsilon^{\mu\nu\xi} q_{\nu} \mathcal{P}_{k,\xi}(q,t);$$

3.2 The local angular momentum current density is the vector cross product of the position and the local momentum current density tensor:

(3.6b) $$\mathcal{J}^{\mu\nu}_{k}(q,t) = \epsilon^{\mu\eta\xi} q_{\eta} \mathcal{P}^{\nu}_{k,\xi}(q,t);$$

3.3 For any process, the total time derivative of the angular momentum density is the vector cross product of the position and the sum of the local body force density and the local surface force density:

(3.6c) $$\frac{d\mathcal{J}^{\mu}_{k}(q,t)}{dt} = \epsilon^{\mu\nu\xi} q_{\nu} [\mathcal{F}_{k,b,\xi}(q,t) + \mathcal{F}_{\partial k,\xi}(q,t)];$$

3.4 The $[t]$ global angular momentum is a system sum function:

(3.6d) $$\mathcal{J}^{\mu}_{[t]}(t) = \sum_{k \in [t]} \mathcal{J}^{\mu}_{k}(t);$$

3.5 The total global angular momentum is conserved in any process:

(3.6e) $$\frac{d\mathcal{J}^{\mu}_{[t]}(t)}{dt} = \frac{d\mathcal{J}^{\mu}_{t}(t)}{dt} = 0.$$

The relation

(3.7) $$\frac{d\mathcal{J}^{\mu}_{k}(q,t)}{dt} = \epsilon^{\mu\nu\xi} q_{\nu} \frac{d\mathcal{P}_{k,\xi}(q,t)}{dt}$$

follows from Axioms 3.1, 3.2 and 2.1 and the fact that $dq/dt = 0$ because q is a fixed point in the macroscopic frame and not a particle coordinate. By (3.7) and Axiom 2, it follows that Axioms 3.3, 3.4, and 3.5 are redundant for the case of central forces used here. It follows also, using (3.6d), (3.6e) and (2.27c), that

(3.8) $$\mathcal{J}^{\mu}_{[t]}(t) = \mathcal{J}^{\mu}_{t}(t) = I^{\mu}_{J} = \text{const.}$$

The system energy density axioms are introduced next. One-half the external potential energy density is used in the expression defining the energy because one-half the external potential energy density is attributed to each of the systems in which the two particles reside.

Axiom 4: Energy.

4.1 The local energy density is the sum of the local kinetic energy density, the local internal potential energy density, and one-half the local external potential energy density:

$$(3.9a) \qquad \mathcal{E}_k(q,t) = \mathcal{K}_k(q,t) + \Phi_{k,n}(q,t) + \tfrac{1}{2}\Phi_{k,x}(q,t);$$

4.2 For any process, the total time derivative of the local energy density is the sum of the local surface heating density, the local body working density, and the local surface working density:

$$(3.9b) \qquad \frac{d\mathcal{E}_k(q,t)}{dt} = \mathcal{Q}_{\partial k}(q,t) + \mathcal{W}_{k,b}(q,t) + \mathcal{W}_{\partial k}(q,t);$$

4.3 The $[t]$ global energy is a system sum function:

$$(3.9c) \qquad \mathcal{E}_{[t]}(t) = \sum_{k\in[t]} \mathcal{E}_k(t);$$

4.4 The total global energy is conserved in any process:

$$(3.9d) \qquad \frac{d\mathcal{E}_{[t]}(t)}{dt} = \frac{d\mathcal{E}_t(t)}{dt} = 0.$$

The relations (3.9c), (3.9d) and (2.27d) are used to obtain

$$(3.10) \qquad \mathcal{E}_{[t]}(t) = \mathcal{E}_t(t) = \mathcal{I}_{H,k} = \text{const.}$$

The k *local Hamiltonian density* is written as $\mathcal{H}_k(q,t) = \mathcal{K}_k(q,t) + \Phi_k(q,t)$.

Axiom 5: Surface Force Density and Surface Energy Flow.

5.1 The total global surface force density vector is zero:

$$(3.11a) \qquad \sum_{k\in[t]} \mathcal{F}_{\partial k}(t) = 0;$$

5.2 The total global surface energy change is zero:

$$(3.11b) \qquad \sum_{k\in[t]} [\mathcal{Q}_{\partial k}(t) + \mathcal{W}_{\partial k}(t)] = 0.$$

It follows from the global version of (3.3b), and from (3.3c), (3.3d), and (3.11a), that

$$(3.12) \qquad \sum_{k\in[t]} \mathcal{F}_{k,b}(t) = 0.$$

The Thermodynamic Axioms

The thermodynamic axioms are concerned with entropy, temperature and pressure. These axioms place constraints on the behavior of the system.

Axiom 6: Entropy.

6.1 The entropy of a system is unbounded below and bounded above by the equilibrium entropy of that system:

$$(3.13a) \qquad -\infty \leq S_k(t) \leq S_k^\epsilon,$$

where S_k^ϵ is the equilibrium entropy; both $S_k(t)$ and S_k^ϵ are evaluated with the same macroscopic constraints on system k;

6.2 (Clausius' inequality) For any process, the local rate of change of the entropy is bounded from below:

$$(3.13b) \qquad \frac{dS_k(q,t)}{dt} \geq k_B \beta_k(q,t) \mathcal{Q}_{\partial k}^*(q,t);$$

6.3 The $[t]$ entropy is a system sum function:

$$(3.13c) \qquad S_{[t]}(t) = \sum_{k \in [t]} S_k(t);$$

The quantity $\mathcal{Q}_{\partial k}^*(q,t)$ that appears in Clausius' inequality is defined in terms of the surface heating $\mathcal{Q}_{\partial k}(q,t)$ and surface working $\mathcal{W}_{\partial k}(q,t)$ as

$$(3.14a) \qquad \mathcal{Q}_{\partial k}^*(q,t) = \mathcal{Q}_{\partial k}(q,t) + \lambda^*(\mathcal{C}_{H,k}) \mathcal{W}_{\partial k}(q,t),$$

where the range of energies available to the k system, $\mathcal{C}_{H,k}$, is defined in (2.30). The function $\lambda^*(\mathcal{C}_{H,k})$ is bounded and small for large systems. It is computed in Chapter 9.

The difference between (3.13b) and the standard version of Clausius' inequality is due to two factors: the boundedness of the range of energies available to the k system and the (possible) motion of the system boundaries. When the boundaries of the system are at rest or the surface force $F_{\partial k}^\mu(q,t)$ vanishes, it follows from (2.19) that $\mathcal{W}_{\partial k}(q,t) = 0$ and $\mathcal{Q}_{\partial k}^*(q,t) = \mathcal{Q}_{\partial k}(q,t)$. If the total k system energy range is unbounded, the result

$$(3.14b) \qquad \lim_{\mathcal{C}_{H,k} \to \infty} \lambda^*(\mathcal{C}_{H,k}) = 0$$

is used to show that $\mathcal{Q}_{\partial k}^*(q,t) = \mathcal{Q}_{\partial k}(q,t)$ in the thermodynamic limit. In either of these cases, (3.13b) reduces to the standard version of Clausius inequality.

Two further definitions will complete the characterization of entropy. A process is said to be *reversible* if, for each $q \in d_k(t)$, the relation

$$(3.15) \qquad \frac{dS_k(q,t)}{dt} = k_B \beta_k(q,t) \mathcal{Q}_{\partial k}^*(q,t)$$

is true. The k system is said to be in *thermodynamic equilibrium* at time t, if its walls are at rest and the global entropy is maximal and constant, i.e.,

$$(3.16a) \qquad S_k(t) = S_k^\epsilon,$$

$$(3.16b) \qquad \frac{dS_k(t)}{dt} = 0.$$

It will be shown subsequently that this definition implies that the system is in both the macroscopic system rest frame and the macroscopic particle rest frame.

Axiom 7: Temperature.
7.1 The local thermature tensor is symmetric:

$$(3.17a) \qquad \beta_k^{\nu\mu}(q,t) = \beta_k^{\mu\nu}(q,t);$$

7.2 The global thermature tensor is an average of the local thermature tensor weighted by the local pressure:

$$(3.17b) \qquad \beta_k^{\mu\nu}(t) = \frac{1}{P_k(t)} \int d^3q\, P_k(q,t)\beta_k^{\mu\nu}(q,t).$$

The definition (3.17b) is a consequence of the facts that the local thermature tensor is defined in terms of the local pressure in Chapter 12 and is an intensive quantity. The local thermature is defined as one-third the trace of the local thermature tensor

$$(3.18) \qquad \beta_k(q,t) = \tfrac{1}{3}\delta_{\mu\nu}\beta^{\mu\nu}(q,t).$$

The relation between the absolute temperature and the system thermature is

$$(3.19) \qquad T_k(t) = \frac{1}{k_B \beta_k(t)}.$$

Axiom 8: Pressure.
8.1 The local pressure tensor is the local momentum current density tensor minus the local internal and external virial tensors:

$$(3.20a) \qquad P_k^{\mu\nu}(q,t) = \mathcal{P}_k^{\mu\nu}(q,t) - \Xi_{k,n}^{\mu\nu}(q,t) - \Xi_{k,x}^{\mu\nu}(q,t).$$

8.2 The k *local pressure* is one third the trace of the pressure tensor:

$$(3.20b) \qquad P_k(q,t) = \tfrac{1}{3}\delta_{\mu\nu}P_k^{\mu\nu}(q,t).$$

The k *pressure* is the average of this quantity over the volume:

$$(3.20c) \qquad P_k(t) = \frac{1}{\delta_k(t)} \int d^3q\, P_k(q,t).$$

8.3 The $[t]$ pressure is a system sum function weighted by the volume:

$$(3.20d) \qquad P_{[t]}(t) = \frac{1}{\delta_t(t)} \sum_{k \in [t]} \delta_k(t) P_k(t).$$

3.3 The Local Version of the Axioms

As noted in the introduction, the local versions of the axioms are used if the underlying particle theory does not contain any action at a distance potentials. The ideal gas is one example of a local theory without any interactions between its particles. Collisions in theories with local interactions between particles either are hard sphere collisions or are mediated by explicit fields that have local field-particle interactions. In either of these cases, the local matter density, momentum, angular momentum, and energy, are locally conserved.

3.4 The Global Version of the Axioms

Except for Axiom 6, the global version of each axiom, expressed in terms of $G_k(t)$ rather than $G_k(q,t)$, parallels the quasilocal version. For Axiom 6, the global version requires the replacements $\beta_k(q,t) \to \beta_k(t)$ and $\mathcal{Q}_{\partial k}(q,t) \to \mathcal{Q}_{\partial k}(t)$ for $q \in \partial d_k(t)$. Using these facts in (3.13b) yields

$$(3.21) \qquad \frac{dS_k(t)}{dt} \geq k_B \beta_k(t) \mathcal{Q}_{\partial k}^*(t).$$

3.5 Comparison with Continuum Thermodynamics

The major differences between the above set of axioms and those of rational continuum thermodynamics as expressed by Coleman and Noll [1963] or Gurtin and Williams [1967] are:

(1) There is no axiom or theorem in this thermodynamics that corresponds to the axiom "heat always flows from hot bodies to cold bodies" of continuum mechanics.[1] In the notation of this theory, this means that neither

$$(3.22a) \qquad \sum_{k \in [t]} \beta_k(q,t) \mathcal{Q}_{\partial k}^*(q,t) \geq 0,$$

---

[1] As the calculation of Thomson [1873], pp. 15–16, implies and Truesdell [1969], p. 35, affirms, heat can flow from cold to hot. Thought experiments of this type were proposed by Toliver Preston and discussed by Clausius [1878], pp. 237–238, in which a cold light gas and a heavy warm one were separated by a porous plug. The same experiments can also be used to show that a gas can flow from a low pressure volume into a high pressure one. This means that formulations of the second law in terms of heat flow and work require a statement concerning the state of the matter, the initial conditions, and the boundary conditions if they are to be correctly stated and precise. See the critiques of Maxwell [1878b] and Boltzmann [1878b] on this.

nor

$$(3.22b) \qquad \sum_{k \in [t]} \int d^3q \, \beta_k(q,t) \mathcal{Q}_{\partial k}^*(q,t) \geq 0,$$

can be proved in general. This, in turn, means that neither

$$(3.23a) \qquad \frac{dS_{[t]}(q,t)}{dt} \geq 0,$$

nor

$$(3.23b) \qquad \frac{dS_{[t]}(t)}{dt} \geq 0,$$

can be demonstrated in the general version of TIS.

(2) The TIS entropy, as opposed to its rate of change, is defined only globally as $S_k(t)$ and there is no local entropy density $S_k(q,t)$ (sometimes called the "specific entropy") and no associated local entropy current density vector $S_k^\mu(q,t)$. However, these quantities do exist in the Boltzmann version of the theory presented in Part III.

(3) Equation (3.1b), with $\mu_k(q,t)$ in place of $\nu_k(q,t)$, and the equations (3.5), (3.6c), and (3.9b), represent the TIS version of the Euler-Cauchy equations.

(4) The pressure tensor formulated here is based on the underlying particle model and is not the stress tensor of continuum mechanics. As will be shown in Chapter 11, the TIS pressure tensor does not play a role in any natural way in the Euler-Cauchy equations of continuum mechanics.

(5) The process of mixing is viewed here as occurring within one system and not as a process in which two systems are combined. The latter process is excluded in this theory by the requirement that each system be closed.

3.6 Observation in Thermodynamics

The local functions themselves may be compared with the results of observation. However, as John Kirkwood [1946] has emphasized, actual experiments used to measure physical quantities have limited resolution in space and time. To reflect this fact, let $\Delta(q,t)$ be a small, but macroscopic, region of spacetime centered at the point (q,t) that will be referred to as a *quasilocal domain*. The size of Δ is determined by the sensitivity of the experimental apparatus and the length of time required to make a measurement. The volume of spacetime associated with the volume set $\Delta(q,t)$ in a given experiment is

$$(3.24) \qquad \mathfrak{d}(\Delta(q,t)) = \int dq' \int dt' \, A_{\Delta(q,t)}(q',t'),$$

where $A_{\Delta(q,t)}(q',t')$ is the *acceptance function* representing the properties of the instruments used to make the observation. In the simplest case, the acceptance function is the indicator function for the set $\Delta(q,t)$, i.e., $A_{\Delta(q,t)}(q',t') = \chi_{\Delta(q,t)}(q',t')$. In the general case, the quasilocal domain volume $\eth$ is used with any local thermodynamic density, $G_k(q,t)$, to define the corresponding *local G_k measurable density* in the form

$$(3.25) \qquad G_k(\Delta(q,t)) = \frac{1}{\eth(\Delta(q,t))} \int dq' \int dt'\, A_{\Delta(q,t)}(q',t') G_k(q',t').$$

A local measurable thermodynamic density $G_k(\Delta(q,t))$ is distinguished from a local thermodynamic density $G_k(q,t)$ by the argument of the function representing it.

For the global observables, the quasilocal domain is the time interval $\Delta(t)$, which is the duration of the observation. The length of this time interval is given by the measure $\mu(\Delta(t))$. To express the G_k measurable density in terms of the G_k local measurable density, the system volume set $d_k(t)$ is also partitioned into a set of non-overlapping volumes $\Delta^a(q_a,t)$ indexed by π with a representative point $(q_a,t) \in \Delta^a(q_a,t)$ such that $\cup_{a\in\pi}\Delta^a(q_a,t) = d_k(t)$.[2] Each $\Delta^a(q_a,t)$ is constructed to cover the same time interval as $\Delta(t)$. These are used in (3.25) to write the G_k *measurable density* in the form

$$(3.26) \qquad G_k(\Delta(t)) = \frac{1}{\mu(\Delta(t))} \sum_{a\in\pi} \eth(\Delta^a(q_a,t)) G_k(\Delta^a(q_a,t)).$$

The use of thermodynamic observable averaging is necessary with local exact densities obtained by phase averaging a thermodynamic function over an exact state. An example of this appears in Section 5.7.

If the measuring instrument has a limited resolution with respect to momentum or energy, an acceptance function A can be defined to account for this. A thermodynamic function $G_k(q,p,t)$, which depends on p as well as q and t, can also be used in place of $G_k(q,t)$. A volume $\Delta^*(q,p,t)$ in (q,p) phase space and an acceptance function of the form $A_{\Delta^*(q,p,t)}(q',p',t)$ can both be defined as well. These can be used in (3.24) and (3.25), along with an additional integration over p', to obtain the local observable average density $G_k(\Delta^*(q,p,t))$.

For future reference, observe that the averaging involved in the definition of a measurable density in (3.25) is applied after phase averaging. It differs from the procedure used in coarse-graining the entropy that involves coarse-graining the particle distribution itself.

3.7 Additional Remarks

The formalism developed in this theory is easily generalizable to systems in which there are several macroscopically distinguishable components. This

---

[2]See Section 6.3 for more details on such a partition.

follows from the fact that many of the formulas of statistical mechanics developed below deal with sums of individual particle functions that can be grouped and summed separately for each component. With a few exceptions, this refinement will not be pursued further.

In the introduction, it was pointed out that the thermodynamics presented here is not closed to new concepts in that new primitive thermodynamic quantities can be added as desired. This is done by adding new analog functions to the statistical mechanics and working out the axioms that govern the behavior of the phase averages of these functions. It is also true that the set of elements in this theory is not closed under the action of the operators of the theory that can be applied to the primitive quantities of the theory. That is, successive applications of the time, volume, temperature, or energy derivative operators will lead to functions that are not included in the primitive set.

The set of relationships between the local nonequilibrium thermodynamic quantities is not as rich as the set for standard equilibrium thermodynamic quantities. This is to be expected due to the generality of the underlying particle distribution. As will be shown in Part II, these relationships are recaptured in the equilibrium version of this theory.

CHAPTER 4

Statistical Mechanics

It is time to introduce the first components of the microscopic side of the Theory of Interacting Systems. Because of the roles played by volumes and boundaries of the subsystems in the theory, the classical Hamiltonian statistical mechanics presented here is extended to include particle systems on a manifold with a boundary.

4.1 Preliminary Remarks

The point of dividing the world into subsystems and selecting one for attention is that this procedure allows us to focus on what is of interest and treat the rest as external. This means that the microscopic coordinates of the particles in external systems are not taken account of directly and their influence is either ignored or accounted for by potentials that represent the average collective effects of the interactions of the external particles with particles in the system under consideration. The implications of this practice will be discussed in Chapter 6 in connection with Maxwell's observation in 1877 that this is the standard practice of physics.

Because a relationship is being established between the macroscopic and microscopic descriptions of the same object, the same partition of the t system volume set that was used for thermodynamics in Chapter 2 will be used for the particle based aspect of the theory. Recall that the symbol $[t]$ represents the system $[t]$, which is the collection of subsystems of t treated together as one system, and the collection of labels $k, l, \ldots$ of the subsystems of t. For convenience, every subsystem discussed below will be assumed to belong to $[t]$ even when this is not stated explicitly. The distinction between the description of a subsystem k and its description as a collection of particles that are part of the isolated system t appears below as the difference between what are called the *system specific Hamiltonian trajectories* for particles in k and the Hamiltonian trajectories associated with the particles $i \in t$ for which $q_i \in d_k(t)$ so that $i \in k$ as well.

The boundary set, $\partial d_k(t)$, is a 2-dimensional dividing set that separates the particles in the k system from the particles in every other system. Each subsystem is assumed to be closed so that no particles enter or leave it. The boundary itself is not allowed to accumulate or dissipate particles, momentum, angular momentum, or energy. With the exception of particles, the boundary does transmit these quantities to and from other systems. This flow will contribute to the change of the k system distribution at the boundary. This change is accounted for in Chapter 5 by showing how the flow of physical quantities through the boundary of k changes the k distribution.

Because each system is closed, the symbol $\mathcal{N}_k$, which is defined as $\nu_k(t)$ in (3.2a), is used to represent the constant number of particles in the system. Every particle in the t system appears in one of the systems $k \in [t]$ and vice versa, so it follows that

$$(4.1a) \qquad \mathcal{N}_t = \sum_{k \in [t]} \mathcal{N}_k.$$

As in the case of the t system, the particles in the k system are indexed and the symbol k represents the set of particle labels in the index as well as representing the system itself. The context will determine which meaning of k is in effect. Using this notation, the particle number in system k is given by

(4.1b) $$\mathcal{N}_k = \sum_{i \in k} 1.$$

Each system, whether isolated or a subsystem, will be described by a system distribution density.

Mechanical results in TIS are required to be independent of, or covariant with, choices made by convention in mechanics and statistical mechanics. This includes the choices of the origin, velocity, and orientation, of the phase space frame of reference for a system. Thermodynamic results, which are computed as phase averages of mechanical analogs, are also required to be independent of these conventional choices for the frame of reference in the same sense. Once a particular frame of reference has been chosen, it is used with all systems.

Let us assume that the k volume set $d_k(t)$ contains $\mathcal{N}_k$ localizable and distinguishable particles that are not chemically reactive and do not have internal energy states. As mentioned before, radiation is also neglected. To simplify calculations, it will be assumed that all interactions can be represented as pair-wise particle-particle interactions and that the total force on a particle is the sum of the pair-wise forces acting on it.[1] It will also be assumed that the forces between particles are central forces. Under these assumptions, the interaction between each pair of particles $i, j \in k$ is described by an interaction potential energy function, $\phi_{ij}(q_i - q_j)$. If there is no interaction between the particles, as in the ideal gas case, $\phi_{ij}(q_i - q_j) = 0$ for each i, j pair.

From the standpoint of statistical mechanics, an isolated system t is one for which (i) there is no interaction between a particle in the system and any particle outside it; (ii) the system is closed (no particles cross $\partial d_t(t)$); and (iii) there is no flow of moments of momentum across the boundary. The consequences of this definition for certain system functions will be explored in succeeding chapters. The macroscopic functions obtained from this formalism by phase averaging will be used to reconcile the microscopic statistical mechanical definition of an isolated system with the macroscopic thermodynamic description of an isolated system given in equations (2.21) of Chapter 2.

4.2 Particle Interaction Potentials

Each member of the set of potential energy functions $\{\phi_{ij}(q_i - q_j)\}$ depends only on the pair of interacting particles, i and j, and the distance

---

[1] For a discussion of this point in classical, atomic, and nuclear systems, see H. Primakoff and T. Holstein [**1939**]. See also H. Margenau and J. Stamper [**1967**] and M. Jammer [**1957**], pp. 251–253, and the references cited in these papers for discussions of the possible failure of this assumption.

$r = |q_i - q_j|$ between them. Because interparticle interactions are tracked individually in the microscopic formalism, the functions ϕ_{ij} are allowed to be unique for each particle pair. The physical model corresponding to this representation of the particle interactions is that of a set of particles without internal degrees of freedom acting on each other at a distance via central forces.

Some realistic and specific assumptions about the behavior of these functions are made next that will facilitate the analysis below. Assume that each function in the set $\{\,\phi_{ij}\,\}$ satisfies the following continuity, boundedness and symmetry conditions:

(4.2a)   (i) $\phi_{ij}(r) \in C^m(\mathbf{R}), \quad m \geq 3, \quad r \geq r_c > 0;$

(4.2b)   (ii) $\phi_{ij}(r) \geq -\phi_0, \quad 0 \leq \phi_0 < \infty;$

(4.2c)   (iii) $\lim_{r \to 0} \phi_{ij}(r) r^a = \infty, \quad a \leq 3;$

(4.2d)   (iv) $\lim_{r \to \infty} \phi_{ij}(r) r^b = 0, \quad b \leq 3;$

(4.2e)   (v) $\dfrac{\partial \phi_{ij}(r)}{\partial q_i} = -\dfrac{\partial \phi_{ij}(r)}{\partial q_j} = (q_i - q_j)\rho_{ij}(r);$

(4.2f)   (vi) $\lim_{r \to 0} \dfrac{d\phi_{ij}(r)}{dr} r^{(a+1)} = -\infty, \quad a \leq 3;$

(4.2g)   (vii) $\lim_{r \to \infty} \dfrac{d\phi_{ij}(r)}{dr} r^{(b+1)} = 0, \quad b \leq 3;$

(4.2h)   (viii) $\|\phi\| = \sup_{i,j \in t} \sup_{1 \leq \mu \leq 3} \sup_{g \leq m} \sup_{r \geq r_c} \left| \dfrac{\partial^g \phi_{ij}(r)}{(\partial q_{i,\mu})^g} \right| < \infty;$

(4.2i)   (ix) $\phi_{ji} = \phi_{ij}, \quad \rho_{ji} = \rho_{ij}.$

As before, the quantity $\mathbf{R}$ represents the set of real numbers. The symbol $C^m(\Upsilon)$ represents the space of m times continuously differentiable functions from $\Upsilon \subset \mathbf{R}$ into $\mathbf{R}$. The quantity r_c, computed in (4.28) below, is the closest any two particles can approach each other. The fact that $r_c > 0$ is a consequence of (4.2c) and the fact that t system total energy is finite and fixed. The quantity $\|\phi\|$ is called the *semi-norm* of ϕ_{ij}. Most of the above assumptions on the interaction potentials are quite standard.

4.3 The System Mechanical State

The quantities q_i and p_i are the real i-particle position and momentum 3-vectors. The ordered pair (q_i, p_i) is the classical *i-particle mechanical state*. In addition to using lower case letters q_i and p_i to represent the position and momentum of particle i, the upper case letters Q_k and P_k are used to represent the $3\mathcal{N}_k$-dimensional direct products

(4.3a) $Q_k = \times_{i \in k} q_i,$

(4.3b) $P_k = \times_{i \in k} p_i.$

The ordered pair (Q_k, P_k) is the k-*system mechanical state*. These variables are the canonical variables in the $6\mathcal{N}_k$-dimensional Hamiltonian description. The phase space for system k is the Euclidean space $\mathbf{R}^{6\mathcal{N}_k}$, which is the $6\mathcal{N}_k$-dimensional direct product of $\mathbf{R}$.

For $i \in k$, let m_i and v_i, where $v_i = \dot{q}_i$, be the mass and velocity of particle i. Since radiation fields have been neglected in these calculations, the relation between the canonical momentum and the velocity is

(4.4a) $$p_i = m_i v_i.$$

The quantity

(4.4b) $$V_k = \times_{i \in k} v_i$$

is the $3\mathcal{N}_k$-dimensional system velocity vector.

4.4 The Hamiltonian Formalism

A *particle sum function* is a phase function G_k that can be decomposed into a contribution from each particle and written in the form

(4.5) $$G_k(Q_k, P_k, t) = \sum_{i \in k} G_i(Q_k, p_i, t).$$

Particle sum functions will be important in what follows because they are the analogs of extensive quantities in thermodynamics. The phase analog $G_i(Q_k, p_i, t)$ refers to a quantity that can be localized at a particle and summed for all the particles. The argument of G_i reflects the properties of a single particle and the possible influences on it. It may depend on the phase variables p_i, (q_i, p_i), (Q_k, p_i), or Q_k, and the time. This analog can depend on Q_k because particle i can potentially interact with all the other particles in k the system. On the other hand, G_i is a function of p_i and not P_k because only the momentum p_i is a property of the particle i. Examples of particle sum functions are the total momentum, total angular momentum, and total energy functions of a system.

The *i-particle kinetic energy* and *k kinetic energy* are expressed in terms of these variables by

(4.6a) $$K_i(p_i) = \tfrac{1}{2} p_i \cdot v_i = \frac{p_i{}^2}{2m_i},$$

(4.6b) $$\mathcal{K}_k(P_k) = \sum_{i \in k} K_i(p_i).$$

The potential energy between particle $i \in k$ and other particles in the k system, called *i-particle internal potential energy*, is

(4.7a) $$\Phi_{i,n}(Q_k) = \sum_{j \in k-i} \phi_{ij}(q_i - q_j),$$

where $k - i$ refers to the set of indices in k with the exception of the index i. Summing this over $i \in k$ and dividing by 2 gives the k *internal potential energy*

$$(4.7b) \qquad \Phi_{k,n}(Q_k) = \tfrac{1}{2} \sum_{i \in k} \Phi_{i,n}(Q_k),$$

where the factor $\tfrac{1}{2}$ is introduced to compensate for double counting of particles.

The representation of the k system must be expressed exclusively in terms of the microscopic particle variables for k system particles. This standard practice is shown in Chapter 6 to be one consequence of requiring the independence of subsystems from each other. It is necessary to express this representation in a way that respects the fact that particles in different subsystems can interact with each other and is consistent with the fact that the subsystems are parts of a larger mechanical system. The k system external potential $\Phi_{k,x}(Q_k, t)$ is defined next as the particle interaction potential that represents the interaction between particles in the k system and those belonging to all other systems. The one-particle version of the k system external potential for particle $i \in k$ is $\Phi_{i,x}(q_i, t)$. It is defined for each external system l and particle $j \in l$ that is interacting with particle $i \in k$ as the phase average of $\phi_{ij}(q_i - q_j)$ over the l system distribution density. Similarly, the external potential $\Phi_{j,x}(q_j, t)$ for particle $j \in l$ due to particle $i \in k$ is the average of $\phi_{ij}(q_i - q_j)$ over the k system distribution density. The phase average at a given time t over the system l will be represented in a convenient bracket notation by $\langle \phi_{ij}(q_i - q_j) \rangle_{lt}$. This notation will be defined explicitly and discussed in Chapter 5.

The potential energy between particle i in k and those in all the systems in $[t]$ except k is called the *i-particle external potential energy*. The sum of this over $i \in k$ is the k *external potential energy*. They are defined by

$$(4.8a) \qquad \Phi_{i,x}(q_i, t) = \sum_{l \in [t] - k} \sum_{j \in l} \langle \phi_{ij}(q_i - q_j) \rangle_{kl},$$

$$(4.8b) \qquad \Phi_{k,x}(Q_k, t) = \sum_{i \in k} \Phi_{i,x}(q_i, t).$$

where $[t] - k$ is the set of subsystems in $[t]$ with the exception of k.

The sum of the internal and external particle potentials are the i *particle phase potential* and k *phase potential* defined as

$$(4.9a) \qquad \Phi_i(Q_k, t) = \Phi_{i,n}(Q_k) + \Phi_{i,x}(q_i, t),$$

$$(4.9b) \qquad \Phi_k(Q_k, t) = \Phi_{k,n}(Q_k) + \Phi_{k,x}(Q_k, t).$$

Because a factor $\tfrac{1}{2}$ appears in the definition of $\Phi_{k,n}(Q_k)$ but not in the definition of the external potential $\Phi_{k,x}(Q_k, t)$, the internal and total k phase potentials are not particle sum functions. This will lead to a distinction below between the system Hamiltonian energy and the system total energy.

One difference between the definitions (4.7)–(4.9) of the potentials in Hamiltonian mechanics and those that are often used in standard thermodynamics is the fact that "constraint parameters", representing macroscopic constraints on the behavior of the particles, are not included. Parameters of this type were included in Hamiltonian functions by Clausius, Boltzmann, and Gibbs, as adjustable parameters that could be used to represent a fixed experimental setup or modified in time to represent the effects of changing macroscopic conditions on the particle system.[2] However, these constraints, which are usually based on the volume and temperature of a system, represent macroscopic choices concerning how an experiment is arranged and are not properly part of the microscopic description of the particle system. From a microscopic perspective, particles feel the pushes and pulls of other particles and nothing more. Thus, if there is an external potential, for example, which confines the particles to the given volume, that potential is already included in the total potential that is part of the mechanical Hamiltonian for the system. Nothing else is needed or allowed.

In sum, macroscopic constraints are not part of the Hamiltonian mechanics and do not belong in the Hamiltonian. The TIS approach to the confinement of particles to a subsystem of an isolated system will be mentioned in Chapter 5 and a macroscopic stochastic boundary formalism supporting it will be introduced in Chapter 6.

Turning now to the k system momenta, the $3\mathcal{N}_k$-dimensional k *phase momentum* vector and the 3-dimensional k *total phase momentum* vector are defined by

(4.10a)
$$P_k^\mu = \times_{i \in k} p_i^\mu,$$

(4.10b)
$$\mathcal{P}_k^\mu(P_k) = \sum_{i \in k} p_i^\mu.$$

The *i-particle angular momentum*, the k *phase angular momentum*, and the k *total phase angular momentum*, are defined by

(4.11a)
$$j_i^\mu(q_i, p_i) = \epsilon^{\mu\nu\xi} q_{i,\nu} p_{i,\xi},$$

(4.11b)
$$J_k^\mu(Q_k, P_k) = \times_{i \in k} j_i^\mu(q_i, p_i),$$

(4.11c)
$$\mathcal{J}_k^\mu(Q_k, P_k) = \sum_{i \in k} j_i^\mu(q_i, p_i).$$

The definition of the $3\mathcal{N}_k$-dimensional k system momentum and angular momentum raises another issue, mentioned above, concerning the choice of a momentum reference frame. A choice is implicit in the use of the vectors

---

[2]See the discussion of the volume as a constraint parameter in Gibbs' work by Bailyn [**1994**], pp. 497–498.

p_i and P_k in (4.10) because these vectors are defined in a frame with an origin, a scale, and an orientation. Momentum based functions, such as the kinetic energy component $\mathcal{K}_k(P_k)$ of system k, are expressed in terms of the absolute value of the momentum in this frame and are not invariant with respect to a Galilean velocity translation (boost) transformation $P_t \to P_t'$ of the whole phase momentum reference frame for all the particles in all subsystems. This behavior is distinct from that of the potential $\Phi_k(Q_k, t)$, which is invariant with respect to a Galilean transformation $Q_t \to Q_t'$ of the whole phase coordinate reference frame.

Consider a Galilean reference frame designated by $\flat$ in which the particles of a system k are all at rest. In this frame, it follows that $p_i^{(\flat)} = 0$ for $i \in k$, the $3\mathcal{N}_k$-dimensional system momentum vector is $P_k^{(\flat)} = \times_{i\in k} p_i^{(\flat)} = 0$, and the total system momentum vanishes: $\mathcal{P}_k^{(\flat)} = \sum_{i\in k} p_i^{(\flat)} = 0$. The kinetic energy also vanishes, $\mathcal{K}_k(P_k^{(\flat)}) = \sum_{i\in k}[p_i^{(\flat)}]^2/2m_i = 0$. Consider next a frame $\sharp$ moving at a velocity $V_a^{(\flat)\mu} = \times_{i\in k} v_a^{(\flat)\mu}$ with respect to the $\flat$ frame. To represent the momentum and kinetic energy of the system in the new frame, the constant momentum vectors $p_{a,i}^{(\flat)} = m_i v_a^{(\flat)}$ for $i \in k$ and $P_a^{(\flat)} = \times_{i\in k} p_{a,i}^{(\flat)}$ will be used. The kinetic energy of the k system in the frame $\sharp$ is then $\mathcal{K}_k(P_k^{\sharp}) = \mathcal{K}_k(P_k^{(\flat)} - P_a^{(\flat)}) = \sum_{i\in k}([p_i^{(\flat)}]^2/2m_i - p_i^{(\flat)} \cdot v_a^{(\flat)} + [p_{a,i}^{(\flat)}]^2/2m_i)$. This shows that the kinetic energy depends on the choice of an origin for the momentum vectors in the phase space coordinate and momentum reference frames. In keeping with the TIS requirement that the theory is independent of the choice of an origin for the phase space coordinate and momentum frames of reference, the vanishing of the k system total kinetic energy cannot be used per se as a particle-based criterion for the classical absolute zero state.

This arbitrariness in the kinetic energy means that thermodynamic quantities that are defined in terms of it have an element of arbitrariness also. To remove this arbitrariness, a special reference frame, called the system rest frame and defined macroscopically as the rest frame of the system boundaries or walls, was introduced in Chapter 2. A second special reference frame is required for equilibrium calculations and will be introduced next. This special reference frame is based on the microscopic center of momentum frame in which both the system total momentum and total angular momentum vanish.

The system total momentum and total angular momentum vectors are the sum functions $\mathcal{P}_k = \sum_{i\in k} p_i$ and $\mathcal{J}_k(Q_k, P_k) = \sum_{i\in k} j_i(q_i, p_i)$ defined in (4.10) and (4.11). The microscopic k *center of momentum frame* is defined for any system k, whether $k \in [t]$, $k = [t]$, or $k = t$, as the frame in which the microscopic conditions $\mathcal{P}_k = 0$ and $\mathcal{J}_k(Q_k, P_k) = 0$ are valid. The thermodynamic total momentum $\mathcal{P}_k(t) = \langle \mathcal{P}_k \rangle_{kt}$ and total angular momentum $\mathcal{J}_k(t) = \langle \mathcal{J}_k(Q_k, P_k) \rangle_{kt}$, obtained by phase averaging the total momentum and total angular momentum, are macroscopically measurable. The *particle rest frame* is defined macroscopically in terms of these quantities by the conditions $\mathcal{P}_k(t) = 0$ and $\mathcal{J}_k(t) = 0$ and coincides with the microscopic center of momentum frame.

The particle rest frame is distinct from the system rest frame in that it is possible for the system to be in one rest frame and not the other. Another distinction is that the particle rest frame has a microscopic analog and the system rest frame does not. The relationship between these two rest frames is important in the definitions of the equilibrium and absolute zero states presented in Chapters 13 and 14.

The *i-particle phase Hamiltonian* is the sum of the kinetic energy defined in (4.6a) and the internal and external potentials defined in (4.7b) and (4.8c),

$$(4.12a) \qquad H_i = H_i(Q_k, p_i, t) = K_i(p_i) + \Phi_i(Q_k, t).$$

The *k phase Hamiltonian* is

$$(4.12b) \qquad \mathcal{H}_k = \mathcal{H}_k(Q_k, P_k, t) = \mathcal{K}_k(P_k) + \Phi_k(Q_k, t).$$

The definition (4.12b) implies that $\mathcal{H}_k$ and $\mathcal{H}_{[t]}$ are not particle sum functions because $\Phi_k(Q_k, t)$ is not. The *i-particle energy* and the *k phase energy* are distinct from the Hamiltonian energy and are defined by

$$(4.13a) \qquad E_i(Q_k, p_i, t) = K_i(p_i) + \tfrac{1}{2}\Phi_i(Q_k, t),$$

$$(4.13b) \qquad \mathcal{E}_k(Q_k, P_k, t) = \sum_{i \in k} E_i(Q_k, p_i, t),$$

$$(4.13c) \qquad \mathcal{E}_{[t]}(Q_t, P_t, t) = \sum_{k \in [t]} \mathcal{E}_k(Q_k, P_k, t).$$

The factor of one-half was used in the definition of the i-particle energy in equation (4.13a) so that both $\mathcal{E}_k$ and $\mathcal{E}_{[t]}$ will be particle sum functions. This means that the single particle energy E_i and the single particle Hamiltonian energy H_i are not equal in general. For a system k that is not isolated, these definitions imply also that $\mathcal{E}_k \neq \mathcal{H}_k$. Similarly, for the $[t]$ system, it is usually the case that $E_{[t]} \neq \mathcal{H}_{[t]}$. However, for the t system, there is no external potential and the definitions (4.12) and (4.13b) imply that $\mathcal{E}_t = \mathcal{H}_t$.

The difference between the system phase energy and phase Hamiltonian lies in the way the potential energy is handled. The job of the energy function is to account for the total energy and assign it to an appropriate entity. Thus, one-half of the mutual potential energy of the interaction between particle i and any other particle is assigned in (4.13a) to particle i and the other half of the potential energy to the other particle. The job of the Hamiltonian function, on the other hand, is to keep track of the interparticle forces correctly. These are computed as the negative gradient of the Hamiltonian with respect to a coordinate vector, so the full interparticle potential is included in the Hamiltonian energy function. Since the sum over $i \in k$ of the single particle internal potential energy counts the particle interaction potential energies twice, this is compensated for by an including a factor of $\frac{1}{2}$ in the definition of the internal potential energy component of the k system phase Hamiltonian.

4.5 The Hamiltonian Evolution Equations

The standard expressions for the Hamiltonian evolution equations in the $3\mathcal{N}_k$-dimensional vector notation are

$$(4.14a) \qquad \dot{Q}_k = \frac{\partial \mathcal{H}_k}{\partial P_k},$$

$$(4.14b) \qquad \dot{P}_k = -\frac{\partial \mathcal{H}_k}{\partial Q_k}.$$

These are expressed in component form by

$$(4.15a) \qquad \dot{q}_i^\mu = \frac{\partial \mathcal{H}_k}{\partial p_{i,\mu}},$$

$$(4.15b) \qquad \dot{p}_i^\mu = -\frac{\partial \mathcal{H}_k}{\partial q_{i,\mu}}.$$

The vectors (4.14) are combined into a $6\mathcal{N}_k$-dimensional vector, V_{2k}, called the k *bivelocity vector*. There is a corresponding $6\mathcal{N}_k$-dimensional gradient operator ∇_{2k}, called the k *bigradient operator*. These operators take the form

$$(4.16a) \qquad V_{2k} = \left(\frac{\partial \mathcal{H}_k}{\partial P_k}, -\frac{\partial \mathcal{H}_k}{\partial Q_k} \right),$$

$$(4.16b) \qquad \nabla_{2k} = \left(\frac{\partial}{\partial Q_k}, \frac{\partial}{\partial P_k} \right).$$

The 3-dimensional dot product or inner product notation can be extended to the $3\mathcal{N}_k$-dimensional vectors of the theory. The $3\mathcal{N}_k$-dimensional inner product for any two $3\mathcal{N}_k$-dimensional vectors G_k and R_k, such that $G_k = \times_{i \in k} g_i$ and $R_k = \times_{i \in k} r_i$ for the 3-dimensional vectors g_i and r_i, is defined by

$$(4.17) \qquad G_k \cdot R_k = \sum_{i \in k} g_i \cdot r_i = \sum_{i \in k} g_i^\mu r_{i,\mu},$$

where the summation convention is used for repeated contravariant and covariant vector or tensor indices. The $6\mathcal{N}_k$-dimensional version of the dot product operator next serves in the definition of the *Poisson bracket operator*, which is expressed in TIS notation in the form

$$(4.18) \qquad V_{2k} \cdot \nabla_{2k} = \frac{\partial \mathcal{H}_k}{\partial P_k} \cdot \frac{\partial}{\partial Q_k} - \frac{\partial \mathcal{H}_k}{\partial Q_k} \cdot \frac{\partial}{\partial P_k}.$$

For future reference, the relations

$$(4.19a) \qquad \nabla_{2k} \cdot V_{2k} = 0,$$

$$(4.19b) \qquad V_{2k} \cdot \nabla_{2k} \mathcal{H}_k = 0,$$

will be important. These are easily obtained from (4.14), (4.16) and (4.18).

Using (4.7b) and (4.7a) with (4.2e), it is also easy to show

$$(4.20\text{a}) \qquad \sum_{i \in k} \frac{\partial \Phi_{k,n}(Q_k)}{\partial q_{i,\mu}} = \sum_{i \in k} \sum_{j \in k-i} \frac{\partial \phi_{ij}(q_i - q_j)}{\partial q_{i,\mu}} = 0,$$

$$(4.20\text{b}) \qquad \sum_{i \in k} \epsilon^{\mu\nu\xi} q_{i,\nu} \frac{\partial \Phi_{k,n}(Q_k)}{\partial q_i^\xi} = 0.$$

These results express the standard facts based on Newton's third law that the sum of the internal forces and the sum of the internal torques of the k system both vanish.

The time derivative of a phase quantity is expressed in the form $\partial / \partial t$ in a reference frame fixed in the t frame or in the form d/dt in a reference frame moving with the system on its trajectory in phase space. The k *system phase total time derivative*, acting on any k phase function $G_k(Q_k, P_k, t)$, is defined by

$$(4.21) \qquad \frac{dG_k(Q_k, P_k, t)}{dt} = \frac{\partial G_k(Q_k, P_k, t)}{\partial t} + V_{2k} \cdot \nabla_{2k} G_k(Q_k, P_k, t).$$

Applying d/dt to $\mathcal{P}_k$, $\mathcal{J}_k$ and $\mathcal{H}_k$, and using (4.19) and (4.18), yields the results

$$(4.22\text{a}) \qquad \frac{d\mathcal{P}_k^\mu(P_k)}{dt} = -\sum_{i \in k} \frac{\partial \Phi_{i,x}(q_i, t)}{\partial q_{i,\mu}},$$

$$(4.22\text{b}) \qquad \frac{d\mathcal{J}_k^\mu(Q_k, P_k)}{dt} = -\sum_{i \in k} \epsilon^{\mu\nu\xi} q_{i,\nu} \frac{\partial \Phi_{i,x}(q_i, t)}{\partial q_i^\xi},$$

$$(4.22\text{c}) \qquad \frac{d\mathcal{H}_k(Q_k, P_k, t)}{dt} = \frac{\partial \Phi_{k,x}(Q_k, t)}{\partial t}.$$

These results show that the total momentum, angular momentum, and Hamiltonian energy are constant if the external potentials satisfy certain conditions so that the external forces, torques, and energy changes, vanish.

4.6 The System Trajectory

The open and bounded k *phase coordinate volume set* is defined as a set of $3\mathcal{N}_k$-dimensional points Q_k by

$$(4.23) \qquad D_k(t) = \times_{i \in k} d_k(t).$$

This is the set of $3\mathcal{N}_k$-dimensional configuration space coordinates, $Q_k = \times_{i \in k} q_i$ that the particles in k can possibly take on at time t. It is bounded because the k volume set $d_k(t)$ is bounded in 3-dimensional coordinate space. In accord with the assumption (4.2c) on the repulsive core of interparticle

potentials, there are points in $d_k(t)$ that are inaccessible to a given particle
when there are other particles in the system.

The t system total energy is fixed and finite. This can be used to derive the finite upper and lower (frame dependent) bounds on the values the
k system Hamiltonian $\mathcal{H}_k$ can take. Let us use $\mathcal{H}_k^l$ to designate the lower
bound and $\mathcal{H}_k^u$ to designate the upper bound of $\mathcal{H}_k$. Let $\Upsilon_{k,\min}(t)$ be the
set of minima of the manifold defined by the function $\mathcal{H}_k(Q_k, P_k, t)$ at time
t. Let us assume for now that these minima are located at nondegenerate
(isolated) points in phase space. The fact that $(Q_k^\circ, P_k^\circ) \in \Upsilon_{k,\min}(t)$ implies
that $\nabla_{2k}\mathcal{H}_k(Q_k^\circ, P_k^\circ, t) = 0$. If U is a suitably small punctured neighborhood of (Q_k°, P_k°) that does not contain this point, it follows that the relation
$\partial^2 \Phi_k(Q_k, t)/\partial q_i \partial q_j > 0$ is valid for $(Q_k, P_k) \in U$ and each $i, j \in k$. The lower
bound $\mathcal{H}_k^l$ of the system Hamiltonian is then defined by

$$(4.24) \qquad\qquad \mathcal{H}_k^l = \inf_{\substack{(Q_k, P_k) \in \Upsilon_{k,\min} \\ t \in [0, \alpha)}} \mathcal{H}_k(Q_k, P_k, t).$$

Next, let $k' = [t] - k$ be the complement of k in $[t]$ and the k' system understood as the system formed by the collection of all systems in $[t]$ with
the exception of k. Also let $\mathcal{H}_{k'}^l$ be the lower limit of the energy in the k'
system as defined by (4.24). The relations $I_{H,t} = \mathcal{H}_t(0)$ and $\mathcal{H}_k(t) + \mathcal{H}_{k'}(t) = \mathcal{H}_t(t) = \mathcal{H}_t(0)$ are then used to show

$$(4.25) \qquad\qquad \mathcal{H}_k^l \leq \mathcal{H}_k(t) \leq \mathcal{H}_t(0) - \mathcal{H}_{k'}^l = \mathcal{H}_k^u.$$

The upper and lower limits on the energy in the k system define the k *system
energy range* as

$$(4.26) \qquad\qquad \mathcal{C}_{H,k} = \mathcal{H}_k^u - \mathcal{H}_k^l.$$

This expresses the energy range that was first stated in (2.30c) in terms of
computable upper and lower bounds for the energy available to the k system.
In some cases, $\mathcal{H}_k^u$ may be chosen to be less than the value stated in (4.25)
and $\mathcal{H}_k^l$ greater than the value given in (4.24). This will be the case when we
know that the system will be confined for a period of time within narrower
limits.

Similar steps are used to define lower and upper bounds on the total k
momentum, $\mathcal{P}_k^l$, $\mathcal{P}_k^u$, and angular momentum, $\mathcal{J}_k^l$, $\mathcal{J}_k^u$, where it is understood
that these are vector quantities.

The lower and upper bounds on $\mathcal{P}_k(P_k)$, $\mathcal{J}_k(Q_k, P_k)$, and $\mathcal{H}_k(Q_k, P_k, t)$ are used to define the bounded, (precompact) phase sets

(4.27a)
$$\Omega_k(t) = \{\, (Q_k, P_k) \in D_k(t) \times \mathbf{R}^{3\mathcal{N}_k} \mid \mathcal{P}_k^\ell \leq \mathcal{P}_k(P_k) \leq \mathcal{P}_k^u,$$
$$\mathcal{J}_k^\ell \leq \mathcal{J}(Q_k, P_k) \leq \mathcal{J}_k^u, \mathcal{H}_k^l \leq \mathcal{H}_k(Q_k, P_k, t) \leq \mathcal{H}_k^u \,\}$$

(4.27b)
$$\Omega_k^\alpha = \cup_{t \in [0, \alpha)} \Omega_k(t).$$

The set $\Omega_k(t)$ is called the k-*system phase volume set* at time t and it will play an important role in what follows. The closure $\overline{\Omega_k}(t)$ of $\Omega_k(t)$, obtained by closing the coordinate set $D_k(t)$, is a compact set in phase space. It is clear that under these definitions, any point (Q_k, P_k) that is a possible mechanical state of the system at time t will fall within the phase set $\Omega_k(t) \subset \Omega_k^\alpha$. Because the upper and lower limits of the momentum, angular momentum, and energy are computed in the t system particle rest frame, this definition of $\Omega_k(t)$ is valid for $k \in [t]$ only in that frame. Its definition in other frames requires a suitable transformation of these limits.

The set $D_k(t)$ is a bounded set in $\mathbf{R}^{3\mathcal{N}_k}$. The set of points (Q_k, P_k) at time t for which $Q_k \in \overline{D_k}(t)$, $\mathcal{P}_k^l \leq \mathcal{P}_k(P_k) \leq \mathcal{P}_k^u$, $\mathcal{J}_k^l \leq \mathcal{J}_k(Q_k, P_k) \leq \mathcal{J}_k^u$, and $\mathcal{H}_k^l \leq \mathcal{H}_k(Q_k, P_k, t) \leq \mathcal{H}_k^u$, is closed and bounded in $\mathbf{R}^{6\mathcal{N}_k}$. It follows that the closure of the set $\Omega_k(t)$ is compact at each time $t \in [0, \alpha)$. Next, by the assumptions in (4.2), it can be shown that the maximum interaction energy available to two particles i and j in the t system is $\mathcal{I}_{H,k} + \frac{1}{2}(\mathcal{N}_k + 1)(\mathcal{N}_k - 2)\phi_0$, when $(Q_t, P_t) \in \Omega_k(t)$.[3] Let the quantity r_c be defined by

(4.28)
$$r_c = \inf_{i, j \in t} \{\, r \mid \phi_{ij}(r) = \mathcal{I}_{H,k} + \tfrac{1}{2}(N_k + 1)(N_k - 2)\phi_0 \,\}.$$

The quantity r_c is called the *distance of closest approach* of the particles and is the closest any two particles can approach each other. By (4.2b) and (4.2c) and the fact that $\mathcal{I}_{\mathcal{H},k} < \infty$, an $r_c > 0$ always exists. If $(Q_k, P_k) \in \Omega_k(t)$, then, for any two particles i and j with $i, j \in t$, the relation $|q_i - q_j| \geq r_c$ is true. This means that $|\phi_{ij}|$ is bounded by the semi-norm $\|\phi\|$ defined in (4.2h).

The restrictions on $\mathcal{H}_k(Q_k, P_k, t)$ are chosen in part so that a unique solution, $(Q_k^{(s)}(t), P_k^{(s)}(t))$, of Hamilton's equations, (4.14), will exist for each $t \in [0, \alpha)$. This implies that one and only one trajectory at time t will pass through each point of $\Omega_k(t)$. The above assumptions also imply that $(Q_k^{(s)}(t), P_k^{(s)}(t)) \in \Omega_k(t)$. The trajectory $(Q_k^{(s)}(t), P_k^{(s)}(t))$ in the system phase space is called the k *system specific trajectory*. As this name implies,

---

[3] If $p_i = 0$ for $i \in t$ in the t frame, it follows that $\frac{1}{2}\sum_{i \in t}\sum_{j \in t-i}\phi_{ij}(q_i - q_j) = I_{\mathcal{H},t}$. By (4.2b) and the assumption that $\phi_{mn}(q_m - q_n) = -\phi_0$ if $m, n \notin \{i, j\}$ and $m \neq n$, it follows that $\phi_{ij} \leq I_{\mathcal{H},t} + \frac{1}{2}(\mathcal{N}_t + 1)(\mathcal{N}_t - 2)\phi_0$.

it is a solution that is specific to the k system and is distinct from the k marginalization of the total system trajectory $(Q_t(t), P_t(t))$. These two trajectories are distinct because the k system specific trajectory depends on the k external potential $\Phi_{k,x}(Q_k, t)$, which is an average potential that is not the same as the sum of individual particle interaction potentials between particles inside k and others. It is the latter exact potential that is used in the computation of the t trajectory. These trajectories are also distinct because the k system evolution equation is not obeyed for particles at the k boundary. The boundary formalism acts at these points to reflect the particles back into $d_k(t)$ and therefore the system trajectory back into set $\Omega_k(t)$. These points mean that the Hamiltonian formalism for a subsystem of a larger isolated system is an approximation and must be understood as such.

The system specific trajectory $(Q_k^{(s)}(t), P_k^{(s)}(t))$ can be expressed in terms of the individual system specific particle trajectories in the form

$$(4.29) \qquad (Q_k^{(s)}(t), P_k^{(s)}(t)) = \times_{i \in k}(q_i^{(s)}(t), p_i^{(s)}(t)).$$

In the case of the isolated t system, the system trajectory $(Q_t(t), P_t(t))$ is assumed to be smooth and lie within the bounded domain $\Omega_t(t)$. For the subsystem k, the system specific coordinate trajectory $Q_k^{(s)}(t)$ is assumed to be continuous and piecewise smooth and the system specific momentum trajectory $P_k^{(s)}(t)$ is assumed to be piecewise continuous and piecewise smooth as a function of t. For the system specific trajectories, the momentum trajectory of the Hamiltonian solution has possible discontinuities. A particle i entering the k system boundary $\partial d_k(t)$ in the mechanical state (q_i, p_i) at time t satisfies the relation $\hat{n}_{\partial k}(q_i, t) \cdot p_i < 0$. It is replaced with a particle in the state (q_i, p_i') that satisfies the relation $\hat{n}_{\partial k}(q_i, t) \cdot p_i' > 0$ for a particle leaving the boundary and returning to the k system. The value of p_i' is defined by the boundary formalism and its determination will be discussed in Chapter 6. As a result, the function $Q_k^{(s)}(t)$ will have a "corner" or kink and the function $P_k^{(s)}(t)$ will not be continuous at certain time points, assumed isolated, for which one or more particles in system k are colliding with the boundary. At other time points, both $Q_k^{(s)}(t)$ and $P_k^{(s)}(t)$ are assumed to be at least once continuously differentiable as functions of the time. These assumptions will be reviewed in Part IV.

As an initial condition, the convention is adopted that the system specific trajectory and the k marginalization of the t trajectory coincide. This means that

$$(4.30a) \qquad (Q_k^{(s)}(0), P_k^{(s)}(0)) = (Q_k(0), P_k(0))$$

at time $t = 0$. This can be written in an equivalent form as the requirement for each $i \in k$ that

$$(4.30b) \qquad (q_i^{(s)}(0), p_i^{(s)}(0)) = (q_i(0), p_i(0)).$$

It is assumed for now that system specific solutions of the equations (4.14) exist and are unique. Because the initial time is arbitrary, this means that for any given time t_0 and point $(Q_k^0, P_k^0) \in \Omega_k(t)$, there is a unique system specific solution, $(Q_k^{(s)}(t), P_k^{(s)}(t))$, such that $(Q_k^{(s)}(t_0), P_k^{(s)}(t_0)) = (Q_k^0, P_k^0)$. Finally, $Q_k^{(s)}(t)$ and $P_k^{(s)}(t)$ are assumed to satisfy the same conditions that $Q_k(t)$ and $P_k(t)$ satisfy. The existence of system specific solutions will be demonstrated in Part IV.

4.7 Isolated Systems

For the isolated t system, it was mentioned above that the system energy and Hamiltonian energy coincide: $E_t(Q_t, P_t) = \mathcal{H}_t(Q_t, P_t)$. The condition that the system be isolated means in addition that

$$(4.31a) \qquad \Phi_{i,x}(q_i, t) = 0,$$
$$(4.31b) \qquad \Phi_{t,x}(Q_t, t) = 0.$$

From this it follows that the formulas in (4.22) become

$$(4.32a) \qquad \frac{d\mathcal{P}_t^\mu(P_t)}{dt} = 0,$$

$$(4.32b) \qquad \frac{d\mathcal{J}_t^\mu(Q_t, P_t)}{dt} = 0,$$

$$(4.32c) \qquad \frac{d\mathcal{H}_t(Q_t, P_t, t)}{dt} = 0.$$

These results mean that the total particle number, total phase momentum, total phase angular momentum, and total phase energy, are constant (conserved) on an isolated system trajectory.

For convenience, it will be assumed temporarily that the t particle rest frame is an inertial frame. Thus, each isolated system t is described by the ordered set of phase constants $\mathcal{I}_t^p = (\mathcal{N}_t, \mathcal{I}_{P,t}^{p,\mu}, \mathcal{I}_{J,t}^{p,\mu}, \mathcal{I}_{H,t}^p)$ such that

$$(4.33a) \qquad \nu_t = \mathcal{N}_t,$$
$$(4.33b) \qquad \mathcal{P}_t^\mu(P_t) = \mathcal{I}_{P,t}^\mu = 0,$$
$$(4.33c) \qquad \mathcal{J}_t^\mu(Q_t, P_t) = \mathcal{I}_{J,t}^\mu = 0,$$
$$(4.33d) \qquad \mathcal{H}_t(Q_t, P_t) = \mathcal{I}_{H,t} = \text{const.}$$

When the quantities in (4.33) are phase averaged, the result $\mathcal{I}_t = \langle \mathcal{I}_t^p \rangle_{tt} = (\mathcal{N}_t, 0, 0, \mathcal{I}_{H,t})$, which is the same as that stated in (2.29), is obtained. The constants in the sets $\mathcal{I}_t$ and $\mathcal{I}_t^p$ are assumed to be fixed from now on. It should be kept in mind that the assumption that the system t particle rest frame is an inertial frame may not be valid. If that is the case, it is not possible to use a Galilean transformation to move from an inertial frame into the t system particle rest frame. This issue, and a similar one associated with a later assumption that the k subsystem particle rest frame is an inertial frame, will be discussed further in Chapter 15.

The relations (4.33) are used to define the t *system phase volume set* for an isolated system by

(4.34)
$$\Omega_t(t) = \{ (Q_t, P_t) \in D_t(t) \times \mathbf{R}^{3\mathcal{N}_t} \mid \mathcal{P}_t^\mu(P_t) = \mathcal{I}_{P,t}^\mu, \mathcal{J}_t^\mu(Q_t, P_t) = \mathcal{I}_{J,t}^\mu,$$
$$\mathcal{H}_t(Q_t, P_t) = \mathcal{I}_{H,t} \}.$$

It is obvious that $\Omega_t(t)$ is a bounded set, so its closure is compact.

CHAPTER 5

The System Distribution and its Evolution Equation

The Hamiltonian mechanics of interacting systems introduced in the last chapter is extended into a general statistical mechanics of interacting systems.

5.1 Probability and Particle Distributions

The central concept in statistical mechanics is the distribution of the mechanical states of the system particles in phase space. It is called the *system distribution density* in TIS and a *density function* in standard statistical mechanics. Both continuous and discrete distributions have played a role in calculations involving statistical mechanics and thermodynamics. A continuous system distribution density $F_k(Q_k, P_k, t)$ is at least once continuously differentiable. It assigns to each mechanical state (Q_k, P_k) the relative probability to finding the system in a neighborhood $dQ_k dP_k$ of this mechanical state. A discrete system distribution, which is often called a *coarse-grained distribution* in statistical mechanics, is based on a partition of phase space and is a function of the cells in the partition. For each cell, the discrete distribution gives the number of particles in the cell.

Discrete distributions are often used in thermodynamic and hydrodynamic calculations. In a discrete distribution, a number $n_a \geq 0$ of particles is assigned to each cell a in a partition π of phase space. Each cell a in this partition is assigned a mechanical state (q_a, p_a), where the phase point (q_a, p_a) is located somewhere in the cell a. The cell itself is treated in calculations as if it were a particle located at q_a with momentum $n_a p_a$. Discrete distributions are easier to work with in some calculations and were originally introduced for this purpose. However, it is easy to see that the evolution equation for the discrete distribution obtained by setting up an evolution equation for the cells in the distribution differs significantly from the Hamiltonian evolution equation for the particles. This means that the description of the system in terms of a discrete distribution and its associated evolution equation is not compatible with the Hamiltonian mechanics of the underlying particles. It has been shown that slight differences between distributions can lead to an exponential divergence of one distribution from another. It follows that discrete distributions lack dynamical significance, so the focus in TIS is on particle distribution densities that are at least once continuously differentiable.[1] Some specific singular and L^2 distribution densities are also part of the theory.

A concrete example of a system distribution is the exponential equilibrium velocity distribution, $f^\epsilon(v) = e^{-mv^2/2}$, discovered by Maxwell [**1860**], [**1866**], and included in his theory of transport equations. It has the form of a normal distribution density in the particle velocities. This equilibrium distribution density was soon generalized by Maxwell [**1873b**] and Boltzmann [**1875**] to an exponential Γ distribution density that includes potential as

---

[1]Discrete distributions play an important role in the discussion of quantum thermodynamics presented in QTS. The possible exponential divergence of "nearby" distributions is discussed there. The formalism employing discrete distributions and the role they play in the relation of thermodynamics to statistical mechanics is presented there as well.

well as kinetic energy in the argument of the exponential.[2] Maxwell's general particle distribution $f(v)$ was a statistical distribution that represented the actual number of particles in a system with velocities between v and $v + dv$. Boltzmann introduced probability distributions in his 1872 work on the Boltzmann equation. Boltzmann's theory was based on the one-particle probability distribution density $f(q, p, t)$, which represents the probability of finding a particle in the mechanical state (q, p). Probability distributions have been a mainstay of statistical mechanics ever since.

Formalizing the probability aspects of the system distribution provides the proper mathematical underpinning for the TIS approach. The components of a probability space are a *universe U* of elementary events, a *collection of sets of events Σ*, and a *normalized probability measure $\boldsymbol{P}$* on Σ. The collection of sets of events Σ is a σ field of subsets of U, which means that Σ is a non-empty class of subsets of U, which includes U, the empty set $\emptyset$, and is closed under countable union and complementation. The normalized measure $\boldsymbol{P}$ satisfies, for any set $R \in \Sigma$, the condition $0 \le \boldsymbol{P}(R) \le 1$. The probability measures of the empty set and the universe are $\boldsymbol{P}(\emptyset) = 0$ and $\boldsymbol{P}(U) = 1$. Elementary events in this setting are individual points in U. The *probability space* is the triple $(U, \Sigma, \boldsymbol{P})$.[3]

The TIS formalism is based on a slightly modified version of this definition of a probability space. The definitions given next will be expressed in terms of a system k, which may be the total system $k = t$, the system represented by the collection of all subsystems $k = [t]$, or a particular subsystem $k \in [t]$. The elementary events are system mechanical states (Q_k, P_k) that are points in $6\mathcal{N}_k$-dimensional phase space. The first extension of the standard probability formalism in TIS is allowing both the universe and the probability measure to be time dependent. The universe for a particular system k at time t is the set $\Omega_k(t)$, which was defined as a bounded subset of the $\mathbf{R}^{6\mathcal{N}_k}$ phase space of system mechanical states in Chapter 4. The time-dependent space of events $\Sigma_k(t)$ is the σ field of subsets on $\Omega_k(t)$. A second extension is the use in TIS of a normed *system measure $\boldsymbol{P}_k(t)$* on $\Omega_k(t)$ in place of a normalized measure. The system mechanical states (Q_k, P_k) are viewed as a random variables on $\Omega_k(t)$ in the probability formalism. Finally, there is an automorphism on phase space determined by the system evolution equation that evolves the measure $\boldsymbol{P}_k(t)$ in time.[4]

The Lebesgue measure $\mu_k(A)$ of a set A in phase space, defined as the phase space version of the measure illustrated in (2.10), is the phase space volume of that set. The *system distribution density $F_k(Q_k, P_k, t)$* is defined as

---

[2]This generalization was originally introduced in the discussion of the thermodynamics of a gas in a gravitational field. The Γ distribution is discussed further in EIS in connection with the work of Khinchin.

[3]For more details on probability theory, see R. G. Laha and Vijay K. Rohatgi [**1979**].

[4]When the system measure $\boldsymbol{P}_k(t)$ is normalized, this is called a *classical dynamical system* by V. I. Arnold and A. Avez [**1968**], p. 1.

the Radon-Nikodym derivative of the system distribution $\mathcal{P}_k(t)$ with respect to the $6\mathcal{N}_k$-dimensional Lebesgue measure μ_k,

$$(5.1) \qquad F_k(Q_k, P_k, t) = \frac{d\mathcal{P}_k(t)}{d\mu_k},$$

where $d\mu_k = dQ_k dP_k$.[5] The system distribution density is called the *state* of the system because it contains all the information available concerning the distribution of the mechanical states for the system and serves as the weighting function for the calculation of system averages.[6] It will play a major role in the formalism to follow.

The *expectation function* $G_k(t)$ of a quantity G_k is computed using its microscopic analog $G_k(Q_k, P_k, t)$, weighted by the system distribution density, and constrained to the k system by the system projection operator. The term *phase average* is generally used in statistical mechanics to designate the expectation function and these terms will be used interchangeably in TIS.

The norm of the system measure $\mathcal{P}_k(t)$, which is written $\|\mathcal{P}_k(t)\|$ and satisfies the bound $\|\mathcal{P}_k(t)\| < \infty$, can be used to normalize $\mathcal{P}_k(t)$. In terms of the system distribution density, this is the *global norm* of F_k defined by

$$(5.2) \qquad Z_k^{(g)}(t) = \int dQ_k \int dP_k \, F_k(Q_k, P_k, t),$$

where the Q_k and P_k integrations are over all of k phase space and not just $\Omega_k(t)$. Because $F_k(Q_k, P_k, t)$ is assumed normable, the global norm is bounded: $Z_k^{(g)}(t) < \infty$.

The particle systems in the classical version of TIS are all required to be closed. This closure is enforced in TIS by a boundary formalism that acts on the system distribution at the boundary in such a way that particles will remain within the system while simultaneously preserving the values of the conserved physical quantities within the system and transmitted between systems.

The system distribution density $F_k(Q_k, P_k, t)$ is required in the literature to be normalized so that it can serve as a probability distribution density and used as a weighting function in the computation of expectation functions of physical quantities. However, this normalization requirement does not stem from the mathematical conditions on the system evolution equation or from requirements for its solution. It is imposed post facto on solutions so that they will serve the macroscopic purpose for which they are intended. The

---

[5] The existence of the Radon-Nikodym derivative is established in the Radon-Nikodym theorem. See Treves [**1967**], p. 211, and Laha and Rohatgi [**1979**], p. 15.

[6] For a discussion of these mathematical concepts in the context of probability theory, and the mathematical form that a probability distribution can take as the unique sum of a step function, a singular function, and an absolutely continuous function, see Laha and Rohatgi [**1979**], Chapter 1. See also R. Kurth [**1960**], Chapter IV.

normalization requirement is therefore a separate and independent condition imposed on solutions of the evolution equation.

There are several problems with requiring normalized system distributions as the solutions of the system evolution equation. First, in light of the statements above, it confounds the mathematics by including an extraneous normalization factor that, for open systems, can change in time. Second, it is shown in Chapter 18 that the normalization of the equilibrium system distribution, provided by the partition function Z_k^ϵ, appears as part of the microscopic system equilibrium entropy analog function in standard equilibrium thermodynamics in the form $S_k^{(n)\epsilon}(Q_k, P_k) = k_B[\beta_k \mathcal{H}_k^\epsilon(Q_k, P_k) + \ln Z_k^\epsilon]$.

There is a problem with this analog because the equilibrium partition function Z_k^ϵ is defined by integrating the equilibrium system distribution density $e^{-\beta_k \mathcal{H}_k^\epsilon(Q_k, P_k)}$ over the phase volume $\Omega_k(t)$ assigned to the k system. The coordinate component of this integration is over those points in $D_k(t) = \times_{i \in k} d_k(t)$ that meet the momentum, angular momentum, and energy, requirements imposed by membership in $\Omega_k(t)$. This means that the domain of the coordinate integration is determined by the macroscopic system volume set $d_k(t)$. This volume set plays no role in the evolution equation of microscopic Hamiltonian mechanics. The problem with including $\ln Z_k^\epsilon$ in the microscopic analog of $S_k^{(n)\epsilon}$ is that doing so violates the TIS principle that macroscopic aspects of the system should not be mixed with microscopic aspects unless they play a role in the microscopic Hamiltonian mechanics.

The computation of expectation functions, known as phase averages in statistical mechanics, requires a normalized weighting function. The normalization of $F_k(Q_k, P_k, t)$ when it is used in the computation of expectation functions is treated as a separate factor in TIS. The importance of this separation will become clearer later when the dependence of the normalization on the time-dependent system volume and its role with respect to the entropy are discussed in more detail. Moreover, for open systems the support of solutions of the k system evolution equation may extend beyond the boundary of the k system. For the TIS formalism to include open systems, as it must for the quantum mechanical version, it follows that the separation between the system distribution density as a solution of its evolution equation and its normalization requirement must be maintained. For these reasons, the primary focus of the TIS classical formalism will be on the k system distribution density $F_k(Q_k, P_k, t)$ rather than the normalized k probability distribution density designated by $F_k^{(n)}(Q_k, P_k, t)$.

The requirement of closure for particle systems in the TIS classical formalism will usually represent a good approximation rather than a strictly enforceable condition. Some particles can leak out of or into a large enough system, and this includes most systems of interest to thermodynamics, without affecting significantly the values of the physical quantities associated with the

system. On the microscopic level, the TIS boundary and wall formalism introduced in the next chapter enforces the idealized closure by using a stochastic
transition probability function, which acts at subsystem boundaries to transform system distribution densities at the boundary. This transformation is
accomplished in a way that accounts for the interactions of system particles
with particles in other systems and the transmission of physical quantities
between them.

From the standpoint of thermodynamics, macroscopic bodies occupy definite volumes and transfer physical quantities of interest between them. A
system projection operator $\chi_{\Omega_k(t)}(Q_k, P_k)$ will be defined below that local-
izes the system expectation function calculations at a given time to phase
points within the k system phase space $\Omega_k(t)$. This operator helps maintain
the connection between the microscopic and macroscopic system definitions.
In conjunction with the boundary transition formalism mentioned above, the
time derivative of the system projection operator is a factor in the representation of the fluxes of physical quantities that pass through the boundary. The
formalism to support these statements will emerge naturally from the fact
demonstrated below that the total time derivative of the system projection
operator is the system flux operator.

While the introduction of a system projection operator may seem a round-
about method of approaching the problem, it is essential in keeping the aspects of a changing situation separate so that the dependence of the phase
averages on the system volume and their changes in time can be made manifest. Simply setting the coordinate and momentum limits of the integrals
to the proper values will not work because the set $\Omega_k(t)$ contains excluded
coordinate volumes within each particle's volume set that correspond to the
inaccessible volumes, with radius greater than or equal to r_c, around each
of the other particles. The size of the volume around particle i from which
particle j is excluded at a given time depends on the ij interaction potential
$\phi_{ij}(q_i - q_j)$ and the relative momentum vector $g_{ji} = p_j - p_i$ of the particles.
Moreover, the system boundaries can change in time, so the limits of the integrals would also need to be time dependent. For these reasons, the system
projection operator offers many benefits over the standard approach.

The system projection operator is used to limit the calculation of the
expectation function of an analog quantity to the coordinate and momentum
sets in phase space assigned to the system. The distinction between the set
of phase points determined in macroscopic terms by the system projection
operator and the set of phase points that make up the support of the microscopic functions, such as the system state $F_k(Q_k, P_k, t)$ or the physical analog
functions, is that neither the system state nor the analogs are required to vanish outside $\Omega_k(t)$. The state $F_k(Q_k, P_k, t)$ is the solution of its Hamiltonian
evolution equation which does not recognize this macroscopic boundary. The
mechanical states that belong to the support of $F_k(Q_k, P_k, t)$ may, of course,

be limited to the $3\mathcal{N}_k$ coordinate volume set $D_k(t)$ by sufficiently powerful forces. In this case, localization is a purely microscopic matter respected by the evolution equation. Another aspect of the formalism mentioned above is the TIS boundary or wall formalism that works in conjunction with the system projection operator. It acts at points $q \in \partial d_k(t)$ that are outside $d_k(t)$ and therefore outside the domain of the system projection operator. Its purpose is to keep particles in the k system. To do so it uses a stochastic operator to determine the particle transitions that reflect them back into the system.

The *system projection operator* $\chi_{\Omega_k(t)}(Q_k, P_k)$ is defined as the indicator function of the set $\Omega_k(t)$, which has the value 1 if $(Q_k, P_k) \in \Omega_k(t)$ and 0 if $(Q_k, P_k) \notin \Omega_k(t)$. It will often be abbreviated as $\chi_{\Omega_k(t)}$. Let us use $\chi_{\Omega_k(t)}(Q_k, P_k)$ with the system distribution density $F_k(Q_k, P_k, t)$ to define the *k partition function* as the *k local norm* $\|F_k\|_k$ of the system distribution in the set $\Omega_k(t)$ by[7]

$$(5.3) \qquad Z_k(t) = \|F_k\|_k = \int dQ_k \int dP_k \, F_k(Q_k, P_k, t)\chi_{\Omega_k(t)}(Q_k, P_k).$$

A requirement for any acceptable solution F_k of the evolution equation is that $Z_k(t) < \infty$. This is implied by the global normability of $F_k(Q_k, P_k, t)$. The k *probability distribution density* is defined in terms of the k system distribution density and its local norm by

$$(5.4) \qquad F_k^{(n)}(Q_k, P_k, t) = [Z_k(t)]^{-1} F_k(Q_k, P_k, t).$$

This formalism generalizes the partition function Z_k^ϵ of equilibrium theory to general statistical mechanics and its associated thermodynamics. In the next few chapters, a formalism associated with the k partition function will be developed for general thermodynamics and statistical mechanics that is very similar to the partition function formalism of equilibrium thermodynamics and statistical mechanics.

The k probability distribution density $F_k^{(n)}(Q_k, P_k, t)$ is interpreted as 'the probability density of finding the k system in a mechanical state that falls within the phase volume $dQ_k dP_k$ centered at (Q_k, P_k) at time t'. The statistical version of this statement is 'the relative frequency with which the state will be found in the neighborhood $dQ_k dP_k$ centered at (Q_k, P_k) at time t'. The statistical viewpoint, as opposed to the probability viewpoint, is reflected in the "ensemble" metaphor used by Gibbs in his approach to the system distribution. From the TIS perspective, the local system distribution

---

[7]The partition function is usually designated by the letter Z because it was introduced by Planck as the *Zustandsumme function* or "state sum function". The local norm $\|F_k\|_k$ of F_k, defined in (5.3) as its norm in $\Omega_k(t)$, is distinct from the definition of its global norm $\|F_k\|$ defined in (5.2), which is computed over all of phase space. This distinction is important for open systems and quantum mechanical systems.

$F_k^{(n)}(Q_k, P_k, t)dQ_k dP_k$ is a probability distribution representing the probability of finding the system in a state (Q_k, P_k) that falls in the neighborhood $dQ_k dP_k$. It is not interpreted as a local population distribution, which is a statistical distribution representing the relative number of systems in an infinite ensemble that have states (Q_k, P_k) that fall within the phase volume $dQ_k dP_k$ at time t. These issues of interpretation will be discussed further in EIS.

5.2 Thermodynamic Observables

The macroscopic functions representing thermodynamic quantities are the *observables* of the theory. The phase functions that represent their associated microscopic analogs are the *observable analogs*. Once a set of observables and their corresponding analogs has been chosen, a function space is chosen in which the analogs and their phase averages are all well defined.

A thermodynamic quantity G_k is defined as the expectation function $G_k(t)$ of the microscopic analog function $G_k(Q_k, P_k, t)$. The system projection operator is used to limit the calculation of the expectation function to the set $\Omega_k(t)$ and the calculation is normalized by dividing it by $Z_k(t)$. At time t, the expectation function associated with the analog $G_k(Q_k, P_k, t)$ is (5.5)

$$G_k(t) = \frac{1}{Z_k(t)} \int dQ_k \int dP_k \, F_k(Q_k, P_k, t)\chi_{\Omega_k(t)}(Q_k, P_k)G_k(Q_k, P_k, t).$$

If $G_k(t)$ is a thermodynamic observable, it is required that $|G_k(t)| < \infty$, that is, that $G_k(t)$ is bounded. For some quantities, $G_k(t)$ may need to be differentiable to some order in the time.

To insure that each thermodynamic observable $G_k(t)$ is mathematically well defined, the strategy is to require that all the system distribution densities that will be used to describe a system belong to a function space that is dual, in the sense of topological vector space methods, to the function space of the set observable analogs for that system. To pursue this idea, let us establish some useful notation. First, as a notational convenience, any phase function $G_k(Q_k, P_k, t)$ is associated with a real-valued time-dependent map from points in the phase space into the real numbers. The phase map is written $\mathfrak{G}_k(t)$ and is defined by $\mathfrak{G}_k(t) : (Q_k, P_k) \to G_k(Q_k, P_k, t)$. This notation will be used for any phase function—including the system distribution.

Let us choose a set $\mathcal{O}_k = \{G_k^s(t)\}$ of *observables* indexed by s and a corresponding set $\mathcal{O}_k^a = \{G_k^s(Q_k, P_k, t)\}$ of *observable analogs* for the k system. The constant function $G_k^1(Q_k, P_k, t) = 1$ is always included in the set of observable analogs because each system distribution is required to satisfy the normalization condition

$$(5.6) \qquad G_k^1(t) = \frac{1}{Z_k(t)} \int dQ_k \int dP_k \, F_k(Q_k, P_k, t)\chi_{\Omega_k(t)}(Q_k, P_k) = 1,$$

which follows in this case from the definitions (5.5) and (5.3). In accord with previous considerations and the intended application of the formalism, it is also required that those phase functions in $\mathcal{O}_k^a$ that represent mechanical properties of the particles, such as the interparticle forces, the momentum, or the energy, are continuous, and therefore uniformly bounded, for $(Q_k, P_k) \in \Omega_k^a$. In addition, those phase analog functions in $\mathcal{O}_k^a$ that depend on the system distribution density, such as those that will be introduced to represent the entropy and thermature, are required to have bounded phase averages.

If each of the observables in $\mathcal{O}_k^a$ belongs to a given space S, then any system distribution density in a topological vector space S' dual to this one will be acceptable. Let us express the duality of S and S' in terms of a bilinear inner product, $(,)_*$, defined for $\mathfrak{F}_k(t) \in S'$ and $\mathfrak{G}_k(t) \in S$ by
(5.7)
$$(\mathfrak{F}_k(t), \mathfrak{G}_k(t))_* = \frac{1}{Z_k(t)} \int dQ_k \int dP_k \, F_k(Q_k, P_k, t) \chi_{\Omega_k(t)} G_k(Q_k, P_k, t).$$

The duality of S and S' implies that this inner product is well defined. If $G_k(Q_k, P_k, t) = 1$, then (5.6) implies $(\mathfrak{F}_k(t), \mathfrak{G}_k^1(t))_* = 1$ as required. Moreover, if G_k is the particle sum function $G_k(Q_k, P_k, t) = \sum_{i \in k} G_i(Q_i, p_i, t)$, the local quantities

(5.8a) $$G_i(q, t) = (\mathfrak{F}_k(t), \delta(q - q_i)\mathfrak{G}_i(t))_*,$$

(5.8b) $$G_k(q, t) = \sum_{i \in k} G_i(q, t),$$

are required to be well defined as well. The functions $G_i(q, t)$ and $G_k(q, t)$, which will be defined as expectation functions below, are also required to be bounded and in some cases are required to be continuously differentiable in q and t to some order.

Let us exclude the entropy and temperature from consideration for the moment and assume that the mechanical functions in $\mathcal{O}_k^a$ belong to the space $S = C^m(\Omega_k(t))$ of m times continuously differentiable functions with support contained in the open, precompact set $\Omega_k(t)$. By condition (4.2a), this means that $m \geq 3$. In this case, for a fixed time t, the global phase averaging procedure acts as a Radon measure that maps functions on $\Omega_k(t)$ into the set of real numbers. The dual space of S is $\mathcal{D}^{m'}(\Omega_k(t))$, the space of distributions of order less than or equal to m with support in the set $\Omega_k(t)$.[8]

Anticipating the definitions of the entropy and thermature in Chapters 9 and 12, the phase averages of $\ln F_k(Q_k, P_k, t)$, $\partial \ln F_k(Q_k, P_k, t)/\partial q_i$, and $\partial \ln F_k(Q_k, P_k, t)/\partial p_i$, for $i \in k$ will be needed. It will be shown in Part IV that it is sufficient for the boundedness and continuity of the thermodynamic entropy and temperature to require that $\mathfrak{F}_k(t)$ belong to a Sobolev space

---

[8] For more on Radon measures and tempered distributions, see Treves [**1967**], Chapters 21 and 25.

$H^s(\Omega_k(t))$, with $s \geq m$. This Sobolev space is a subspace of $\mathcal{D}^{m\prime}(\Omega_k(t))$. Some of the system distributions needed, such as the system exact distribution discussed below, belong to $\mathcal{D}^{m\prime}(\Omega_k(t))$ but not to a Sobolev space. This means, as we shall see, that when the system is described by one of these distributions, the entropy is not defined.

5.3 Phase Averaging

The TIS approach to probability measures on phase space and the associated phase averaging procedure is more general than the version employed in many treatments of equilibrium thermodynamics. These other versions are based on the relative Lebesgue measures of certain sets in phase space. This form for the probability measure was introduced in the work of Poincaré and is closely associated with the Liouville theorem in physics. To illustrate this measure, suppose the set V in phase space is a universe of possible events (possible mechanical states) for a system. In the usual probability measure, the probability that the system phase point will fall in the phase set $M \subset V$ is the ratio of the Lebesgue measure of M to that of V written as $\mathcal{P} = \mu(M)/\mu(V)$. This definition of the probability plays an important role, for example, in the work of Garrett D. Birkhoff on ergodic theory, the work Alexsandr Khinchin on asymptotic distributions, and in the dispute between Ernst Zermelo and Boltzmann. It will be shown below that using it corresponds to the computation of expectation functions using Gibbs' microcanonical state as the system distribution.

A more general concept is needed in TIS because of the requirements of both the modern theory of probability and of the nonequilibrium version of thermodynamics developed here.

The expectation function or phase average of the thermodynamic quantity G_k is defined in parallel to (5.5) using a convenient bracket notation by

(5.9)
$$G_k(t) = \langle G_k(Q_k, P_k, t) \rangle_{kt}$$
$$= \frac{1}{Z_k(t)} \int dQ_k \int dP_k \, F_k(Q_k, P_k, t) \chi_{\Omega_k(t)}(Q_k, P_k) G_k(Q_k, P_k, t).$$

For quantities G_k that are particle sum functions, the *i-particle local G density* and the *i-particle local G current density vector* are defined in accord with (5.8) by

(5.10a) $G_i(q, t) = \langle \delta(q - q_i) G_i(Q_k, p_i, t) \rangle_{kt}$,

(5.10b) $G_i^\mu(q, t) = \langle \delta(q - q_i) v_i^\mu G_i(Q_k, p_i, t) \rangle_{kt}$.

Particle sum functions represent quantities that can be localized at the particles in the system. The argument of the i-particle component of a particle sum function may depend on q_i, or even Q_k because particle i may interact with all other particles, and on p_i and the time, but not on p_j for $j \neq i$. Summing the quantities in (5.10) over $i \in k$ yields the k *local G density* and k *local G current density vector*

$$(5.11a) \qquad G_k(q,t) = \sum_{i \in k} G_i(q,t),$$

$$(5.11b) \qquad G_k^\mu(q,t) = \sum_{i \in k} G_i^\mu(q,t).$$

Integrating the local density and the local current density over $q \in d_k(t)$ gives the i-particle global phase average of that density:

$$(5.12a) \qquad G_i(t) = \int d^3q \, G_i(q,t) = \langle G_i(Q_k, p_i, t) \rangle_{kt},$$

$$(5.12b) \qquad G_i^\mu(t) = \int d^3q \, G_i^\mu(q,t) = \langle v_i^\mu G_i(Q_k, p_i, t) \rangle_{kt}.$$

Summing (5.12) over $i \in k$ results in the function $G_k(t)$ defined in (5.11) above and $G_k^\mu(t)$. Finally, setting $k = t$ and $G_t = \mathcal{I}_t$ in (5.11) leads to the result

$$(5.13) \qquad \langle (\mathcal{N}_t, \mathcal{P}_t(P_t), \mathcal{J}_t(Q_t, P_t), \mathcal{H}_t(Q_t, P_t)) \rangle_{tt} = \mathcal{I}_t.$$

5.4 The System Distribution

Let us turn now to a more detailed look at the system distribution density, F_k. The localization of the system distribution to $\Omega_k(t)$ is expressed in the form $F_k(Q_k, P_k, t) \chi_{\Omega_k(t)}(Q_k, P_k)$. The support of this localized distribution in phase space and time falls within the space $\Theta \times \theta$. The set Θ is an open, connected, bounded subset of $\mathbf{R}^{6\mathcal{N}_k}$, with $\Omega_k^a \subset \Theta$; the set θ is an open, bounded, connected set in $\mathbf{R}$, with $[0, \alpha) \subset \theta$. Because $\Omega_k(t)$ is precompact, the continuity of the mechanical observables in the set $\mathcal{O}_k^a$ on the compact set $\overline{\Omega_k}(t)$ guarantees the uniform boundedness of these observables as functions of the system mechanical state (Q_k, P_k).

The definition (5.7) guarantees that the global phase average integrals of these quantities will converge for each system distribution, $\mathfrak{F}_k(t)$, for which $\mathfrak{F}_k(t) \in \mathcal{D}^{m'}(\Omega_k(t))$. The boundedness conditions (2.15) then follow from a computation of the local phase averages using the fact that both $G_k(Q_k, P_k, t)$ and v_i are bounded for $(Q_k, P_k) \in \Omega_k(t)$.

The conditions on $F_k(Q_k, P_k, t)$ can be summarized as

$$(5.14a) \qquad F_k(Q_k, P_k, t) \geq 0,$$
$$(5.14b) \qquad Z_k(t) = Z_k < \infty.$$

The first condition must be interpreted in the distributional sense, namely, if $\mathfrak{G}_k(t) \geq 0$, then $(\mathfrak{F}_k(t), \mathfrak{G}_k(t))_* \geq 0$. Because the integrand in (5.9) vanishes outside the precompact set $\Omega_k(t)$, the divergence theorem can be used to show for any observable analog $G_k(Q_k, P_k, t)$ that

(5.15)
$$\frac{1}{Z_k(t)} \int dQ_k \int dP_k \, \nabla_{2k} \cdot [V_{2k} F_k(Q_k, P_k, t) \chi_{\Omega_k(t)}(Q_k, P_k) G_k(Q_k, P_k, t)] = 0,$$

where the integrations over Q_k and P_k are over all of their respective $\mathbf{R}^{3\mathcal{N}_k}$ spaces. The relation (5.15) allows integration by parts freely.

5.5 The System Evolution Equation

The motions of the individual particles in system k are determined by the Hamiltonian equations. The evolution of $F_k(Q_k, P_k, t)$ in time can be visualized as a flow of points on trajectories representing these particles that are determined by the system evolution equation. This representation in terms of the particle trajectories is called the *Hamiltonian flow* of the system.

The boundary formalism will be introduced formally in Chapter 6, but it will be explained briefly here to show how it works. The system boundary is located at $\partial d_k(t)$ outside the open k volume set $d_k(t)$. The boundary formalism replaces the k evolution equation at this boundary by an operator that reflects particles back into the k system while transmitting other physical quantities through the boundary. In this formalism, when particle i in state (q_i, p_i) enters the boundary at $q_i \in \partial d_k(t)$, a stochastic operator replaces it with a particle in state (q_i, p_i') exiting the boundary and moving back into the system. This description is an approximate description of the encounters of particles with the system boundary when there is not sufficient information to determine the exact particle trajectories. This picture of the collision of a particle with the wall of a system implies that the i-particle trajectory at q_i will have a corner in its 3-dimensional trajectory and a discontinuity from the right in the 3-dimensional trajectory of its momentum variable p_i.

Because the number of particles is finite and their locations cannot coincide, the set of locations of particles at the boundary at time t, $\{ q_i \mid q_i \in \partial d_k(t) \}$, is a set isolated points of measure 0 in $\partial \Omega_k(t) \subset \mathbf{R}^{6\mathcal{N}_k - 1}$ and therefore in $\mathbf{R}^{6\mathcal{N}_k}$. The singular points of $\nabla_{2k} F_k$ due to particle collisions with the wall are therefore assumed to be isolated so that they belong to a set of measure 0 in $\mathbf{R}^{6\mathcal{N}_k}$.

Consider now the evolution of an isolated system t of particles in terms of the Hamiltonian flow of the particles introduced above. There is no boundary formalism in this case because there are no particles outside the system with which the particles in the system can interact. Let A_0 be a bounded phase set in system t at time $t = 0$ and define the bounded phase set A_1 at time $t = 1$ as the collection of all particles with trajectories that originated in A_0. An important property of the Hamiltonian flow, which is central to later work, is that it preserves the Lebesgue measure of volumes in phase space. This means that the Lebesgue measure $\mu(A_0)$ in phase space is preserved under the

Hamiltonian flow, so that $\mu(A_1) = \mu(A_0)$. This flow is therefore a measure preserving automorphism of phase space that transforms it onto itself.[9]

For points of a subsystem k not at the system boundary, that is, for $Q_k \notin \partial D_k(t)$, the total time derivative of F_k is required to be 0 by the equation of continuity for a distribution. Because the boundary neither accumulates nor creates particles, the total time derivative of F_k can be set to 0 for $Q_k \in \partial D_k(t)$ as well. As a consequence, the TIS evolution equation for F_k is written[10]

$$(5.16) \qquad \frac{dF_k(Q_k, P_k, t)}{dt} = 0 \qquad \text{(a.e.)},$$

where (a.e.) means *almost everywhere*, that is, everywhere except possibly on a set of Lebesgue measure zero in $\mathbf{R}^{6\mathcal{N}_k}$. The 'almost everywhere' qualification on the phase analogs is understood but will rarely be mentioned explicitly. Thermodynamic quantities obtained from analogs that differ on a set of Lebesgue measure zero are the same because sets of measure zero do not contribute to the value of the phase average. This means that the definition (5.16) of the total time derivative of the system distribution density is consistent with the transitions at the wall described above because these transitions take place on sets of measure zero in $\Omega_k(t)$.

The system projection operator limits the allowed phase points to $\Omega_k(t)$ in the phase averaging integral. Because the set $\Omega_k(t)$ depends on the time, the total time derivative of the system projection operator is needed. For convenience, this time derivative is abbreviated as

$$(5.17)$$
$$\dot{\chi}_{\Omega_k(t)}(Q_k, P_k) = \frac{d\chi_{\Omega_k(t)}(Q_k, P_k)}{dt} = \frac{\partial\chi_{\Omega_k(t)}(Q_k, P_k)}{\partial t} + V_k \cdot \nabla_k \chi_{\Omega_k(t)}(Q_k, P_k).$$

The important fact that about $\dot{\chi}_{\Omega_k(t)}(Q_k, P_k)$, demonstrated below, is that it is zero except at the boundary of $\Omega_k(t)$, where it is singular. It follows that if the state (Q_k, P_k) is not at the boundary of $\Omega_k(t)$, the total time derivative of $\chi_{\Omega_k(t)}(Q_k, P_k)F_k(Q_k, P_k, t)$ vanishes. To compute $\dot{\chi}_{\Omega_k(t)}(Q_k, P_k)$, the definition of $\Omega_k(t)$ stated in (4.27a) is used first to write the function $\chi_{\Omega_k(t)}(Q_k)$ in the form

$$(5.18)$$
$$\chi_{\Omega_k(t)}(Q_k, P_k) = \chi_{D_k(t)}(Q_k)\theta(\mathcal{P}_k^u - \mathcal{P}_k)\theta(\mathcal{P}_k - \mathcal{P}_k^l)\theta(\mathcal{J}_k^u - \mathcal{J}_k)\theta(\mathcal{J}_k - \mathcal{J}_k^l)$$
$$\times \theta(\mathcal{H}_k^u - \mathcal{H}_k)\theta(\mathcal{H}_k - \mathcal{H}_k^l).$$

---

[9]This was proved first by Joseph Liouville in 1838. See the discussion of Liouville's theorem and subsequent related work by Karl G. J. Jacobi along with and references to both in Boltzmann [**1896a**], pp. 274–290.

[10]Because of Liouville's demonstration (in modern terminology) that Hamiltonian flows are measure preserving, this evolution equation is often called the *Liouville equation* in statistical mechanics. However, Liouville did not write an equation of this form, so that name will not be used here.

The indicator function $\chi_{D_k(t)}(Q_k)$ is next decomposed as

$$(5.19) \qquad \chi_{D_k(t)}(Q_k) = \prod_{i \in k} \chi_{d_k(t)}(q_i).$$

It follows from its definition as an indicator function that the rate of change of $\chi_{d_k(t)}(q_i)$ as a function of time is zero except at the boundary. Anticipating the formula (6.34) of Chapter 6, the i-particle boundary flux is

$$(5.20) \quad \dot{\chi}_{d_k(t)}(q_i, v_i) = \frac{d\chi_{d_k(t)}(q_i)}{dt} = \int_{\partial k} d^2r \, \delta(q_i - r)\hat{n}_{\partial k}(r, t) \cdot (v_i - v_{\partial k}(r, t)),$$

where $\hat{n}_{\partial k}(r, t)$ is the inward-pointing normal unit vector at the boundary and $v_{\partial k}(r, t)$ is the boundary velocity vector. Integrating this over q_i, gives

$$(5.21) \qquad \int_k d^3q_i \, \dot{\chi}_{d_k(t)}(q_i, v_i) = \int_{\partial k} d^2r \, \hat{n}_{\partial k}(r, t) \cdot (v_i - v_{\partial k}(r, t)).$$

This illustrates how volume integrals are converted into surface integrals by the singular operator $\dot{\chi}_{d_k(t)}(q_i, v_i)$. Finally, using (5.20) and (5.19), the time derivative of $\chi_{D_k(t)}(Q_k)$ is obtained

$$(5.22) \qquad \frac{d\chi_{D_k(t)}(Q_k)}{dt} = \chi_{D_k(t)}(Q_k) \sum_{i \in k} \frac{\dot{\chi}_{d_k(t)}(q_i, v_i)}{\chi_{d_k(t)}(q_i)}.$$

The total time derivative of the energy bounds is written

$$(5.23) \qquad \frac{d}{dt}\theta(\mathcal{H}_k^u - \mathcal{H}_k)\theta(\mathcal{H}_k - \mathcal{H}_k^l) = -\frac{d\mathcal{H}_k}{dt}[\delta(\mathcal{H}_k^u - \mathcal{H}_k) - \delta(\mathcal{H}_k - \mathcal{H}_k^l)],$$

with similar relations for $\mathcal{P}_k$ and $\mathcal{J}_k$. Assume now that at some time t_0, $\mathcal{H}_k$ takes an extreme value, i.e., either $\mathcal{H}_k = \mathcal{H}_k^u$ or $\mathcal{H}_k = \mathcal{H}_k^l$. Assume also that none of the particles is at the system boundary $\partial d_k(t)$. The system specific trajectory will satisfy $(Q_k^{(s)}(t_0), P_k^{(s)}(t_0)) \in \partial\Omega_k(t_0)$ if the system trajectory is in $\mathrm{supp}F_k \cap \Omega_k(t_0)$, where $\mathrm{supp}F_k(Q_k, P_k, t)$ is the *support* of F_k defined in this case as the set of points (Q_k, P_k) for which $F_k(Q_k, P_k, t) > 0$. There is no trajectory in $\mathrm{supp}F_k \cap \Omega_k(t_0)$ that will increase $\mathcal{H}_k$ above $\mathcal{H}_k^u$ or decrease it below $\mathcal{H}_k^l$. It follows that $d(F_k\mathcal{H}_k)/dt = 0$ at time t_0 in this case, so the energy flux through the boundary vanishes.

The total time derivative of a system projection operator gives rise to a component that is a singular function at the system boundary. It also gives rise to singular components at the energy-dependent boundaries around each particle in the system due to the excluded volumes surrounding them. The previous result $d(F_k\mathcal{H}_k)/dt = 0$ for the energy flux, along with similar results of the form $d(F_k\mathcal{P}_k)/dt = 0$ and $d(F_k\mathcal{J}_k)/dt = 0$ for the momentum and angular momentum fluxes through the boundaries, are consistent with the fact that there is no particle flux into or out of these excluded volumes around each particle.

Suppose now that one or more particles are at the boundary. Because of the discontinuity in the momentum or velocity of these particles, $F_k(Q_k, P_k, t)$ is not continuously differentiable in this case. However, as pointed out above, $dF_k(Q_k, P_k, t)/dt$ is not bounded on the set of points in $\Omega_k(t)$ for which particles are at the boundary. This is a set of measure zero in phase space that will not effect those thermodynamic phase averages that have $dF_k(Q_k, P_k, t)/dt$ in the integrand. It is possible to set $dF_k(Q_k, P_k, t)/dt = 0$ on this set of points without changing measurable quantities, so this convention will be adopted. It is important to remember that the issue of discontinuity in F_k is connected with the employment of the wall transition approximation on the boundary $\partial\Omega_k(t)$ and does not imply that there are discontinuities in the system distribution viewed as a solution of the Hamiltonian evolution equation within $\Omega_k(t)$.

By the above extension of equation (5.16) to the boundary $\partial\Omega_k(t)$, the result $d(F_k H_k)/dt = 0$ is equivalent to

$$(5.24) \qquad F_k \left.\frac{d\mathcal{H}_k}{dt}\right|_{t=t_0} = 0 \qquad \text{(a.e.)}.$$

This relation means that no energy is being transported into or out of domains in phase space forbidden to the particles, i.e., into $\mathbf{R}^{6\mathcal{N}_k} - \Omega_k(t)$. Using these results in (5.23) gives us the result

$$(5.25) \qquad F_k \frac{d}{dt}\theta(\mathcal{H}_k^u - \mathcal{H}_k)\theta(\mathcal{H}_k - \mathcal{H}_k^l) = 0.$$

There are similar arguments and similar results concerning the vanishing of the time derivatives of the product of F_k and the time derivative of the components of the projection operator concerned with $\mathcal{P}_k$ and $\mathcal{J}_k$. It follows that no momentum and no angular momentum are being transported into or out of the phase space volume $\mathbf{R}^{6\mathcal{N}_k} - \Omega_k(t)$. It follows from these results and (5.22) that

$$(5.26) \quad \dot{\chi}_{\Omega_k(t)}(Q_k, P_k) = \frac{d\chi_{\Omega_k(t)}(Q_k, P_k)}{dt} = \chi_{\Omega_k(t)}(Q_k, P_k) \sum_{i\in k} \frac{\dot{\chi}_{d_k(t)}(q_i, v_i)}{\chi_{d_k(t)}(q_i)}.$$

It is possible to include the system projection operator in the definition of the system probability distribution density. That is, the system distribution density can be defined as $F_k^{(d)}(Q_k, P_k, t) = \chi_{\Omega_k(t)}(Q_k, P_k)F_k(Q_k, P_k, t)$, where the d refers to the confinement of the system to the volume set $d_k(t)$. Distributions similar to $F_k^{(d)}(Q_k, P_k, t)$ have been used in quantum mechanics in attempts to define localized states. The evolution equation for this state is written

$$(5.27) \qquad \frac{dF_k^{(d)}(Q_k, P_k, t)}{dt} = \frac{\dot{\chi}_{\Omega_k(t)}(Q_k, P_k)}{\chi_{\Omega_k(t)}(Q_k, P_k)} F_k^{(d)}(Q_k, P_k, t).$$

The form $F_k^{(d)}(Q_k, P_k, t)$ will not be used here. As discussed above, it is important that the macroscopic concerns with volumes and boundaries are separated from the microscopic evolution equations and their solutions because these quantities play no role in the evolution equations. For this reason, it is necessary to use F_k instead of $F_k^{(d)}$ to represent the system state. As an additional benefit, the F_k form generalizes better than $F_k^{(d)}$ to the quantum case.

For each $i \in k$, the form of (5.20) shows that $\langle \dot{\chi}_{d_k(t)}(q_i, v_i) \rangle_{kt}$ is the net i-particle probability flux into the boundary at q_i. Since each system is closed and no particles cross the boundary, the net i-particle flux into the boundary is required to vanish for each $i \in k$ and for any distribution of particles in the volume. As will be shown in Chapter 7, this requirement will be satisfied if, for $q_i \in \partial d_k(t)$, F_k satisfies the stronger closure condition

$$(5.28) \qquad \int dP_k\, F_k(Q_k, P_k, t)[v_i^\mu - v_{\partial k}^\mu(q_i, t)] = 0.$$

It follows from this relation that no potential energy or other quantity dependent only on particle position will be transferred across the boundary. That is, for any continuous phase function such as $G_i(Q_k, t)$, $G_k(Q_k, t)$, $G_i(p_j)$, or 1, which does not depend on p_i, the *closure conditions*

$$(5.29a)$$
$$\langle \delta(q - q_i)\dot{\chi}_{d_k(t)}(q_i, v_i)\{G_i(Q_k, t), G_i(p_j), 1\} \rangle_{kt} = \{0, 0, 0\}$$

and

$$(5.29b) \qquad \langle \dot{\chi}_{\Omega_k(t)}(Q_k, P_k)G_k(Q_k, t) \rangle_{kt} = 0,$$

are satisfied.

For an isolated system t, no matter, moments of momentum, or position dependent quantities, are transferred across the boundary. This condition is met by choosing the boundary so that $F_t(Q_t, P_t, t) = 0$ if $Q_t \in \partial D_t(t)$. This has the consequence

$$(5.30) \qquad \dot{\chi}_{\Omega_t(t)}(Q_t, P_t)F_t(Q_t, P_t, t) = 0.$$

For any single particle or system phase functions, $G_i(Q_t, p_i, t)$ or $G_t(Q_t, P_t, t)$, it follows for an isolated system that

$$(5.31a) \qquad \langle \delta(q - q_i)\dot{\chi}_{d_t(t)}(q_i, v_i)G_i(Q_t, p_i, t) \rangle_{tt} = 0,$$
$$(5.31b) \qquad \langle \dot{\chi}_{\Omega_t(t)}(Q_t, P_t)G_t(Q_t, P_t, t) \rangle_{tt} = 0.$$

The relations (5.29) and (5.31) will be very useful in what follows.

To ascertain the consistency of the evolution of the system distribution described by (5.16) with the conditions (5.14) it is established next that $F_k(t)$

will meet the conditions (5.14) if $F_k(0)$ meets them. To verify the finite local norm condition (5.14b), it is assumed to be true at $t = 0$. The relation is differentiated with respect to t for any time t. The result $dF_k/dt = 0$ is expanded using the definition (5.16) with (4.21) for the total time derivative operator to obtain the useful relation

$$(5.32) \qquad \frac{\partial F_k(Q_k, P_k, t)}{\partial t} = -V_{2k} \cdot \nabla_{2k} F_k(Q_k, P_k, t).$$

The time partial derivative of the relation (5.9) with $G_k = 1$ is taken to show

$$(5.33)$$
$$\frac{\partial \langle 1 \rangle_{kt}}{\partial t} = \frac{1}{Z_k(t)} \int dQ_k \int dP_k \, \frac{\partial [F_k(Q_k, P_k, t) \chi_{\Omega_k(t)}(Q_k, P_k)]}{\partial t} - \frac{\partial \ln Z_k(t)}{\partial t}$$
$$= \frac{1}{Z_k(t)} \int dQ_k \int dP_k \left[\frac{\partial \chi_{\Omega_k(t)}}{\partial t} - \chi_{\Omega_k(t)} V_{2k} \cdot \nabla_{2k} \right] F_k - \frac{\partial \ln Z_k(t)}{\partial t}$$
$$= \left\langle \dot{\chi}_{\Omega_k(t)}(Q_k, P_k) \right\rangle_{kt} - \frac{\partial \ln Z_k(t)}{\partial t}.$$

In this calculation, integration by parts and (4.19a) were used. This result is next set to 0 because $\langle 1 \rangle_{kt} = 1$ is a constant independent of the time. By (5.33) and (5.29b), it follows that

$$(5.34) \qquad \frac{\partial \ln Z_k(t)}{\partial t} = \left\langle \dot{\chi}_{\Omega_k(t)}(Q_k, P_k) \right\rangle_{kt} = 0.$$

This leads immediately to the results

$$(5.35a) \qquad \frac{\partial Z_k(t)}{\partial t} = 0,$$
$$(5.35b) \qquad Z_k(t) = Z_k = \text{const.}$$

It also follows that if $\langle 1 \rangle_{kt} = 1$ at $t = 0$, it is also true at each $t \in [0, \alpha)$.

There is a second consistency condition associated with time derivatives. This is the preservation of the closure condition (5.29). The change in the system projection operator in time is associated with the movement of the system boundaries. It is required that the closure condition is preserved as the boundaries move. Although the solutions of the k system evolution equation are not limited to $\Omega_k(t)$, the system distribution density $F_k(Q_k, P_k, t)$ outside $\Omega_k(t)$ is used only with a boundary or wall transition formalism to compute the change in F_k due to particles reaching the boundary that are reflected back into the k system. This formalism acts to maintain the relation $\hat{n}_{\partial k}(q, t) \cdot (\bar{v}_i(q, t) - v_{\partial k}(q, t)) = 0$ for $q \in \partial d_k(t)$ that is the basis for the closure condition. In other words, the k system boundary formalism does not change the probability of finding a particle in system k.

Because an imposed formalism is introduced for these particles that is not subject to the evolution equation, it follows that the total time phase derivative of $\dot{\chi}_{\Omega_k(t)}(Q_k, P_k)$, which, as the flux operator, represents the boundary formalism, can be set to zero by convention as long as that is consistent with the behavior of the rest of the formalism. This condition on $\dot{\chi}_{\Omega_k(t)}(Q_k, P_k)$ is expressed as

(5.36)
$$\frac{d\dot{\chi}_{\Omega_k(t)}(Q_k, P_k)}{dt} = 0.$$

Using (5.35), (5.32), and integration by parts, it is easy to demonstrate the required consistency condition

(5.37)
$$\frac{\partial \langle \dot{\chi}_{\Omega_k(t)}(Q_k, P_k) \rangle_{kt}}{\partial t} = \frac{1}{Z_k(t)} \int dQ_k \int dP_k \left[\frac{\partial \dot{\chi}_{\Omega_k(t)}}{\partial t} F_k + \dot{\chi}_{\Omega_k(t)} \frac{\partial F_k}{\partial t} \right]$$
$$= \left\langle \frac{d\dot{\chi}_{\Omega_k(t)}(Q_k, P_k)}{dt} \right\rangle_{kt} = 0.$$

Demonstrating condition (5.14a) requires the formalism for the change in the distribution at the wall in the general interacting systems case and it will be proved in Chapter 26. For the isolated system case, condition (5.14a) is standard and (5.14b) follows easily.[11]

5.6 The Evolution of Thermodynamic Quantities

The formulas for the evolution equation of the system distribution and the time derivative of the system projection operator can be used now to compute the change in time of a phase averaged quantity. Computing the time partial derivative of $G_i(q, t)$ yields[12]

(5.38)
$$\frac{\partial G_i(q, t)}{\partial t} = \frac{1}{Z_k(t)} \int dQ_k \int dP_k \, \delta(q - q_i) \frac{\partial [F_k \chi_{\Omega_k(t)}(Q_k, P_k)]}{\partial t} G_i(Q_k, p_i, t)$$
$$+ \left\langle \delta(q - q_i) \frac{\partial G_i(Q_k, p_i, t)}{\partial t} \right\rangle_{kt} - \frac{d \ln Z_k(t)}{dt} G_i(q, t).$$

The last term on the right is removed using (5.35a). The relation (5.32) is then used to replace $\partial F_k / \partial t$ with $-V_{2k} \cdot \nabla_{2k} F_k$ in (5.38). Integrating the

---

[11]See Kurth [**1960**], p. 93, for example.

[12]This computation is justified by the distribution version of Leibniz' rule for differentiating an integral that depends on a parameter. See Dieudonné [**1960**], p. 172, for the standard case. In Part IV, it will be shown that the integrals in (5.38) all exist.

result by parts and using (5.15) and (4.19a) leads to the result

$$(5.39) \quad \frac{1}{Z_k(t)} \int dQ_k \int dP_k \, \delta(q - q_i) G_i(Q_k, p_i, t) \frac{\partial[F_k \chi_{\Omega_k(t)}(Q_k, P_k)]}{\partial t}$$
$$= \langle V_{2k} \cdot \nabla_{2k}(\delta(q - q_i)G_i) \rangle_{kt} + \langle \delta(q - q_i)\dot{\chi}_{\Omega_k(t)}(Q_k, P_k)G_i \rangle_{kt} \,.$$

The relation

$$(5.40) \qquad V_{2k} \cdot \nabla_{2k}\delta(q - q_i) = v_i \cdot \frac{\partial}{\partial q_i}\delta(q - q_i) = -v_i \cdot \frac{\partial}{\partial q}\delta(q - q_i)$$

is used next to rewrite the first term on the right in (5.39). This result is combined with (5.38) and the second term on the right in (5.39) to obtain (5.41a)

$$\frac{dG_i(q, t)}{dt} = \left\langle \delta(q - q_i)\frac{dG_i(Q_k, p_i, t)}{dt} \right\rangle_{kt} + \langle \delta(q - q_i)\dot{\chi}_{\Omega_k(t)}G_i(Q_k, p_i, t) \rangle_{kt} \,,$$

where $dG_i(q, t)/dt = \partial G_i(q, t)/\partial t + \partial G_i^\mu(q, t)/\partial q^\mu$ is the thermodynamic total time derivative defined in (2.12). As a notational point, it should be observed that although the time derivative in (5.41a) is designated as the total time derivative, it refers only to aspects of the time derivative and to quantities, which may include $\delta_k(t)$ and $\beta_k(t)$, in which the time is an explicit parameter. If $G_i(q, t)$ does depend on $\delta_k(t)$ or $\beta_k(t)$ as an aspect of the phase averaging, the phase partial derivative of the analog with respect to the volume or thermature is expressed in the form $\dot{\delta}_k(t)\partial/\partial\delta_k$ or $\dot{\beta}_k(t)\partial/\partial\beta_k$, where the derivatives are with respect to the parameters δ_k or β_k and do not refer to the thermodynamic partial derivatives with respect to these quantities.

The quantity $\dot{\chi}_{\Omega_k(t)}(Q_k, P_k)G_i(Q_k, p_i, t)$ that appears in (5.41a) is the local G_i surface phase flux density term. The phase average of this flux term is singular, but provides a finite result for the average of G_i over a quasilocal spacetime volume associated with an observation. Summing (5.41a) over $i \in k$ yields

$$(5.41b) \qquad\qquad \frac{dG_k(q, t)}{dt} = \sum_{i \in k} \frac{dG_i(q, t)}{dt}.$$

Integrating (5.41) over $q \in \mathbf{R}^3$ gives

(5.42a)
$$\frac{\partial G_i(t)}{\partial t} = \left\langle \frac{dG_i(Q_k, p_i, t)}{dt} \right\rangle_{kt} + \langle \dot{\chi}_{\Omega_k(t)}(Q_k, P_k)G_i(Q_k, p_i, t) \rangle_{kt} \,,$$

(5.42b)
$$\frac{\partial G_k(t)}{\partial t} = \sum_{i \in k} \frac{\partial G_i(t)}{\partial t},$$

where the partial time derivative notation for thermodynamic quantities that depend only on the time is explicitly retained to indicate that these quantities may depend on the volume and thermature.

The time partial derivative of any global quantity expressed as a phase average, whether $G_k(t)$ or $G_i(t)$, can be computed directly using the definition (5.9) or (5.10) with the appropriate analog. The result for $G_k(t)$ is[13]

$$(5.43) \qquad \frac{\partial G_k(t)}{\partial t} = \left\langle \frac{dG_k(Q_k, P_k, t)}{dt} \right\rangle_{kt} + \left\langle \dot{\chi}_{\Omega_k(t)}(Q_k, P_k) G_k(Q_k, P_k, t) \right\rangle_{kt}.$$

5.7 The System State

Any system distribution that satisfies (5.14), (5.15), (5.16), and (5.29) or (5.31), is called a *system phase state* or simply a *system state*. The term 'state' in this context means the information necessary and sufficient for a full characterization of the system at a given point in time from the standpoint of the theory. The distinction between the system state, which is the system distribution $F_k(Q_k, P_k, t)$, and the system mechanical state, which is (Q_k, P_k), is important. The paradoxes that have resulted from confusing the two will be discussed in a later chapter.

Two particular limiting states play an important role in the TIS version of thermodynamics. The first is defined in terms of the k system specific trajectory in phase space $(Q_k^{(s)}(t), P_k^{(s)}(t))$, which was presented before as the solution of the k system Hamiltonian equations. It is the singular *k-system exact state*

$$(5.44) \qquad F_k^\delta(Q_k, P_k, t) = \delta(Q_k - Q_k^{(s)}(t))\delta(P_k - P_k^{(s)}(t)),$$

where

$$(5.45a) \qquad \delta(Q_k - Q_k^{(s)}(t)) = \prod_{i \in k}\prod_{\mu=1}^{3} \delta(q_i^\mu - q_i^{(s)\mu}(t)),$$

$$(5.45b) \qquad \delta(P_k - P_k^{(s)}(t)) = \prod_{i \in k}\prod_{\mu=1}^{3} \delta(p_i^\mu - p_i^{(s)\mu}(t)).$$

It also follows from the definitions (5.3) and (5.44) that

$$(5.46) \qquad Z_k^\delta(t) = \chi_{\Omega_k(t)}(Q_k^{(s)}(t), P_k^{(s)}(t)).$$

Note that $Z_k^\delta(t)$ is a projection operator because $Z_k^\delta(t) = 1$ or $Z_k^\delta(t) = 0$ depending on whether $(Q_k^{(s)}(t), P_k^{(s)}(t)) \in \Omega_k(t)$ or not.

---

[13]The flux term on the right hand side of this equation is essentially that derived by Maxwell [**1866**], equation (58). This equation also appears on p. 57 of the reprint in Brush [**1965**], Vol. 2.

The bracket notation $< \cdot >_{k\delta t}$ will be used to indicate a phase average over the F_k^δ state. Then, because of the relation

(5.47)
$$Z_k(t) \left\langle F_k^\delta(Q_k, P_k, t) \right\rangle_{kt} = \left\langle F_k(Q_k, P_k, t) \right\rangle_{k\delta t} = Z_k^\delta(t) F_k(Q_k^{(s)}(t), P_k^{(s)}(t), t),$$

which is an immediate consequence of the definitions, it follows that the motion of F_k^δ is compatible with the motion of F_k, when F_k is considered as a probability density defined on the F_k^δ states.[14] In terms of the present classical point of view, the system is always in some F_k^δ state; the state F_k represents the probability of finding the system in a given F_k^δ state.

It is easy to show that F_k^δ is a positive Radon measure. If $G_k \geq 0$ almost everywhere, it follows that $(F_k^\delta(t), G_k(t))_* = Z_k^\delta(t) G_k(Q_k^{(s)}(t), P_k^{(s)}(t), t) \geq 0$, a.e., because $Z_k^\delta(t) \geq 0$. This means that F_k^δ is non-negative and (5.14a) is fulfilled. The fact that $(F_k^\delta(t), 1)_* = \chi_{\Omega_k(t)}(Q_k^{(s)}(t), P_k^{(s)}(t)) = Z_k^\delta(t)$ means that the state F_k^δ is also normalized if $(Q_k^{(s)}(t), P_k^{(s)}(t)) \in \Omega_k(t)$. This implies in turn that (5.14b) holds. The validity of the conditions (5.15) and (5.29) follows easily for the F_k^δ state as well. As previously established, the relation (5.16) for an F_k^δ state is equivalent to Hamilton's equations.

The particles in system k are also members of system $[t]$. It follows immediately that the $[t]$ exact state can be expressed in terms of the set of subsystem exact states as

(5.48)
$$F_{[t]}^\delta(Q_t, P_t, t) = \prod_{k \in [t]} F_k^\delta(Q_k, P_k, t).$$

The t system exact state, on the other hand, is based on the trajectory $(Q_t(t), P_t(t))$ that is the solution of Hamilton's equations for the t system. Because it is usually the case that $\mathcal{H}_t(Q_t, P_t, t) \neq \sum_{k \in [t]} \mathcal{H}_k(Q_k, P_k, t)$, it follows that $F_{[t]}^\delta(Q_t, P_t, t) \neq F_t^\delta(Q_t, P_t, t)$ in general. It also follows that the trajectory obtained by selecting the trajectories of particles $i \in k$ from the t system trajectory $(Q_t(t), P_t(t))$ to form the trajectory $(Q_k^{(t)}(t), P_k^{(t)}(t))$ is not the same as the k system specific trajectory $Q_k^{(s)}(t), P_k^{(s)}(t)$.

The quantity $G_k^\delta(t)$ is defined in terms of this formalism by

(5.49)
$$G_k^\delta(t) = Z_k^\delta(t) G_k(Q_k^{(s)}(t), P_k^{(s)}(t), t).$$

This quantity is finite as long as $G_k(Q_k, P_k, t)$ is bounded for $(Q_k, P_k) \in \Omega_k(t)$ and it vanishes if $(Q_k, P_k) \notin \Omega_k(t)$.

Let us next form the local i-particle phase average of $G_i(Q_k, p_i, t)$ for the F_k^δ state. Using (5.44) in (5.10a) with the state F_k^δ gives the singular result

(5.50)
$$G_i^\delta(q, t) = \nu_i^\delta(q, t) G_i(Q_k^{(s)}(t), p_i^{(s)}(t), t),$$

---

[14]Gibbs [1902], p. 118, expressed a similar perspective with regard to the relation of microcanonical states to canonical states.

immediately, where

(5.51) $\qquad \nu_i^\delta(q,t) = \langle \delta(q - q_i) \rangle_{k\delta t} = Z_k^\delta(t) \delta(q - q_i^{(s)}(t)).$

In this case, the i-particle density is singular and the thermodynamic observable average (3.25) must be used to obtain a useful result. Setting $A_{\Delta(q,t)} = \chi_{\Delta(q,t)}$ in (3.24) and (3.25) for simplicity, a finite result is easily obtained for the observable average of the local density:

(5.52a)
$$\nu_i^\delta(\Delta(q,t)) = \frac{1}{\eth(\Delta(q,t))} \int d^3q' \int dt' \, \chi_{\Delta(q,t)}(q',t') \nu_i^\delta(q',t')$$
$$= \frac{1}{\eth(\Delta(q,t))} \int dt' \, \chi_{\Delta(q,t)}(q_i(t'),t') Z_k^\delta(t'),$$

(5.52b)
$$\nu_k^\delta(\Delta(q,t)) = \sum_{i \in k} \nu_i^\delta(\Delta(q,t)).$$

These results are used to construct the local observable average G_k from G_i in the form

(5.53a)
$$G_i^\delta(\Delta(q,t)) = \frac{1}{\eth(\Delta(q,t))} \int dt' \chi_{\Delta(q,t)}(q_i(t'),t') Z_k^\delta(t') G_i(Q_k^{(s)}(t'), p_i^{(s)}(t'), t'),$$

(5.53b)
$$G_k^\delta(\Delta(q,t)) = \sum_{i \in k} G_i^\delta(\Delta(q,t)).$$

The second limiting state is the singular k *system microcanonical state*, F_t^γ, which is defined only for an isolated system. The microcanonical state is expressed in terms of the set of mechanical constants $\mathcal{I}_t$ describing the t system as

(5.54a)
$$F_t^\gamma(Q_t, P_t; \mathcal{I}_t) = [Z_t^\gamma(\mathcal{I}_t)]^{-1} \chi_{\Omega_t(t)}(Q_t, P_t) \delta(\mathcal{P}_t(P_t) - \mathcal{I}_{P,t})$$
$$\times \delta(\mathcal{J}_t(Q_t, P_t) - \mathcal{I}_{J,t}) \delta(\mathcal{H}_t(Q_t, P_t) - \mathcal{I}_{H,t}),$$

where $\delta(\mathcal{P}_t(P_t) - \mathcal{I}_{P,k}) = \prod_\mu \delta(\mathcal{P}_t^\mu(P_t) - \mathcal{I}_{P,k}^\mu)$, etc., and

(5.54b)
$$Z_t^\gamma(\mathcal{I}_t) = \int dQ_t \int dP_t \, \chi_{\Omega_t(t)}(Q_t, P_t) \delta(\mathcal{P}_t(P_t) - \mathcal{I}_{P,t})$$
$$\times \delta(\mathcal{J}_t(Q_t, P_t) - \mathcal{I}_{J,t}) \delta(\mathcal{H}_t(Q_t, P_t) - \mathcal{I}_{H,t}).$$

$Z_t^{\gamma}(\mathcal{I}_t)$ is the area of a $(6\mathcal{N}_k - 7)$-dimensional surface in phase space with I_t fixed. Equation (4.32) is used to show that it is also true that

$$(5.55) \qquad \frac{d}{dt}\delta(\mathcal{P}_t(P_t) - \mathcal{I}_{P,t}) = \delta'(\mathcal{P}_t(P_t) - \mathcal{I}_{P,t})\frac{d\mathcal{P}_t(P_t)}{dt} = 0$$

with similar relations for $\delta(\mathcal{J}_t - \mathcal{C}_{J,t})$ and $\delta(\mathcal{H}_t - \mathcal{I}_{H,t})$. Let us set $F_t^{\gamma} = 0$ for $Q_t \in \partial D_t(t)$ because the boundary of an isolated system can be chosen outside the region accessible to the particles. Then, computing dF_t^{γ}/dt and using (5.55), it can be verified that (5.16) and (5.31) hold for F_t^{γ}. As in the case of F_t^{δ}, it is easy to verify that (5.14) and (5.15) hold for F_t^{γ} as well.

The version of the microcanonical state presented here differs considerably in mathematical form, but not in intention, from the version introduced by Gibbs [**1902**]. Because mathematics in Gibbs' time was not comfortable with singular distributions, Gibbs defined the energy of a microcanonical state in terms of an energy shell of width ΔE around a particular energy E rather than as a precisely determined surface at energy $E = I_{H,t}$ as was done in (5.54).[15]

There are two major problems with using an energy shell ΔE in the definition of the microcanonical state. The first is that for an isolated classical system the energy is fixed at a single value E and does not occupy a shell of width ΔE about E. There is therefore no physical justification for this definition. Even if the first problem is ignored, the second problem is that the width ΔE is completely arbitrary. It is often carried along for a while in the calculations and then simply disappears.

Careful modern treatments of microcanonical calculations use Gibbs definition of the function $\delta_k^{(G)}(E_k)$, which represents the phase volume with energy less than or equal to the energy E_k, as the basis for calculations. A time-dependent TIS version of Gibbs' phase volume function is $\delta_k^{(G)}(E_k, t) = \int dQ_k \int dP_k \chi_{\Omega_k(t)}(Q_k, P_k)\theta(E_k - H_k(Q_k, P_k, t))$. Bailyn [**1994**], pp. 501–503, computed the spread in $\delta_k^{(G)}(E_k, t)$ associated with the energy spread ΔE_k in the definition of the microcanonical state as (TIS notation)

$$(5.56) \qquad \Delta\delta_k^{(G)}(E_k, t) = \frac{\partial \delta_k^{(G)}(E_k, t)}{\partial E_k}\Delta E_k.$$

Computing the energy derivative of the function $\delta_k^{(G)}(E_k, t)$ results in a δ measure in the integrand due to the differentiation of $\theta(E_k - \mathcal{H}_k(Q_k, P_k, t))$ with respect to E_k. This yields a finite result for $\partial\delta_k^{(G)}(E_k, t)/\partial E_k$ when the integration is performed, but the problem with the arbitrary size of ΔE_k remains. Bailyn observed in connection with this calculation that while Gibbs [**1902**] had not introduced a function equivalent to a δ measure, he had

---

[15]See Bailyn [**1994**], Chapter 11, for a recent treatment of Gibbs' formalism.

introduced a sharply peaked function with a peak width of about ΔE_k that is equivalent to a δ measure in the limit $\Delta E_k \to 0$.

5.8 Stationary States

Let us turn to a more detailed examination of the nature of the stationary states of the isolated t system and their functional description. Since a stationary state of the system satisfies the equation

$$(5.57) \qquad \frac{\partial F_t^{(s)}(Q_t, P_t)}{\partial t} = 0,$$

by (5.16) and (5.32) it must also satisfy the equation

$$(5.58) \qquad V_{2t} \cdot \nabla_{2t} F_t^{(s)}(Q_t, P_t) = 0.$$

Every solution of this equation is a differentiable function of the set of constants of the motion of the system. This set is called a *complete invariant of the motion* by Truesdell [**1960**], p. 31. These constants themselves are a consequence of the symmetry properties of the system Hamiltonian (Goldstein [**1950**], p. 261). Since t is isolated, $\mathcal{H}_t(Q_t, P_t)$ is independent of the time and is invariant under translations and rotations. It follows immediately that $\mathcal{N}_t$, $\mathcal{P}_t^\mu(P_t)$, $\mathcal{J}_t^\mu(Q_t, P_t)$ and $\mathcal{H}_t(Q_t, P_t)$ are constants of the motion. If these symmetries exhaust the symmetries of the Hamiltonian, then the ordered set $\mathcal{I}_t$ defined in Chapter 4 is a complete invariant of the motion.

The set $\Omega_t(t)$ is defined in terms of the set $\mathcal{I}_t$ of constants of the motion in (4.33). A value of $\mathcal{I}_t$ is called a *regular value* if the set $\Omega_t(t)$ associated with it is actually a manifold in the t system phase space.[16] When this is the case, the state $F_t^\gamma(Q_t, P_t; \mathcal{I}_t)$ satisfies (5.58). Suppose the function $G_t(Q_t, P_t)$ is constant over the entire manifold defined by $\mathcal{I}_t$. Then it can be written in the form $A_M F_t^\gamma(Q_t, P_t; \mathcal{I}_t)$, where A_M is a constant. From this, it follows that F_t^γ is the unique stationary state of the t system with support covering this manifold. If there are other constants of the motion of the system, the set $\mathcal{I}_t$ can be expanded to $\mathcal{I}_t'$ and the state $F_t^{\gamma\prime}$ defined accordingly on the $\mathcal{I}_t'$ submanifold of the manifold defined by $\mathcal{I}_t$.

Consider next the *time-averaged distribution*, $\overline{F_t}(Q_t, P_t; \alpha)$ obtained from the state $F_t(Q_t, P_t, t)$ by

$$(5.59) \qquad \overline{F_t}(Q_t, P_t; \alpha) = \frac{1}{\alpha} \int_0^\alpha dt\, F_t(Q_t, P_t, t).$$

It is easy to show that this quantity has all the properties required of a state because F_t is a state. Furthermore, if the $\alpha \to \infty$ limit exists for

---

[16] See Ralph Abraham and Jerrold Marsden [**1978**], pp. 204–207.

$\overline{F_t}(Q_t, P_t; \alpha)$, then $\lim_{\alpha \to \infty} \overline{F_t}(\alpha)$ is a stationary state.[17] To show this, first assume that $\partial F_t(Q_t, P_t, t)/\partial t \in L^1(\Omega_t(t))$ and $\nabla_{2t} F_t(Q_t, P_t, t) \in L^1(\Omega_t(t))$, where $L^1(\Omega_t(t))$ is the space of (Q_k, P_k) integrable functions on $\Omega_t(t)$. Then use (5.32) followed by Fubini's theorem to write

$$(5.60) \quad \frac{1}{\alpha}[F_t(Q_t, P_t, \alpha) - F_t(Q_t, P_t, 0)] = \frac{1}{\alpha} \int_0^\alpha dt \frac{\partial F_t}{\partial t} = -V_{2t} \cdot \nabla_{2t} \overline{F_t}(\alpha).$$

The fact that the t system Hamiltonian is independent of the time was also used in this calculation. Next, phase average the relation

$$(5.61) \quad |F_t(Q_t, P_t, \alpha) - F_t(Q_t, P_t, 0)| \leq F_t(Q_t, P_t, \alpha) + F_t(Q_t, P_t, 0)$$

and use the result with (5.60) to obtain

$$(5.62) \quad \frac{1}{Z_k(t)} \int dQ_t \int dP_t \; |V_{2t} \cdot \nabla_{2t} \overline{F_t}(Q_t, P_t; \alpha)| \leq \frac{2}{\alpha}.$$

Under the TIS assumptions, $|V_{2t}|$ is bounded and continuous on $\overline{\Omega_t}(t)$ and $\overline{\Omega_t}(t)$ is compact, so $|V_{2t}|$ is uniformly bounded on $\Omega_t(t)$. It follows that $V_{2t} \cdot \nabla_{2t} F_t(Q_t, P_t, t) \in L^1(\Omega_t)$. It also follows from (5.62) that $V_{2t} \cdot \nabla_{2t} \overline{F_t}(Q_t, P_t; \alpha) \in L^1(\Omega_t)$ for $\alpha > 0$, so it can be concluded that

$$(5.63) \quad \lim_{\alpha \to \infty} V_{2t} \cdot \nabla_{2t} \overline{F_t}(Q_t, P_t; \alpha) = 0 \quad \text{a.e.}$$

This means that the $\alpha \to \infty$ limit of $\overline{F_t}(\alpha)$, if it exists, satisfies (5.58). Let us assume that this limit exists and that the support of $\overline{F_t}(Q_t, P_t; \alpha)$ remains bounded as $\alpha \to \infty$. Let us then set $\Omega_\alpha = \text{supp}\, \overline{F_k}(\alpha)$ and $\Omega_\infty = \lim\sup_{\alpha \to \infty} \Omega_\alpha$. With these assumptions and definitions, it follows that

$$(5.64) \quad \lim_{\alpha \to \infty} \overline{F_t}(Q_t, P_t; \alpha) = A_M \chi_{\Omega_\infty}(Q_t, P_t) F_t^\gamma(Q_t, P_t; \mathcal{I}_t),$$

for some constant A_M. Note that, in ergodic theory terms, this result is limited to a single metrically transitive domain. It is also different from that needed by ergodic theory which is concerned with the equivalence of the $\alpha \to \infty$ limit of time averaged thermodynamic functions defined on an exact state, designated by $G_k^\delta(\alpha)$, and the ordinary phase average $G_k(t)$. These definitions and ergodic theory issues will be discussed in EIS.

---

[17]See Kurth [1960], pp. 111–119, for an approach to this problem via Birkhoff's theorem and a discussion of the conditions under which this limit exists and the limiting state is normalized.

CHAPTER 6

Macroscopic Approximations:
Independence, Quasilocality, and the Wall

The standard procedure in statistical mechanics is to study a single macroscopic system of microscopic particles and represent the effects of all others as boundary conditions and external potentials. Justifying this arbitrary division of the world into parts and examining the consequences of doing so is the subject of this chapter.

6.1 Subject and Object

In the beginning of Maxwell's [**1877**] book *Matter and Motion* he defined a material system, as the subject of mechanics, by stating, p.2, that

> In all scientific procedure we begin by marking out a certain region or subject as the field of our investigations. To this we must confine our attention, leaving the rest of the universe out of account till we have completed the investigation in which we are engaged. In physical science, therefore, the first step is to define clearly the material system that we make the subject of our statements. This system may be of any degree of complexity. It may be a single material particle, a body of finite size, or any number of such bodies, and it may even be extended so as to include the whole universe.

This led Maxwell immediately to discuss the distinction between relations that are internal to the material system chosen and those that are between the whole or any part of the material system chosen and any external system. With reference to the external systems he noted that "These we study only so far as they affect the system itself, leaving their effect on external bodies out of consideration. Relations and actions between bodies not included in the system are to be left out of consideration. We cannot investigate them except by making our system include these other bodies."

These commonplace observations by Maxwell actually reflect an implicit epistemological stance concerning the division of the world into parts for separate study. While most physicists who select a system for study and treat the rest as external do not give this step much thought, Bohr realized the choices made by experimenters can have a profound influence on how experiments are represented theoretically and the interpretation of what is measured. To make sense of the situation in quantum mechanics, he set out to make the procedures used in setting up experiments, representing them theoretically, and interpreting measurements, manifest and explicit. One of the issues that Bohr focused on was what he called the "unambiguous use of the formalism" in his terminology. With respect to the quantum formalism, he concluded that a clear separation is required between the representations of the observing subject and the observed object. He formalized this requirement as the Subject/Object Distinction and made it one of the basic elements of his quantum epistemology.

There is a similar epistemological issue at work in classical theory. The purpose for separating a system for study from the rest of the universe, as Maxwell described it, is to allow the representation of the physical aspects of the system by phase functions that depend exclusively on quantities that are part of the system and the time. This means that the arguments of the functions representing physical quantities associated with particles in the system, the functions connected with the system boundary formalism, and the formula representing the external potential, are expressed exclusively in terms of the variables representing the particles in the system and possibly the time. An example is the Hamiltonian energy function defined in (4.12). It is written $\mathcal{H}_k(Q_k, P_k, t) = \mathcal{K}_k(P_k) + \Phi_k(Q_k, t)$ and its argument list is limited to the k system variables Q_k, P_k, and t. This Hamiltonian function takes part in the k system specific equation of motion for the particles, so solutions $F_k(Q_k, P_k, t)$ of this equation of motion are also expressed exclusively in terms of variables belonging to the particles in the k system and the time.

In this chapter, the TIS formalism needed to support this division of the world into parts will be presented and the implications and consequences of this standard practice in physics will be discussed. The position adopted in TIS is that the proper understanding of one system in a collection of interacting systems requires that all the systems be treated in a symmetric and consistent way that respects the fact that all the systems are composed of microscopic interacting particles. This requirement stems from the fact that, from the microscopic point of view, the mechanical formalism may be applied to describe a collection of particles either in the context of the subsystem it belongs to or in the context of the larger isolated system to which it also belongs. The independent systems approximation, introduced in this chapter as part of a formalism that treats all systems on an equal footing, addresses the issue of the relation between these different viewpoints as a matter of mathematics and physics. The assumption of quasilocal forces and the wall formalism, also treated in this chapter, follow from this general perspective as well.

6.2 Independence

Formally, the world is represented in TIS in terms of whole and part: the total system t and the collection $[t]$ of its subsystems. Two approximations are necessary when representing the total system as a collection of interacting subsystems in this way. These are called the *independent systems approxima-tion,* which means treating each subsystem as independent of the others in the sense of probability theory, and the *quasilocal forces approximation,* in which forces are assumed to have a finite range that is very small by macroscopic standards.

The goal of the study of a physical system is to obtain objective infor-mation that is independent of the both the experimenter and the rest of the universe. In RSO, the procedure of selecting a system for study and dealing with it separately from the environment that surrounds it was examined in the light of these goals. It was mentioned there that in order to "disambiguate" the values of the variables representing a system from the values of variables representing quantities in other systems when these systems are interacting, it is necessary to consider how the part is related to the whole in both the macroscopic formalism of thermodynamics and the microscopic formalism of statistical mechanics. In considering how the decomposition of the t system into the [t] system is to be represented in the microscopic formalism, it is clear that the central intent of this division is to make the representation of one subsystem independent of the microscopic variables describing any other system.

Alexandr Khinchin raised an apparent problem with the use of probability distributions in this way to represent physical systems in connection with the issue of decomposing a system into components for analysis. Khinchin [**1943**], pp. 41–43, presented a "methodological paradox" associated with such decompositions. He argued that if the particles in different systems are to be treated as independent in the sense of probability, their joint probability distribution must factor and remain factored in time. This means that the particle interaction energies between particles in different systems must be neglected. However, this means neglecting the exchange of energy between this system and other systems, which Khinchin had previously stated is the basis both for calculating the pressure and for the theory that an exchange of energy between systems is required for them to go to equilibrium with each other. Khinchin justified the use of probability theory and factored distributions in this context by stating that the interactions are weak and short range and can be neglected in most cases without appreciable error. He did not prove this claim, however, and it cannot be supported.

In the TIS formalism, the *independent systems approximation* will be adopted. In this approximation, the [t] system distribution density is defined as the product of the subsystem distribution densities for $k \in [t]$. The formalism is constructed so that these distribution densities for the subsystems do remain factored during the course of time. As Khinchin observed, this does imply that correlations between individual particles that appear in the t system distribution density are neglected in the factored [t] system distribution density when they are between particles in one subsystem and particles in another. It does not follow, however, that the interactions between systems must be neglected in order to use this formalism. The interactions between systems are included in the TIS formalism in the external potentials. These potentials were defined in Chapter 4 as averages of the interactions between particles in a selected subsystem and particles in all the other subsystems. Using external potentials in this way is therefore an approximation of the physical situation required by the independent systems approximation. The popular "weakly interacting systems approximation", which involves neglecting the interactions between subsystems, is not used in TIS.

The independent systems approximation is supported by a boundary formalism that maintains the separation between the systems while allowing the flow of physical quantities between them. The boundary formalism accounts quantitatively for these flows of conserved quantities between subsystems and maintains the validity of the conservation laws. By using the boundary formalism and external potentials, the independence approximation respects the fact that all interactions are ultimately between individual particles. This means that the macroscopic decomposition is consistent with the system specific mechanics of the underlying particle system.

It is important to note that the independence and quasilocal forces approximations are of a different kind than approximations made for mathematical tractability, such as the approximations of the stress tensor by various authors discussed in Chapter 11. The approximations introduced in this chapter are required for epistemological reasons and, as mentioned above, are an aspect of Bohr's analysis of the necessity of the subject/object distinction in physics.

To assess the mathematical aspects and implications of independence, consider a joint probability distribution density for the occurrence of two events. If the two events are independent in the sense of probability theory, the joint probability distribution is the product of the probability distributions for each event. This means that there is no correlation between the events that would allow the occurrence of one event to influence the probability computed for the other. In keeping with previous definitions, the event space of the t system at time t is defined as the set $\Omega_t(t)$ in t system phase space. Similarly, the event space for each subsystem $k \in [t]$ is $\Omega_k(t)$. The $[t]$ system event space is the direct product of the event spaces for each subsystem $k \in [t]$: $\Omega_{[t]}(t) = \times_{k \in [t]} \Omega_k(t)$.

Consider an event $A_k \subset \Omega_k(t)$, which is a set of elementary events (Q_k, P_k) for the k system state at time t. The event in $\Omega_{[t]}(t)$ that corresponds to the event A_k in $\Omega_k(t)$ is $A_k^t = \times_{m \in [t]-k} \Omega_m(t) \times A_k$. The conditional probabilities associated with the event A_k are defined in this reduced event space. Next, let $A_{kl}^t = A_k^t \cap A_l^t = \times_{m \in [t]-k-l} \Omega_m(t) \times A_k \times A_l$ and let $\mathcal{P}_t(A_k^t)$ represent the probability that $(Q_t, P_t) \in A_k^t$ is true. Then the events A_k^t and A_l^t are independent if and only if $\mathcal{P}_t(A_{kl}^t) = \mathcal{P}_t(A_k^t)\mathcal{P}_t(A_l^t)$.

These aspects of probability can be expressed in terms of system distribution densities or probability distribution densities. Let F_{kl} represent the joint probability distribution density of systems k and l. When the systems are independent in the sense of probability, these quantities satisfy the relation $F_{kl} = F_k F_l$. It is important to remember that independence is a property of the descriptions of systems and does not correspond to some physical aspects of those systems. It is the epistemological situation that both requires and warrants the replacement of a joint distribution by a product of distributions.

Consider next the implications of the fact that the system distribution density that can legitimately be assigned to a system is arbitrary as long at it reflects the macroscopic information we have about the system. Let $t = k + l$, where t is an isolated total system and k and l are its two subsystems. The k and l systems are assumed to be large enough so that we do not have full information concerning the particles that belong to them.

Let $F_k^{(i)}(Q_k, P_k, t)$ and $F_l^{(i)}(Q_l, P_l, t)$, for $i \in \{1, 2\}$, be two acceptable descriptions of the k and l systems. It is possible to describe the $[t]$ system by the distribution densities $F_{[t]}^{(1)}(Q_t, P_t, t) = F_k^{(1)}(Q_k, P_k, t)F_l^{(1)}(Q_l, P_l, t)$ and $F_{[t]}^{(2)}(Q_t, P_t, t) = F_k^{(2)}(Q_k, P_k, t)F_l^{(2)}(Q_l, P_l, t)$ at the same time even though $F_{[t]}^{(1)}(Q_t, P_t, t) \neq F_{[t]}^{(2)}(Q_t, P_t, t)$. However, this macroscopic choice of 1 or 2 for the $[t]$ system distribution density and the corresponding subsystem distribution densities can be made only once for the initial states of the systems involved. After that, the laws of motion determine the states that follow. For the thermodynamics obtained from the theory to be determinate and reproducible, these facts require that the thermodynamic quantities and relations obtained from the theory cannot depend significantly on the microscopic details of the initial state assigned to a system.

Assume that the events in a set of systems are independent at time $t = 0$. If these systems are mutually isolated, then the events in these systems will always remain independent. To see this formally, assume that a total system is composed of two subsystems, k and l, which are isolated from each other. Because these systems are mutually isolated, the external potentials representing forces between the systems vanish, i.e., $\Phi_{k,x}(Q_k, t) = 0$ and $\Phi_{l,x}(Q_l, t) = 0$ are true. It is not hard to show that the total time phase derivative operator of the system is then the sum of operators of the form (4.21) for the isolated k and the l subsystems:

$$(6.1) \qquad\qquad \frac{d_t}{dt} = \frac{d_k}{dt} + \frac{d_l}{dt},$$

where d_t/dt is the total time derivative applied to phase functions of the t system and d_k/dt and d_l/dt are defined similarly for the k and l systems. Let us apply this to the collection $[t]$ of subsystems that satisfy the evolution equation (5.16). Then, any joint distribution $F_{[t]}$ that factors into $F_{[t]}(0) = F_k(0)F_l(0)$ at time $t = 0$, will factor into the product $F_k(t)F_l(t)$ at any time $t > 0$. It follows that independent systems that are isolated from each other will remain independent.

The concepts of isolation and independence are distinct. First, isolation does not imply independence. That is, if k and l are mutually isolated and the initial choice $F_{[t]}(0) \neq F_k(0)F_l(0)$ is made, then an examination of the evolution equations shows that it will also be true in general that $F_{[t]}(t) \neq F_k(t)F_l(t)$. Independence does not imply isolation either. To show this, consider events in one system k and represent the influence of a second system l on the particles of k by an external potential function. The k external potential is expressed solely in terms of variables of the k system and no correlations are set up between the k particle variables and the l particle variables. For this representation, the events in the k system are independent of events in the l system even though the boundary conditions of the first system are determined as an average of these events in the l system.

The *independent systems approximation* provides a formalism in which the t system, the $[t]$ system, and each of the subsystems $k \in [t]$, has its own evolution equation and its own representation by a system distribution. The *independent systems distribution density expresses the relation between these descriptions.* It is defined by

$$(6.2) \qquad F_{[t]}(Q_t, P_t, t) = \prod_{k \in [t]} F_k(Q_k, P_k, t).$$

This was called the $[t]$ system distribution density above. Since the number of systems is finite and each system distribution density F_k is required to be non-negative and normable, $F_{[t]}$ is non-negative and normable. Moreover, the support of $F_{[t]}$ is easily seen to fall within $\Omega_t(t)$, so $F_{[t]}$ satisfies the conditions (5.14).

It is of interest next to compare the time evolution of F_t and $F_{[t]}$. To begin with, the subsystem total time phase derivative (4.21), with the help of (6.2) and (5.16), can be used to show

$$(6.3a) \qquad \frac{dF_t(Q_t, P_t, t)}{dt} = 0,$$

$$(6.3b) \qquad \frac{dF_{[t]}(Q_t, P_t, t)}{dt} = 0.$$

The k marginalization of (6.3b) is the k system evolution equation, (5.16). Let us compare this with the k marginalization of (6.3a).[1] To do this, define the k marginalization of F_t by

$$(6.4) \qquad F_k^t(Q_k, P_k, t) = \left[\prod_{s \in [t]-k} \int dQ_s \int dP_s \right] F_t(Q_t, P_t, t)$$

and the kl marginalization by

$$(6.5) \qquad F_{kl}^t(Q_k, P_k, Q_l, P_l, t) = \left[\prod_{s \in [t]-k-l} \int dQ_s \int dP_s \right] F_t(Q_t, P_t, t).$$

The global norms of these marginalizations are

$(6.6a)$

$$Z_t^{(g)} = \int dQ_k \int dP_k \, F_k^t(Q_k, P_k, t),$$

$(6.6b)$

$$Z_t^{(g)} = \int dQ_k \int dP_k \int dQ_l \int dP_l \, F_{kl}^t(Q_k, Q_l, P_k, P_l, t).$$

---

[1] A similar analysis for the isolated system case appears in Kurth [**1960**], pp. 95–96.

Let us also set

(6.7) $$\Phi_{kl}(Q_k, Q_l) = \sum_{i \in k} \sum_{j \in l} \phi_{ij}(q_i - q_j).$$

It follows immediately from this definition that

(6.8) $$\frac{\partial \Phi_{ks}(Q_k, Q_s)}{\partial Q_k} = -\frac{\partial \Phi_{ks}(Q_k, Q_s)}{\partial Q_s}.$$

With this notation, the t system Hamiltonian can be written in the form

(6.9) $$\mathcal{H}_t(Q_t, P_t) = \sum_{k \in [t]} \mathcal{H}_k(Q_k, P_k) + \frac{1}{2} \sum_{k \in [t]} \sum_{l \in [t] - k} \Phi_{kl}(Q_k, Q_l).$$

Now, by the distribution version of Leibniz's rule for differentiating under the integral sign, it follows that

(6.10) $$\left[\prod_{s \in [t] - k} \int dQ_s \int dP_s \right] \frac{\partial F_t}{\partial t} = \frac{\partial}{\partial t} \left[\prod_{s \in [t] - k} \int dQ_s \int dP_s \right] F_t = \frac{\partial F_k^t}{\partial t}.$$

The relation $\int dP_s \, (\partial F_{ks}^t / \partial P_s) G_{ks}(Q_k, Q_s) = 0$ is established for any G_{ks} using integration by parts. This fact leads to the result

(6.11) $$\left[\prod_{s \in [t] - k} \int dQ_s \int dP_s \right] \frac{\partial \Phi_{t,n}(Q_t)}{\partial Q_t} \cdot \frac{\partial F_t}{\partial P_t} = \frac{\partial \Phi_{k,n}(Q_k)}{\partial Q_k} \cdot \frac{\partial F_k^t}{\partial P_k}$$
$$+ \sum_{s \in [t] - k} \int dQ_s \int dP_s \frac{\partial \Phi_{ks}(Q_k, Q_s)}{\partial Q_k} \cdot \frac{\partial F_{ks}^t}{\partial P_k}.$$

This is used with the similar relation $\int dQ_s \, (\partial F_{ks}^t / \partial Q_s) \cdot V_s = 0$, which is valid for any V_s, to show

(6.12) $$\left[\prod_{s \in [t] - k} \int dQ_s \int dP_s \right] V_{2t} \cdot \nabla_{2t} F_t = V_k \cdot \frac{\partial F_k^t}{\partial Q_k} - \frac{\partial \Phi_{k,n}(Q_k)}{\partial Q_k} \cdot \frac{\partial F_k^t}{\partial P_k}$$
$$- \sum_{s \in [t] - k} \int dQ_s \int dP_s \frac{\partial \Phi_{ks}(Q_k, Q_s)}{\partial Q_k} \cdot \frac{\partial F_{ks}^t}{\partial P_k}.$$

Let us now complete the k marginalization of dF_t/dt by assembling these parts. For the k-marginalization of dF_t/dt, using the formulas above yields the result

(6.13) $$\frac{\partial F_k^t}{\partial t} + V_k \cdot \nabla_k F_k^t - \frac{\partial \Phi_{k,n}(Q_k)}{\partial Q_k} \cdot \frac{\partial F_k^t}{\partial P_k}$$
$$- \sum_{s \in [t] - k} \int dQ_s \int dP_s \frac{\partial \Phi_{ks}(Q_k, Q_s)}{\partial Q_k} \cdot \frac{\partial F_{ks}^t}{\partial P_k} = 0.$$

This is the equation to be approximated.

To approximate (6.13), the following replacements are made:

(6.14a)   (i) F_k in place of $\dfrac{Z_k^{(g)}}{Z_t^{(g)}} F_k^t,$

(6.14b)   (ii) $F_k F_s$ in place of $\dfrac{Z_k^{(g)} Z_l^{(g)}}{Z_t^{(g)}} F_{ks}^t,$

Multiplying equation (6.13) by $Z_k^{(g)}/Z_t^{(g)}$, using these substitutions, and rearranging, yields

(6.15) $\dfrac{\partial F_k}{\partial t} + V_k \cdot \nabla_k F_k - \dfrac{\partial \Phi_{k,n}(Q_k)}{\partial Q_k} \cdot \dfrac{\partial F_k}{\partial P_k} - \dfrac{\partial \Phi_{k,x}(Q_k, t)}{\partial Q_k} \cdot \dfrac{\partial F_k}{\partial P_k} = 0,$

where
(6.16)

$$\Phi_{k,x}(Q_k, t) = \sum_{s \in [t]-k} \frac{1}{Z_s^{(g)}} \int dQ_s \int dP_s \, F_s(Q_s, P_s, t) \Phi_{ks}(Q_k, Q_s) = 0.$$

The approximation (6.15) is the k system evolution equation (5.16).

In summary, the *independent systems approximation* consists of approximating system t by system $[t]$, that is,

(a) replacing F_t by $F_{[t]}$,
(b) replacing the t system evolution equation (6.3a) by the $[t]$ system evolution equation (6.3b), and
(c) adopting, for each $k \in [t]$, the initial condition

(6.17) $F_k(Q_k, P_k, 0) = F_k^t(Q_k, P_k, 0).$

By this choice of an initial condition, as will be shown below, the $[t]$ system has the same conserved quantities as the t system.

6.3 The Quasilocal Forces Approximation

The quasilocal forces approximation was introduced above as the assumption that the forces between particles are very short range compared to macroscopic distances. Long-range electromagnetic and gravitational potentials do not fit the quasilocal model and are either ignored or represented as ordinary external potential fields.

Under the quasilocal forces approximation, local processes are not affected by distant particles. The issues connected with locality are important in the quasilocal version of the TIS theory because the standard action at a distance potentials used to represent the interactions of particles act instantaneously over finite distances, which means that local conservation laws are violated. Retarded potentials would not help unless used in combination with energy carrying fields to account for the energy in transit. Because of the

problem of the instantaneous transmission of momentum, angular momentum, and energy, quasilocal conservation laws will be constructed to replace
the local conservation laws for this version of the theory.

The approximations introduced in this section and the last one are not
new. As Brush [1970] pointed out, Newton used a form of the quasilocal
forces assumption in some of his discussions. Implicit forms of both the
independence and the quasilocality approximations were essential to the work
of Maxwell [1866].[2] The quasilocal forces approximation was used by Boltzmann [1868] and is often used as one of the assumptions in the approximation
procedure used to "derive" Boltzmann's equation.[3] Assumptions similar to
these have been used, usually implicitly and without notice, ever since. From
an empirical perspective, the quasilocal forces assumption is justified by the
saturation properties of electronic bond interactions between atoms, which
means that a given electron in an atom can interact significantly with only a
few other electrons in other nearby atoms.

To develop the quasilocal conservation laws, first assume that at each time
t the particles in the total system, t, can be divided into mutually exclusive
sets such that particles in any one set do not interact with particles in any
other set. Let $\pi(t)$ be a finite partition at time t of $d_t(t)$ into open, connected
sets $\Delta_t^a(t)$, $a \in \pi(t)$, called *quasilocal domains*. This partition is defined
at time t so that no particle is on a partition boundary and each group of
particles is contained in one and only one set $\Delta_t^a(t)$ for some $a \in \pi(t)$. From
these definitions, the relations

(6.18a) $$\cup_{a \in \pi(t)} \overline{\Delta_t^a}(t) = \overline{d_t}(t),$$

(6.18b) $$\Delta_t^a(t) \cap \Delta_t^b(t) = \emptyset, \quad \text{if } a \neq b,$$

follow. The $\pi(t)$ partition is used to partition the sets $d_k(t)$ and $\partial d_k(t)$. In
accord with this, the k/a *quasilocal domain* and the k/a *boundary domain*
are defined by

(6.19a) $$\Delta_k^a(t) = \Delta_t^a(t) \cap d_k(t),$$

(6.19b) $$\partial \Delta_k^a(t) = \overline{\Delta_t^a}(t) \cap \partial d_k(t).$$

It follows easily that

(6.20a) $$\cup_{a \in \pi(t)} \overline{\Delta_k^a}(t) = \overline{d_k}(t),$$

(6.20b) $$\cup_{a \in \pi(t)} \partial \Delta_k^a(t) = \partial d_k(t).$$

---

[2]These are often characterized as the 'short-range forces approximation' or 'short-
range forces assumption'. See, for example, Maxwell [1879a](in Maxwell [1890], p. 706).

[3]See Henry W. Watson [1876] on this and other approximation schemes presented as
derivations of Boltzmann's equation.

For each $k \in [t]$ or for $k = t$, let us define an index set for the particles in the quasilocal domain a by

(6.21a) $$\gamma_k^a = \gamma_k^a(t) = \{ i \in k \mid q_i \in \Delta_k^a(t) \}.$$

The union over $a \in \pi$ of these particle indexes yields the set k of the particle indices in the k system:

(6.21b) $$k = \cup_{a \in \pi(t)} \gamma_k^a(t).$$

The direct product of the position and momentum vectors for those particles in a is designated by

(6.22a) $$Q_k^a = \times_{i \in \gamma_k^a} q_i,$$
(6.22b) $$P_k^a = \times_{i \in \gamma_k^a} p_i.$$

Using this formalism, the a domain total momentum, angular momentum, and kinetic energy, are

(6.23a) $$\mathcal{P}_k^{a,\mu}(P_k^a) = \sum_{i \in \gamma_k^a} p_i^\mu,$$

(6.23b) $$\mathcal{J}_k^{a,\mu}(Q_k^a, P_k^a) = \sum_{i \in \gamma_k^a} j_i^\mu(q_i, p_i),$$

(6.23c) $$\mathcal{K}_k^a(P_k^a) = \sum_{i \in \gamma_k^a} \mathcal{K}_i(p_i).$$

Because of the assumption that at time t the particles in a given quasilocal domain are interacting only with other particles in that domain, it follows that in each domain $\Delta_k^a(t)$, the momentum, angular momentum and energy are conserved. At the boundary of the k and l systems, suppose $\Delta_k^a(t) \cap w_{kl}(t) \neq \emptyset$. The boundary transition $(q_i, p_i') \to (q_i, p_i)$ due to the interaction of particle i at $q_i \in w_{kl}$ in domain a for the case of a domain that crosses the kl boundary must be accompanied by a boundary transition for particles in the l system that maintains the conservation laws. Assuming that the boundary formalism introduced below has been constructed so that this is true, the relations valid for the a domain total momentum, angular momentum, and kinetic energy, are

(6.24a) $$\mathcal{P}_k^{a,\mu}(P_k^{a\prime}) + \mathcal{P}_l^{a,\mu}(P_l^{a\prime}) = \mathcal{P}_k^{a,\mu}(P_k^a) + \mathcal{P}_l^{a,\mu}(P_l^a),$$

(6.24b)
$$\mathcal{J}_k^{a,\mu}(Q_k^a, P_k^{a\prime}) + \mathcal{J}_l^{a,\mu}(Q_l^a, P_l^{a\prime}) = \mathcal{J}_k^{a,\mu}(Q_k^a, P_k^a) + \mathcal{J}_l^{a,\mu}(Q_l^a, P_l^a),$$

(6.24c) $$K_k^a(P_k^{a\prime}) + K_l^a(P_l^{a\prime}) = K_k^a(P_k^a) + K_l^a(P_l^a).$$

The global case is used when the division into quasilocal domains is not possible or not of interest. In this case, the systems are not partitioned into subdomains, so π has one element, $\pi = \{1\}$, and it follows that $\Delta_k^1(t) =$

$d_k(t)$, $\partial \Delta_k^1(t) = \partial d_k(t)$, and $\gamma_k^1(t) = k$. This means that the global case fits smoothly into this formalism and requires no essential modifications in the previous calculations.

A *quasilocal function* is a step function $b_k^a(t)$ that is constant on a single quasilocal domain a and zero outside it. In most cases of interest, a quasilocal density function is needed and written as $b_k^a(q, t)$. A quasilocal density is defined by requiring that $q \in \Delta_k^a(t)$ and dividing the value $b_k^a(t)$ by the volume $\delta_a(t) = \mu(\Delta_k^a(t))$ of the quasilocal domain a:

$$(6.25) \qquad b_k^a(q, t) = \frac{1}{\delta_a(t)} \chi_{\Delta_k^a(t)}(q) b_k^a(t).$$

Any local function, $b_k(q, t)$, can be approximated as a sum of quasilocal functions by

$$(6.26) \qquad b_k(q, t) \approx \sum_{a \in \pi(t)} b_k^a(q, t).$$

The global version of $b_k^a(q, t)$ is obtained by integrating it over $q \in d_k(t)$. This yields the approximation of the global function $b_k(t)$ in the form

$$(6.27) \qquad b_k(t) = \int d^3q \, b_k(q, t) \approx \sum_{a \in \pi} \int d^3q \, b_k^a(q, t) = \sum_{a \in \pi} b_k^a(t).$$

Quasilocal functions will be used in the discussion of the temperature in connection with the definition of the equilibrium entropy.

6.4 The Thermodynamic Limit

The thermodynamic limit was mentioned briefly above in a discussion of other approaches to the relation of statistical mechanics to thermodynamics. It treats each system as though the number of particles in the system and its volume are infinite, while the particle density and energy per particle are required to be finite. Formally, the thermodynamic limit takes the form $\mathcal{N}_k \to \infty$, $E_k \to \infty$, and $\delta_k \to \infty$, with $\mathcal{N}_k/\delta_k \to \rho_k < \infty$ and $E_k/\delta_k \to e_k < \infty$.

The thermodynamic limit was contemplated by Maxwell [**1878b**] (see Maxwell [**1890**], p. 670) and Gibbs [**1902**], p. 166–167, and employed by Boltzmann [**1896a**]. However, Boltzmann [**1896a**], pp. 61–62, felt that it was a "dubious" assumption.[4] It has since come to be thought of by many as a necessary step in obtaining thermodynamic formulas from statistical mechanical ones.

The standard justifications for using the thermodynamic limit have been reviewed by Joel Lebowitz [**1999**], pp. 589–590. He stated that this limit represents an "idealization of a macroscopic physical system whose spatial

---

[4]Boltzmann, nevertheless, depended on this procedure to evade the consequences of Poincaré's theorem. See Boltzmann [**1897**] (in Brush [**1965**], Vol. 2, p. 245).

extension, although finite, is very large on the microscopic scale of interparticle distances or interactions." He noted that one advantage of this limit is that the boundary and finite-size effects present in real systems are eliminated. Lebowitz observed also that particle sum functions in this limit take on determinate values with zero variance and are independent of the volume δ_k and the nature of the boundaries of the volume.

The thermodynamic limit simplifies some calculations and some formulas of standard thermodynamics. Another purpose for using it is to define a system that will show irreversible behavior and not exhibit Poincaré recurrences. It follows immediately that the thermodynamic formalism based on the thermodynamic limit is not compatible with the mechanics of the underlying particle system. There are also other serious mathematical and physical problems with this procedure that make any claims based on this limit indeed dubious. From the TIS standpoint, the thermodynamic limit does not make mathematical sense because quantities crossing the boundaries of systems are important to a comprehensive theory and thermodynamic states obtained by using it do not have any dynamical significance.

In TIS notation, the thermodynamic limit of a formula that is expressed in terms of the number of particles and their energies is obtained by applying to it the following limiting procedure:

$$(6.28a) \qquad \mathcal{C}_{H,t} \to \infty, \quad \mathcal{N}_t \to \infty \quad \text{with } \mathcal{C}_{H,t}/\mathcal{N}_t = \text{const.}$$

The other limit, outlined in Chapter 1, is expressed in terms of the number of particles, the volume δ_k of the particle system, and the density of particles. It takes the form

$$(6.28b) \qquad \delta_k \to \infty, \quad \mathcal{N}_t \to \infty \quad \text{with } \mathcal{N}_t/\delta_k = \text{const.}$$

In other cases, the energy per particle is also held constant. The asymptotic limits of a number of thermodynamic quantities are computed in TIS using the central limit theorem of probability in EIS, Part V.

6.5 The System Wall

The physical boundary l around a system k of particles, such as the container of a gas, is itself a material system. The boundary system l is often called the *wall* of the k system in thermodynamics. From the standpoint of the TIS formalism, a wall system is just another particle subsystem in the $[t]$ collection. Because subsystems are closed, the concept of a wall plays a special role in the TIS formalism. As mentioned briefly in Chapter 5, the system wall in TIS is represented by a stochastic operator that reflects particles back into their respective systems while allowing the transmission of physical quantities between systems and keeping proper account of the conserved quantities transmitted between them.

Longer-range interactions between particles in different systems that are not at the boundary between systems are accounted for by the external potentials for each system and not by the wall formalism.

The TIS stochastic wall transition probability operator is defined by its action on the system distributions for two systems sharing a boundary. Aspects of this boundary formalism were introduced in Chapter 5 in the form of the boundary flux operator.

The use of the term 'wall' to refer to the systems bounding a given system is in accord with common usage. Maxwell [**1879a**] was the first to introduce consideration of the physical, as opposed to idealized, wall into statistical mechanics. He based his discussion of the wall on several physical models. His most important innovation in this regard was the introduction of explicit transition probabilities for the momentum change as a particle rebounds from the wall. After Maxwell's work, models of the wall have been proposed by a number of authors. This was a popular topic in the middle of the twentieth century, when walls and wall formalisms were discussed by Peter G. Bergmann and J. L. Lebowitz [**1955**], Lebowitz and H. L. Frisch [**1957**], and Lebowitz and Bergmann [**1957**]. Twenty years after that, wall formalisms were discussed by Carlo Cercignani [**1975**], J. R. Dorfman and H. van Beijeren [**1977**], and Truesdell and Muncaster [**1980**].

Many of the theories connecting thermodynamics and statistical mechanics avoid the issue of accounting for the wall by considering only isolated systems or systems at equilibrium. In these cases, there is no flow of moments of momentum across the boundary. In other cases, the wall is assigned an infinite potential to keep particles from penetrating it.[5] Theories that are more general allow systems to interact, but impose unrealistic assumptions on these interactions. Theories that speak in terms of "weakly interacting" systems and claim that this means that systems other than the one under investigation can be ignored—except when they function as "heat baths" for the purpose of bringing that system into equilibrium at a given temperature—fit into this category. The main purpose for assuming that two systems k and l are weakly interacting is to allow the total energy of the two systems to be written as $E_t = E_k + E_l$ and ignore the potential energy E_{kl} due to the interactions between particles in different systems.[6] The only reason for discussing interacting systems at all in most cases has been to consider formally how heat flows from one system to another or how a system will go to equilibrium when connected to a heat bath. The question of reconciling the assumption

---

[5]See, for example, Gibbs' [**1902**], pp. 163-164, assumption of infinite wall potentials and Tolman's [**1938**], pp. 505, 541–547, criticism of it. Almost all the 'particles in a box' calculations make this assumption.

[6]This approach is taken in many treatments of thermodynamics and statistical mechanics. See Jordan [**1933**], p. 29, for one example among many of how the weakly interacting assumption is used. Recent versions of this assumption appear in numerous textbooks.

that an interaction is strong enough to bring a system to equilibrium with the assumption that it is weak enough to be negligible is not explored, however.[7]

In most standard treatments of interacting systems, the boundary conditions for the system under study have properties that are inconsistent with the fact that the boundary system is itself made up of particles and has changing values for its own energy, momentum, angular momentum, pressure, temperature, and entropy.[8] Another missing component in these theories is the fact that either the motion of the boundary is neglected or the effects of this motion are not fully explored. In each of the cases cited, an essential component of the analysis of momentum and energy flow is missing and compatibility with the underlying particle model is lost. One consequence of these problems is that a logically acceptable analysis of entropy and the second law of thermodynamics in nonequilibrium settings that is based on the underlying particle dynamics has not been possible.

The theoretical account of the wall developed here will differ somewhat from Maxwell's. The essential notion of a transition probability for particles at the wall is retained, but will not be connected with any particular physical model of the wall. The reason for this difference is that the relationships stated in the TIS axioms, including Clausius' inequality, depend only on the stochastic character and some very general properties of the wall transition probability and not on the details of a particular physical model. A specific physical model is required, of course, if the actual rate at which certain quantities change at the wall is to be computed. The central point is that the wall is a conceptual construct that facilitates, at a certain level of approximation, the description of the effects two systems have on each other at their common boundary.

6.6 The TIS Approach to the Wall

The role of a boundary formalism is to enforce the closure condition on each system, maintain system independence, and to account for the transmission of conserved quantities between systems. The two-dimensional boundary surface associated with these considerations is a macroscopic aspect of the situation, a conventional choice not determined by the particle physics, that we impose for the purpose of organizing our theoretical relations and interpreting experiments.

---

[7]The TIS interacting systems approximation, introduced previously in this chapter, includes the energy of interaction between systems. The justification for writing $E_t = E_k + E_l$ for two interacting systems in TIS is that half of the interaction energy for each pair of particles in different systems is assigned to the energy of each system in the form of the external potential energy and included in E_k and E_l.

[8]Examples of these theories are those that assume special reflection laws at the wall. The assumed wall reflection laws have included specular, reverse, and diffuse reflection, or some combination of these, for particles reaching the boundary.

One consequence of the need to divide the world into parts is that the motion of the particles in the subsystems in $[t]$ is expressed theoretically in terms of system specific trajectories. This means, as noted above, that particle trajectories when they are viewed as belonging to a member of the $[t]$ system are not the same as the trajectories of those of the same particles when they are considered to be components of the t system because the evolution equations are different.

It is useful to collect these aspects of the system boundaries into a formalism that corresponds to our macroscopic experience of the relations between macroscopic objects. The basic concept to be introduced is the *interface transition probability*. The interface transition probability W_{kl} between systems k and l specifies the probability of each specific transition from the joint mechanical state $(Q_k^\natural, P_k^\natural) \times (Q_l^\natural, P_l^\natural)$ entering the kl boundary to the joint mechanical state $(Q_k, P_k) \times (Q_l, P_l)$ exiting the boundary. At this level, we are dealing with the bare physics of the multiparticle interactions at the boundary and the system distributions play no role. Momentum, angular momentum, and energy are simply transferred across the boundary from particles to particles and no physical quantities are created, destroyed, or retained, at the boundary.

The purpose of replacing the exact particle trajectories determined by the equations of motion with those obtained from the approximate interface transition probability is to make the interacting subsystems k and l independent and allow their separate description in the formalism. The local *wall transition probability* for a system k is constructed next in accord with this by phase averaging the interface transition probability at a given spacetime point (q, t) with the distribution for the wall system l at that point. The wall transition probability expresses the probability that the k system mechanical state $(Q_k^\natural, P_k^\natural)$ will make the transition into the state (Q_k, P_k) due to an interaction between particles in the k system and particles in the wall. This notion of a wall is symmetric in that if the l system is (part of) a wall for the k system, then the k system is (part of) a wall for the l system.

In the final step, the wall transition probability is used to fashion an operator, called the *wall operator*. This operator maps an incoming k system state $F_k(Q_k^\natural, P_k^\natural, t)$ to an outgoing state $F_k(Q_k, P_k, t)$. Expressing the system evolution in terms of a wall operator makes explicit the causal assumptions of the theory that are implicit in the calculation of $d(\chi_{\Omega_k(t)} F_k)/dt$ in Chapter 5. This concept of a wall operator is consistent with the conception of Truesdell and Muncaster [**1980**], who have emphasized that the wall operator serves as the general boundary condition for kinetic theory.

6.7 The Interface Transition Probability

A mechanical state (q_i, p_i) for particle i with $q_i \in \partial d_k(t)$ at time t is said to be an *incoming state* if the particle is entering the boundary, that is, if $\hat{n}_{\partial k}(q_i, t) \cdot (v_i - v_{\partial k}(q_i, t)) < 0$. It is said to be in an *outgoing state* if it is leaving the boundary, i.e., if $\hat{n}_{\partial k}(q_i, t) \cdot (v_i - v_{\partial k}(q_i, t)) > 0$. The basic TIS assumption is that a particle that enters the boundary at time t in state $(q_i^\natural, p_i^\natural)$ is replaced by an identical particle leaving the boundary at time t in state (q_i, p_i). Before and after the transition at the boundary, the motion of a particle is governed by its evolution equation. For particles not at the boundary or particles in an outgoing state at the boundary, the particle state at time t is unmodified by this transition formalism.

To simplify notation, the symbol G_k will be used in place of $G_k(Q_k, P_k, t)$ and $G_k^\natural$ in place of $G_k(Q_k^\natural, P_k^\natural, t)$. The physical quantity of most concern is the rate at which transitions occur at the wall since this is needed in the evolution equations. This rate is determined by the flux of particles into the wall. Since particles carry the mechanical quantities of interest, the flux into the wall can be expressed as the flux of particles, momentum, angular momentum, or energy. To represent these fluxes of system quantities into the wall, some notation must be added. An auxiliary function, for $q_i \in \partial d_k(t)$, is defined that is called the *incoming* $(-)$ or *outgoing* $(+)$ *i-particle state projection operator* by

$$(6.29) \qquad \theta^\pm(q_i, v_i) = \theta(\pm\hat{n}_{\partial k}(q_i, t) \cdot (v_i - v_{\partial k}(q_i, t))),$$

where $\theta^-(q_i, v_i) = 1$ if (q_i, p_i) is an incoming state and 0 otherwise, and $\theta^+(q_i, v_i) = 1$ if (q_i, p_i) is an outgoing state and 0 otherwise. This function is used in the expression of the *incoming* $(-)$ and *outgoing* $(+)$ *i-particle relative velocity*, defined for q_i at the boundary as

$$(6.30) \qquad u_i^{\pm,\mu}(q_i, v_i) = \theta^\pm(q_i, v_i)(v_i^\mu - v_{\partial k}^\mu(q_i, t)).$$

The incoming and outgoing flux notation is used to represent formally the mechanical state transition at the kl boundary. The *interface transition probability*, $W_{kl}(Q_k, Q_l, Q_k^\natural, Q_l^\natural, P_k, P_l, P_k^\natural, P_l^\natural)$, is the quantity that describes this transition. It will consistently be abbreviated as W_{kl}. It represents the probability of a transition between the flux of the incoming joint mechanical state $(Q_k^\natural, P_k^\natural) \times (Q_l^\natural, P_l^\natural)$ and the flux of the outgoing joint mechanical state $(Q_k, P_k) \times (Q_l, P_l)$. W_{kl} is required to map incoming fluxes with mechanical states that fall within $\Omega_k(t) \times \Omega_l(t)$ to outgoing fluxes with mechanical states that fall within $\Omega_k(t) \times \Omega_l(t)$. For $i \in k$ or $i \in l$, the connection between the flux of the incoming joint kl system state, $u_i^- F_{kl}$, and the flux of the outgoing joint kl system state, $u_i^+ F_{kl}$, is expressed in terms of W_{kl} by

$$(6.31) \quad u_i^{+,\mu}(q_i, v_i)F_{kl} = -\int dQ_k^\natural \int dP_k^\natural \int dQ_l^\natural \int dP_l^\natural \, W_{kl} u_i^{-,\mu}(q_i^\natural, v_i^\natural)F_{kl}^\natural.$$

The quantity $u_i^\pm$ is used next to define the *incoming* $(-)$ and *outgoing* $(+)$ *i-particle boundary flux operator* by

(6.32) $$\dot{\chi}_{d_k(t)}^\pm(q_i, v_i) = \int_{\partial k} d^2r\, \delta(q_i - r)\hat{n}_{\partial k}(r, t) \cdot u_i^\pm(r, v_i).$$

In this formula, r is a three-dimensional vector that is integrated over the two dimension boundary of the k system. The vector inner product of the operator $\int_{\partial k} d^2r\, \delta(q - r)\hat{n}_{\partial k}(q, t)$ with both sides of (6.31) is computed next followed by the use of the definition (6.32). The resulting relation expresses the outgoing i-particle flux in terms of the incoming i-particle flux:
(6.33)

$$\dot{\chi}_{d_k(t)}^+(q_i, v_i)F_{kl} = -\int dQ_k^\natural \int dP_k^\natural \int dQ_l^\natural \int dP_l^\natural\, W_{kl}\dot{\chi}_{d_k(t)}^-(q_i^\natural, v_i^\natural)F_{kl}^\natural.$$

Summing this over $i \in k$ or $i \in l$ yields the incoming and outgoing total boundary fluxes.

It is also easy to use this formalism to show

(6.34a) $$u_i(q_i, v_i) = u_i^+(q_i, v_i) + u_i^-(q_i, v_i),$$
(6.34b) $$\dot{\chi}_{d_k(t)}(q_i, v_i) = \dot{\chi}_{d_k(t)}^+(q_i, v_i) + \dot{\chi}_{d_k(t)}^-(q_i, v_i).$$

By the definition (5.20) and the results (5.29), (6.30), and (6.32) above, the *i-particle boundary flux* and *k boundary flux* can be written

(6.35a)

$$\dot{\chi}_{d_k(t)}(q_i, v_i) = \frac{d\chi_{d_k(t)}(q_i)}{dt} = \int_{\partial k} d^2r\, \delta(q_i - r)\hat{n}_{\partial k}(r, t) \cdot (v_i - v_{\partial k}(r, t)),$$

(6.35b) $$\dot{\chi}_{\Omega_k(t)}(Q_k, P_k) = \chi_{\Omega_k(t)}(Q_k, P_k)\sum_{i \in k} \frac{\dot{\chi}_{d_k(t)}(q_i, v_i)}{\chi_{d_k(t)}(q_i)}.$$

As a transition probability, W_{kl} must meet certain conditions. In addition to these, there are conditions based on its role in determining physical transitions between particle states. These include maintaining the validity of the closure conditions (5.29) and (5.31) as well as the conservation of particles, momentum, angular momentum, and energy. It is also required that W_{kl} map a system state into a system state. In accord with these needs, the following conditions are placed on W_{kl}:

(6.36a)   (i) $W_{kl} \geq 0,$

(6.36b)   (ii) $\int dP_k \int dP_l \, W_{kl} = 1,$

(6.36c)   (iiia) $V_{2k} \cdot \nabla_{2k} W_{kl} = -V_{2k}^\natural \cdot \nabla_{2k}^\natural W_{kl},$

(6.36d)   (iiib) $V_{2l} \cdot \nabla_{2l} W_{kl} = -V_{2l}^\natural \cdot \nabla_{2l}^\natural W_{kl},$

(6.36e)   (iva) $W_{kl} = W_{kl}^{(\text{unb})} * W_{kl}^{(\text{demon})},$

(6.36f)   (ivb) $W_{kl} = W_{kl}^{(\text{det})},$

(6.36g)   (ivc) $W_{kl} = W_{kl}^{(\delta)}.$

The *unbiased interface transition probability* $W_{kl}^{(\text{unb})}$, the *demon interface transition probability* $W_{kl}^{(\text{demon})}$, the *deterministic interface transition probability* $W_{kl}^{(\text{det})}$, and the *exact interface transition probability* $W_{kl}^{(\delta)}$, are special forms of the interface transition probability and $*$ is the law of composition for the interface transition probabilities.

Because of the properties (i) - (iii), it is not hard to see that the law of composition of the W_{kl} is a convolution. To show this, observe that conditions (6.36c) and (6.36d) imply that the interface transition probabilities are velocity frame independent because the transition depends only on the difference between the incoming and outgoing momentum vectors. This means that the variables of W_{kl} are defined so that if the law of composition is written in the form

(6.37) $W_{kl}^3 = W_{kl}^1 * W_{kl}^2$

$$= \int dQ_k' \int dP_k' \int dQ_l' \int dP_l' \, W_{kl}^1(Q_k, Q_l, Q_k', Q_l', P_k, P_l, P_k', P_l')$$

$$\times W_{kl}^2(Q_k', Q_l', Q_k^\natural, Q_l^\natural, P_k', P_l', P_k^\natural, P_l^\natural),$$

it is clear that $*$ is a convolution and W_{kl}^3 will satisfies conditions (6.36). It follows that (6.31) and (6.33) are also convolutions.

To understand the working of the interface transition probability, consider first a collision between two particles at a wall. The conservation of momentum, angular momentum, and energy together constrain the outcome of this collision so a definite result is determined by the initial conditions with respect to the transmission of conserved quantities across a boundary due to this collision Let the wall be represented by a plane through the center of momentum of this collision With respect to this plane, the net momentum, angular momentum, and kinetic energy, moving into the plane, will be preserved by the collision and will move out of the plane in the same direction.

This means that momentum, for example, will be transferred by this collision from one system to another if the net normal momentum of the two particles entering the plane does not vanish with respect to the plane. The component of the net momentum entering the plane that is parallel to the plane will also be preserved and transmitted from one system to another. For a multiparticle collision on both sides of the boundary, the situation is more complex but the net conserved quantities will be transmitted across the plane of the boundary just as for a two particle collision

Consider now two systems sharing a common boundary such that one has a higher total energy than the other does. If particles are presented to the wall from both systems in proportion to their representation in each of the systems, more high energy particles will be presented by the higher energy system than the other. If the particle energies presented to the wall are generally in proportion to their presence in a system, the net effect of a wall transition probability that is not biased toward one system or the other will be to reduce the difference in the energies of the systems in the long run. If the two systems are isolated from other influences, this will lead them eventually to equilibrium concerning the transfer of energy between them. However, it is not assumed in TIS that particle quantities presented to the wall must be in proportion to their presence in a system because this is not implied or guaranteed by the equations of motion.

To illustrate the workings of the second law and explore the conceptual issues surrounding it, Maxwell [1872], pp. 308–309, proposed that the eventual equilibrium of two systems could be evaded by an intelligent being, the Maxwell demon. This demon has the power to select particles based on some criterion and control which particles get to or through the wall.[9] The mechanism the demon uses to achieve this result is a massless door that closes the hole between the systems when a particle is rejected and opens it for those particles allowed to pass. While the idealized door is massless, and can be moved without the expenditure of work, it is not physically irrelevant in that it must absorb the momentum change of each rejected particle and dump it to a fixed laboratory reference system.

In the classical version of the TIS theory, the systems are closed, so a demon will be used that works for closed systems. The TIS version of Maxwell's demon for closed systems acts as an asymmetric band pass filter for particles of various energies or momenta approaching the boundary from either side. This filter is able to prevent particles with energies or momenta in a certain range from reaching the boundary at all and returns the momentum

---

[9]For a discussion of Maxwell's demon, its history, its properties, the connection between entropy and information, the connection between entropy and computation, and other extensions of the ideas behind it into other disciplines, see Harvey Leff and Andrew Rex [1990], pp. 2–32, and the reprints of individual papers collected in Leff and Rex [1990]. See also John Earman and John D. Norton [1998] for a brief history of the introduction of the demon and subsequent developments. Klein [1970] is of interest as well.

normal to the wall absorbed from particles in this range with a reversed sign back to the system the particle belongs to. As mentioned, the conservation laws can be satisfied by the operation of this demon only if account is taken of the momentum relations of the door and its laboratory support. This version of the filter demon will allow only slow particles to reach the wall in one system and only fast particles to reach the wall in the other. Particles reaching the wall are transformed into particles leaving the wall in accord with the standard wall transition probability. This arrangement will work as a kind of pump to transfer momentum, angular momentum, and energy between systems and create a gradient between them of the density of momentum, angular momentum, energy, or some combination of these.

Let us consider next the unbiased interface transition probability $W_{kl}^{(\text{unb})}$. It is a stochastic interface transition probability that acts on the average to transmit faithfully the spectrum of conserved quantities through the wall in each direction. It does not selectively retain low energy in one system and high energy in the other. As with all interface transition probabilities except the demon, any energy entering the wall that is not transmitted is reflected back into the original system.

The deterministic interface transition probability $W_{kl}^{(\text{det})}$ connects each incoming mechanical state in both systems with an outgoing mechanical state in both systems that is determined by some particular law of reflection at the wall. The exact interface transition probability $W_{kl}^{(\delta)}$ connects particular incoming and outgoing trajectories for particles approaching the wall.

Specific forms for these specialized interface transition probabilities will be given in Section 6.8.

The interface transition probabilities are all required to satisfy conditions (6.36a)–(6.36d). Because the systems are closed, the interface transition probabilities are defined in such a way that every incoming particle that reaches the interface is reflected back into its system of origin. Requirements (i) and (ii) are standard for probabilities. Requirement (iii) expresses velocity frame independence and Newton's third law. As will be demonstrated shortly, requirement (iii), with the help of (6.37), also insures that states will map into states. Requirement (iv) makes explicit the conservation of momentum, angular momentum, and energy at the wall by requiring that W_{kl} be expressed in either the special form (6.36e), (6.36f), or (6.36g).

Assuming for now $W_{kl}^{(\text{demon})} = 1$, the next step is to show that the requirements (6.36) serve their purpose, namely, that incoming states are mapped into outgoing states by W_{kl}. Recall that a state is a distribution that meets conditions (5.14), satisfies the evolution equation (5.16), and satisfies (5.29) or (5.31). The first step is to show that conditions (5.14) are satisfied by F_{kl} in (6.31) if they are satisfied by $F_{kl}^{\natural}$. Let us begin with condition (5.14a) for F_{kl}. Because $W_{kl} \geq 0$, condition (5.14a) follows from (6.28) because $F_{kl}^{\natural} \geq 0$.

For (5.14b), assume next that $F_{kl}^{\natural}$ is normalizable, i.e., that $Z_{kl}(t) < \infty$, and for simplicity use the normalized states $F_{kl}^{(n)\natural}$ in place of $F_{kl}^{\natural}$ and $F_{kl}^{(n)}$ in place of F_{kl} in (6.31). To show that $F_{kl}^{(n)}$ of the left side of (6.31) is normalized if $F_{kl}^{(n)\natural}$ is normalized, integrate both sides of (6.31) over $(Q_k, P_k) \times (Q_l, P_l)$, interchange the order of the integrations by Fubini's theorem, and use (6.36b) to obtain the result from the normalization of $F_{kl}^{(n)\natural}$. It follows from (6.36c), (6.36d), (6.33), (6.37), and setting $F_{kl}^{(n)} = F_k^{(n)} F_l^{(n)}$ (in accord with the independent systems approximation), that F_k and F_l are normalizable, so that $Z_k(t)$ and $Z_l(t)$ are both finite, and that the outgoing states F_k and F_l are solutions of their respective evolution equations if the incoming states $F_k^{\natural}$ and $F_l^{\natural}$ are solutions of the same evolution equations.

It is not hard to show that the fundamental relation (5.28) is valid here for $i \in k$. A parallel result is true for $i \in l$. To do this, first replace F_{kl} by $F_k F_l$ in (6.31) and then integrate both sides of the result over (Q_l, P_k, P_l). Next, interchange the (P_k, P_l) and $(P_k^{\natural}, P_l^{\natural})$ integrations by Fubini's theorem and then use (6.36b) followed by (5.3) for the local norm of F_l. These steps yield

$$(6.38) \qquad Z_l(t) \int dP_k \, u_i^+(q_i, v_i) F_k = -Z_l(t) \int dP_k^{\natural} \, u_i^-(q_i, v_i^{\natural}) F_k^{\natural}.$$

The dummy integration variable $P_k^{\natural}$ on the right side is then changed to P_k. Using (6.34a) leads to (5.28). Next, (5.28) is multiplied by $\chi_{\Omega_k(t)}(Q_k, P_k)\delta(q - q_i)$ and integrated over Q_k. This gives, for $q \in \partial d_k(t)$,

$$(6.39a) \qquad \langle \delta(q - q_i)[v_i^{\mu} - v_{\partial k}^{\mu}(q_i, t)] \rangle_{kt} = 0,$$

which can be written, using the definitions (8.1) and (8.5) in Chapter 8 below, as

$$(6.39b) \qquad \bar{v}_i^{\mu}(q, t) = v_{\partial k}^{\mu}(q, t) \quad \text{for } i \in k, \ q \in \partial d_k(t).$$

This relation implies that the i-particle stream velocity is equal to the boundary velocity at the boundary as it should. Condition (5.29) is met by the relation (6.39a). If G_i is set to $G_i(Q_k, t) = 1$ in (5.29), the left-hand side of the resulting equation represents the net i-particle flux into the wall. This quantity is 0 as required since the wall is not allowed to create, absorb or accumulate particles.

6.8 The Wall Transition Probability

As discussed above, the probability that the k and l systems are in the state $(Q_k, P_k) \times (Q_l, P_l)$ is $F_k(Q_k, P_k, t) F_l(Q_l, P_l, t)$ in the independent systems approximation. Therefore, to express the above formulas in a form appropriate to the independent systems case, each occurrence of F_{kl} above is replaced with $F_k F_l$. Since W_{kl}, as an energy and momentum conserving

transition, does not factor into a product of independent parts, let us operate in the spirit of the independent systems approximation and average W_{kl} over F_l. This leads to the definition of the kl *wall transition probability* as
(6.40)
$$W_k(Q_k, P_k, Q_k^\natural, P_k^\natural; F_l) = \frac{1}{Z_l(t)} \int dQ_l \int dP_l \int dQ_l^\natural \int dP_l^\natural \, W_{kl} F_l(Q_l^\natural, P_l^\natural, t).$$

Summing $W_k(F_l)$ over $l \in [t] - k$ gives the k *wall transition probability*

(6.41) $$W_k(Q_k, P_k, Q_k^\natural, P_k^\natural; F) = \sum_{l \in [t] - k} W_k(Q_k, P_k, Q_k^\natural, P_k^\natural; F_l).$$

Equation (6.31) can also be expressed in terms of F_k and W_k. Choose $i \in k$, integrate both sides of (6.4) over (Q_l, P_l) and sum over $l \in [t] - k$. Using the definitions (6.40) and (6.41), the relation for the i-particle flux in terms of W_k is obtained in the form

(6.42) $$u_i^{+,\mu}(q_i, v_i) F_k = - \int dQ_k^\natural \int dP_k^\natural \, W_k u_i^{-,\mu}(q_i, v_i) F_k^\natural.$$

By (6.36b), (6.31), (6.40) and (6.41), it is easy to demonstrate that W_k satisfies the conditions

(6.42a) (i) $W_k \geq 0,$

(6.42b) (ii) $\int dP_k \, W_k = 1,$

(6.42c) (iii) $V_{2k} \cdot \nabla_{2k} W_k = -V_{2k}^\natural \cdot \nabla_{2k}^\natural W_k,$

Next, the relation (6.33) is used with the definitions (6.34), (6.35), and (6.42) to show
(6.43)
$$\dot{\chi}_{\Omega_k(t)}(Q_k, P_k) F_k = - \int dQ_k^\natural \int dP_k^\natural \, \dot{\chi}_{\Omega_k(t)}^-(Q_k^\natural, P_k^\natural) F_k^\natural$$
$$\times \, [W_k(Q_k, P_k, Q_k^\natural, P_k^\natural; F) + \delta(Q_k - Q_k^\natural)\delta(P_k - P_k^\natural)].$$

This equation demonstrates, as promised, that the total flux is computed solely from the incoming flux. It also makes explicit the causal relationship (in the stochastic sense) between the incoming state and the outgoing state at the wall in the evolution equation.

By (6.43), (6.42b), and (5.35), it is easy to show that $dZ_k(t)/dt = 0$, so the local norm of F_k is preserved by the wall transition. Moreover, the support of F_k will remain within $\chi_{\Omega_k(t)}(Q_k, P_k)$ because the fluxes (6.41) and (6.43) are required to do so. Although it has been shown that the wall transition preserves the non-negativity of a non-negative state F_k, it has not been shown that F_k is non-negative. The non-negativity of $F_k(Q_k, P_k, t)$, as a solution of the evolution equation, is addressed in Chapter 25.

6.9 Special Interface Transition Probabilities

To express the unbiased interface transition probability formally, the definition (6.29) is extended to

$$(6.45) \qquad \theta^{\pm}(Q_k, P_k) = \prod_{i \in k} \theta^{\pm}(q_i, p_i/m_i).$$

The *unbiased interface transition probability* is then defined by

(6.46a)
$$
\begin{aligned}
W_{kl}^{(\mathrm{unb})} = {} & [Z_{kl}^{(\mathrm{unb})}(Q_k^{\natural}, Q_l^{\natural}, P_k^{\natural}, P_l^{\natural})]^{-1} \theta^{-}(Q_k^{\natural}, P_k^{\natural}) \theta^{+}(Q_k, P_k) \\
& \times \theta^{-}(Q_l^{\natural}, P_l^{\natural}) \theta^{+}(Q_l, P_l) \delta(Q_k^{\natural} - Q_k) \delta(Q_l^{\natural} - Q_l) \\
& \times \delta(\mathcal{P}_k(P_k^{\natural}) + \mathcal{P}_l(P_l^{\natural}) - \mathcal{P}_k(P_k) - \mathcal{P}_l(P_l)) \\
& \times \delta(\mathcal{J}_k(Q_k, P_k^{\natural}) + \mathcal{J}_l(Q_l, P_l^{\natural}) - \mathcal{J}_k(Q_k, P_k) - \mathcal{J}_l(Q_l, P_l)) \\
& \times \delta(K_k(P_k^{\natural}) + K_l(P_l^{\natural}) - K_k(P_k) - K_l(P_l)),
\end{aligned}
$$

where

(6.46b)
$$
\begin{aligned}
Z_{kl}^{(\mathrm{unb})}(Q_k^{\natural}, Q_l^{\natural}, P_k^{\natural}, P_l^{\natural}) = {} & \int dQ_k \int dP_k \int dQ_l \int dP_l \, \theta^{-}(Q_k^{\natural}, P_k^{\natural}) \theta^{+}(Q_k, P_k) \\
& \times \theta^{-}(Q_l^{\natural}, P_l^{\natural}) \theta^{+}(Q_l, P_l) \delta(Q_k^{\natural} - Q_k) \delta(Q_l^{\natural} - Q_l) \\
& \times \delta(\mathcal{P}_k(P_k^{\natural}) + \mathcal{P}_l(P_l^{\natural}) - \mathcal{P}_k(P_k) - \mathcal{P}_l(P_l)) \\
& \times \delta(\mathcal{J}_k(Q_k, P_k^{\natural}) + \mathcal{J}_l(Q_l, P_l^{\natural}) - \mathcal{J}_k(Q_k, P_k) - \mathcal{J}_l(Q_l, P_l)) \\
& \times \delta(K_k(P_k^{\natural}) + K_l(P_l^{\natural}) - K_k(P_k) - K_l(P_l)).
\end{aligned}
$$

An examination of the form of the integration shows that $Z_{kl}^{(\mathrm{unb})}$ is really a function of $\mathcal{P}_k(P_k^{\natural}) + \mathcal{P}_l(P_l^{\natural})$, $\mathcal{J}_k(Q_k, P_k^{\natural}) + \mathcal{J}_l(Q_l, P_l^{\natural})$ and $K_k(P_k^{\natural}) + K_l(P_l^{\natural})$, so $W_{kl}^{(\mathrm{unb})}$ simply connects states with the same total momentum, angular momentum and kinetic energy at the boundary without preferring any. Recall also that the wall transition probability acts only on particles at the wall of a system.

It is not hard to show by the properties of the δ measure that, if the domain conservation relations (6.24) are valid, $W_{kl}^{(\mathrm{unb})}$ will factor into a product of interface transition probabilities with one factor for each of the quasilocal domains that intersects the kl interface. On the other hand, it is easy to see that $W_{kl}^{(\mathrm{unb})}$ does not factor into a product $W_k W_l$, with W_k depending only on the k variables and W_l depending only on the l variables.

The *deterministic interface transition probability* is defined by

(6.47)
$$W_{kl}^{(\text{det})} = \theta^-(Q_k^\natural, P_k^\natural)\theta^+(Q_k, P_k)\theta^-(Q_l^\natural, P_l^\natural)\theta^+(Q_l, P_l)\delta(Q_k^\natural - Q_k)$$
$$\times \, \delta(Q_l^\natural - Q_l)\delta(P_k^\natural - P_k^\star(P_k, P_l))\delta(P_l^\natural - P_l^\star(P_k, P_l)),$$

where $P^\star(P_k, P_l)$ is a unimodular, invertible function of P_k and P_l defined so that the pairs $(P_k^\star(P_k, P_l), P_l^\star(P_k, P_l))$ and (P_k, P_l) both have the same total momentum, angular momentum and energy for fixed (Q_k, Q_l). The deterministic form for the interface transition probability includes commonly used deterministic reflection laws such as the specular reflection law and the reverse reflection law.

The *exact interface transition probability* is a specialized form of the deterministic interface transition probability. It is based on the fact that the closure of system means that the exact state representing the particle trajectories that are the solutions of the system evolution equations will not reach or cross the boundary. This interface transition probability simply sets $P_k^\star(P_k, P_l) = P_k$ and $P_l^\star(P_k, P_l) = P_l$ and connects the path of each particle approaching the wall with its path leaving the wall. If the exact deterministic interface transition probability is used, the details of the exact particle interactions near the wall are preserved. Then, if both systems k and l are described by an exact state, it follows that no microscopic information will be lost at the system boundaries. This means that two or more systems can be described equivalently by two exact system states $F_k^\delta F_l^\delta$ with an exact deterministic wall between them or by one exact state F_{kl}^δ for the combined systems.

6.10 On Walls and Demons

As mentioned above, the conditions (6.36e), (6.36f), and (6.36g), include the conservation laws that the interface transition probability must meet. The interface transition distribution $W_{kl}^{(\text{demon})}$ has been dubbed the "demon" because it can be used to mimic the behavior of a Maxwell demon. While the unbiased wall makes the transition from a given state into any of the allowed final states equally probable, the demon can skew this in particular ways to express specific properties for each wall. Since systems are closed, the demon is not allowed to transfer particles, but it can be used to pump momentum, angular momentum, or energy, across the boundary.

As mentioned, this demon must make use of the laboratory reference frame in its manipulation of energy and momentum in order to achieve its goals. This means that a third system capable of emitting and absorbing momentum, angular momentum, and energy, must be included if the conservation laws are to be upheld. The demon may be implemented formally as a sum of parts such that each limited by orthogonal projection operators to one range of the net energy or momentum coming into the wall. Because the conservation laws are not required to be obeyed by the two systems managed by the demon when the fixed laboratory system is not taken account of, this demon may selectively use a specular reflection law for both sets of particles reaching the wall in the two systems when the net momentum meets one criterion and the unbiased wall for both sets of particles reaching the wall in the two systems that meet another criterion.

For future reference, a property of the unbiased wall that is not shared by those with a demon (i.e., assume $W_{kl}^{(\text{demon})} = 1$) is stated now. Assume that the product of the k and l states can be written in the form

$$(6.48) \quad F_k(Q_k, P_k^\natural, t) F_l(Q_l, P_l^\natural, t) =$$
$$C_{kl}(\mathcal{P}_k(P_k^\natural) + \mathcal{P}_l(P_l^\natural), \mathcal{J}_k(Q_k, P_k^\natural) + \mathcal{J}_l(Q_l, P_l^\natural), K_k(P_k^\natural) + K_l(P_l^\natural)),$$

where C_{kl} is a function of the total momentum, angular momentum and energy of systems k and l. Using the definition (6.47) of the unbiased interface transition probability, it is easy to show by (6.31), with the $F_k F_l$ of (6.48) in place of F_{kl}, that the product of the F_k and F_l outgoing states is

$$(6.49) \quad F_k(Q_k, P_k, t) F_l(Q_l, P_l, t) =$$
$$C_{kl}(\mathcal{P}_k(P_k) + \mathcal{P}_l(P_l), \mathcal{J}_k(Q_k, P_k) + \mathcal{J}_l(Q_l, P_l), K_k(P_k) + K_l(P_l)).$$

This result will be important in the discussions of Clausius' inequality and the equilibrium state. In general, the transition $F_k^\natural F_l^\natural$ to $F_k F_l$ does not preserve the form (6.48) when a non-trivial demon ($W_{kl}^{(\text{demon})} \neq 1$) is operating. It is shown in Chapter 13 that there is a unique form for products of states satisfying (6.48) or (6.49). This implies that $W_{kl}^{(\text{unb})}$ is the unique interface transition probability for which the transition from (6.48) to (6.49) holds for all products of states $(Q_k, P_k) \times (Q_l, P_l)$ and $(Q_k', P_k') \times (Q_l', P_l')$ satisfying (6.24) in each quasilocal domain at each time t.

In spite of the emphasis on unbiased walls, no claim is made here that any real wall system must be unbiased or exhibit a particular form for its interface transition probability. Nor does the theory presented in this chapter depend on any special form for the interface transition probability. The important aspects of an interface transition probability from the standpoint of the Theory of Interacting Systems are that it represents a stochastic transition and that it satisfies conditions (6.36) and (6.39). What does depend on the specific form for the interface transition probability are the nature of the equilibrium state preserved by that wall and the actual rates at which particular momentum and energy transfers will occur.

CHAPTER 7

The Mechanical Axioms

The formalism developed in Chapters 4 and 5 will be used now to show that the mechanical axioms of Chapter 3 are a consequence of statistical mechanical relationships and the phase averaging procedure. In this and the next few chapters, the formulas derived will be expressed in terms of the k system, but they are valid for any $k \in [t]$, $k = t$ or $k = [t]$ unless otherwise noted.

7.1 Density

The *i-particle local matter density* and the *k local matter density* for $i \in k$ are defined by

(7.1a)
$$\nu_i(q,t) = \langle \delta(q - q_i) \rangle_{kt},$$

(7.1b)
$$\nu_k(q,t) = \sum_{i \in k} \nu_i(q,t),$$

(7.1c)
$$\nu_k(t) = \sum_{i \in k} 1 = \mathcal{N}_k,$$

where the fact that $\nu_i(t) = \langle 1 \rangle_{kt} = 1$ was used in (7.1c). If G_i is set to $G_i(Q_k, p_i, t) = 1$ in the definition (5.10a) of the local phase average, then (5.14a) can be used to obtain

(7.2a)
$$\nu_i(q,t) \geq 0,$$

(7.2b)
$$\nu_k(q,t) \geq 0.$$

Integrating $\nu_i(q,t)$ in (7.1a) over $q \in d_k(t)$ and using the definition of the phase average shows that $\nu_i(t) = 1$ as stated above.

The current densities associated with the particle matter densities are the *i-particle local matter current density vector* and the *k local matter current density vector* defined by[1]

(7.3a)
$$\nu_i^\mu(q,t) = \langle \delta(q - q_i) v_i^\mu \rangle_{kt},$$

(7.3b)
$$\nu_k^\mu(q,t) = \sum_{i \in k} \nu_i^\mu(q,t).$$

Combining (7.1b) with (7.2b) confirms Axiom 1.1. In addition, the *i-particle local stream velocity* can be defined in terms of these quantities by

(7.4)
$$\bar{v}_i^\mu(q,t) = [\nu_i(q,t)]^{-1} \nu_i^\mu(q,t).$$

The *i-particle mass density*, *k mass density*, and *k mass* are defined similarly as:

(7.5a)
$$\mu_i(q,t) = m_i \nu_i(q,t),$$

(7.5b)
$$\mu_k(q,t) = \sum_{i \in k} \mu_i(q,t),$$

(7.5c)
$$\mathcal{M}_k = \sum_{i \in k} \mu_i(t) = \sum_{i \in k} m_i,$$

where $\mu_i(t) = m_i \nu_i(t) = m_i$ was used. Because $m_i \geq 0$ for $i \in k$, it follows from (7.2) that $\mu_i(q,t)$ and $\mu_k(q,t)$ are non-negative. The mass current density is the same as the momentum density defined below.

---

[1]See Maxwell [**1866**] equations (74) and (75).

Setting $G_i(Q_k, p_i, t) = 1$ in (5.41a) and then using (5.29a) leads immediately to

(7.6a)
$$\frac{d\nu_i(q,t)}{dt} = \frac{\partial \nu_i(q,t)}{\partial t} + \frac{\partial \nu_i^\mu(q,t)}{\partial q^\mu} = 0,$$

(7.6b)
$$\frac{d\nu_k(q,t)}{dt} = \frac{\partial \nu_k(q,t)}{\partial t} + \frac{\partial \nu_k^\mu(q,t)}{\partial q^\mu} = 0.$$

These equations express the conservation of particles. It follows from (7.5) there are similar equations for $\mu_i(q,t)$ and $\mu_k(q,t)$ that express the conservation of mass.

Equation (7.6b) confirms Axiom 1.2 and (7.1c) gives (3.2a). Equation (3.2b) is obtained in a similar fashion. Axiom 1.3 is obtained by summing $d\nu_i(t)/dt = 0$ over $i \in k$. Finally, the sum of $\mathcal{N}_k$ over $k \in [t]$ leads to $\mathcal{N}_t$. These results confirm Axiom 1.4.

7.2 Momentum and Force

To demonstrate Axiom 2, let us define the 3-dimensional *i-particle local momentum density* vector, the *k local momentum density*, which is a $3\mathcal{N}_k$-dimensional vector, and the 3-dimensional *k local total momentum density* vector by

(7.7a) $p_i^\mu(q,t) = \langle \delta(q - q_i) p_i^\mu \rangle_{kt}$,

(7.7b) $P_k^\mu(q,t) = \times_{i \in k} p_i^\mu(q,t),$

(7.7c) $\mathcal{P}_k^\mu(q,t) = \sum_{i \in k} p_i^\mu(q,t).$

These in turn are used to define the *i-particle local stream momentum* and the *k local stream momentum* as

(7.8a) $\bar{p}_i^\mu(q,t) = [\nu_i(q,t)]^{-1} p_i^\mu(q,t),$

and

(7.8b) $\bar{\mathcal{P}}_k^\mu(q,t) = [\nu_k(q,t)]^{-1} \sum_{i \in k} \nu_i(q,t) \bar{p}_i^\mu(q,t) = [\nu_k(q,t)]^{-1} \mathcal{P}_k^\mu(q,t).$

Since the energy of the t system is fixed and finite, the kinetic energy and momentum available to the k system are bounded. This can be used to show that the bounds (2.15) are valid.

The *k global total momentum* is the phase average of the total phase momentum $\mathcal{P}_k^\mu$ and is expressed in the form

(7.9) $\mathcal{P}_k^\mu(t) = \langle \mathcal{P}_k^\mu \rangle_{kt}.$

Because $\mathcal{P}_{[t]}^{\mu} = \sum_{k \in [t]} \mathcal{P}_k^{\mu}$, the definition (6.2) of the $[t]$ state and its partition function $Z_{[t]}(t) = \prod_{k \in [t]} Z_k(t)$ are used to show that the global system momentum is a system sum function in accord with Axiom 2.3:

$$(7.10) \qquad\qquad \mathcal{P}_{[t]}(t) = \sum_{k \in [t]} \mathcal{P}_k(t).$$

The symmetric *i-particle phase momentum current tensor* is

$$(7.11) \qquad\qquad p_i^{\mu\nu} = p_i^{\mu} v_i^{\nu}.$$

The *i-particle local momentum current density tensor* is defined in terms of $p_i^{\mu\nu}$ by

$$(7.12a) \qquad\qquad p_i^{\mu\nu}(q,t) = \langle \delta(q - q_i) p_i^{\mu\nu} \rangle_{kt} .$$

The *k local momentum current density tensor* and the *k local total momentum current density tensor* are

$$(7.12b) \qquad\qquad P_k^{\mu\nu}(q,t) = \times_{i \in k} p_i^{\mu\nu}(q,t),$$

$$(7.12c) \qquad\qquad \mathcal{P}_k^{\mu\nu}(q,t) = \sum_{i \in k} p_i^{\mu\nu}(q,t).$$

Because electromagnetism is not included in TIS until EIS, it follows that $p_i = m_i v_i$, so $\mathcal{P}_k^{\mu\nu}(q,t)$ is obviously symmetric and Axiom 2.1 is proved.

The total thermodynamic force on particle i in the t inertial frame at (q,t) is equal to $dp_i(q,t)/dt$. This leads to the definitions of the *i-particle local body force density*, the *k local body force density*, and the *k local total body force density* as

$$(7.13a) \qquad\qquad F_{i,b}^{\mu}(q,t) = - \left\langle \delta(q - q_i) \frac{\partial \Phi_i(Q_k,t)}{\partial q_{i,\mu}} \right\rangle_{kt},$$

$$(7.13b) \qquad\qquad F_{k,b}^{\mu}(q,t) = \times_{i \in k} F_{i,b}^{\mu}(q,t),$$

$$(7.13c) \qquad\qquad \mathcal{F}_{k,b}^{\mu}(q,t) = \sum_{i \in k} F_{i,b}^{\mu}(q,t).$$

The relation (4.9) is used to decompose this into the *i-particle local internal body force density*, the *k local internal body force density*, and the *k local total internal body force density*

$$(7.14a) \qquad\qquad F_{i,n}^{\mu}(q,t) = - \left\langle \delta(q - q_i) \frac{\partial \Phi_{i,n}(Q_k)}{\partial q_{i,\mu}} \right\rangle_{kt},$$

$$(7.14b) \qquad\qquad F_{k,n}^{\mu}(q,t) = \times_{i \in k} F_{i,n}^{\mu}(q,t),$$

$$(7.14c) \qquad\qquad \mathcal{F}_{k,n}^{\mu}(q,t) = \sum_{i \in k} F_{i,n}^{\mu}(q,t).$$

It also leads to the *i-particle local external body force density*, the *k local external body force density*, and the *k local total external body force density*

(7.15a) $$F_{i,x}^{\mu}(q,t) = -\left\langle \delta(q-q_i)\frac{\partial\Phi_{i,x}(q_i,t)}{\partial q_{i,\mu}} \right\rangle_{kt},$$

(7.15b) $$F_{k,x}^{\mu}(q,t) = \times_{i\in k}F_{i,x}^{\mu}(q,t),$$

(7.15c) $$\mathcal{F}_{k,x}^{\mu}(q,t) = \sum_{i\in k}F_{i,x}^{\mu}(q,t).$$

It is easy to see, using the decomposition of $\Phi_k(Q_k,t)$, that the body forces are the sums of the internal and external forces

(7.16a) $$F_{i,b}^{\mu}(q,t) = F_{i,n}^{\mu}(q,t) + F_{i,x}^{\mu}(q,t),$$

(7.16b) $$F_{k,b}^{\mu}(q,t) = F_{k,n}^{\mu}(q,t) + F_{k,x}^{\mu}(q,t),$$

(7.16c) $$\mathcal{F}_{k,b}^{\mu}(q,t) = \mathcal{F}_{k,n}^{\mu}(q,t) + \mathcal{F}_{k,x}^{\mu}(q,t).$$

There are similar relations for the global forms of these forces.

The singular *i-particle local surface force density*, the *k local total surface force density*, and the *k local surface force density* at the boundary are

(7.17a) $$F_{\partial i}^{\mu}(q,t) = \langle \delta(q-q_i)\dot{\chi}_{\Omega_k(t)}(Q_k,P_k)p_i^{\mu} \rangle_{kt},$$

(7.17b) $$F_{\partial k}^{\mu}(q,t) = \times_{i\in k}F_{\partial i}^{\mu}(q,t),$$

(7.17c) $$\mathcal{F}_{\partial k}^{\mu}(q,t) = \sum_{i\in k}F_{\partial i}^{\mu}(q,t).$$

To obtain the macroscopic form of Newton's second law, (7.7a) is differentiated by d/dt. The total phase time derivative of p_i^{μ} is easily shown to be $dp_i^{\mu}/dt = -\partial\Phi_i(Q_k,t)/\partial q_{i,\mu}$. This is used with the definition $dp_i^{\mu}(q,t)/dt = \partial p_i^{\mu}(q,t)/\partial t + \partial p_i^{\mu\nu}(q,t)/\partial q^{\nu}$ obtained from (2.12), (5.41), and the definitions (7.13) and (7.17) to obtain the results

(7.18a) $$\frac{dp_i^{\mu}(q,t)}{dt} = F_{i,b}^{\mu}(q,t) + F_{\partial i}^{\mu}(q,t),$$

(7.18b) $$\frac{d\mathcal{P}_k^{\mu}(q,t)}{dt} = \mathcal{F}_{k,b}^{\mu}(q,t) + \mathcal{F}_{\partial k}^{\mu}(q,t),$$

which demonstrate Axiom 2.2.

Let us now turn to the conservation of momentum. The quasilocal and global versions of the theory are non-local and allow the instantaneous transfers of momentum and energy over finite distances. For this reason, momentum is not conserved locally, which would be written as $d\mathcal{P}_k(q,t)/dt = 0$, in these versions of the theory. However, in the quasilocal or global cases, (6.24a) will be used for the conservation of momentum in each quasilocal domain of each system, including those that intersect the system boundary, to show that Axiom 2.4 is valid. To do this requires using the incoming and outgoing state projection operator defined in (6.32) and averaging the momentum flux at the boundary with the help of (6.23a).

For the first step in proving Axiom 2.4, integrate (7.14a) over $q \in d_k(t)$ and sum over $i \in k$ to obtain

$$(7.19) \qquad \mathcal{F}^{\mu}_{k,b}(t) = -\sum_{i \in k} \left\langle \frac{\partial \Phi_k(Q_k, t)}{\partial q_{i,\mu}} \right\rangle_{kt}.$$

The definitions (4.9) and (4.8) are used to decompose the body force as

$$(7.20) \qquad \mathcal{F}^{\mu}_{k,b}(t) = \mathcal{F}^{\mu}_{k,n}(t) + \mathcal{F}^{\mu}_{k,x}(t).$$

The result (4.20a) is then used to show that

$$(7.21) \qquad \mathcal{F}^{\mu}_{k,n}(t) = 0.$$

These results allows us to write (7.19) in the form

$$(7.22) \qquad \mathcal{F}^{\mu}_{k,b}(t) = \mathcal{F}^{\mu}_{k,x}(t) = -\sum_{i \in k} \sum_{l \in [t]-k} \sum_{j \in l} \left\langle \left\langle \frac{\partial \phi_{ij}(q_i - q_j)}{\partial q_{i,\mu}} \right\rangle_{lt} \right\rangle_{kt}.$$

Summing this over $k \in [t]$ and using the fact that the sum is symmetric in i and j while, by (4.2e), the integrand is antisymmetric, it follows that the sum of the external forces and the sum of the body forces over all the systems in $[t]$ vanishes:

$$(7.23a) \qquad \sum_{k \in [t]} \mathcal{F}^{\mu}_{k,x}(t) = 0,$$

$$(7.23b) \qquad \sum_{k \in [t]} \mathcal{F}^{\mu}_{k,b}(t) = 0.$$

The second step in demonstrating Axiom 2.4 is to show that Axiom 5.1 is true:

$$(7.24) \qquad \sum_{k \in [t]} \mathcal{F}^{\mu}_{\partial k}(t) = \sum_{k \in [t]} \left\langle \dot{\chi}_{\Omega_k(t)}(Q_k, P_k) \mathcal{P}^{\mu}_k \right\rangle_{kt} = 0.$$

To prove this relation, observe first that by (6.34b) and (6.35), the quantity $\dot{\chi}_{\Omega_k(t)}(Q_k, P_k)$ can be written in the form

$$(7.25)$$

$$\dot{\chi}_{\Omega_k(t)}(Q_k, P_k) = \chi_{\Omega_k(t)}(Q_k, P_k) \sum_{i \in k} \left[\frac{\dot{\chi}^+_{d_k(t)}(q_i, v_i)}{\chi_{d_k(t)}(q_i)} + \frac{\dot{\chi}^-_{d_k(t)}(q_i, v_i)}{\chi_{d_k(t)}(q_i)} \right]$$

$$= \dot{\chi}^+_{\Omega_k(t)}(Q_k, P_k) + \dot{\chi}^-_{\Omega_k(t)}(Q_k, P_k).$$

Because each quasilocal domain conserves its momentum and energy separately, those quasilocal domains that fall within a given system will conserve momentum and energy within that system. For a domain a that intersects the kl boundary, it is necessary to show

$$(7.26) \qquad \langle \dot{\chi}_{\Omega_k(t)} \mathcal{P}_k^{a,\mu} \rangle_{kt} + \langle \dot{\chi}_{\Omega_l(t)} \mathcal{P}_l^{a,\mu} \rangle_{lt} = 0.$$

This relation is then summed over $a \in \pi(t)$ followed by a sum over all systems sharing a common boundary to obtain (7.24). To demonstrate (7.26), rewrite it in the form

$$(7.27) \qquad \langle \langle [\dot{\chi}_{\Omega_k(t)} + \dot{\chi}_{\Omega_l(t)}][\mathcal{P}_k^{a,\mu} + \mathcal{P}_l^{a,\mu}] \rangle_{lt} \rangle_{kt} = 0.$$

To prove the validity of (7.27), let us begin by substituting $F_{kl} = F_k F_l$ in (6.33) and then sum the result over $i \in k$. This yields

$$(7.28) \qquad \chi_{\Omega_k(t)}^+(Q_k, P_k) F_k F_l = - \int dP_k' \int dP_l' \, W_{kl} \, \chi_{\Omega_k(t)}^-(Q_k, P_k') F_k' F_l'.$$

From this it follows

(7.29)

$$\int dP_k \int dP_k \, [\mathcal{P}_k^{a,\mu} + \mathcal{P}_l^{a,\mu}][\dot{\chi}_{\Omega_k(t)}^+ + \dot{\chi}_{\Omega_l(t)}^+] F_k F_l$$

$$= - \int dP_k \int dP_k \, [\mathcal{P}_k^{a,\mu} + \mathcal{P}_l^{a,\mu}]$$

$$\times \quad \int dP_k' \int dP_l' \, W_{kl}[\dot{\chi}_{\Omega_k(t)}^{-\prime} + \dot{\chi}_{\Omega_l(t)}^{-\prime}] F_k' F_l'$$

$$= - \int dP_k \int dP_k \int dP_k' \int dP_l' \, W_{kl}$$

$$\times [\mathcal{P}_k^{a,\prime\mu} + \mathcal{P}_l^{a,\prime\mu}][\dot{\chi}_{\Omega_k(t)}^{-\prime} + \dot{\chi}_{\Omega_l(t)}^{-\prime}] F_k' F_l'$$

$$= - \int dP_k' \int dP_l' \, [\mathcal{P}_k^{a,\prime\mu} + \mathcal{P}_l^{a,\prime\mu}]$$

$$\times [\dot{\chi}_{\Omega_k(t)}^{-\prime} + \dot{\chi}_{\Omega_l(t)}^{-\prime}] F_k' F_l'$$

To obtain this result, (6.24a) for the conservation of momentum in the a domain and (6.36b) were both used. The dummy integration variables (P_k', P_l') are changed to (P_k, P_l) on the right-hand side of (7.29). The relation (7.25) is used with the result (7.29) to prove (7.27) and therefore that (7.26) is valid. Summing over $a \in \pi(t)$ and taking note of the systems involved, it is not hard to show that (7.24) is also true. This confirms Axiom 5.1.

Axiom 2.4 follows from Axiom 2.2, (7.23) and (7.24).

7.3 Angular Momentum

The i-particle phase angular momentum and k phase angular momentum are defined by equations (4.11). The i-particle *local angular momentum density,* the *k local angular momentum density,* and the *k local total angular momentum density* are given in terms of these quantities by

(7.30a) $$j_i^\mu(q,t) = \langle \delta(q - q_i) j_i^\mu(q_i, p_i) \rangle_{kt}\,,$$

(7.30b) $$J_k^\mu(q,t) = \times_{i \in k} j_i^\mu(q,t),$$

(7.30c) $$\mathcal{J}_k^\mu(q,t) = \sum_{i \in k} j_i^\mu(q,t).$$

Using (4.11a) in (7.30a) gives

(7.31) $$j_i^\mu(q,t) = \epsilon^{\mu\nu\xi} q_\nu p_{i,\xi}(q,t).$$

This, with (7.30c), confirms Axiom 3.1.

The *i-particle local angular momentum current density,* the *k local angular momentum current density,* and the *k local total angular momentum density* are

(7.32a) $$j_i^{\mu\nu}(q,t) = \langle \delta(q - q_i) j_i^\mu(q_i, p_i) v_i^\nu \rangle_{kt}\,,$$

(7.32b) $$J_k^{\mu\nu}(q,t) = \times_{i \in k} j_i^{\mu\nu}(q,t),$$

(7.32c) $$\mathcal{J}_k^{\mu\nu}(q,t) = \sum_{i \in k} j_i^{\mu\nu}(q,t).$$

Substituting (4.11a) into (7.32a), followed by an application of the definitions (7.12) and (7.32c), yields Axiom 3.2. As noted in Chapter 3, Axioms 3.3, 3.4 and 3.5 are redundant for systems that allow only central forces between particles.

7.4 The Diffusion and Viscosity Equations

For completeness, the transport of matter and momentum will be discussed briefly. The concepts of diffusion and viscosity are associated with phenomenological equations of mass and momentum transport. The viscosity, for example, appears in the phenomenological analysis of shear forces in fluid flow.[2] To give a rigorous treatment of either topic on the particle level and make a connection with the phenomenological analysis, explicit nonequilibrium solutions of the evolution equation are required. In the absence of these explicit solutions, various approximation methods have been used with some success.[3]

The diffusion equation is concerned with the transport of particles. The basic equation for this transport is the continuity equation (7.6). To make a connection with the standard approach, let us assume the system is described

---

[2] See Section 11.8

[3] For a discussion of previous work based on the study of mean free paths and random walks in Brownian motion, and a calculation of the diffusion constant, see Bailyn [**1994**], Chapter 10.

by a special distribution, so that Fick's law of mass transport for the i-particle current,

$$(7.33) \qquad \nu_i^\mu(q,t) \approx -D_i \frac{\partial \nu_i(q,t)}{\partial q_\mu},$$

with D_i constant, is approximately true. This approximation is then used in (7.6) to obtain the i-particle diffusion equation

$$(7.34a) \qquad \frac{\partial \nu_i(q,t)}{\partial t} \approx D_i \nabla^2 \nu_i(q,t).$$

The quantity $D_i > 0$ appearing in these equations is called the i-particle diffusion coefficient. Assuming $D_i = D_k$ for each $i \in k$, which would be the case for a collection of identical particles, and summing this equation over $i \in k$ yields the TIS version of Fick's diffusion equation

$$(7.34b) \qquad \frac{\partial \nu_k(q,t)}{\partial t} \approx D_k \nabla^2 \nu_k(q,t).$$

The next step from the TIS perspective is to calculate the coefficient of diffusion D_k in terms of the particle properties. This calculation will show what approximations are needed and thereby exhibit the limits of validity for Fick's equation and the viscosity current equation are. The calculation of these coefficients in terms of the particle properties requires the specification of one of some special distribution for which (7.33) is approximately true.

A calculation of this type will be illustrated by Maxwell's [1860] mean-free path calculations.[4] Maxwell assumed a local steady state, so that the local density does not depend on the time: $\nu_k(q,t) \approx \nu_k(q)$. He calculated the average magnitude of the particle velocity, which is expressed in TIS notation as $\overline{|v_k|} = [\mathcal{N}_k]^{-1} \sum_{i\in k} \overline{|v_i|}$, where $\overline{|v_i|} = \langle |v_i| \rangle_{kt}$. For the diffusion, Maxwell calculated the passage of particles or physical quantities across a plane placed at an arbitrary location in a gas. Let us focus on a point q_0 in this plane. Let λ be the mean free path for particles in the gas. Maxwell employed Clausius' result $e^{-r/\lambda}$ for the probability that a particle will survive a distance r from its last scattering before being scattered again in his calculation.[5] He treated the particle density as locally constant near any point in the plane and eventually obtained for the normal current of particles entering one side of the plane at the point q_0 the result

$$(7.35) \qquad \nu_k^{(\perp)}(q_0) = \hat{n}_\mu \nu_k^\mu(q_0) = \tfrac{1}{4}\overline{|v_k|}\nu_k(q_0),$$

---

[4] This summary is adapted from Bailyn's [1994], pp. 431–435, presentation of the mean free path calculations of Clausius and Maxwell [1860].

[5] One of the important assumptions underlying the use of the mean free path formalism is that the distribution of particles is the same from the point of view of either the rest frame of the system or the rest frame of any pair of particles chosen as the basis for the calculation. See Bailyn [1994], p. 428. This assumption is dubious, however, and is not used in TIS.

where $\hat{n}_\mu$ is a unit normal vector pointing into the plane. The result (7.35) is independent of the mean free path. For the pressure, Maxwell obtained the approximation

$$(7.36) \qquad \qquad P_k^{(M)} = \tfrac{1}{3}m \left[\overline{|v_k|}\right]^2 \nu_k(q_0),$$

where $m = m_i$ for $i \in k$ is the mass of a particle.

To compute the coefficient of diffusion, Maxwell allowed the density to vary near the plane and expanded it in a Taylor's series. For the current, he obtained the result

$$(7.37) \qquad \qquad \nu_k^\mu(q_0) = -\tfrac{1}{3}\lambda\overline{|v_k|}\,\frac{d\nu_k(q)}{dq_\mu}\bigg|_{q=q_0}.$$

This implies immediately by (7.33) and $D_i = D_k$ that[6]

$$(7.38) \qquad \qquad D_k = \tfrac{1}{3}\lambda\overline{|v_k|}.$$

An examination of the work in kinetic theory shows that the mean free path calculations concerned with diffusion, viscosity, and heat flow, implicitly or explicitly assume special distributions such as those Maxwell used in the mean free path calculations. The same is true for the Boltzmann equation based calculations of David Enskog that appeared in 1911, 1912, and in his dissertation of 1917, which was published as a paper in 1921. The work on these issues by Sidney Chapman in 1912, 1916, and 1917, was based originally on Maxwell's transport equations. The approaches of both Enskog and Chapman assume system states that are "near" equilibrium.[7]

Viscosity is an aspect of the transport of momentum through a substance. It reflects the resistance to this transport due to interparticle interactions and the conversion of some of the transported energy associated with the flow of momentum into heat. The concept of viscosity was given a mathematical expression in Newton's *Principia* in terms of the force F required to maintain a difference $\Delta\bar{v}$ in the average velocity over a distance Δz in laminar fluid flow. Newton's formula was

$$(7.39) \qquad \qquad F = \eta\frac{\Delta\bar{v}}{\Delta z},$$

---

[6] For an improved calculation of the diffusion constant based on the analysis of Brownian motion by Einstein, Langevin, and others, and the theory of random walks, see Bailyn [**1994**], pp. 481–491.

[7] The presentation of Enskog's theory in S. Chapman and T. G. Cowling [**1939**] includes some historical information on the development of the theory and references to original papers.

where η is the coefficient of viscosity. This formula is used today to measure the viscosity of a fluid when it is in the 'Newtonian flow' regime.[8]

Versions of the viscosity later appeared in the equations of fluid motion developed by Navier and Stokes. They analyzed the way stresses could be represented in small portions of a fluid in motion and how these stresses would affect the rate of change of the average stream velocity in time. This work was not particle based and depended on the average stream velocity and force density averages for particles in the neighborhood of the macroscopic point (q, t) rather than on the precise details in terms of particle coordinates. This approach will be discussed in more detail in Chapter 11.

The transport of momentum is used to calculate the viscosity of a substance. For the i-particle current in a viscous substance, the *i-particle momentum current* for the momentum component μ in the direction ν is written

$$(7.40) \qquad p_i^{\mu\nu}(q, t) \approx -\eta_i \frac{\partial v_i^{\mu}(q, t)}{\partial q_\nu},$$

where η_i is the *i-particle coefficient of viscosity*. Assuming as before that the particles are identical so that $\eta_i = \eta_k$ and using this in the local equation for conservation of momentum yields

$$(7.41a) \qquad \frac{\partial p_i^{\mu}(q, t)}{\partial t} \approx \eta_i \nabla^2 v_i^{\mu}(q, t).$$

$$(7.41b) \qquad \frac{\partial p_k^{\mu}(q, t)}{\partial t} \approx \eta_k \nabla^2 v_k^{\mu}(q, t).$$

Returning the Maxwell's calculations, Bailyn [**1994**], p. 434, has shown that the general formula for the current C_A^{μ} of the quantity $A_k(q, t)$ into the plane is expressed in terms of the unit vector $\hat{n}_\nu$ normal to the plane by

$$(7.42) \qquad C_A^{(\perp)}(q_0) = -b_A \lambda v_k(q_0)\hat{n}_\nu (dA_k(q)/dq_\nu)|_{q=q_0},$$

where b_A is a constant that depends on the geometry. For the momentum transport into the plane, $p_k^{\mu}(q)$ is used in place of $A_k(q)$ to obtain the current $p_k^{(\perp)\mu}(q) = -b_p \lambda v_k(q_0)\hat{n}_\nu \partial p^{\mu}(q)/\partial q_\nu$ of $p_k^{\mu}(q)$ normal to the plane. This is used with the same procedure as above and the appropriate calculation of the flux of momentum into the plane to obtain the result

$$(7.43) \qquad \eta = \tfrac{1}{3}\lambda m v_k(q_0)\overline{|v_k|}$$

for the coefficient of viscosity. The value constant $b_p = 1/3$ follows from the geometry of the calculation. Because $\lambda \sim 1/v_k(q_0)$ for a wide range of densities, η is independent of the density when the density is in this range.

---

[8]Newton's approach was criticized by Stokes [**1849**], pp. 303–304. Newton's assumption that a fluid under this stress can be in a steady state is incorrect. For a brief history of viscosity and the concepts associated with it, see Brush [**1962**].

More recent work is based on correlations between different portions of a fluid in motion. This work also requires the use of specialized distributions to compute results and will not be reviewed here.

Maxwell's calculations of the coefficients of diffusion and viscosity have been presented to show how an approximation procedure, in this case the mean free path approximation, can be used with the TIS formalism to obtain concrete results. The assumptions needed to calculate these coefficients are too specialized for the results to be part of the general theory.

7.5 Energy

The *i-particle local energy density*, the *k local energy density*, and the *k local total energy density* are defined by

(7.44a) $$E_i(q,t) = \langle \delta(q - q_i) E_i(Q_k, p_i, t) \rangle_{kt} \,,$$
(7.44b) $$E_k(q,t) = \times_{i \in k} E_i(q,t),$$
(7.44c) $$\mathcal{E}_k(q,t) = \sum_{i \in k} E_i(q,t),$$

where $E_i(Q_k, p_i, t)$ is defined in (4.13a). The *i-particle local energy current density*, the *k local energy current density*, and the *k local total energy current density* are also defined as

(7.45a) $$E_i^{\mu}(q,t) = \langle \delta(q - q_i) v_i^{\mu} E_i(Q_k, p_i, t) \rangle_{kt} \,,$$
(7.45b) $$E_k^{\mu}(q,t) = \times_{i \in k} E_i^{\mu}(q,t),$$
(7.45c) $$\mathcal{E}_k^{\mu}(q,t) = \sum_{i \in k} E_i^{\mu}(q,t).$$

The *i-particle local kinetic energy density*, the *k local kinetic energy density*, and the *k local kinetic energy density* are next defined as

(7.46a) $$K_i(q,t) = \langle \delta(q - q_i) K_i(p_i) \rangle_{kt} \,,$$
(7.46b) $$K_k(q,t) = \times_{i \in k} K_i(q,t),$$
(7.46c) $$\mathcal{K}_k(q,t) = \sum_{i \in k} K_i(q,t).$$

The *i-particle local internal potential energy density* and the *k local internal potential energy density* are defined by

(7.47a) $$\Phi_{i,n}(q,t) = \langle \delta(q - q_i) \Phi_{i,n}(Q_k) \rangle_{kt} \,,$$
(7.47b) $$\Phi_{k,n}(q,t) = \tfrac{1}{2} \sum_{i \in k} \Phi_{i,n}(q,t),$$

and, finally, the *i-particle local external potential energy density* and the *k local external potential energy density* by

(7.48a)
$$\Phi_{i,x}(q,t) = \langle \delta(q - q_i)\Phi_{i,x}(q_i,t)\rangle_{kt} = \nu_i(q,t)\Phi_{i,x}(q,t),$$

(7.48b)
$$\Phi_{k,x}(q,t) = \sum_{i \in k} \Phi_{i,x}(q,t).$$

These quantities are used to define the *i-particle local potential energy density* and the *k local potential energy density*

(7.49a) $\Phi_i(q,t) = \Phi_{i,n}(q,t) + \Phi_{i,x}(q,t),$

(7.49b) $\Phi_k(q,t) = \Phi_{k,n}(q,t) + \Phi_{i,x}(q,t).$

By the definitions of E_i, E_k, H_i and $\mathcal{H}_k$, it is obvious that

(7.50a) $E_i(q,t) = K_i(q,t) + \frac{1}{2}\Phi_i(q,t),$

(7.50b) $\mathcal{E}_k(q,t) = \mathcal{K}_k(q,t) + \Phi_{k,n}(q,t) + \frac{1}{2}\Phi_{k,x}(q,t),$

(7.50c) $H_i(q,t) = K_i(q,t) + \Phi_i(q,t),$

(7.50d) $\mathcal{H}_k(q,t) = \mathcal{K}_k(q,t) + \Phi_{k,n}(q,t) + \Phi_{k,x}(q,t).$

Axiom 4.1 is obtained from (7.50b).

7.6 Energy Flow

The global energy $\mathcal{E}_k(t)$ is defined by

(7.51) $$\mathcal{E}_k(t) = \langle E_k(Q_k, P_k, t)\rangle_{kt}.$$

By (4.13c), this leads to

(7.52) $$\mathcal{E}_{[t]}(t) = \sum_{k \in [t]} \mathcal{E}_k(t),$$

which confirms Axiom 4.3.

The *i-particle local body working density* and *k local body working density*, which is the working done on particle i by other particles in k and by particles in other systems, are next defined as[9]

(7.53a) $$W_{i,b}(q,t) = \left\langle \delta(q - q_i)\frac{dE_i(Q_k, p_i, t)}{dt}\right\rangle_{kt},$$

(7.53b) $$\mathcal{W}_{k,b}(q,t) = \sum_{i \in k} W_{i,b}(q,t).$$

---

[9]The terms working and heating for the rate of change of the work and the heat in a system are discussed below.

Using the formula (5.41a) with these definitions gives the result

$$(7.54) \qquad \frac{dE_i(q,t)}{dt} = W_{i,b}(q,t) + \left\langle \delta(q-q_i)\dot{\chi}_{\Omega_k(t)}(Q_k, P_k)K_i(p_i) \right\rangle_{kt},$$

where the fact based on (5.29) that

$$(7.55) \qquad \left\langle \delta(q-q_i)\dot{\chi}_{\Omega_k(t)}(Q_k, P_k)\Phi_i(Q_k, t) \right\rangle_{kt} = 0$$

was also used.

The flow of energy through the boundary is next separated into a part identified with the rate of doing work, called *working* in TIS, and a part identified with the rate of the flow of heat, called *heating*.[10] The macroscopic boundary velocity vector, $v_{\partial k}(q,t)$, is used in making this separation. The singular *i-particle local surface working density* and singular *k local surface working density* are defined with reference to (2.19) by

$$(7.56a) \qquad W_{\partial i}(q,t) = F_{\partial i}(q,t) \cdot v_{\partial k}(q,t),$$

$$(7.56b) \qquad \mathcal{W}_{\partial k}(q,t) = \sum_{i \in k} W_{\partial i}(q,t).$$

By (7.17), the equations (7.56) can be put in the form

$$(7.57a)$$
$$W_{\partial i}(q,t) = \left\langle \delta(q-q_i)\dot{\chi}_{\Omega_k(t)}(Q_k, P_k)p_i \cdot v_{\partial k}(q_i, t) \right\rangle_{kt},$$

$$(7.57b)$$
$$\mathcal{W}_{\partial k}(q,t) = \sum_{i \in k} \left\langle \delta(q-q_i)\dot{\chi}_{\Omega_k(t)}(Q_k, P_k)p_i \right\rangle_{kt} \cdot v_{\partial k}(q, t).$$

Finally, the singular *i-particle local surface heating density* and *k local surface heading density* are defined at the boundary for $q \in \partial d_k(t)$ as

$$(7.58a)$$
$$Q_{\partial i}(q,t) = \left\langle \delta(q-q_i)\dot{\chi}_{\Omega_k(t)}(Q_k, P_k)K_i(p_i - \bar{p}_i(q_i, t)) \right\rangle_{kt},$$

$$(7.58b)$$
$$\mathcal{Q}_{\partial k}(q,t) = \sum_{i \in k} Q_{\partial i}(q,t).$$

where $\bar{p}_i(q_i, t)$ is the local stream momentum at the boundary. It follows from (5.29) that

$$(7.59) \qquad \left\langle \delta(q-q_i)\dot{\chi}_{\Omega_k(t)}(Q_k, P_k)\bar{p}_i(q_i, t) \cdot \bar{v}_i(q_i, t) \right\rangle_{kt} = 0.$$

To compute the total time derivative of the energy, first use the relation

$$(7.60) \qquad K_i(p_i) = K_i(p_i - \bar{p}_i(q_i, t)) + p_i \cdot \bar{v}_i(q_i, t) - \tfrac{1}{2}\bar{p}_i(q_i, t) \cdot \bar{v}_i(q_i, t),$$

---

[10]For caveats on the separation of energy into heat and work, see the remarks in Maxwell [**1878b**] (Maxwell [**1890**], p. 669). See also the discussion in Klein [**1970**], pp. 89–91.

in (7.54). Next use the definitions (7.57) and (7.58) followed by an application of the condition (5.28) to replace $\bar{v}_i(q, t)$ by $v_{\partial k}(q, t)$ at the boundary. These steps give the result

(7.61a)
$$\frac{dE_i(q, t)}{dt} = Q_{\partial i}(q, t) + W_{i,b}(q, t) + W_{\partial i}(q, t).$$

Summing (7.61a) over $i \in k$ yields

(7.61b)
$$\frac{d\mathcal{E}_k(q, t)}{dt} = \mathcal{Q}_{\partial k}(q, t) + \mathcal{W}_{k,b}(q, t) + \mathcal{W}_{\partial k}(q, t),$$

which is the same as (3.9b) and proves Axiom 4.2.

7.7 The Conservation of Energy

The time derivative of $E_i(Q_k, p_i, t)$ is obtained by applying the operator d/dt defined in (4.21) to the definition (4.13a) of $E_i(Q_k, p_i, t)$. This leads by way of (4.18) to

(7.62)
$$\frac{dE_i(Q_k, p_i, t)}{dt} = \tfrac{1}{2} V_k \cdot \frac{\partial \Phi_{i,n}(Q_k)}{\partial Q_k} - v_i \cdot \frac{\partial \Phi_{k,n}(Q_k)}{\partial q_i}$$
$$+ \tfrac{1}{2} \left[\frac{\partial \Phi_{i,x}(q_i, t)}{\partial t} - v_i \cdot \frac{\partial \Phi_{i,x}(q_i, t)}{\partial q_i} \right].$$

This is used to define the *i-particle local internal working* and *k local internal working* by

(7.63a)
$$W_{i,n}(q, t) = \left\langle \delta(q - q_i) \left[\tfrac{1}{2} V_k \cdot \frac{\partial \Phi_{i,n}(Q_k)}{\partial Q_k} - v_i \cdot \frac{\partial \Phi_{k,n}(Q_k)}{\partial q_i} \right] \right\rangle_{kt},$$

(7.63b)
$$\mathcal{W}_{k,n}(q, t) = \sum_{i \in k} W_{i,n}(q, t),$$

and the *i-particle local external working* and *k local external working* by

(7.64a)
$$W_{i,x}(q, t) = \tfrac{1}{2} \left\langle \delta(q - q_i) \left[\frac{\partial \Phi_{i,x}(q_i, t)}{\partial t} - v_i \cdot \frac{\partial \Phi_{i,x}(q_i, t)}{\partial q_i} \right] \right\rangle_{kt},$$

(7.64b)
$$\mathcal{W}_{k,x}(q, t) = \sum_{i \in k} W_{i,x}(q, t).$$

The decomposition of $\Phi_k(Q_k, t)$ is used again to show for the body working that

(7.65a) $$W_{i,b}(q, t) = W_{i,n}(q, t) + W_{i,x}(q, t),$$
(7.65b) $$\mathcal{W}_{k,b}(q, t) = \mathcal{W}_{k,n}(q, t) + \mathcal{W}_{k,x}(q, t).$$

Integrating these equations over q gives the global forms of the working

(7.66a) $$W_{i,b}(t) = W_{i,n}(t) + W_{i,x}(t),$$
(7.66b) $$\mathcal{W}_{k,b}(t) = \mathcal{W}_{k,n}(t) + \mathcal{W}_{k,x}(t).$$

Let us consider now the proof of the conservation of energy. Because of non-locality involved in the transfer of momentum, angular momentum, and energy, it follows that in the general case

(7.67) $$\sum_{i \in k} W_{i,b}(q, t) \neq 0.$$

However, in the global case, where

(7.68) $$\mathcal{W}_{k,b}(t) = \sum_{i \in k} \int d^3q \, W_{i,b}(q, t),$$

the definition (7.8) and (4.13b) can be used to obtain

(7.69) $$\mathcal{W}_{k,b}(t) = \left\langle \frac{dE_k(Q_k, P_k, t)}{dt} \right\rangle_{kt}.$$

Summing (7.62) over $i \in k$, gives

(7.70) $$\frac{dE_k(Q_k, P_k, t)}{dt} = \frac{1}{2}\left[\frac{\partial \Phi_{k,x}(Q_k, t)}{\partial t} - V_k \cdot \frac{\partial \Phi_{k,x}(Q_k, t)}{\partial Q_k} \right].$$

This implies that $\mathcal{W}_{k,b}(t) = \mathcal{W}_{k,x}(t)$, which represents the work done by external forces on the system.

To analyze (7.70) further, the time partial derivative of $\Phi_{k,x}(Q_k, t)$ is taken with the help of the definition (4.8). The relation (5.32) is used next to compute $\partial F_l / \partial t$ followed by integration by parts and the use of (5.29) to obtain

(7.71) $$\frac{\partial \Phi_{k,x}(Q_k, t)}{\partial t} = \sum_{i \in k} \sum_{l \in [t]-k} \sum_{j \in l} \left\langle v_j \cdot \frac{\partial \phi_{ij}(q_i - q_j)}{\partial q_j} \right\rangle_{lt}.$$

The second term on the right in (7.70) can immediately be put in the form

(7.72) $$V_k \cdot \frac{\partial \Phi_{k,x}(Q_k, t)}{\partial Q_k} = \sum_{i \in k} \sum_{l \in [t]-k} \sum_{j \in l} \left\langle v_i \cdot \frac{\partial \phi_{ij}(q_i - q_j)}{\partial q_i} \right\rangle_{lt}.$$

Next, use (7.71) and (7.72) in (7.70) and the result in (7.69). Replacing the $\langle\langle \cdot \rangle_{lt} \rangle_{kt}$ average with the equivalent $\langle \cdot \rangle_{[t]t}$ average gives the result

(7.73) $$\mathcal{W}_{k,b}(t) = -\frac{1}{2} \sum_{i \in k} \sum_{l \in [t]-k} \sum_{j \in l} \left\langle \left(v_i \cdot \frac{\partial}{\partial q_i} - v_j \cdot \frac{\partial}{\partial q_j} \right) \phi_{ij}(q_i - q_j) \right\rangle_{[t]t}.$$

Summing this over $k \in [t]$, the antisymmetry of i and j in the summand of (7.73) leads to the conclusion that

$$(7.74) \qquad \sum_{k \in [t]} \mathcal{W}_{k,b}(t) = 0.$$

When the quantities $W_{\partial i}(q, t)$ and $Q_{\partial i}(q, t)$ are integrated over $q \in d_k(t)$ and summed over $i \in k$, a computation similar to (7.21) shows that

$$(7.75)$$
$$\sum_{k \in [t]} [\mathcal{Q}_{\partial k}(t) + \mathcal{W}_{\partial k}(t)] = \sum_{k \in [t]} \left\langle \dot{\chi}_{\Omega_k(t)}(Q_k, P_k) \mathcal{K}_k(P_k) \right\rangle_{kt}$$
$$= 0,$$

in accord with Axiom 5.2. Axiom 4.4 follows immediately from Axioms 4.2 and 4.3 and the results (7.74) and (7.75).

7.8 The Center of Momentum

Two system rest frames were introduced in Chapters 2 and 4. It was pointed out there that these rest frames are important to the definitions of some thermodynamic quantities. The macroscopic system rest frame is defined as the rest frame of the system boundaries, if one exists. The macroscopic particle rest frame is based on the microscopic center of momentum frame as its particle analog.

To define the center of momentum of the k system, the location of the k *phase center of mass* is first computed by the formula

$$(7.76) \qquad \mathcal{R}_k^{(cm)\mu}(Q_k) = \frac{1}{\mathcal{M}_k} \sum_{i \in k} m_i q_i^{\mu},$$

where $\mathcal{M}_k$ was defined in (7.5c) as the total mass. The phase average of this is the k *center of mass* that is written

$$(7.77) \qquad \mathcal{R}_k^{(cm)\mu}(t) = \left\langle \mathcal{R}_k^{(cm)\mu}(Q_k) \right\rangle_{kt}.$$

The macroscopic k *center of mass frame* is defined as the frame in which the origin is at the center of mass. This frame is defined by the relation $\mathcal{R}_k^{(cm)\mu}(t) = 0$.

The total phase momentum and the total system momentum can be expressed in terms of the phase center of mass by

$$(7.78a) \qquad \mathcal{P}_k^{\mu} = \mathcal{M}_k \frac{dR_k^{(cm)\mu}(Q_k)}{dt},$$

$$(7.78b) \qquad \mathcal{P}_k^{\mu}(t) = \mathcal{M}_k \left\langle \frac{dR_k^{(cm)\mu}(Q_k)}{dt} \right\rangle_{kt}.$$

The macroscopic k center of linear momentum frame was defined as the frame in which $\mathcal{P}_k^\mu(t) = 0$. Similarly, the macroscopic k *center of angular momentum frame* was defined as the frame in which $\mathcal{J}_k^\mu(t) = 0$. Finally, the macroscopic k *particle rest frame* was defined in Chapter 4 as the center of momentum frame in which both $\mathcal{P}_k^\mu(t) = 0$ and $\mathcal{J}_k^\mu(t) = 0$.

The k *center of momentum phase velocity* and the k *center of momentum velocity* at time t are defined in terms of the phase center of mass by

$$(7.79a) \qquad v_k^{(cm)}(Q_k, V_k) = \frac{dR_k^{(cm)}(Q_k)}{dt} = \frac{V_k}{\mathcal{M}_k} \cdot \frac{\partial R_k^{(cm)}(Q_k)}{\partial Q_k},$$

$$(7.79b) \qquad v_k^{(cm)}(t) = \frac{dR_k^{(cm)}(t)}{dt} = \left\langle \frac{dR_k^{(cm)}(Q_k)}{dt} \right\rangle_{kt} = \frac{\mathcal{P}_k(t)}{\mathcal{M}_k}.$$

The relation (7.78) is used next with the integral of (7.18b) over $q \in d_k(t)$ to show

$$(7.80)$$
$$\mathcal{M}_k \frac{d^2 R_k^{(cm)}(t)}{dt^2} = \frac{d\mathcal{P}_k(t)}{dt} = \left\langle \frac{d\mathcal{P}_k}{dt} \right\rangle_{kt} + \left\langle \dot{\chi}_{\Omega_k(t)}(Q_k, P_k)\mathcal{P}_k \right\rangle_{kt}$$
$$= \mathcal{F}_{k,x}^\mu(t) + \mathcal{F}_{\partial k}^\mu(t).$$

To reach the k particle rest frame, the origin is first transformed to the center of mass and the velocity frame is then transformed into the center of linear momentum frame so that $\mathcal{P}_k(t) = 0$. The angular momentum $\mathcal{J}_k$ is computed in this new frame and then, by rotating the system with a uniform angular velocity using the vector $\mathcal{J}_k(t)$ as an axis, the system frame is transformed in the second step so that $\mathcal{J}_k(t) = 0$. This transformation into the center of angular momentum frame does not change the value of the total momentum so that $\mathcal{P}_k(t) = 0$ remains valid. Note that the second step is not a Galilean transformation and, if the system was in an inertial frame, this transformation will introduce a frame-dependent non-inertial force field in the final frame. The second step leaves the k system in the k particle rest frame.

If the k system is subject to arbitrary external forces, (7.80) and a similar formula for the angular momentum indicate that a particular transformation into the center of momentum will be valid only at a given instant.

7.9 The Enthalpy

The enthalpy plays an important role in calculations that involve heat changes at constant pressure. The thermodynamic definition of the enthalpy is expressed in terms of the Hamiltonian energy and the pressure by $\mathcal{H}_k^{(en)} = \mathcal{H}_k + \delta_k \mathbf{P}_k$. The TIS version of the enthalpy is given in terms of the i-*particle local enthalpy density,* the k *local enthalpy density,* and the k *enthalpy* by

(7.81a) $$H_i^{(\mathrm{en})}(q,t) = H_i(q,t) + \mathrm{P}_i(q,t),$$

(7.81b) $$\mathcal{H}_k^{(\mathrm{en})}(q,t) = \mathcal{H}_k(q,t) + \mathrm{P}_k(q,t),$$

(7.81c) $$\mathcal{H}_k^{(\mathrm{en})}(t) = \mathcal{H}_k(t) + \delta_k(t)\mathrm{P}_k(t).$$

The pressure will be discussed in Chapter 11. Recall that $\mathcal{H}_k^{(\mathrm{en})} \neq \sum_{i \in k} H_i^{(\mathrm{en})}$ because the system Hamiltonian energy $\mathcal{H}_k$ is not a particle sum function.

It is easy to see that the enthalpy is a true density. The definition (3.20c) can be written in the form $\delta_k(t)\mathrm{P}_k(t) = \int d^3q \, \mathrm{P}_k(q,t)$. It follows that

(7.82) $$\mathcal{H}_k^{(\mathrm{en})}(t) = \int d^3q \, \mathcal{H}_k^{(\mathrm{en})}(q,t),$$

so $\mathcal{H}_k^{(\mathrm{en})}(q,t)$ is a density in the TIS sense. While the enthalpy is often classified as a thermodynamic potential, these results imply that it is a thermodynamic density and its classification as a thermodynamic potential is incorrect. Thermodynamic potentials will be discussed in Chapter 19.

CHAPTER 8

Entropy and the Second Law

In this chapter, I begin an inquiry into the second law of thermodynamics, entropy, and their relation to particle mechanics. This is the first step in a progressively deeper look into these fundamental issues that will be continued in subsequent volumes.

8.1 Introduction

Since the introduction of the steam engine in the late seventeenth century, observations by engineers had indicated that every heat engine requires an upper and lower working temperature and that certain properties of the engine depend on the difference between these temperatures. Other investigations had indicated that certain circumstances would decrease the efficiencies of the machines. In addition, it was clear that these machines always expend more energy than can be used to perform useful work. This extra energy is dissipated as heat into the environment during the operation of the machine.

Sadi Carnot [1824] used caloric theory, which at that time was the dominant theory of heat, to investigate this situation theoretically. He provided a graphical description of the thermal behavior of idealized engines as processes represented by trajectories on a plane spanned by two physical quantities, such as the temperature and volume of the working substance used by the engine. His graphical methods went beyond previous treatments in introducing the important concepts of *cyclic processes* and *efficiency*. This led Carnot to the notion of a *reversible engine,* which can traverse a given cyclic diagram in either direction by converting heat to work or converting work to heat.

William Thomson [1851] built on the work of Carnot, Émile Clapeyron, and James Prescott Joule by stating two fundamental propositions on which to build thermodynamics. The first proposition was the recently developed law of the conservation of energy. It had been quickly adopted and was not controversial because its employment in mechanics could be understood in terms of visualizable physical models. Thomson's second proposition was concerned with the efficiency of heat engines. He associated Carnot's limits of efficiency with a dissipation of useable energy in the cyclic operation of steam engines. The focus of this second proposition soon became a concern with the irreversibility of certain macroscopic processes involving heat flow. The second proposition lacked the clear mechanical foundation enjoyed by the first and subsequently spawned a host of alternative formulations and interpretations. Thomson's two propositions became codified as the two fundamental laws of thermodynamics and a formalism began to develop around them.

The discovery of thermoelectricity in the early nineteenth century and its explanation by William Thomson in the period 1851–1853 provided a significant challenge to the developing thermodynamics. Thomson was able to express thermoelectric phenomena in a circuit in terms of Carnot cycles. In order to apply Carnot's theory of reversible cyclic processes, Thomson divided the energy flow in the circuit into a reversible energy flow carried by the circuit and an irreversible energy loss due to Joule heating. He then simply ignored the irreversible Joule heating and successfully computed the thermoelectric coefficients using Carnot's theory for reversible cycles. This work and its implications for thermodynamics are discussed in detail in EIS and will not be pursued further here.

The question of the proper formulation and interpretation of the second law became one of the main issues that concerned researchers. An important step in this direction was the introduction by Clausius [**1865**] of the concept of entropy, which as Bailyn [**1994**], pp. 113–114, has pointed out, was already present, although unnamed, for reversible heat flows in Clausius [**1854**]. Clausius expressed the change in entropy of a body in terms of the heat flow through its boundary divided by its temperature. Clausius also showed in 1865 that irreversible macroscopic processes, in the sense introduced by Carnot, led to an increase in entropy, but that reversible processes did not. He expressed this in the form of an inequality for the entropy that provides a lower bound on the rate of entropy change in a process. Clausius felt that he had shown that this inequality implied that certain processes must be irreversible and that the entropy could not decrease. It was therefore natural to associate entropy with the second law.

After the advent of kinetic theory in the 1850s, the work of Maxwell and Boltzmann in the 1860s and 1870s combined statistical and mechanical ideas in the form of particle distributions and equations for their evolution in time. The association of the entropy with the macroscopic second law led a number of authors during this period, beginning with Boltzmann [**1866**], pp. 23–33, and later including Clausius, C. Szily, and Hermann von Helmholtz, to attempt to demonstrate that the non-decrease of the entropy function (TIS sign convention) is a mathematical consequence of the evolution of the underlying microscopic particle system in accord with its mechanics. This requires showing that there is a mechanical quantity defined in terms of the particle variables that has the properties associated with Clausius' macroscopic entropy. These attempts to reduce thermodynamics to statistical mechanics all failed.

Macroscopic thermodynamics continued to develop in ways that were independent of the attempted reductions to statistical mechanics. In extensions of the macroscopic theory, the entropy and other thermodynamic quantities were assigned various properties depending on their role in describing the system. Certain concepts such as momentum, angular momentum, and energy, were considered extensive, which means that they are proportional in some sense to the number of particles or size of the system. Other quantities such as the temperature and pressure were considered independent of the size of the system and called intensive quantities.

The entropy in particular was considered to be an extensive quantity. This made the entropy proportional to the size of the system and an additive quantity both within and between systems. It was sometimes also considered a homogeneous function of its arguments. These considerations determined in part the form of the thermodynamic functions that were used to represent the entropy.

Pinning down the entropy proved to be especially difficult because there is no associated mechanical concept to guide the analysis. How, when, and where entropy is produced was a particularly vexing problem. Clausius considered two forms of entropy change. The first was an external change d_eS due to the flow of heat into or out of a system. The second was an internal change d_iS associated with the production of entropy within a body. Although Clausius explicitly rejected the production of internal entropy at one point, it is needed if his famous 1865 maxim,

> The energy of the universe is constant.
> The entropy of the universe tends to a maximum.

is to be valid for isolated systems.

Because of the work of Lars Onsager in the 1930's, the entropy became associated with various macroscopic fluxes and flows. Onsager's work led directly to a concept of entropy that is very similar to a fluid and described by a balance equation at equilibrium. These properties are also very similar to the properties that had been assigned to the caloric in caloric theory.

From around the 1860s, through the development of quantum mechanics, and into the 1940s, phenomenological thermodynamics was felt by many scientists to be much more secure than the statements obtained from the microscopically based theories. For this reason, conclusions concerning the evolution of a microscopic particle system were often based on the implications of the second law of macroscopic phenomenological thermodynamics that were felt to be valid in that situation.[1] This viewpoint appears in the work examined in QTS of Planck, Einstein, von Neumann, and many others.

In discussions of the entropy, we often seem to be in a different world than the one we occupy during the study of the energy and momentum relationships. In a detailed review article on the entropy, Wehrl [1978], p. 221, remarked that, in spite of the passage of more than a century since it was introduced, it is "astonishing to note that most of the main features of entropy are unknown to physicists and that many problems in connection with entropy are still open or have been solved only in the past few years." He attributed this to the concern of physicists with special cases rather than the general features of entropy and to the requirement of some "so-to-say unusual mathematics" in the proofs of statements about it. Wehrl's statements about entropy reflect those generally accepted in the literature, so the remarks made here and in the next chapter are not directed specifically against his work.

---

[1] Arthur Eddington [1935], p. 74, expressed this feeling about the second law in the form "The law that entropy always increases—the second law of thermodynamics—holds, I think, the supreme position among the laws of nature." This is typical of many stated views on the second law at that time.

Wehrl, p. 222, began his survey with the statements "it has to be stressed that the concept of entropy is not at all unclear but a very well defined one. ... Entropy relates macroscopic and microscopic aspects of nature and determines the behavior of macroscopic systems, i.e., real matter, in equilibrium (or close to equilibrium). Why this is true unfortunately is not yet understood in full detail, in spite of a century's efforts of thousands and thousands of physicists." Some of the reasons will emerge below.

8.2 The Basic Issues

Although there are differences between Clausius' concept of entropy and some of the formulations of the second law, these are not germane to this discussion, so the Clausius' form for the lower bound on the change of his thermodynamic entropy will be treated from now on as an expression of the second law in terms of measurable physical quantities.

As G. Bierhalter [1992] pointed out in a study of nineteenth century attempts to provide a mechanical basis for the second law, most physicists in the nineteenth century felt that the theorems and principles of physics would ultimately be derived from mechanical principles. This meant that the lack of a mechanical proof of the second law left a serious gap. The unsuccessful attempts by prominent physicists to provide mechanical proofs led them and others to seek the foundation of the second law in other directions.[2]

The question of how to interpret the entropy and the macroscopic second law in terms of a particle model became a concern in the second half of the nineteenth century. Because there are many possibilities for defining the entropy in terms of particle properties and behavior, and quite a few of these possibilities have appeared in the literature, the important question is how to evaluate the various proposals and to find criteria for choosing between them.

Each different model or explanation proposed for the entropy in terms of particles has in fact spawned a new form of entropy to contend with. Some textbooks list five or more distinct forms of entropy. One of the goals of this chapter is to decide (1) if any of these particle analogies are valid models for the entropy, and (2) whether Axiom 6, with an appropriate analog function for computing the entropy, is the mathematical expression of what is meant by the thermodynamic formulations of the second law. It goes without saying that the relationship between the entropy, entropy analogs, entropy models, and the various versions of the second law, has been a tangled one.

---

[2] The failure of attempts at a mechanical derivation of the second law is pursued further in EIS. For a good general discussion of the attempts in this era to provide a mechanical foundation for thermodynamics, see Bailyn [1994], Chapter 10.

8.3 Carnot and the Second Law

In Carnot's [1824] study of heat engines, he considered idealized engines that take in heat from a reservoir at a high temperature, convert some of that heat into an equivalent amount of work, and then discharge the remaining part of the heat into a reservoir operating at a lower temperature. These steps are carried out in the engine by means of a working substance, such as steam, which transfers heat and transforms some of it into work by pressing on a cylinder.[3] Carnot's important theoretical innovation, which reflected the design of steam engines, was to consider the cyclic operation of an engine in which the working substance at the end of a cycle is in the same macroscopic state it was in at the beginning of the cycle. Moreover, Carnot observed that if such an engine can be operated in reverse it will take in heat at a lower temperature, combine it with a certain amount of work and discharge heat at a higher temperature. He defined a *reversible cycle* as a cyclic process for which the amount of heat transformed into work by the engine in the forward operation equals the amount of work consumed by the engine operating in reverse between the same temperature extremes. Carnot and his successors could show for these engines that the upper limit on the efficiency at which heat engines can operate is always less than 1 if the two operating temperatures both fall within the range $0 < T_k < \infty$. In addition, Carnot showed that reversible engines achieve the maximum efficiency possible for a cyclic process. Because of the loss of energy due to friction and heat loss into the environment, it was clear that no real macroscopic engine is reversible so that real engines cannot operate at the theoretical maximum efficiency in converting heat to work.[4]

An engine that operates at less than maximal efficiency will produce less work for a given amount of heat consumed than is produced by the reversible engine. Some of the heat energy "disappears" during the operation of an irreversible engine and is dissipated into the environment.[5] Émile Clapeyron [1834], followed by Thomson [1848] and Clausius [1850] (who also removed the theory from the context of caloric theory) were responsible for the conceptual and mathematical development of the theory and for making it the foundation of macroscopic thermodynamics.

Since the operation of all real machines is at less than the theoretical maximum efficiency, Thomson and Clausius attempted to codify this fact by stating two laws that governed the conversion of energy from heat into work. Thomson's [1851] first proposition, mentioned above, treated heat as a form of energy. He attributed this first proposition to Joule, who had measured the

---

[3]For the history of this subject, see Truesdell [1980], Brush [1976], and Partington [1949]. For a discussion of the roots of Carnot's thinking in engineering studies, see Thomas Kuhn [1960].

[4]See Chapter 17 for a more detailed treatment of Carnot's work.

[5]See the discussion in Brunold [1930], pp. 45–50.

mechanical equivalent of heat. The second proposition, which he attributed to Carnot and Clausius, stated that a reversible Carnot engine produces as much mechanical effect as any thermodynamic engine can.

In order to prove the second proposition, Thomson [**1851**] presented an axiom on which he based his demonstration:[6]

> *It is impossible, by means of inanimate material agency, to derive mechanical effect from any portion of matter by cooling it below the temperature of the coldest of the surrounding objects.*

Thomson also stated Clausius' 1850 version of this axiom as follows:[7]

> *It is impossible for a self-acting machine, unaided by any external agency, to convey heat from one body to another at a higher temperature.*

In another paper at about the same time, Thomson [**1852**], (see also Thomson [**1854**]) referred to the "dissipation" of energy as the important factor:

> "When heat is created by an unreversible process (such as friction), there is a *dissipation* of mechanical energy, and a full *restoration* of it to its primitive condition is impossible."

Clausius [**1862**], p. 81, recalled a proposition he had published in 1854.[8] This proposition expressed the second law as:

> *Heat cannot of itself pass from a colder into a warmer body.*

An influential version of the second law was later stated in 1897 by Planck [**1922**], p. 89, in the form[9]

> "In connection with what has just been said we shall now base the general proof of the second law on the following empirical law: 'it is impossible to construct an engine which will work in a complete cycle, and produce no effect except the raising of a weight and the cooling of a heat reservoir.'"

It has also been given many other formulations.[10]

8.4 Maxwell's Perspective

Several authors have pointed out the indebtedness of the work of Maxwell [**1860**], [**1866**], which extended kinetic theory by introducing more powerful statistical methods, to the writings of Adolphe Quetelet on social statistics

---

[6]Thomson's statement of the axiom is reprinted in Kestin [**1976**], p. 179. For a discussion of some issues surrounding these axioms, see Daub [**1970**].

[7]See Kestin [**1976**], p. 181.

[8]See also the reprint in Kestin [**1976**], p. 133.

[9]See also the formulation in Planck [**1932**], p. 52.

[10]It is not the intention here to give a complete account of formulations of the second law. The above statements will suffice for the current discussion. For further information and a critical analysis, see Truesdell [**1980**], Chapter 11.

and the law of errors.[11] Part of Maxwell's reasoning concerning the equilibrium state was based on thinking about the independent accumulation of small random errors and their description by the law of errors distribution. He went on to develop his transport equations in accord with these statistical conceptions. He was aware from almost the beginning of the tension between the statistical approach and the dynamical approach to understanding the mechanics of a large number of particles.

Maxwell [1870] addressed the British Association on the Advancement of Science on current issues. In reference to kinetic theory and thermodynamic phenomena, he, pp. 225–226, stated:

"One of the most remarkable results of the progress of molecular science is the light it has thrown on the nature of irreversible processes—processes, that is, which always tend towards and never away from a certain limiting state."

As examples, Maxwell stated that two gases in a vessel will always become more uniformly mixed and heat differentials between parts of a substance will always decrease. He cited Thomson's theory of the irreversible dissipation of energy as the basis for the latter phenomenon and stated that Thomson's view was equivalent to Clausius' doctrine of the growth of entropy. He also mentioned Fourier's theory of the conduction of heat as another irreversible theory.

As a consequence of these ideas, Maxwell [1870], p. 226, drew the conclusion

"But if we attempt to ascend the stream of time by giving to its symbol continually diminishing values, we are led up to a state of things in which the formula has what is called a critical value; and if we inquire into the state of things the instant before, we find that the formula becomes absurd.

We thus arrive at the conception of the state of things which cannot be conceived as the physical result of a previous state of things, and we find that this critical condition actually existed at an epoch not in the utmost depths of past eternity, but separated from the present time by a finite interval.

This idea of a beginning is one which the physical researches of recent times have brought home to us, more than any observer of the course of scientific thought in former times would have had reason to expect."[12]

---

[11]See Theodore Porter [1981], for example.

[12]This conclusion was based on the theory of irreversible thermodynamic phenomena. This is very likely a reference to Thomson's [1852] calculations of the rate of cooling of the earth and the maximum length of time the earth could have been in existence in a habitable form. There is also a striking parallel between this statement by Maxwell and

In 1872, Maxwell [**1872**], pp. 308–309, discussed limitations on the second law of thermodynamics. Because we do not perceive individual molecules, Maxwell stated that we are compelled to adopt a statistical method of calculation. He [**1873a**] expanded on this point in a further discussion of the recent introduction of the methods of statistics into mathematical physics. He contrasted the "historical method," embodied in the equations of dynamics, which requires a perfect knowledge of the data, and the "statistical method," which is our only recourse in studying large numbers of molecules.

Maxwell [**1878a**] (Maxwell [**1890**], p. 644) again associated an increase in entropy with "a diminution of the available energy of the system, that is to say, the total quantity of work which can be obtained from the system." He also maintained that the entropy of a material system is the sum of the entropies of its parts. Proceeding further, Maxwell [**1878b**] ([**1890**], pp. 668–670) examined various formulations of the second law of thermodynamics and the conceptual difficulties surrounding them in his review of Tait's *Thermodynamics*. On this he stated that ([**1890**], p. 669), "It is probably impossible to reduce the second law of thermodynamics to a form as axiomatic as that of the first law, for we have reason to believe that though true, its truth is not of the same order as that of the first law." He noted in this regard, as mentioned in Chapter 7, that the distinction between heat and work would vanish if we could follow the individual molecules in their paths. Thus, p. 670, "The truth of the second law is therefore a statistical not a mathematical truth, for it depends on the fact that the bodies we deal with consist of millions of molecules and that we can never get hold of single molecules." A consequence of this is that[13]

> "the second law of thermodynamics is continually being violated, and that to a considerable extent, in any sufficiently small group of molecules belonging to a real body. As the number of molecules in the group is increased, the deviations from the mean of the whole become smaller and less frequent, and when the number is increased till the group includes a sensible portion of the body, the probability of a measurable variation from the mean occurring in a finite number of years becomes so small that it may be regarded as practically an impossibility."[14]

In accord with this statistical point of view, Maxwell stated further that no deduction of the second law from dynamical principles alone "can be a sufficient explanation of the second law."

---

current thinking on the "big-bang" theory of the origin of the universe, which is supported in part by the thermodynamic study of the background radiation pervading the universe.

[13]See also Gibbs 1875 statement below on the change of the second law from a statement of impossibility to a statement of improbability in relation to this.

[14]This is an application of the statistical law of large numbers.

8.5 The Emergence of Entropy

An important formal step leading to the definition of thermodynamic entropy was taken by Thomson [**1854**], pp. 126–127, when he extended Carnot's work on cyclic processes. For a reversible cyclic process, Thomson presented the formula equivalent to (TIS notation)

$$(8.1) \qquad \sum_{l \in A} \frac{\Delta Q_l}{T_l} = 0,$$

where A is a set of segments of the closed cycle on each of which the temperature T_l has a specific value and the net heat flow is ΔQ_l.[15]

Clausius [**1854**] extended his account of the operation of Carnot cycles and defined the equivalence value of a flow of heat between bodies at different temperatures. He recognized that dQ/T is an exact differential for these processes. Clausius [**1865**] later called it the differential of the entropy and designated it by dS. The integral of the entropy around a closed cycle is written $S = \oint dQ/T$. The second law was expressed in Clausius' work in terms of this quantity (using both $+$ and $-$ sign conventions at different points) as $\oint dQ/T \geq 0$.[16]

As irreversible phenomena, Clausius cited those for which energy is actually lost into the environment. He also maintained that the expansion of a gas into an evacuated chamber without gain or loss of heat and without the transformation of heat into work is an irreversible phenomenon.[17] Clausius based this view on the fact that work is required to reestablish the original state of the gas. However, there is no heat flow into or out of the system during the expansion process, so $S = \int dQ/T = 0$. This implies that Clausius was not entitled to conclude that this is an example of irreversibility on the grounds of an increase in entropy. Similar problems of interpretation appear in many other cases and will be discussed further below.

Thomson in 1855 made a distinction between reversible and irreversible processes. After Clausius introduced the entropy, reversible processes were identified as processes that do not change the entropy and irreversible processes were identified as those that do. This distinction subsequently played an important role in understanding the entropy and its changes. The work

---

[15]For more details, see the account in Partington [**1949**], pp. 169–172.

[16]The translation between Carnot's theory, rooted in caloric theory, and the theory Clausius developed out of it was discussed by Charles Brunold [**1930**], pp. 84–86. In reference to the entropy, Truesdell [**1980**] has argued that Ferdinand Reech instead of Carnot deserves credit for discovering it. In support of this, Truesdell [**1986**] observed that Gibbs [**1875**] had mentioned Reech. Clausius is accepted as the discoverer of the entropy here because (1) Reech's version was developed in the context of the theory of vortex atoms; (2) Reech did not derive the formula for the entropy explicitly before 1850; and (3) Reech did not exploit it in the way Clausius did. See Hutchinson [**1981**], pp. 76–80, on this.

[17]This free expansion of a gas is called Joule's experiment—although Planck [**1922**], p. 49, maintained that it was first performed by Gay-Lussac.

in Planck [**1887**], for example, is based on this distinction. Planck divided all processes into "reversible processes" and "natural processes", which are irreversible. He stated that the difference between these two types of processes rests only on the beginning and final states and not on the path between them. He stated that nature has "more desire" for the end state than the beginning state in a natural process and concluded that this makes these processes irreversible. As natural processes, Planck mentioned the flow of heat from a hot body to a cold one, the generation of heat by friction or collision among other processes. He associated the states with larger entropy as the more desired states and natural processes as those for which $dS > 0$. He applied these ideas soon afterward to his short-lived theory of natural radiation.

8.6 Boltzmann's Analog

Boltzmann [**1872**], in his groundbreaking work on the Boltzmann equation, presented the natural logarithm of the one-particle distribution as the statistical mechanical analog for the entropy. He defined the quantity $\mathcal{H}(t)$ as the phase average of his analog over the 6-dimensional phase space of a particle in the form[18]

$$(8.2) \qquad H(t) = -k_B \int d^3q \int d^3p \, f(q,p,t) \ln f(q,p,t).$$

Boltzmann showed in his H theorem that $H(t)$ never decreases.

After criticism by Loschmidt [**1876**] concerning issues of reversibility, and in response to later criticism by Zermelo [**1896a**] regarding recurrence, Boltzmann [**1895**], [**1896b**], claimed that the H theorem is a "statistical theorem" and that its truth depends on the initial state of the system under investigation. In more than one paper, Boltzmann attempted to justify this claim and deal with the further difficulty that entropy seemed to go up and down in some cases. He approved of Culverwell's [**1890**] description of the curve associated with the H theorem as a 'flattened, inverted tree' with discontinuities at its maxima. However, as will be shown in Chapter 22, these ideas do violence to both the mathematics behind the H theorem and the conceptual basis of Boltzmann's equation.[19]

One significant aspect of Boltzmann's definition of $H(t)$ was his use of an analog in his H theorem. This extended to a thermodynamic quantity Maxwell's idea of defining mechanical quantities in his transport equations by mechanical phase analogs. The logarithmic form $\ln f(q,p,t)$ of this analog was another important innovation. These ideas were later taken up by Planck and Gibbs and developed in a broader context. The formalism associated with

---

[18]Boltzmann's original analog is $\ln f(q,p,t)$. The TIS version of Boltzmann's analog for a one-particle probability distribution density $f(q,p,t)$ is $S = -k_B \ln f(q,p,t)$.

[19]These ideas are also discussed in more detail in EIS.

Boltzmann's definition of an entropy analog in connection with the Boltzmann equation will be discussed in Chapter 22.

As a technique of working with distributions and finding those that maximize the entropy, Boltzmann introduced discrete methods in his 1868 paper on equilibrium thermodynamics and his 1872 paper presenting the Boltzmann equation. He divided the total system energy into a collection of discrete units and assigned these units of energy to the particles in the system. His combinatorial formalism yielded a count of the number of ways in which this could be done. Subsequently, Boltzmann [1877b] significantly enlarged his conception of entropy compared with his work in 1868 and 1872 by extending the discrete methods of analyzing distributions he had introduced in 1872 and identifying the number of different microscopic states in a system that correspond to a given macroscopic state as the entropy of that system. Pursuing a direction indicated in his 1872 paper, he identified this entropy as the probability of a given macroscopic state. Boltzmann selected the most probable state, which is the distribution that maximizes this entropy, as the equilibrium state.

Boltzmann's 1877 formalism was used by Planck in a derivation of the blackbody radiation law. The success of Planck's theory gave a strong boost to Boltzmann's views regard the relation of entropy and probability. Coupled with the other problems associated with the classical definitions of the entropy, Planck's success also gave rise to the idea that entropy cannot be properly defined in classical theory. This issue will be mentioned again in Chapter 9, but a full analysis requires discussion of discrete distributions, Planck's work on blackbody radiation, and the conception of the entropy as the probability of a state. This analysis will be taken up in QTS.

8.7 Gibbs' Approach

Gibbs [1875] worked on the thermodynamics of multicomponent systems at equilibrium. In this work, he encountered a problem in the definition of the equilibrium entropy of a substance. This problem, now called the *Gibbs paradox*, concerns the increase in entropy due to the mixing of two gases predicted by standard thermodynamics when these gases diffuse into each other. According to the standard equilibrium theory, the increase in the equilibrium entropy when two quantities of two substances are allowed to mix in this way does not depend on the degree of difference between them. However, when the two substances are the same, an increase in entropy for the diffusion of two portions into each other does not make sense thermodynamically.[20] In addition, since mixed substances can also be separated, Gibbs [1875], p. 167, echoed Maxwell's statement quoted above that the association of entropy with mixing means that "the impossibility of an uncompensated decrease of entropy seems to be reduced to improbability."

---

[20]See Gibbs [1875], pp. 165–168, and Gibbs [1902], Chapter XII.

In 1902, Gibbs generalized the statistical mechanical approach of Maxwell and Boltzmann. He worked with an $\mathcal{N}$ particle probability distribution density rather than the one-particle version that Boltzmann had used and emphasized the importance of analog functions in defining the thermodynamic functions associated with the statistical mechanics. Gibbs [**1902**], p. 16, called the natural logarithm of the $\mathcal{N}$ particle system probability distribution density the "index of probability" and identified it as the entropy analog. He, p. 33, also chose the "most simple case conceivable" as the index of probability for the equilibrium case, namely, that it is linear in the energy. He noted that "when the system consists of parts with separate energies, the laws of distribution in phase of the separate parts are of the same nature." Mandelbrot [**1964**] presented this in modern terminology as the statement that the total energy is a sufficient statistic for the equilibrium (Maxwell-Boltzmann) distribution.

The problems with mixing and the Gibbs paradox led Gibbs [**1902**], p. 188, to introduce a distinction between what he called "specific phases" and "generic phases" in the statistical mechanical description of a system. A specific phase is represented by either a point in the $6\mathcal{N}$-dimensional phase space of the system (an exact state) or a probability distribution density over these points. Generic phases are sums over all possible combinations of specific phases obtained by interchanging identical particles. An example of a generic phase function is the discrete particle distribution employed by Boltzmann in his combinatorial calculations. Generic phase functions are divided by the quantity $[\mathcal{N}!]^{-1}$, which represents the number of distinct arrangements of $\mathcal{N}$ distinguishable particles, to maintain the proper normalization. The inclusion of this factor in his calculations was justified by Gibbs on the grounds of the practical indistinguishability of the particles. It was later given a post-hoc justification by others on the grounds of quantum mechanics.

Gibbs' resolution of the Gibbs paradox by introducing generic phases to replace specific phases for describing the entropy of a system and dividing the phase average of the entropy by $\mathcal{N}!$ means physically that the system distribution is seen as describing a superposition or mixture of all possible microscopic classical states at any given time. Using this as a general procedure for computing the phase averages of physical quantities has the effect of making the time derivatives of the thermodynamic functions representing these quantities meaningless except, perhaps, at equilibrium. From the standpoint of classical physics, the system trajectory in phase space represents a unique history for the particles in the system and a factor of $[\mathcal{N}!]^{-1}$ cannot be included in classical thermodynamic calculations without violating this unique aspect of the system dynamics and its history. Furthermore, even if the fact that a system is actually in one specific phase is ignored, the system distribution cannot simply be divided by $\mathcal{N}!$ unless there is an explicit sum over the $\mathcal{N}$! specific phases as well.[21] This approach will therefore be rejected and only specific phases will be used in the theory presented here.

In the later chapters of his "Elementary Principles," Gibbs [**1902**] took up the questions of entropy and the approach to equilibrium. He first gave an informal proof of the Poincaré recurrence theorem for an isolated system in Chapter XII. He then observed that the distribution that maximizes (TIS sign convention) his entropy analog is unique and is a "permanent" (i.e., time independent) distribution. On the other hand, he found that the phase average of his general (nonequilibrium) entropy analog is constant in time for an isolated system when that system is described by a distribution for which the entropy is defined. This means, as Gibbs noted, that an isolated system shows "no approach toward statistical equilibrium in the course of time." This fact called into question for Gibbs the validity of his choice for a statistical mechanical analog to represent the entropy.

Gibbs came to doubt that this phase average of the index of probability for systems not at equilibrium, which is constant in time, is a proper analog for thermodynamic entropy. His doubts stemmed from the observation that two initially separate quantities of colored liquid in contact with each other will diffuse into each other until the whole quantity of liquid is uniformly colored. This process can be speeded up by stirring the liquid. In the face of such "irreversible processes," Gibbs abandoned the phase average of his index of probability as the thermodynamic entropy and put in its place a quantity that does increase in time. This was soon called (Poincaré [**1906**], p. 373) the "coarse-grained entropy."

Overall, Gibbs' approach to the concept of entropy as the phase average of the logarithm of a probability distribution density defined in $6\mathcal{N}$-dimensional phase space was an important conceptual advance but did not solve the problems surrounding it. In addition, as will be discussed in Chapter 9, his coarse-grained entropy was a step backwards.

8.8 Thermodynamic Entropy and Statistical Analogs

In the application of his entropy formula $\oint dQ/T = 0$ to a reversible Carnot cycle, Clausius [**1854**], pp. 94–95, argued that the temperature refers to the "momentary temperature of the changing body" rather than to the reservoir to which it is attached when heat is being exchanged. By implication, this applies also to the use of the formula $\int dQ/T$ to compute the

---

[21]The question of whether to divide distributions by $\mathcal{N}$! was the subject of an extended debate in the physics community between 1900 and 1926. For more details on the controversy concerning the $[\mathcal{N}!]^{-1}$ factor, see the discussion in Epstein [**1937**], Section 16, and the critical discussion in Ehrenfest and Trkal [**1920**], Sections 1 and 8. See also the remarks in Tisza and Quay [**1963**], pp. 85–86. The $\mathcal{N}$! controversy, discrete distributions, and Boltzmann's Principle will be discussed in QTS and further references will be given there.

entropy when the process is not a reversible one.[22] The problem came up
again towards the end of the nineteenth century, when a number of authors
became interested in generalizing thermodynamics to the nonequilibrium domain.[23] Subsequently, Brunold [1930], pp. 82–83, for example, argued that
T in $\int dQ/T$ represents the temperature of the reservoir not that of the system. This issue cannot be resolved using a formalism that is valid only for
reversible processes because, as Clausius pointed out, the two approaches are
equivalent in that case. It has subsequently been ignored in the literature
without being resolved.

In the derivation of Clausius' inequality in the next chapter, the temperature associated with Clausius' calculation is that of the boundary of the
system itself. This choice is consistent with the overall approach of the Theory
of Interacting Systems, in which each system is treated on an equal footing.
There is no *in principle* distinction between a reservoir and any other system
that would warrant giving primacy to the temperature of the reservoir system over the actual temperature of a system boundary in the calculation of
a thermodynamic quantity for that system.

In discussions of the thermodynamic entropy, the equilibrium entropy S
is shown to be a "state function" of two variables taken from the set $\{P, V, T\}$.
This is another way of stating Clausius' point that dS is an exact differential
of the independent variables used to describe the system. The state of a
system at equilibrium can also be represented as a point on a plane with the
independent macroscopic variables as axes. In the nonequilibrium case, on the
other hand, the thermodynamic entropy is not a state function in some space
like this with macroscopic variables as coordinate axes. The reason is that the
set of microscopic configurations, each of which represents a different possible
point in phase space, far exceeds what using a set of macroscopic variables
as axes in a space can represent. This means that two systems characterized
by the same values for a set of macroscopic variables can eventually have
very divergent histories. Because of this, discussions of the thermodynamic
entropy based on the formulas and relations valid in the equilibrium case
cannot be carried over to the nonequilibrium realm. This is also true for
Carathéodory's approach to thermodynamics.[24]

An acceptable analysis of a thermodynamic concept in terms of a microscopic analog or interpretation, such as the association of the thermodynamic

---

[22]See also Clausius [1856], p. 246, footnote, where the treatment is less clear. In
addition, there is a discussion of this point in Partington [1949], p. 192, footnotes 2 and 3.

[23]See the references in Partington [1949], p. 193, footnote 7.

[24]See Carathéodory [1909], [1919]. See also Born [1921] for a recasting of thermodynamics into Carathéodory's form and the discussion of Carathéodory's work in Jordan
[1933]. For a critical discussion of Carathéodory's work, see Truesdell [1986], pp. 114–123.
Connections between the work of Carathéodory's and that of Helmholtz [1887] have been
traced by Bailyn [1994], p. 425, footnote 7.

concept of entropy with the microscopic "order" or "arrangement" of the particles, for example, must meet two tests:

(i) (Analog Test) An explicit analog function for the thermodynamic concept must exist and be expressed as a function of the canonical particle position and momentum variables and the interparticle potential energy functions of Hamiltonian mechanics and nothing else;

(ii) (Validity Test) A justification must be given for using this function to represent the thermodynamic concept.

If the further statement is made that this thermodynamic concept helps determine the evolution of a process, a claim that is often made with respect to various forms of the entropy,[25] a third test is appended:

(iii) (Relevance Test) It must be shown how the analog function plays a non-vacuous role in the evolution equation of the underlying mechanics.

The third requirement is a crucial one. If the particle analog function representing the thermodynamic concept under consideration does not appear as a component of the evolution equation of a system, thereby modifying the set of solutions that represent possible trajectories of the system in phase space, then that concept can play no essential role in determining the future of the system.

Note that these tests apply only to specific interpretations of a thermodynamic concept in particle terms and specific claims as to how this concept determines the microscopic evolution of a system during processes. However, if no analog can be found for a thermodynamic concept on the microscopic level or the concept cannot be expressed in terms of thermodynamic quantities that do have microscopic analogs, that concept cannot play a role in the general nonequilibrium theory.

8.9 Order and Disorder

Let us consider now the most common proposal for the interpretation of entropy in particle terms. This is the proposal that entropy can be understood in terms of the concept of 'order' on the microscopic level. This model of the entropy has almost attained the status of dogma in thermodynamics. The association of entropy with order originated early with Maxwell and Thomson and has appeared in many discussions of entropy since then, including a recent one by Penrose [1989], Chapter 7. These models usually relate the changes in entropy to changes in the spatial arrangement of the underlying particles.[26] In other cases the entropy is divided into contributions of various types corresponding to different measures of disorder expressed in

---

[25]See, for example, Dickerson [1969], p. 206, or Bent [1965], Chapter 25.

[26]See, for example, Dickerson [1969], pp. 8–9: "Energy is a measure of molecular motion; entropy is a measure of molecular arrangement. ... Entropy, as defined, is a measure of disorder."

terms of different attributes associated with the particles. These measures have included distributions of mass in a volume, distributions of particles in coordinate space, etc.[27] It is often claimed, for example, that certain chemical reactions proceed in the direction they do because of such configurational considerations.[28]

In support of the view that order or disorder is the measure of entropy, the equilibrium state could be considered as the most disordered state because all temperatures are uniform and pressure differences are related only to outside forces. This means that there are no peaks and valleys in the values of these macroscopic parameters over the system volume that is not associated with the external forces. The fact that the entropy of the equilibrium state is maximal then suggests that the statistical mechanical analog of the entropy is proportional to a measure of the microscopic order in the system. While plausible to a degree, this point of view requires an independent justification. In the case of Boltzmann's entropy analog, for example, this justification would amount to showing that $\ln f(q, p, t)$ is a measure of microscopic order that corresponds in some specific way to the particle arrangement in space.

The original justification given for the correspondence between order and entropy was stated by Boltzmann [**1868**] using a discrete (coarse-grained) measure obtained by partitioning a collection of particles randomly into cells and counting them. In the limit $\mathcal{N}_k \to \infty$, Boltzmann used this measure to show that the state with particles distributed uniformly is the "most probable state" in the sense that the distribution has a single very high peak so that most particle configurations with fixed total energy, momenta, and number of particles will correspond closely to this situation. This may be considered as the most disordered state in some sense of that word because it is uniform and has no localized structures in phase space. On the other hand, this situation may correspond to a highly ordered crystal structure for the particles when viewed in coordinate space. Clearly, something more definite is needed as a definition of 'order' or 'disorder' for a collection of particles if it is to be used to define the entropy. The danger is that it will turn into a circular reference such that the most disordered state is subsequently defined as the one configuration that corresponds to a state with maximal entropy.

There is another problem in using combinatorial calculations of the type Boltzmann introduced in demonstrating properties of the system distribution or in justifying the H theorem. This has to do with the fact that the coarse-grained distributions treated in combinatorial calculations are not equivalent to the continuous distributions used in calculation of the H theorem or in the computation of thermodynamic values in the case of Gibbs' theory. Moreover,

---

[27]See, for example, the discussion of "configurational entropy" in Bent [**1965**], pp. 179–181. There are many other examples in thermodynamic texts, monographs, papers, and popular books.

[28]Dickerson [**1969**], p. 206.

the point is made in Chapter 19 that the procedure used to coarse-grain the particle distribution does not commute with the total time derivative operator used in the system evolution equation. This means that Boltzmann's discrete measure is not a dynamically evolving physical quantity and is not compatible with the underlying mechanics.[29] Because of these problems, Boltzmann's argument for defining the entropy by means of coarse-grained distributions is rejected.

Crystalline states are often cited as low entropy states because they have the most (spatial) order. However, the expression of the macroscopic equilibrium entropy in terms of the energy, given in TIS as $S_k^\epsilon = k_B \beta_k H_k^\epsilon$ in equation (18.14) below, or in the standard version of the equilibrium entropy as $S_k^{(n)\epsilon} = k_B[\beta_k \mathcal{H}_k^\epsilon + \ln Z_k^\epsilon]$, means that these are the lowest entropy states because they are the lowest energy states. The spatial ordering is simply coincidental to that.[30]

The concept of order, as such, has not been shown to play any role in the classical description of the mechanical interaction between the particles. It may express symmetries in the spatial distribution of the forces between them, but there are no *a priori* symmetry requirements imposed on the spatial ordering of the particles or on the forces between them that appear in the classical Hamiltonian of a system. Since, on the mechanical level, the interparticle forces, particle masses, and particle kinetic energies are the only physical agents that determine the motion of particles or the outcome of a chemical reaction, the concept of order is superfluous. It may be at best a calculating tool and allow a shorthand method of computing the outcome, but it imposes no restriction on the underlying mechanical equations and adds nothing necessary to the mechanical determination of the outcome of a reaction.[31]

8.10 Dissipation of Energy

Except for popular accounts of the entropy, the concept of the dissipation of energy, introduced in 1852 by Thomson [**1852**], [**1891**], has not remained a serious contender as either an analog or explanation for entropy. In spite of the intuitive appeal of this point of view, it became clear rather early that while the dissipation of energy may be of crucial importance on the scale of human activities, it has no analog on the microscopic level, where energy moves from one place to another, but is not created or destroyed.

---

[29]The mathematical assumptions underlying the use of $f(q, p, t)$ are discussed in Chapters 22, 24 and 25. Various methods that have been used for coarse-graining a distribution are discussed in Chapter 23.

[30]The TIS definition of the entropy analog and the difference between the standard definition and the TIS definition in both the nonequilibrium and equilibrium cases are discussed further in Chapter 9.

[31]It is shown in QTS that the situation is the same with regard to order and entropy in quantum mechanics.

The mathematical connection between Thomson's dissipation of heat and the loss of available work was made by Clausius. Clausius [1856], pp. 245–249, showed that the loss of work in an irreversible cyclic process is $T_0 \oint dQ/T$, where T_0 is the temperature of the environment into which the heat is lost. Maxwell [1878a] (Maxwell [1890], p. 644) expressed this by stating (TIS notation) "the quantity of energy which is dissipated in a given process is equal to $T_0(S_2 - S_1)$, where S_1 is the entropy at the beginning, S_2 is the entropy at the end of the process and T_0 is the temperature of the system in its ultimate state, when no more work can be gotten out of it."[32] Because the mechanical energy obtained from a system can be stored and made use of, it is the heat dissipated rather than the energy dissipated that is the quantity of interest. The heat dissipated is the heat that leaves a system through a boundary, which, in TIS notation, is measured by the quantity $\Delta Q = \int_{t_0}^{t_1} dt\, Q_{\partial k}(t)$. The important issue is that of choosing the appropriate boundary on which to analyze the dissipation of heat.

In this same paper, Maxwell [1878a] (Maxwell [1890], p. 646) discussed issues concerned with the interpretation of the dissipation of energy

"It follows from this that the idea of dissipation of energy depends on the extent of our knowledge. Available energy is energy we can direct into any desired channel. Dissipated energy is energy which we cannot lay hold of and direct at pleasure, such as the energy of confused agitation of molecules which we call heat. Now, confusion, like the correlative term order, is not a property of material things in themselves, but only in relation to the mind which perceives them. A memorandum book does not, provided it is neatly written, appear confused to an illiterate person, or to the owner who understands it thoroughly, but to any other person able to read it appears to be inextricably confused. Similarly the notion of dissipated energy could not occur to a being who could turn any of the energies of nature to his own account, or to one who could trace the motion of every molecule and seize it at the right moment. It is only to a being in the intermediate stage, who can lay hold of some forms of energy while others elude his grasp, that energy appears to be passing inevitably from the available to the dissipated state."

---

[32]See Daub [1970] for a further discussion of these issues. For the case of heat flow without doing work, he showed on pp. 338–339 that the entropy change when an amount of heat ΔQ leaves a body at temperature T and is absorbed by its surroundings at temperature T_0, where $T \geq T_0$, is $\Delta S = \Delta Q \left(\frac{1}{T_0} - \frac{1}{T} \right)$. Looking ahead to equation (17.18), the theoretical maximum work that could have been obtained from ΔQ using a reversible engine operating between these temperatures is $\Delta W = \Delta Q \left(\frac{T - T_0}{T} \right) = T_0 \Delta S$. This last relation holds generally, so the notion of the dissipation of energy can be reduced to that of the increase in entropy.

This is the clearest and most explicit statement Maxwell made on the subject of our knowledge and the role it plays in the second law.

8.11 Mixing

A process in which two systems become mixed, and the "irreversibility" associated with it, has often been cited as a paradigmatic case of an entropy increasing process. Thomson [1873] and Maxwell [1878a], for example, both used examples of mixing in their discussion of the increase in entropy of a system made from two systems. Maxwell observed, along with Thomson, that it is extremely improbable that a quantity of gas containing a substantial number of oxygen and nitrogen molecules, say, would "unmix" into separate volumes of oxygen and nitrogen. For Maxwell, this illustrated the statistical nature of the second law.[33] As stated above, Maxwell [1878b] (Maxwell [1890], pp. 669–671) observed that if we could follow each molecule, there would be no distinction between work and heat and no second law. The second law in this view is a consequence of the enormous number of degrees of freedom in a molecular system and the minuteness of the molecules compared to human sizes. However, Maxwell subsequently had serious misgivings on this point.[34] It was also pointed out above that Gibbs had considered mixing in both his earlier work on the equilibrium of heterogeneous systems (discovering the Gibbs paradox in the process) and in his later book on statistical mechanics and thermodynamics.

There are two important reasons why the mixing of particles in two separate systems is not allowed in the TIS formalism. The first is that the concept of a system boundary breaks down in a situation in which the mixing of two systems is being considered and two systems become one system. When, for example, does the boundary between two systems disappear as the particles of the two systems are allowed to mix? If we imagine that a hole appears in the mathematical boundary surface between the systems, so that particles can leave or enter the volume that was occupied initially by one system, the two systems must be described microscopically as one system from that point on. If we do not imagine that a hole or holes will develop in the boundary between the systems, there is no clear time at which we can say the boundary disappears. In any case, the mathematical description of either of these approaches to removing a boundary during mixing is problematic. This does

---

[33]For Maxwell's views on mixing and entropy, see Maxwell [1878b], pp. 665–671. See also Brush [1976], pp. 587–593, and Klein [1970], pp. 74–77.

[34]The claim that entropy is a measure of mixing does not solve the problem of finding an objective, physically based, definition of entropy that does not depend on the state of our knowledge. Maxwell's [1878a], ([1890], pp. 645–646) counterexample to the claim that entropy is a measure of mixing was a process in which two gases are mixed that we think have indistinguishable molecules but later find that we can distinguish and separate them. See also the discussion in Brush [1976], pp. 592–593. For other objections to the association of mixing with entropy, see Lewis [1930], pp. 572–573.

not limit the generality of the theory because the two systems can always be treated as one system from the microscopic point of view and the subsequent amalgamation of the systems and mixing of the particles are then internal to that system. Internal boundaries may be defined within that system to reflect the initial situation and calculations can be made of quantities crossing this boundary. Internal boundaries such as these allow the passage of particles and do not meet the conditions required by the classical TIS boundary formalism for a boundary between independent systems.

The second, and most telling, argument against mixing comes again on the microscopic level. As far as the particles are concerned, our labeling of a collection of particles as two systems with a boundary between them rather than one system is a macroscopic matter. The particles, whatever their material identity, make no such distinctions. They simply interact mutually with those particles around them in a way appropriate to the kinds of particles they are. The boundaries we choose for the systems and the labels we attach to the particles, such as assigning them to one system or the other, have no effect on their mechanical behavior. Mixing, therefore, does not pass the Relevance Test proposed above as one of the determining factors in the mechanical evolution of the system.

8.12 Complexity and Chaos

The concept of "complexity" has been invoked by a number of workers to explain the difference between thermodynamic irreversibility and the reversibility of the underlying particle mechanics. This complexity is itself a consequence of the enormously large number of microscopic particles or degrees of freedom in a macroscopic system. Similar arguments, based on the concept of "integrability", have been put forth by chaos theorists.

Although complexity has lately become the explanation of choice by post chaos theorists, the physical mechanism by which it works its magic has never been elucidated. At just what point a system has become complex enough to exhibit irreversible behavior, and why this should be so, has not been discussed. Complexity shares with the concepts of order and disorder the attribute of residing in the eye of the beholder. Moreover, to meet the tests stated above, any claim that the complexity of a situation has dynamical consequences must show how these consequences emerge from the particular form in which this complexity is expressed.

Attempts to justify the thermodynamic limit by referring to complexity fail for these reasons. There is no physical reason for equating the behavior of a large number of particles to the behavior of an infinite number of them and attempting to justify this by referring to our ignorance of the details. The thermodynamic limit, for example, clearly changes the dynamics of a particle system in several ways. While taking the asymptotic limit of a physical quantity may simplify its calculation, these approximate limiting quantities lack dynamical significance. Moreover, when the evolution equation for a system is examined, there is nothing in it that indicates that a radically different behavior, expressed by a change in the evolution equation, will result at some point when the number of particles, or the number of degrees of freedom, grows large but remains finite while the particle density remains constant.

Chaos theory, with its bifurcating sets and fractal scaling, is sometimes offered as an explanation. But chaos theory in these settings is a macroscopic theory that is not concerned with the equations of motion of particles but with macroscopic phenomenological equations or with equations based on some arbitrary generating function, such as the Lorentz equation. Others have invoked "the instabilities of microscopic trajectories induced by chaotic dynamics" to obtain irreversibility.[35] Such statements concerning "instabilities of trajectories" in relation to "chaotic dynamics" are not aspects of the microscopic Hamiltonian flow of particles—although collections of these particle flows can exhibit bifurcations in chaotic regimes. This emergent chaotic behavior may show up in the flows of thermodynamic quantities, but it is simply a consequence of the Hamiltonian mechanics of the particles in that circumstance. It is not caused by some physical aspect or special feature that is added to the evolution equation describing a system of particles obeying reversible microscopic Hamiltonian particle mechanics. This means that neither chaos nor complexity can lead to irreversibility or an increase in the entropy. In short, complexity and chaos, whether based on the number of particles or some other measure of the number of degrees of freedom in a system, fail the three tests stated above.

8.13 Entropic Forces

Entropy has played a concrete role in the analysis of experiments on mixtures of colloidal particles.[36] Lars Onsager [1949] introduced these ideas. He discussed an isotropic fluid of long thin rods and argued that at a high enough concentration, this fluid must undergo a phase transition from a disordered to an ordered, or nematic phase, with a preferred orientation. The transition from a disordered phase to an ordered phase involves, according to standard theory, a loss of entropy. On the other hand, ordering the rods is supposed to lead to a gain in entropy due to the increase in the free volume of the ordered state. As the concentration of rods increases, Onsager maintained that this transition to order will occur because the loss of entropy associated with becoming oriented is more than compensated by the gain in entropy due to the increase in the free volume brought about by the ordering.

---

[35]See Lebowitz [1999], p. 588, for example.

[36]I will cite only selected recent articles from a series of articles on these topics in the literature. See the references cited in these articles for more details and earlier work. The work of Onsager is discussed in detail in EIS in connection with thermoelectricity.

Building on Onsager's work, S. Asakura and F. Oosawa [1958] predicted an attractive "depletion" or "excluded volume" force that depends on the ratio of the number of large spheres to small spheres in a colloidal solution and their relative sizes.

In the last ten years, there have been a number of experiments that involve observing the behavior of large particles in a mixture of small and large particles in a solution. Experiments performed using containers with grooves or a step in the wall, showed that the large particles in these mixtures became ordered in relation to the grooves or step. The experimenters interpreted this ordering as an effect of entropic forces acting on the large particles. In one such analysis, A. D. Dinsmore, A. G. Yodh and D. J. Pine [1996] prepared an aqueous solution containing large spherical particles ($\sim$ 500 nanometers in diameter), and small spherical particles ($\sim$ 83 nanometers), which are present in a ratio 1 to 3000, respectively.[37] They measured the probability that a large particle would jump between bins in a partition of the volume over a surface and used this to determine the free energy $A_k^{(fe)}$ of the system as a function of the position of the large sphere in one of the bins. They assumed that the probability of this transition is proportional to $e^{-A_i^{(fe)}/k_B T}$, where $A_i^{(fe)}$ is the free energy of the i-th bin. For comparison, the experimenters also computed the free energy using the *excluded volume theory* that is based on the fact that the large spheres exclude the centers of smaller spheres from a volume that is equal to the radius of the larger sphere plus the radius of the smaller sphere. This volume is greater than that of the large spheres alone, so the theory predicts that minimizing the configurational entropy will result in aligning or packing the large spheres to minimize this lost volume.

Dinsmore, Yodh and Pine, p. 240, quantified the force acting on a large sphere to move it into bin i as the gradient of the bin i free energy: $\mathcal{F}_i = -\nabla A_i^{(fe)}$. The experiment observed large spheres at a step boundary formed by the edge of a glass terrace in the solution. In the direction parallel to the step boundary, they found no forces. However, perpendicular to the step boundary, they found a free energy barrier about $2k_B T_k$ in magnitude. The force on the large particles due to this free energy barrier was computed to be about 0.04×10^{-12} newtons.

The explanation offered for these observations was that "A mixture of hard spheres maximizes its entropy by maximizing the volume accessible per particle." The theory predicts that attractive forces between the large spheres and between a large sphere and the wall can arise due to this configurational aspect of the entropy even when the particles are all hard spheres with repulsive interactions. Quantitatively, moving a large sphere from the bulk to the surface decreases the free energy of the mixture by an amount computed by the authors to be $3(r_L/r_S)\phi_S k_B T_k$, where r_L is the radius of the large sphere,

---

[37]See also the work in A. D. Dinsmore, D. T. Wong, P. Nelson and A. G. Yodh [1998].

r_S is the radius of the small sphere, $\phi_S = \mathcal{N}_S \pi r_S^3 / (6V)$, $\mathcal{N}_S$ is the number of small spheres, V is the volume of the mixture, and T is the temperature. Dinsmore, Yodh and Pine stated that this free energy gradient leads to a force of about 10^{-12} newtons pushing the large particles to the wall. This force is also called the *volume depletion force.*

In an article commenting on a similar experiment, H. N. W. Lekkerkerker and A. Stroobants [**1998**], p. 306, observed that computer experiments have shown that highly ordered smectic and crystalline phases in systems of rods can be produced with spheres that have purely repulsive interactions. They concluded that this "again leaves entropy as the driving force." They also stated that the novel structures formed by mixtures of rods and spheres in the experiments of M. Adams, Z. Dogic, S. L. Keller, and S. Fraden [**1998**] "provide further challenges to understanding the way in which entropy produces such remarkable ordering." Indeed, Adams, Dogic, Keller, and Fraden, p. 351, concluded their analysis of the behavior of rod-sphere mixtures by stating that "effective attractive potentials of surprising complexity can be constructed through the choice of the shape, size, and number of molecules in a suspension, indicating the possibility of manipulating these variables to 'engineer with entropy,' building order from disorder."

Let us review the ideas expressed in these articles and examine the theoretical basis proposed for entropic forces. In the proposal given above that the entropic force law is the gradient of the free energy, this gradient was interpreted as $F_i^{(S)\mu} = -\nabla_i^\mu A_i^{(\text{fe})}$, where i refers one "bin". This bin is part of a collection of bins formed by an arbitrary macroscopic partition of the volume over a surface. The definition of the free energy in the form $A_i^{(\text{fe})} = H_i - T_i S_i$ is used with the gradient above to express the entropic force in the form $F_i^{(S)\mu} = -[\nabla_i^\mu H_i - T_i \nabla_i^\mu S_i]$, where the temperature is assumed to be spatially uniform. To analyze this force further, it is necessary to know how the Hamiltonian energy H_i is defined and how to compute the gradient of the entropy S_i.

No formula was given that shows how H_i depends on the velocities and coordinates of the particles in the bin and on the macroscopic coordinates q that characterizes points in the bin. The interpretation of $\nabla_i H_i$ is therefore not obvious given that ∇_i is presumably a macroscopic bin gradient operator acting on the macroscopic coordinate q. Similar concerns hold for the entropy term. So that the argument can proceed, each bin will be interpreted as a system $k \in [t]$, where $[t]$ is the set of bins, the entropic force will be defined as the local macroscopic relation

$$(8.3) \qquad F_k^{(S)\mu}(q,t) = -\nabla_k^\mu \mathcal{H}_k(q,t) - T_k \nabla_k^\mu S_k(q,t).$$

The Hamiltonian and the entropy may be treated as step functions on the bins and the gradient operator as a difference operator between the bins. However, in spite of a concrete interpretation of the components of the theory, this approach breaks down because, as will be discussed in the next chapter, the entropy cannot be expressed in this way as a local function $S_k(q, t)$. Passing over this problem and considering entropy as a step function on the bins and the gradient of the entropy as a difference function, the next problem is that the entropy is not uniquely determined by the macroscopic information we have on the system. Further steps may be taken in this direction, but the upshot is that there is no suitable interpretation for the "entropic force" defined in (8.3) that connects it with the mechanics of the underlying particles.

While this theory of entropic forces is not precise enough to criticize in further detail, any theory that depends on the localization of the entropy will fail because the entropy in a general $\mathcal{N}$ particle theory, as opposed to Boltzmann's one-particle theory, is not associated with individual particles and cannot be expressed as a local particle sum function.

The accounts of entropic forces in other papers are even sketchier and are not amenable to precise analysis.

There are alternative explanations for the observed colloidal phenomena. From a terminological perspective, these phenomena are known in biological research as consequences of "macromolecular crowding". This term is much more felicitous from the TIS point of view than "entropic forces" because it focuses properly on the effects of crowding in producing the observed behavior. The effects of depletion forces at equilibrium, for example, can be understood in terms of minimizing the total system energy over the set of possible particle configurations with fixed boundary conditions of the kinds described in the experiments. Maximizing the entropy then produces the equilibrium relation between the entropy and the energy as $S_k = k_B \beta_k \mathcal{H}_k$, so the concept of entropy does not add anything further to the situation obtained by minimizing the energy. Furthermore, the concept of crowding and the fact that ordered structures in crowded situations, even when the particles have purely repulsive potentials, often have less energy tied up in the particle interaction potential energies than disordered ones. These facts can explain depletion forces and how crystals can form from particles that have purely repulsive interaction forces. In addition, there is an explanation of how the small particles enforce an order on the larger particles that is similar to one explanation in Descartes' time of how the force of gravity works. The attractive force of gravity was explained then as due to the fact that the large particles in an isotropic sea of small particles will shadow each other as they come closer together, so the number of small particles pushing each large particle is greater on the hemisphere opposite to the other large particle. This results in a net attractive force along the line joining the centers of the large particles.

8.14 The Second Law and Life

It has sometimes been claimed that life somehow evades the second law of thermodynamics. Living beings maintain their order and functionality in the face of heat and motion, which usually degrade and erode macroscopic structures.[38] Typical views in the 1930's on this subject were expressed in an exchange of letters to the journal *Nature* between F. G. Donnan, a professor of physical and inorganic chemistry, and the physicist James H. Jeans. Donnan [**1934**] challenged the view expressed by Jeans in his book *The New Background of Science* that organisms must have some means for evading the second law of thermodynamics. Jeans [**1934**] replied that a person steering a large steamship adds little entropy in the activities of steering, but the "positional entropy" of the cargo is changed considerably. This is due to moving the cargo from a location where it has a high density ν_1 to a location where it has a low density ν_2, thereby changing the positional entropy by $\Delta S = k_B \mathcal{N}[\ln \nu_1 - \ln \nu_2]$. In this case, the intervention of an intelligent being is incidental to the entropy calculation.

In spite of the entry of the physicist E. A. Guggenheim on Donnan's side, and the exchange of several more letters,[39] this controversy died without being resolved and without, apparently, leaving any further trace. Jeans' positional entropy is only one example of the types of entropy proposed during this period. Many other forms of the second law and the entropy proliferated during this time and there was no clear way or criteria to distinguish legitimate forms for these quantities, if there are such, from those that are not.

Later discussions of entropy and life were initiated by the publication of Schrödinger's [**1943**] book *What is Life?* in which he spoke of living organisms that "feed upon negative entropy" so that they can maintain their order and complexity against thermodynamic degradation and dissipation. In spite of sporadic exchanges on these matters between the 1930's and 1990's, this topic has not received much attention or led to new insights in either physics or biology.

From the TIS standpoint, the analysis of life and evolution in terms of entropy is both misleading and fruitless. Animals require energy to maintain life, their internal organization, and operations. Consuming food to obtain this energy implies that an animal cannot be isolated. The assumption that life or living things must violate the second law of thermodynamics, and require the consumption of negative entropy to maintain themselves, depends on the notion of entropy in terms of order, complexity, or something else. There is no explanation of why this presumed consumption of negative entropy is required, over and above the consumption of food to maintain activity and repair structure. It follows that these statements concerning entropy and life are not supportable and not valid.

---

[38] For more information on these ideas in the larger perspective of entropy and evolution, see Bailyn [**1994**], pp. 171–178, and the recent references given there.

[39] See the sequence of letters listed in the citations F. G. Donnan and E. A. Guggenheim [**1934**] and Jeans [**1934**].

8.15 The Second Law Revisited

The statements of the second law presented above and the particle models associated with them do appeal to our sense of how the world usually operates. We cannot construct perpetual motion machines. We never observe macroscopic quantities of a gas spontaneously unmixing into its components or heat flowing from a cold body to a hot one, and so on. These observations give a strong experiential base to the formulations of the second law based on mixing and heat flow. However, as Maxwell, Boltzmann, and Gibbs were well aware, this does not constitute a proof that these processes must operate in this way. The expression of the second law in terms of heat flow between bodies of different temperatures, for example, is thus an empirical generalization that is not a necessary consequence of the particle dynamics. Moreover, as Maxwell stated several times, situations can be imagined using the particle model in which each version of the second law is violated. On the other hand, we cannot arrange a measurable violation in a large collection of particles simply by manipulating macroscopic boundary conditions.

The statements made in this chapter concerning the entropy will be supported by the definition of the entropy in Chapter 9 and the analysis of its behavior. The resulting entropy is not a property of the particles, and is therefore not a particle sum function, but is based on the collection of particles in a system taken as a whole. The TIS version of the second law, Axiom 6, is a macroscopic law, but not a statistical law in the sense that Maxwell meant. For Maxwell, the second law is sometimes violated by the behavior of the underlying particles. He called the second law a "statistical law" because the entropy computed from it will sometimes decrease. Nevertheless, this statement is not a mathematical statement based on a computation and is not related to theorems of statistics.[40]

The macroscopic notion of irreversibility is quite different from the microscopic notion. Reversing a macroscopic process means bringing a system back along a macroscopic path to a state that is macroscopically indistinguishable from the original state. Reversing the microscopic particle paths is only one way to do this. The original microscopic system state forms a subset of Lebesgue measure zero in the set of all microscopic states that cannot be distinguished from each other by macroscopic means. Thus macroscopic irreversibility does not imply microscopic irreversibility. Because these ideas are different, a macroscopic concept called *unreturnability* will be introduced in the discussion of the second law of thermodynamics in EIS, Chapter 11, to replace the misleading aspects of the term irreversibility in reference to the evolution of macroscopic bodies in time. The old thermodynamic concept of *unrecoverability* will be reintroduced and discussed there as well.

---

[40] This issue will be pursued in EIS.

CHAPTER 9

Entropy

The relationships between the macroscopic second law of thermodynamics, the macroscopic entropy, and their interpretations in terms of particle models were explored in the last chapter. The attempts at interpreting the entropy in terms of these models and as a driving force shaping future particle distributions were rejected as inadequate to the task. From the TIS perspective, the entropy is a thermodynamic quantity that reflects a property of a collection of particles as a whole and does not play a role in the particle evolution equations.

177

9.1 Basic Issues

The ongoing difficulties in interpreting the entropy have led some to conclude that it can be properly defined only in quantum mechanics. A. Wehrl [1978], p. 222, for example, stated a viewpoint on entropy that is currently considered obvious by many physicists: "Of course, a correct definition is only possible in the framework of quantum mechanics, whereas in classical mechanics entropy can only be introduced in a somewhat limited and artificial manner." The quantum mechanical definition he had in mind is a discrete one based on the work of von Neumann [1932c]. By "limited and artificial manner" Wehrl meant that the classical distribution must be coarse-grained so that the associated classical entropy will fit the discrete quantum model.[1]

Two definitions of the entropy are widely used in quantum mechanics. The first is contained in Planck's expression for the entropy in terms of a discrete combinatorial model of a particle system. Planck stated that this conception of the entropy was the foundation for his groundbreaking work on the spectrum of blackbody radiation in 1900. His [1913] retrospective justification for using this form of the entropy in his work was based on Boltzmann's [1877b] paper proposing a connection between entropy and probability.

The second definition was presented in von Neumann's [1927], [1932c], codification of the formalism of quantum mechanics. Von Neumann extended the quantum formalism he was presenting to include a quantum thermodynamics that was based on results obtained from an adaptation of the procedures of macroscopic phenomenological thermodynamics. Von Neumann defined his entropy in terms of the representation of a quantum system in terms of its statistical matrix or density matrix. In the context of the measurement of a particular physical quantity, the statistical matrix is expressed in terms of the set of eigenfunctions representing the possible outcomes for this kind of measurement. An important feature of von Neumann's entropy is that it changes only during a measurement and not during the evolution of the quantum system in accord with Schrödinger's equation.

One aspect of the entropy that has created significant problems for those using it in thermodynamic formulas is a lack of an absolute definition for it. An absolute definition fixes its value at some thermodynamic state and allows the entropy of other states to be computed using the thermodynamic calculations for entropy differences. It is true that in most applications of the thermodynamic formalism it is the entropy change dS or ΔS during a processes, rather than the absolute entropy S itself, that is needed. However, the formula for the change in the free energy, $dF = dU - T dS - S dT$, requires the knowledge of the value of S and not just dS as long as the temperature change dT does not vanish. This problem was recognized by Hermann von

---

[1]The concept of entropy in quantum mechanics, and the details of Wehrl's review article, will be discussed in QTS.

Helmholtz [1882] and by Henri Louis Le Chatelier in 1888.[2] The need for S and not just dS implies that calculations with the free energy contain an irremovable ambiguity due to an unknown constant in the definition of the entropy. In the early twentieth century, this issue stimulated a search, primarily by Planck and Walter Nernst, for a definition of the absolute entropy in terms of other physical quantities that does not contain an undetermined constant.

The claim that the entropy cannot be defined properly in classical physics rests on a misunderstanding concerning the definitions of the quantum entropy. This misunderstanding stems from the beliefs that the treatment of the entropy by Planck or von Neumann is adequate for thermodynamics and that the use of discrete or eigenfunction methods is necessary to obtain proper results in calculations involving the entropy. To begin with, there are significant differences between the versions of the entropy presented by Planck and von Neumann. This is shown in QTS by a detailed comparative analysis of the different classical and quantum definitions of the entropy by Boltzmann, Gibbs, Planck, von Neumann, Einstein, and others. More to the point here, however, the work in this chapter shows that the second of these beliefs is incorrect and that discrete methods are not required to define the entropy.

After a review of the major classical analogs that have been proposed for the entropy, a TIS analog is introduced that is suitable for a general thermodynamics that includes both nonequilibrium and equilibrium thermodynamics. The formal TIS definition of the entropy analog presented in this chapter is essentially an extension and refinement of the formalism originally proposed by Boltzmann, Gibbs, and Planck. However, the TIS definition of the entropy is made within the TIS framework and this imposes several further requirements on it and its interpretation. In particular, there is a concern with the proper definition of entropy in the nonequilibrium domain that satisfies the requirements on the entropy when it is specialized to equilibrium theory.

Many versions and definitions of the entropy have been introduced over the years in the contexts of both classical and quantum mechanics.[3] Because the entropy is one of the central aspects of thermodynamics investigated in this series, each important definition by an author will be labeled for future reference in other volumes. The first version of Gibbs' entropy, for example, is designated by $S_k^{(G1)}$. The purpose of the labeling is conceptual, rather than historical, so these labeled versions may not reflect the historical order in which they were introduced.

---

[2]See the discussion of this in Bailyn [1994], pp. 248, 346–348.

[3]The detailed survey of thermodynamics by Bailyn [1994] will be used as one point of reference to other author's work.

The primary goal of the work in this chapter is to find a suitable analog for the general spacetime version of the TIS entropy and work out its properties. To assist in this effort, some of the classical versions of the entropy by various authors will be introduced and evaluated as candidates. A classical entropy analog suitable for TIS is identified and used with the TIS boundary formalism in a proof of Clausius' inequality.

9.2 Boltzmann's Entropy Analogs

Boltzmann [1872] presented an analog for the entropy in the form of the logarithm of the one-particle probability distribution density for the system. This introduced an element into physics that was foreign to previous mechanical thought. In making the entropy dependent on a one-particle probability distribution rather than a quantity possessed by the particle itself, Boltzmann, whether he was aware of it or not, took a step away from the materialist worldview. Some of the consequences of this step will emerge later in the discussion of the reconciliation of thermodynamic irreversibility with the reversibility of the underlying particle mechanics.

The formal expression of Boltzmann's 1872 entropy analog is expressed in terms of the normalized one-particle probability distribution $f^{(n)}(q,p,t)$, in TIS notation, by[4]

$$(9.1a) \qquad S^{(B2)}(q,p,t) = -k_B \ln f^{(n)}(q,p,t).$$

For consistency with the other work below, and at the cost of some historical accuracy in writing the formulas, Boltzmann's entropy has been expressed using Planck's formulation, which includes the Boltzmann constant k_B that Boltzmann never used. Boltzmann designated the phase average of this analog by H, which is written here as $H(t)$ and defined by

$$(9.1b) \qquad H(t) = S^{(B2)}(t) = \int d^3q \int d^3p\, f^{(n)}(q,p,t) S^{(B2)}(q,p,t).$$

The phase averaging procedure that Boltzmann used is extended over all phase space. He showed in his H theorem that the rate of change of $H(t)$ is never negative (TIS sign convention) and vanishes only when the one-particle probability distribution density is the equilibrium Maxwell-Boltzmann probability distribution density. Boltzmann identified this non-decreasing quantity as the thermodynamic entropy.

Joseph Loschmidt [1876] challenged the irreversibility of Boltzmann's equation as established in the H theorem. He pointed out the contradiction between the irreversible behavior of a system implied by the H theorem with the reversibility of the equations of motion of the underlying particle system. In the light of these criticisms, Boltzmann began changing his conception of what the H theorem proves and repositioning his statements about it.

---

[4]Both the 1868 and 1872 versions of Boltzmann's entropy are analyzed in Chapter 22.

One of Boltzmann's responses to the challenge of Loschmidt was to generalize his approach to the entropy in Boltzmann [**1877b**]. In this approach, he represented the state of a system k, with $\mathcal{N}_k$ particles and a fixed total energy E_k, by a discrete coarse-grained combinatorial measure $\mathfrak{P}_k$, which he called the *permutability*. It represents the number of "microscopic complexions" or distributions of particles into phase cells, obtained by a partition of the system phase space, for a system with the fixed total energy E_k. He defined the probability that the system has a particular coarse-grained distribution of particles into cells as the number of arrangements of particles with this configuration having the total energy E_k divided by the number $\mathfrak{N}_k$ of arrangements in all possible configurations at this energy. This statement of a connection between entropy and probability is often called *Boltzmann's principle*.

Planck in 1901 gave formal expression to Boltzmann's principle in connection with his work on the blackbody radiation spectrum. He ignored the division of $\mathfrak{P}_k$ by $\mathfrak{N}_k$ and called the permutability, which he denoted by W, the "thermodynamic probability". In these terms, Planck's first version of Boltzmann's entropy took the form $S_k^{(\mathrm{P1})} = \mathrm{k}_B \ln \mathfrak{P}_k + \mathrm{const}$. Soon after he had expressed the entropy in this form, Planck made a significant change and wrote it as $S_k^{(\mathrm{P2})} = \mathrm{k}_B \ln \mathfrak{P}_k$ without the arbitrary constant—thereby providing a definition for an *absolute entropy*. Almost immediately after Planck's work, a number of authors, including Walter Nernst, O. Sakur, H. Tetrode, Otto Stern, Planck, Einstein, and others attempted to compute the absolute entropy. Planck's definition of the absolute entropy and Nernst's work will be taken up in Chapter 14 below.

9.3 Gibbs' Entropy Analogs

A more general entropy analog was presented by Gibbs [**1902**]. This analog, which he called the *index of probability*, was based on the $\mathcal{N}_k$ particle probability distribution density. It is expressed in terms of the normalized system probability distribution density $F_k^{(\mathrm{n})}$ in the form (TIS notation)

$$(9.2) \qquad S_k^{(\mathrm{G1})}(Q_k, P_k, t) = -\mathrm{k}_B \ln F_k^{(\mathrm{n})}(Q_k, P_k, t).$$

Gibbs went on to show for an isolated system t that the rate of change of the phase average of this entropy analog vanishes. To show this using the TIS formalism, assume the system t is isolated and that F_t vanishes when $Q_t \in \partial d_t(t)$. In other words, the system boundary is chosen outside the set of points at which the system particles are located. Next, the total time derivative of the Gibbs' entropy analog $S_t^{(\mathrm{G1})}(Q_t, P_t, t)$ is computed using (5.16) with the result

$$(9.3) \qquad \frac{dS_t^{(\mathrm{G1})}(Q_t, P_t, t)}{dt} = -\frac{\mathrm{k}_B}{F_t(Q_t, P_t, t)} \frac{dF_t(Q_t, P_t, t)}{dt} = 0.$$

By (5.43), (9.3), and (5.31), the time derivative of $S_t^{(G1)}(t)$ is

$$(9.4) \qquad \frac{dS_t^{(G1)}(t)}{dt} = \left\langle \frac{dS_t^{(G1)}(Q_t, P_t, t)}{dt} \right\rangle_{tt} + \left\langle \dot{\chi}_{\Omega_t(t)} S_t^{(G1)}(Q_t, P_t, t) \right\rangle_{tt} = 0,$$

which is the result Gibbs obtained in his computation.

The standard result that the entropy has a unique maximum when the system is in an equilibrium state $F_t^\epsilon(Q_t, P_t)$ will be demonstrated in Chapter 18. The result (9.4) implies that the initial system distribution $F_t(Q_t, P_t, 0)$ for an isolated system will not approach $F_t^\epsilon(Q_t, P_t)$ as time goes on unless $F_t(Q_t, P_t, 0) = F_t^\epsilon(Q_t, P_t)$ to begin with.

This result posed a severe problem for Gibbs who believed, as did everyone else, that an isolated system should go to equilibrium and that this should be reflected in the entropy.[5] Gibbs therefore rejected $-k_B \ln F_t^{(n)}(Q_t, P_t, t)$ as the proper system entropy analog and replaced the t probability distribution density $F_t^{(n)}(Q_t, P_t, t)$ by a coarse-grained version of the system distribution in the definition of the entropy analog. It is not hard to show, because of the fact that information is discarded when the system probability distribution is coarse-grained, that the entropy associated with the resulting coarse-grained distribution always increases to a maximum.

Gibbs [**1902**] also defined two other entropy analogs that represent the entropy of systems described by the microcanonical state. These analogs were based on the phase volume accessible to a system with energy less than or equal to E_k. Gibbs, p. 87, defined the quantity that he called the 'extension-in-phase', which is called the *accessible phase volume* in TIS and expressed in TIS notation using the system projection operator that Gibbs did not use, by

$$(9.5) \qquad \delta_k^{(G)\epsilon}(E_k) = \int dQ_k \int dP_k \, \chi_{\Omega_k(t)}(Q_k, P_k)\theta(E_k - \mathcal{H}_k^\epsilon(Q_k, P_k)),$$

where $H_k^\epsilon(Q_k, P_k)$ is the equilibrium Hamiltonian function defined in Chapter 13.[6] Gibbs used this quantity to define, pp. 88, 128, 170–171, two equilibrium entropy analogs that are appropriate for a system described by the microcanonical probability distribution in the form

$$(9.6a) \qquad\qquad S_k^{(G2)\epsilon}(E_k) = k_B \ln \delta_k^{(G)\epsilon}(E_k),$$

$$(9.6b) \qquad\qquad S_k^{(G3)\epsilon}(E_k) = k_B \ln \left[\frac{d\delta_k^{(G)\epsilon}(E_k)}{dE_k} \right],$$

---

[5] Bailyn [**1994**], pp. 525–530, has presented a brief but thoughtful discussion of how (9.4) follows from the Liouville theorem, which states that the phase volume of a set is preserved under evolution in time in accord with Hamilton's equations.

[6] The nonequilibrium generalizations of the equilibrium accessible phase volume and its derivative with respect to the energy play an important role in EIS.

where Boltzmann's constant has been included for dimensional reasons and consistency with other TIS formulas.

Gibbs [1902], p. 183, stated that these analogs are asymptotically the same as the number of particles increases. This was soon echoed by others, such as Planck [1904], who presented a proof that the analogs (9.6) and Boltzmann's 1877 version of the entropy all give the same asymptotic results when $\mathcal{N}_k \gg 1$. Planck concluded that Boltzmann's 1877 approach is superior because it refers only to the complexions or particle arrangements of the actual system while Gibbs' formalism requires reference to the whole imaginary ensemble. However, this difference is only apparent and vanishes when one compares the probability space underlying Boltzmann's complexions and Gibbs' coarse-grained ensembles and sees that they are the same. Similarly, Planck's statement that Boltzmann's theory is valid in the general case while Gibbs' theory is only valid for equilibrium states is also seen to be incorrect because Gibbs' did define a nonequilibrium version of the entropy, which is expressed in TIS notation in (9.2).[7]

An extension of Gibbs' work was made by Bailyn [1994], pp. 501–503. He computed a version of the entropy associated with the energy shell at E_k with the spread ΔE_k as (TIS notation)[8]

$$(9.7) \qquad S_k^{(\mathrm{Ba})}(E_k, t) = \mathrm{k}_B \ln[\Delta \delta_k^{(\mathrm{G})}(E_k, t)] = S_k^{(\mathrm{G2})}(E_k, t) + \ln[\Delta E_k],$$

where, with the addition of the system projection operator as before, the nonequilibrium extension of Gibbs' accessible phase volume is written

$$(9.8\mathrm{a}) \quad \delta_k^{(\mathrm{G})}(E_k, t) = \int dQ_k \int dP_k\, \chi_{\Omega_k(t)}(Q_k, P_k)\theta(E_k - \mathcal{H}_k(Q_k, P_k, t)),$$

the phase volume associated with the energy spread ΔE_k is $\Delta \delta_k^{(\mathrm{G})}(E_k, t) = (\partial \delta_k^{(\mathrm{G})}(E_k, t)/\partial E_k)\Delta E_k$, and the extension of Gibbs' entropy to the general case that is associated with this is

$$(9.8\mathrm{b}) \qquad S_k^{(\mathrm{G3})}(E_k, t) = \mathrm{k}_B \ln\left[\frac{\partial \delta_k^{(\mathrm{G})}(E_k, t)}{\partial E_k}\right].$$

The entropy $S_k^{(\mathrm{Ba})}(E_k, t)$ suffers from the problem mentioned above that $\Delta \delta_k^{(\mathrm{G})}(E_k, t)$ is arbitrary because ΔE_k is arbitrary.

---

[7]For other discussions of Gibbs' analogs (9.6) and their relation to Boltzmann's 1877 conception of the relation between entropy and probability, see Haas [1936a], [1936b], [1936c], Epstein [1931], and Klein [1963], pp. 89–96. See also the detailed analysis of the connection claimed between entropy and probability in QTS, Part III.

[8]To fit in with the TIS approach and future discussions, Bailyn's version of this entropy calculation has been specialized to associate the variable y used in his Hamiltonian with the time t.

Gibbs second and third equilibrium entropy analogs are concerned with isolated systems at a definite energy and based on the microcanonical assumption that each particle configuration with the system energy has the same probability. This assumption is too restrictive to represent general $\mathcal{N}_k$ particle states and Gibbs' entropy analogs are not generalizable to the nonequilibrium theory, so they will not be used in TIS. They will be discussed again briefly in EIS. This method of defining the entropy will also be examined critically in QTS and compared to other approaches. The inclusion of an arbitrary and indeterminate factor, such as ΔE_k in Bailyn's extension of Gibbs' ideas, is found to be unacceptable there.

Gibbs did not pursue the nonequilibrium aspects of his formalism. He moved quickly to the equilibrium case and the canonical distribution that represents it in his formalism. In this case, the index of probability is expressed as a function of the energy in the "simplest way possible" as a linear function.

9.4 Other Aspects of Entropy

Bailyn went on to discuss the coarse-grained entropy theory, which leads to the result $dS_k/dt > 0$ for all nonequilibrium states, and predicts the evolution of the system to equilibrium as expected in macroscopic thermodynamics. In this discussion, Bailyn, pp. 528–529, presented the objection of Terrell Hill [**1958**], p. 95, that the shape of the domains that are the basis of the coarse-graining has an effect on the values computed. Hill observed that the arbitrariness in the choice of the shape of the domain at any given time and using that shape in the ensuing calculation is "tantamount to saying that disorder lies in the eye of the beholder and has no objective significance." Hill [**1958**], p. 95, repeated this criticism and stated that the coarse-grained entropy depends in a "critical and artificial way" on the shape of the cells. However, Bailyn rejected that point of view and felt that some special choice of a "simple" shape could save the version of the coarse-grained theory presented by Richard Tolman [**1938**].[9]

While coarse-graining a distribution and then taking the limit as the size of the domains tends to zero can reproduce the results of a calculation with a continuous function, this procedure cannot be used when a time derivative is required. In formal terms, the coarse-graining of a distribution does not commute with the time derivative operation. This is, of course, why a

---

[9]Hill's objection concerning the arbitrariness of the elementary domains in reference to Boltzmann's entropy was raised first by O. Sakur in 1911 who felt that it left a logical gap in the sequence of consequences leading from kinetic theory to the equation of state of a gas. This issue of the arbitrariness of the size and shape of the domains in coordinate space or phase space in the definition of the entropy is similar to the objection discussed above concerning Gibbs' definition of the microcanonical state in terms of an energy spread ΔE_t. Discrete distributions, their entropies, and the controversy concerning elementary domains are discussed for both the classical and quantum cases in QTS.

coarse-grained distribution gives a different result for its time derivative than the continuous version of that distribution. Moreover, it was maintained in the last chapter that there is no physical reason for coarse-graining the system distribution apart from our thermodynamic preconvictions about how systems should evolve. Finally, the arguments concerning the evolution of systems based on Boltzmann's H theorem and Gibbs' notion of mixing presented in Bailyn [1994], pp. 525–530, are not mathematical consequences of the theory and are therefore unconvincing.

From the TIS point of view, the increase in entropy in time obtained by coarse-graining the system distribution is a mathematical artifact of the coarse-graining procedure and invalidates the spacetime approach. For this reason and others mentioned above, coarse-grained distributions will not be used in TIS. It should be kept in mind that the formalism for representing completed observations, introduced in Chapter 3, reflects the limits in the resolution of a macroscopic instrument used for observation, and is unrelated to coarse-graining the system distribution.

Another postulated aspect of the entropy that plays an important role in many treatments of the standard theory is an entropy balance equation. Clausius' had provided a balance equation for the entropy change during a reversible process that took the form $dS = dQ/T$. A more ambitious entropy balance equation for a time-dependent entropy was introduced in the work of Lars Onsager [1931]. Onsager included an entropy current term at the boundary and a quantity representing the rate of internal entropy creation in a system. In TIS notation, Onsager's balance equation is written

$$(9.9) \qquad \frac{dS_k(t)}{dt} = \frac{\partial S_k(t)}{\partial t} + \oint_{\partial k} d^2r \, n_{\partial k,\mu}(r) S_k^\mu(r,t) = \int_k d^3q \, \sigma_k(q,t),$$

where $S_k(t)$ is the entropy, $S_k^\mu(q,t)$ is the local entropy current, and $\sigma_k(q,t)$ is the local rate of internal creation of the entropy.[10] The integration over ∂k is over the surface of the body and the integration over k is over the whole system volume. This equation is one example of an extensive body of theoretical work that has been developed out of Onsager's notions of thermodynamic fluxes and forces.

Attempts to define the entropy have continued unabated. One such attempt tries to define thermodynamics and the entropy based on the assumption that the particles can be described by a state that is a collection of

---

[10]This version is a modification of an equation in Bailyn [1994], pp. 168–171, 306. See Bailyn for a further discussion of entropy balance equations in standard thermodynamics. Onsager's work is studied and critiqued in the context of thermoelectricity in EIS. For more detailed discussions of entropy balance equations and the internal and external generation of entropy, see Wolfgang Yourgrau, Alwyn van der Merwe and Gough Raw [1966], Chapter 1, and a number of other texts.

components that are in Local Thermodynamic Equilibrium (LTE).[11] A version of thermodynamics based on a mixture of microscopic and macroscopic physical variables, called Irreversible Thermodynamics (IT), has also occupied a number of workers for some time.[12] Some of the recent work on defining the entropy has gone further a field and introduced concepts of chaos theory, contractions of the phase space due to the loss of particles from a system, the inclusion of a velocity dependent "dissipation" as a modification of the particle dynamics, and defined the production of "irreversible entropy" in terms of "losses along the information dimension of the stable manifold associated with a chaotic saddle."

One example of the use of ideas from chaos theory in the definition of the entropy appears in the paper Wolfgang Breymann, Tamás Tél, and Jürgen Vollmer [**1996**]. They were concerned with using the ideas connecting chaos theory and entropy stated above to extend to open dynamical systems their previous work on the entropy production in closed system. They based their work on what they called the "single particle picture" for low dimensional open chaotic systems in an external field and assumed a velocity dependent energy dissipation to "ensure energy conservation". Because particles are allowed to escape from an open system in their approach, the phase space of the system shrinks. The authors assumed that this shrinkage is proportional to $e^{-\sigma(x)t}$, where $\sigma(x)$ is a smooth function of the coordinates. The rate of change of the entropy was then deemed to be $\dot{s} = -\bar{\sigma}$, which is the average rate of shrinkage of the phase space. Breymann, Tél, and Vollmer computed the Lyapunov exponents associated with both the shrinkage and the dissipation and then, p. 2947, associated the average of these Lyapunov exponents with the asymptotic rate of change of the irreversible component of the entropy: "$s_{\mathrm{irr}} = -\dot{s} = \bar{\sigma} = -\sum_i \bar{\lambda}_i + \kappa > 0$". In conjunction with the Kolmogorov-Sinai (KS) entropy, they, p. 2947, claimed that "irreversible specific entropy production is proportional to the deviation from unity of the partial information dimension of the stable manifold." Finally, they considered the irreversible increase in entropy due to the loss of information concerning the system as being due to the finite resolving power of the observer and the consequent coarse-graining of the system.[13]

It is difficult to associate the concepts presented in this paper with the underlying particle dynamics. Several assumptions are made concerning rates of change of various quantities and their asymptotic properties that are not justified by further argument or analysis. The association of an increase in entropy with the "shrinkage" of the phase space is not physically motivated and the division of the entropy change into that due to particle escape and

---

[11]A number of theorems in Bailyn [**1994**] depend on the assumption of LTE.

[12]See Chapter 12 for a discussion of attempts to define the temperature using an extended version of irreversible thermodynamics designated here by EIT.

[13]For more information, see Breymann, Tél, and Vollmer [**1996**] and the references cited there.

that due to "dissipation" is also simply an assumption without physical basis in the particle mechanics. Breymann, Tél, and Vollmer, p. 2948, stated simply that "The essential effect leading to entropy production is the contraction of the phase space volume in an ever refining fractal manner." This is the justification for their earlier conclusion, p. 2495, that "The fractal structure of the underlying invariant chaotic set is connected in both open and closed systems with the entropy production by the fact that it contains an infinite amount of information on arbitrarily fine scales that cannot be extracted through any observation."

The major problems with this theory, and many others in this genre, are the lack of grounding of the concepts introduced in particle mechanics and the unjustifiable mixing of microscopic and macroscopic concepts. Chaos theory is a macroscopic theory concerned with collections of particle orbits. It is usually based on some assumed, hopefully plausible, generating function that is not derived from particle mechanics. In addition, Breymann, Tél, and Vollmer bring in formulas from many different approaches to defining the entropy and try to show that these are all encompassed by their work, but some of these approaches seem to be based on assumptions or viewpoints that are incompatible with their own or are out of place in their approach. The result on diffusion, for example, is brought in from their discussion of Fokker-Plank theory in another paper, and it is stated that $\dot{s}_{\mathrm{irr}} = \kappa = j^2/(2D)$, where j is the particle current and D is the coefficient of diffusion of Fokker-Planck theory. An increasing entropy based on the Fokker-Planck theory has been constructed by a number of authors, but these have not been based on a current of escaping particles. Finally, the use of information theory with physics must be done carefully or nonsense results. The loss of information in a physical situation is a macroscopic issue concerning observers and experiments that does not affect the microscopic physical equations or the manifold of physical possibilities.

The TIS approach to the entropy in relation to classical mechanics avoids these problems by maintaining a close connection to both classical Hamiltonian statistical mechanics and the macroscopic functions in thermodynamics that represent physical quantities obtained from the underlying mechanics. No assumptions concerning rates of change of Lyapunov exponents or their asymptotic values are needed, nor are any assumptions concerning local thermodynamic equilibrium required. All aspects of the fundamental TIS theory either follow directly from the particle mechanics without approximations or are based on clear approximations that meet specific epistemological goals such as the independence and quasilocality assumptions. Approximations introduced for specific purposes, such as those associated with phase boundary theory, for example, are introduced as theoretical approximations and not as assumptions concerning how the underlying particle mechanics will function.

The TIS system entropy for the general theory introduced below is identified with the microscopic particle description in terms that would be immediately recognizable to Maxwell, Boltzmann, and Gibbs. The difference between the TIS approach and the previous work lies in part in the status of the entropy and its relation to information. The TIS version of the entropy analog is not a particle sum function and cannot be given a local interpretation. This implies that there is no local internal entropy current $S_k^\mu(q, t)$ and no local internal entropy production $\sigma_k(q, t)$. These statements are supported by the calculation below of the time derivative of the entropy that shows that any change in the entropy is due to entirely to individual particle encounters with the system boundary.

The TIS thermodynamic entropy, which is the phase average of the TIS entropy analog, differs in some important ways from the standard definitions. It is not based on a normalized probability distribution density $F_k^{(n)}(Q_k, P_k, t)$ but on the system distribution density $F_k(Q_k, P_k, t)$. Because the TIS analog is not a particle sum function, the thermodynamic entropy obtained from it is therefore not extensive. The dependence of the resulting general macroscopic entropy function on macroscopic arguments is indirect and due to the inclusion of the system projection operator in phase averaging calculations. The question of the homogeneity of its dependence on its macroscopic arguments does not come up in this context. By design, on the other hand, the independent systems assumption insures that thermodynamic entropy of a collection of systems is the sum of the entropies for each system.

9.5 The TIS Entropy Analog

The TIS classical entropy is defined in the same way as Planck's version of Gibbs' entropy except that the system probability distribution density is not normalized. The k *phase entropy* is defined by

$$(9.10) \qquad S_k(Q_k, P_k, t) = -k_B \ln F_k(Q_k, P_k, t).$$

The important aspect of this definition is that the logarithm is a convex function. Paul and Tatiana Ehrenfest pointed out that any convex function of F_k, such as $[F_k]^n$, will do in place of $\ln F_k$ in the proof of Clausius' inequality.[14] However, the logarithmic form has the virtue that the phase entropy of the total system is the sum of the phase entropies of its component systems when the total system distribution factors into a product of the component distributions. This is the case when the Independence Approximation is used to approximate F_t by $F_{[t]}$. Similarly, within a system, the system phase entropy is a particle sum function if and only if the system distribution factors into a product of single particle distributions. That is the case for the TIS version of Boltzmann's equation which is presented in Part III.

---

[14]See P. and T. Ehrenfest [**1912**], pp. 51. See also the function $\chi^2(F_k, F_k^\epsilon)$ defined in Section 26.2 below.

Because the system phase entropy depends on the system probability distribution density in this formulation, it is clearly not a mechanical quantity. In addition, Boltzmann's constant k_B establishes a specific relation between the temperature scale and the energy scale, and so depends on both of these scales. Another significant consequence of the definition (9.10) is that the entropy is undefined for a system described by an exact state $F_k^\delta(Q_k, P_k, t)$. This fact will be used in several places below as part of the reconciliation of the differences between the claims of irreversibility in the thermodynamic description of a system and the reversibility in the mechanical description of its underlying particle system.

There are some mathematical aspects of this analog for the entropy that will be important in future work. First, it will be assumed for now, and demonstrated in Part IV, that integrals involving the entropy analog (9.10) are well defined when they are used. Second, the logarithmic form for the entropy presented in (9.10) is not the most general logarithmic form. In the relativistic version of the theory, which employs a group of transformations wider than the Galilean transformations used in the classical case (for which the Jacobian is always 1), the entropy analog can be generalized to $-k_B \ln \mathfrak{F}_k(Q_k, P_k, t)$, where $\mathfrak{F}_k$ is both a distribution density and a tensor density in the sense of Schouten [**1954**], p. 12. This is justified on dimensional grounds and is required to give the argument of the logarithm the proper behavior under transformations.

In the independent systems approximation, factorization is explicitly introduced. The approximation of F_t by $F_{[t]}$ leads by (6.2) and (9.10) to

$$(9.11) \qquad S_{[t]}(Q_t, P_t, t) = \sum_{k \in [t]} S_k(Q_k, P_k, t).$$

The phase average of $S_k(Q_k, P_k, t)$ is written

$$(9.12) \qquad S_k(t) = \langle S_k(Q_k, P_k, t) \rangle_{kt}.$$

Phase averaging both sides of (9.11) by $\langle \cdot \rangle_{[t]t}$, and using the fact that this phase average is normalized by $Z_{[t]}(t) = \prod_{k \in [t]} Z_k(t)$, so that the expectation function is computed with $F_{[t]}^{(n)}$, yields

$$(9.13) \qquad S_{[t]}(t) = \sum_{k \in [t]} S_k(t)$$

in accord with Axiom 6.3.

Differentiating $S_k(Q_k, P_k, t)$ by d/dt, (5.16) is used to show

$$(9.14) \qquad \frac{dS_k(Q_k, P_k, t)}{dt} = 0.$$

The time partial derivative of the thermodynamic entropy $S_k(t)$ is computed by replacing $G_k(Q_k, P_k, t)$ with $S_k(Q_k, P_k, t)$ in the formula (5.43) and then using (9.14) to obtain

$$(9.15) \qquad \frac{\partial S_k(t)}{\partial t} = \left\langle \dot{\chi}_{\Omega_k(t)}(Q_k, P_k) S_k(Q_k, P_k, t) \right\rangle_{kt}.$$

For the isolated t system, the result $\partial S_t(t)/\partial t = 0$ was obtained previously in (9.4). The decomposition of $\dot{\chi}_{\Omega_k(t)}(Q_k, P_k)$ in (5.26) allows $\partial S_k(t)/\partial t$ to be written in the form of a particle sum function with the local i-particle contribution defined in parallel with (5.41a) as

$$(9.16) \qquad \frac{dS_i(q, t)}{dt} = \left\langle \delta(q - q_i) \frac{\dot{\chi}_{d_k(t)}(q_i, v_i)}{\chi_{d_k(t)}(q_i)} S_k(Q_k, P_k, t) \right\rangle_{kt}.$$

This implies that the local individual particle encounters with the boundary are the source of the change in the entropy. In spite of this, (9.16) is not a true particle sum function because $S_k(Q_k, P_k, t)$ itself cannot in general be expressed as a particle sum function.

There is another important aspect of the system entropy $S_k(t)$ that is not related to the dynamics of the underlying particle system. It is connected with the difference between the TIS and standard definitions of the entropy that was mentioned in connection with the equilibrium entropy in Chapter 8. For normalized distributions, the expectation function $S_k^{(n)}(t)$ is based on $F_k^{(n)}(Q_k, P_k, t)$ and easily seen to be

$$(9.17) \qquad S_k^{(n)}(t) = \langle S_k(Q_k, P_k, t) + k_B \ln Z_k(t) \rangle_{kt} = S_k(t) + k_B \ln Z_k(t).$$

The entropy $S_k^{(n)}(t)$ depends on the system volume set $d_k(t)$ and $S_k(t)$ does not. This is a consequence of the fact that the partition function $Z_k(t)$ defined in (5.3) depends on the phase volume set $\Omega_k(t)$ in the TIS formalism due to the presence of $\chi_{\Omega_k(t)}(Q_k, P_k)$ in its integrand, so it ultimately depends on the system coordinate volume set $d_k(t)$. This dependence of $Z_k(t)$ on $d_k(t)$ means that $F_k^{(n)}(Q_k, P_k, t)$ depends on $d_k(t)$ as well. In the TIS formalism, the system distribution and the system entropy analog are functions only of the system state (Q_k, P_k) and the time and do not depend on macroscopic aspects of the situation such as $d_k(t)$. The system state is therefore connected only with the microscopic Hamiltonian formalism and not with any of the macroscopic parameters associated with the system.

The review above of entropy analogs just scratches the surface of this topic. This review of versions of the entropy will be extended in QTS, Part III, and the work of Boltzmann, Gibbs, Planck, Lorentz, Poincaré, Einstein, Sakur, Tetrode, Natanson, Fowler, and many others will be analyzed. The conclusions reached there concerning the proper way to represent the entropy in thermodynamics are consistent with the points of view expressed here.

9.6 The Reference Distribution Inequality

Let us proceed now to the mathematical aspects of the entropy analog that will be used in the classical TIS theory. This consists of a demonstration of Clausius' inequality, Axiom 6.2, using the system evolution equation and the wall formalism developed in Chapter 6.

Although $\partial S_t(t)/\partial t = 0$ is always true for the thermodynamic entropy of an isolated system t, it is not true in general for the rate of change of the entropy of the subsystems $k \in [t]$ computed in (9.15) using the standard TIS formalism for computing the time derivative of a macroscopic thermodynamic quantity. This formalism connects the rate of change of the entropy to the total time derivative of the entropy analog and to boundary terms. To measure the change in a distribution density such as the phase entropy $S_k(Q_k, P_k, t) = -k_B \ln F_k(Q_k, P_k, t)$ due to the evolution in time of the underlying particle system, another distribution density is needed for comparison along with a method for comparing them. The distribution introduced in this section to use for this comparison is called the *reference distribution*. The method for comparing its rate of change with that of the system distribution is introduced next. This reference distribution will be specialized into a useful form afterward.

The proof of Clausius' inequality consists of a significantly modified version of Carlo Cercignani's version of a proof employed by Jean-Sylvestre Darrozès and Jean-Pierre Guiraud [**1966**].[15] It is related to work by Joel Louis Lebowitz and Harry Lloyd Frisch [**1957**], Section 2, and Jay Robert Dorfman and Henk van Beijeren [**1977**], pp. 82–92. This prior work will be discussed briefly using the TIS formalism. Without loss of generality, let us consider the case in which system k is enclosed by system l, which will play the role of the boundary system or wall.

Darrozès and Guiraud were the first to approach the problem of taking account of the wall in computing Clausius' inequality. They considered a wall system with a fixed wall transition probability W_k and analyzed how this will modify the k system particle distribution. While this is a crucial step, there is a significant problem with the proofs of Clausius' inequality based on it. A central assumption made in these proofs is that any two different k system states, F_k and L_k, say, have the same k wall transition probability, W_k, at any point in time. However, the wall transition probability depends on the state of the particles in the wall. For the wall transition probability to be independent of the k system distribution, the evolution l system distribution must be independent of the evolution of the k system distribution. From the point of view of particle mechanics, such an assumption is not physically reasonable because the particles in the k and l systems are interacting. They

---

[15]See Cercignani [**1975**], Chapter III, Section 4.

evolve together, so the state of the wall also depends on the state of the system enclosed by it.

This critique of the approach of Darrozès and Guiraud and Cercignani means that both the k and l systems must be included in the calculations in a symmetric manner. This is done in the TIS formalism where the evolution of the k system is determined by its wall transition probability W_k, which, by (6.40) and (6.41), is an average over F_l. Similarly, the k system is part of the wall of the l system. The wall transition probabilities W_k and W_l depend in turn on the interface transition probability, W_{kl}, which depends only on incoming and outgoing particle mechanical states and not on particle distributions. The TIS calculations will therefore be based on a symmetric treatment of the two systems together and on the properties interface transition probability W_{kl}.

Two special distribution densities, $L_k = L_k(Q_k, P_k, t)$ for system k and $L_l = L_l(Q_l, P_l, t)$ for system l, will be defined below and called *reference distributions*. It is not necessary that these reference distributions be system states, but the reference distribution is required to factor into components when the system is divided at a given time into quasilocal domains. This reflects the fact that each quasilocal domain at that time can be treated as an independent system. Let us assume then that the reference distributions taken in pairs satisfy the following conditions:

1. The reference distribution for each system is the product of reference distributions for each quasilocal domain that intersects $d_k(t)$ at time t. For each system $k \in [t]$, this is written

$$(9.18a) \qquad\qquad L_k = \prod_{a \in \pi(t)} L_k^a.$$

2. For each $k \in [t]$, the k reference distribution is bounded below and above and is integrable in $\Omega_k(t)$:

$$(9.18b) \qquad 0 < L_k(Q_k, P_k, t) < \infty \qquad \text{for all } (Q_k, P_k) \in \Omega_k.$$

$$(9.18c) \qquad \int dQ_k \int dP_k \, L_k(Q_k, P_k, t) < \infty;$$

3. The products $F_k F_l$ and $L_k L_l$ satisfy (6.31) with the same interface transition probability, W_{kl}.

An explicit reference distribution L_k will be displayed in Section 9.4 and assumption (3) will be discussed in Section 9.6.

As before, G_k will often be used in place of $G_k(Q_k, P_k, t)$ and G_k' in place of $G_k(Q_k, P_k', t)$. The quantity $\hat{n}_{\partial k}(q_i, t) \cdot u_i^{\pm}(q_i, v_i)$ will be abbreviated as $\hat{n}_{\partial k} \cdot u_i^{\pm}$ for $i \in k$ and $\hat{n}_{\partial l}(q_i, t) \cdot u_i^{\pm}(q_i, v_i)$ as $\hat{n}_{\partial l} \cdot u_i^{\pm}$ for $i \in l$. Following Cercignani, these elements will be used to define a non-negative i-particle weight function for $i \in k$ and $q_i \in d_k(t)$ by

$$(9.19) \qquad U_i(P_k', P_l') = \frac{W_{kl}|\hat{n}_{\partial k} \cdot u_i'^-|L_k'L_l'}{|\hat{n}_{\partial k} \cdot u_i^+|L_kL_l}.$$

For $i \in l$, there is a similar weight function expressed in terms of $\hat{n}_{\partial l} \cdot u_i^{\pm}$. The distribution U_i is well defined almost everywhere because of (9.18b) and the fact based on (6.31) that $\int dP_k' \int dP_l' W_{kl}\hat{n}_{\partial k} \cdot u_i'^- L_k'L_l' = 0$ if $\hat{n}_{\partial k} \cdot u_i^+ = 0$.

The form chosen for U_i is a modified version of a function used by Cercignani that has been adapted to this context. It is easy to demonstrate that the replacements $|\hat{n}_{\partial k} \cdot u_i^-| = -\hat{n}_{\partial k} \cdot u_i^-$ and $|\hat{n}_{\partial k} \cdot u_i^+| = \hat{n}_{\partial k} \cdot u_i^+$ are valid using the above definitions. Since both L_kL_l and F_kF_l satisfy (6.31), taking the vector dot product of both sides of (6.31) with $\hat{n}_{\partial k}$ or $\hat{n}_{\partial l}$ and using the appropriate replacement relation in the result shows for $i \in k$ or $i \in l$, respectively, that

$$(9.20a) \qquad \int dP_k' \int dP_l' \, U_i(P_k', P_l') = 1,$$

$$(9.20b) \qquad \int dP_k' \int dP_l' \, U_i(P_k', P_l') \frac{F_k'F_l'}{L_k'L_l'} = \frac{F_kF_l}{L_kL_l}.$$

Following Cercignani [**1975**], 115–118, let us define $C(G)$ as the convex function $C(G) = G \ln G$.[16] By Jensen's inequality,[17] it follows that
$$(9.21a)$$
$$C\left(\int dP_k' \int dP_l' \, U_i(P_k', P_l') \frac{F_k'F_l'}{L_k'L_l'} \right) \leq \int dP_k' \int dP_l' \, U_i(P_k', P_l') C\left(\frac{F_k'F_l'}{L_k'L_l'} \right)$$

and, by (9.20b),

$$(9.21b) \qquad C\left(\frac{F_kF_l}{L_kL_l} \right) \leq \int dP_k' \int dP_l' \, U_i(P_k', P_l') C\left(\frac{F_k'F_l'}{L_k'L_l'} \right).$$

There is equality in (9.21) if and only if
1. it is the case that

$$(9.22a) \qquad U_i(P_k', P_l') = \delta(P_k' - P_k^\star(P_k, P_l))\delta(P_l' - P_l^\star(P_k, P_l))$$

for some map $P^\star$, which, by virtue of the equations of motion, must be invertible, unimodular, and proper as well as conserve momentum, angular momentum and kinetic energy; or

---

[16]Observe that $C(F)$ is continuous for $F > 0$, that $\lim_{F \to 0+} C(F) = 0$, and that $C''(F) \geq 0$. In addition, the current general assumptions on F_k imply that the integrals used exist whenever $S_k(t)$ exists.

[17]For further information on Jensen's inequality, see, for example, Laha and Rohatgi [**1979**], p. 368.

2. it is the case that

$$F_k(Q_k, P_k, t) = M_k(Q_k, t)L_k(Q_k, P_k, t) \qquad \text{and}$$

(9.22b) $\qquad F_l(Q_l, P_l, t) = M_l(Q_l, t)L_l(Q_l, P_l, t)$

for some functions $M_k(Q_k, t)$, $M_l(Q_l, t)$.

The next step is to multiply both sides of (9.21) by the non-negative function $|\hat{n}_{\partial k} \cdot u_i^+|L_kL_l$. This is followed by the use of the definition (9.19) of $U_i(P_k', P_l')$ and the definition of $C(G)$ to expand the result. This gives for $q_i \in \partial d_k(t)$ the result

(9.23)

$$|\hat{n}_{\partial k} \cdot u_i^+| F_k F_l \ln\left(\frac{F_k F_l}{L_k L_l}\right) \leq \int dP_k' \int dP_l' W_{kl} |\hat{n}_{\partial k} \cdot u_i'^-| F_k' F_l' \ln\left(\frac{F_k' F_l'}{L_k' L_l'}\right),$$

with a similar result for $i \in l$. The next step is to integrate both sides of (9.23) over P_k and P_l and use the relation (6.36b) on the right-hand side. The dummy integration variables P_k' and P_l' are then changed to P_k and P_l on the right-hand side. After this, the right side is moved to the left of the inequality and the relation $|\hat{n}_{\partial k} \cdot u_i^+| - |\hat{n}_{\partial k} \cdot u_i^-| = \hat{n}_{\partial k} \cdot u_i$ is used to obtain

(9.24) $\qquad \displaystyle\int dP_k \int dP_l\, \hat{n}_{\partial k} \cdot u_i\, F_k F_l \ln\left(\frac{F_k F_l}{L_k L_l}\right) \leq 0.$

The non-negative singular operator $\delta(q - q_i) \int_{\partial k} d^2r\, \delta(q_i - r)$ is applied to both sides of (9.24) and the integrations over Q_k and Q_l are performed. The definition (6.35) of $\dot{\chi}_{d_k(t)}(q_i, v_i)$ is used to write the result in the form

(9.25) $\qquad \left\langle \left\langle \delta(q - q_i)\dfrac{\dot{\chi}_{d_k(t)}(q_i, v_i)}{\chi_{d_k(t)}(q_i)} \ln\left(\dfrac{F_k F_l}{L_k L_l}\right)\right\rangle_{lt}\right\rangle_{kt} \leq 0.$

Expanding the logarithm into a sum of logarithms and using (5.26a) to annihilate terms depending on $\ln F_l$ and $\ln L_l$, the relation (9.16) expresses for $q_i \in \partial d_k(t)$ the rate of change of the entropy for particle i in terms of the *reference distribution inequality* as

(9.26) $\qquad \dfrac{dS_i(q, t)}{dt} \geq -k_B \left\langle \delta(q - q_i)\dfrac{\dot{\chi}_{d_k(t)}(q_i, v_i)}{\chi_{d_k(t)}(q_i)} \ln L_k\right\rangle_{kt}$

with a similar result for $i \in l$.

9.7 The Reference Distribution

The next step requires the choice of a reference distribution. It is clear from the conditions (9.18) that a wide range of distributions can play the role of a reference distribution. In particular, any actual states of the k and l systems at a given time that satisfy (9.18) may be used in this way. However, it is not necessary at this stage that the reference distribution be a possible state of the system.

The most important aspect in the choice of a reference distribution is that it remains unchanged by an interaction at the wall. This allows the possibility of a steady state solution of the system equations of motion. Each choice of an interface transition probability and a stationary reference distribution leads to a corresponding version of Clausius' inequality. Many different reference distributions meet this requirement for a given interface transition probability, so additional criteria must be supplied to single out a reference distribution that has independent physical significance. In the light of the previous discussion and (9.18), four natural requirements on the interface transition probability and on the reference distributions are:

1. The reference distribution for the system factors into a product of the reference distributions for each independent quasilocal domain that intersects $d_k(t)$:

(9.27a)
$$L_k = \prod_{a \in \kappa} L_k^a,$$

where

(9.27b)
$$\kappa = \{\, a \in \pi(t) \mid \Delta_k^a(t) \neq \emptyset \,\}.$$

2. The reference distribution preserves its form under transitions at the wall governed by the interface transition probability chosen.
3. The reference distributions for systems k and l have local momentum, angular momentum and energy averages that equal those of the states, F_k and F_l, being compared to them.
4. The reference distributions are extrema of the entropy (within the limits of the constraints imposed by (3)).

Properties (1), (3) and (4) will be pursued now and (2) will be examined in the next section.

Choosing a reference distribution that is an extremum of $S_k(t)$ means that $dS_k(q,t)/dt = 0$ for this distribution. In addition, as will be shown in Chapter 13, if the constraints listed in (3) are independent of q and t and the reference distribution is also a state, it will be an equilibrium state of the system. Furthermore, if (9.22b) applies and the product of distributions $M_k(Q_k, t)L_k(Q_k, P_k, t)$ is a state, it also satisfies the criterion (3.14) for a reversible process state. Reversible processes and the consequences of requiring that the product $M_k L_k$ be a state will be examined in Chapter 13.

Let us now use requirements (3) and (4) to further analyze L_k. Assume that $p_i^{L,\mu}(q,t)$ and $K_i^L(q,t)$ are the local macroscopic momentum and local kinetic energy densities for particle i in the k system described by the L

distribution. They are given by

(9.28a) $$\langle \delta(q - q_i)p_i^{\mu}\rangle_{Lt} = p_i^{L,\mu}(q,t),$$

(9.28b) $$\langle \delta(q - q_i)K_i(p_i)\rangle_{Lt} = K_i^{L}(q,t).$$

Assume also that $p_i^{L}(q,t)$ and $K_i^{L}(q,t)$ are bounded and that $K_i^{L}(q,t) \geq 0$. In addition, for $q \in \partial d_k(t)$, let $[m_i\nu_i(q,t)]^{-1}p_i^{L,\mu}(q,t) = v_{\partial k}^{L,\mu}(q,t)$ in accord with the boundary condition (6.39b).

It is not necessary to include a separate local angular momentum condition in (9.28) because the exclusive use of central forces in the classical versions of TIS makes it redundant.

Let us now compute the maximum entropy state using the first order calculus of variations. To do so, the integral $\langle \ln L \rangle_{lt}$ is maximized subject to the constraints (9.28). Let δL_k be an arbitrary variation of L_k with respect to the momentum variables that leaves Q_k and the momentum limits fixed. The system projection operator is therefore fixed as well, which leads to

(9.29) $$\int dP_k \, \delta L_k = 0.$$

Then, in accord with the first order calculus of variations, it is sufficient for each $i \in k$ to use four phase functions $\{\, b_i(q_i,t), -g_i^{\mu}(q_i,t) \,\}$ as the undetermined multipliers. Let $R_k(Q_k,t)$ be an arbitrary function of Q_k and t, which is bounded for $Q_k \in D_k(t)$, and vary $\int dP_k \, L_k \ln L_k$ to obtain

(9.30)
$$\int dP_k \, \delta L_k \{\ln L_k + 1 + \sum_{i\in k}[b_i(q_i,t)K_i(p_i) - g_i(q_i,t) \cdot p_i] + R_k(Q_k,t) \,\} = 0.$$

This has the solution

(9.31a) $$L_k(Q_k, P_k, t) = e^{-\left[\sum_{i\in k}(b_i(q_i,t)K_i(p_i) - g_i(q_i,t)\cdot p_i) + R_k(Q_k,t)\right]}.$$

The partition function for L_k is

(9.31b)
$$Z_k^{L}(t) = \int dQ_k \int dP_k \, \chi_{\Omega_k(t)}(Q_k, P_k)e^{-\left[\sum_{i\in k}(b_i(q_i,t)K_i(p_i) - g_i(q_i,t)\cdot p_i) + R_k(Q_k,t)\right]}.$$

The fact that $L_k > 0$ implies $\partial^2(L_k \ln L_k)/\partial L_k^2 = L_k^{-1} > 0$. This means that the choice (9.30) for L_k minimizes the integral and maximizes the associated entropy. The relations (9.31) will be cast in a form below that meets requirement (1) above as stated in (9.27).

In Section 13.2 it will be shown that when L_k is the state of the k system the i-particle local heating $\mathcal{Q}_{\partial i}^{*}(q,t)$ vanishes in the particle rest frame of the k system. This means that the rate of change of the i-particle entropy vanishes, i.e., $dS_i(q,t)/dt = 0$. The positive semi-definiteness of $K_i^{L}(q,t)$ implies that $b_i(q,t)$ is also positive semi-definite, which implies in turn that $b_i(q,t) \geq 0$ for $q \in d_k(t)$.

The relation (9.31) is next expressed in a more convenient form. The definitions

(9.32a) $$v_i^{*,\mu}(q,t) = [b_i(q,t)]^{-1} g_i^\mu(q,t),$$

which is valid when $b_i(q,t) > 0$, and

(9.32b) $$p_i^{*,\mu}(q,t) = m_i v_i^{*,\mu}(q,t)$$

will be useful. The fact that $R_k(Q_k,t)$ is still arbitrary allows the absorption of the term $-K_i(p_i^*)$ into $R_k(Q_k,t)$. Then L_k can be written in the form

(9.33) $$L_k(Q_k, P_k, t) = e^{-\left[\sum_{i \in k}(b_i(q_i,t)K_i(p_i - p_i^*)) + R_k(Q_k,t)\right]}.$$

9.8 The Unbiased Wall

The choice (9.33) for L_k and L_l clearly satisfies (9.18a)–(9.18c) because the phase energy and momentum are bounded for both systems. Let us consider now the question of whether a wall exists that meets the third condition on reference distributions. This condition requires that there is an interface transition probability W_{kl} such that (6.31) is satisfied with $L_k L_l$ in place of F_{kl}. By this is meant that the special forms for L_k and L_l exhibited in (9.33) are preserved under transitions at the boundary. It will be shown next that the unbiased wall transition probability $W_{kl}^{(\text{unb})}$ has this property for the k and l reference distributions defined by (9.33).

In parallel with the factorization in (9.27), the factorization mentioned in Chapter 6 of the δ measures in $W_{kl}^{(\text{unb})}$ into a product of δ measures representing the conservation of momentum, angular momentum, and energy will be used in each quasilocal domain separately. If it can be shown that the product $L_k L_l$ has the form (6.49), then L_k and L_l will preserve their forms at an unbiased wall. Equation (6.49) is valid in either the quasilocal or global case if, for each $a \in \pi(t)$, it is the case that

(9.34) $$\sum_{i \in \gamma_k^a} b_i(q_i,t)K_i(p_i - p_i^*(q_i,t)) + \sum_{j \in \gamma_l^a} b_j(q_j,t)K_j(p_j - p_j^*(q_j,t))$$
$$= \sum_{i \in \gamma_k^a} b_i(q_i,t)K_i(p_i' - p_i^*(q_i,t)) + \sum_{j \in \gamma_l^a} b_j(q_j,t)K_j(p_j' - p_j^*(q_j,t))$$

for $q_i, q_j \in w_{kl}(t)$. Since this must be true for all pairs of momentum vectors $(P_k^{a'}, P_l^{a'})$ and (P_k^a, P_l^a) that satisfy the conservation relations (6.24), and be true for all (Q_k, Q_l), it will be shown next that it is necessary and sufficient that (1) $b_i(q_i,t)$ and $v_i^*(q_i,t)$ are quasilocal functions that are continuous across the boundary, (2) $b_i(q_i,t)$ is independent of i, and (3) $v_i^*(q_i,t)$ takes a particular form.

By the independence of $b_i(q_i, t)$ from i and the continuity of $b_i(q_i, t)$ across the boundary for $i \in k$ are meant

(9.35a)   (i) $b_i(q_i, t) = b_k^a(t),$

(9.35b)   (ii) $b_l^a(t) = b_k^a(t).$

This implies that $b_k^a(t)$ depends only on the quasilocal domain a and not on i, k or l. The special form for the velocity $v_i^*(q_i, t)$ is

(9.35c)   (iii) $v_i^{*,\mu}(q_i, t) = \epsilon^{\mu\nu\xi} s_{k,\nu}(t) q_{i,\xi} + c_k^\mu(t);$

If (9.35c) is true, then $v_i^{*,\mu}(q_i, t)$ depends on i only through its argument, q_i. Assuming this is the case, $v_i^{*,\mu}(q_i, t)$ will be written as $v_k^{*,\mu}(q_i, t)$ from now on. By the continuity of $v_k^{*,\mu}(q_i, t)$ across the boundary is meant

(9.35d)   (iv) $v_l^{*,\mu}(q_i, t) = v_k^{*,\mu}(q_i, t)$ for $q_i \in w_{kl}(t).$

In terms of (9.35c), this implies

(9.35e) $s_l(t) = s_k(t)$ $c_l(t) = c_k(t).$

The special form for $v_k^{*,\mu}(q_i, t)$ is determined by the fact that

(9.36) $v_k^*(q_i, t) \cdot p_i = s_k(t) \cdot j_i(q_i, p_i) + c_k(t) \cdot p_i.$

Summing (9.36) over $i \in \gamma_k^a$ and similarly summing $v_l^*(q_i, t) \cdot p_i$ over $i \in \gamma_l^a$ and adding the results yields $[s_k(t) \cdot \mathcal{J}_k^a(Q_k, P_k) + s_l(t) \cdot \mathcal{J}_l^a(Q_l, P_l)] + [c_k(t) \cdot \mathcal{P}_k^a(P_k) + c_l(t) \cdot \mathcal{P}_l^a(P_l)]$. It follows from (6.24) that this quantity is conserved if $s_k(t) = s_l(t)$ and $c_k(t) = c_l(t)$.

Using the definition of $K_i(p_i)$ to expand $K_i(p_i - p_i^*(q_i, t))$ into a sum of components shows that the above relations, in conjunction with the conservation of the total kinetic energy $K_k(P_k^a) + K_l(P_l^a)$ in a quasilocal domain, are sufficient for (9.34) to hold. To show necessity, a variety of momentum vectors that satisfy (6.24) can be chosen. For fixed $b_i(q_i, t)$ and $b_j(q_j, t)$, which do not satisfy (9.35a) and (9.35b), it can then be shown that there is no solution for (9.34). A similar result holds if it is assumed that $s_k(t)$ and $c_k(t)$ are not continuous across the boundary. These steps are left to the reader.

The relations (9.35) then yield the result that the functional form (9.33) for L_k will be preserved at the wall when (a) the wall is described in terms of an unbiased interface transition probability, $W_{kl}^{(\text{unb})}$, (b) the k and l system reference distributions are both written in the form (9.33), (c) the functions b_k and g_k defining L_k are quasilocal, and (d) b_k and g_k are continuous across all interfaces.

The deterministic and exact interface transition probabilities, W_{kl}^{det} and $W_{kl}^{(\delta)}$, do not preserve the form of L_k and L_l in general. This is due to the fact that multiparticle transformations $P_k^*(P_k', P_l')$, $P_l^*(P_k', P_l')$ may be constructed such that for some (P_k, P_l) there is no pair of momentum vectors

(P'_k, P'_l) that meet the condition $(P_k, P_l) = (P^*_k(P'_k, P'_l), P^*_l(P'_k, P'_l))$. This shows that the deterministic interface transition probability may not lead to an equilibrium state.

The essential aspect of the interface transition probability is that it reflects our macroscopic ignorance of the detailed processes at the wall. As macroscopic beings, we cannot track this information nor do we want to. The exact $W^{(\delta)}_{kl}$ interface transition probability, which is valid for each individual transition, is therefore replaced with a general interface transition probability W_{kl}, which is valid over a large enough collection of transitions in the sense of probability. It is W_{kl}, and more precisely $W^{(\text{unb})}_{kl}$, that leads to a steady state solution that reflects our experience of equilibrium systems. For these reasons, $W^{(\text{unb})}_{kl}$ is used instead of $W^{(\delta)}_{kl}$ as the interface transition probability in the proof of Clausius' inequality.

This distinction between $W^{(\text{unb})}_{kl}$ and $W^{(\delta)}_{kl}$ with regard to a steady state is the crux of the epistemological issue that must be faced in any attempt to reconcile the irreversible aspects of thermodynamics with statistical mechanics. No purely mechanical proof of Clausius' inequality or the second law of thermodynamics can be given. If the exact description of the microscopic system is retained, including the interactions of its particles across any macroscopically chosen interface, the system does not show movement to equilibrium. This is consistent with the reversibility of the particle mechanics and the Poincaré recurrence theorem for the total system containing systems k and l. If, on the other hand, the exact interface transition probability is replaced by the unbiased interface transition probability $W^{(\text{unb})}_{kl}$, the system will show movement to equilibrium. This reflects the fact that the general interface transition probability summarizes the average behavior of the particles in a system at the interface, so it is a good guide to the behavior of the system in most cases for some period of time. However, the equilibrium state for the system that results from these calculations does not imply that the system is in a genuine time-independent state because it is still appropriate to describe the system by some unknown exact state $F^\delta_k(Q_k, P_k, t)$ that may move away from quiescent behavior at any time. Finally, the use of a wall demon may lead to a steady state that is not an equilibrium state. This possibility will not be investigated further.

9.9 Clausius' Inequality

The next task is to use (9.28) to compute the undetermined multipliers $b_i(q_i, t)$ and $g^\mu_i(q_i, t)$, which will complete the definition of L_k. When this is done, it will be a simple matter to compute Clausius' inequality.

In standard equilibrium theory, where $\Omega^{(\text{std})}_k = D_k \times \mathbf{R}^{3\mathcal{N}_k}$, momentum integrations are extended to infinity. With the assumption that the range of momentum integration is infinite, the form of p^{a*}_i is determined by (9.35) and

(9.36). The definition (9.32a) can then be used with (9.35a) in the computation of the integral (9.28). The definition $f_i^{(g)}(q, p, t) = \langle \delta(q - q_i)\delta(p - p_i)\rangle_{Lt}^{(g)}$ of the global phase average, in which the system distribution is L_k, the projection operator is not used, and the integration is over all phase space, leads to the result

(9.37) $$f_i^{(g)}(q, p, t) = e^{-b_k(t)K_i(p - p_i^*(q,t))}\nu_i^{(g)*}(q, t),$$

where

(9.38)

$$\nu_i^{(g)*}(q, t) = \frac{1}{Z_k^{L(g)}(t)} \int dQ_k \int dP_k \, \delta(q - q_i)\delta(p - p_i)$$
$$\times e^{-b_k(t)\sum_{j \in k-i} K_j(p_j - p_j^*(q_j, t)) - R_k(Q_k, t)}.$$

The global partition function $Z_k^{L(g)}(t)$ is obtained by integrating both sides of (9.37) over $q \in \mathbf{R}^3$ and $p \in \mathbf{R}^3$ and setting the result equal to 1. It follows from (9.37) and (9.38) that

(9.39a) $$\int d^3p \, f_i^{(g)}(q, p, t) = \nu_i(q, t) = \left[\frac{2\pi m_i}{b_k(t)}\right]^{\frac{3}{2}} \nu_i^{(g)*}(q, t),$$

(9.39b) $$\left[\frac{2\pi m_i}{b_k(t)}\right]^{\frac{3}{2}} \int d^3q \, \nu_i^{(g)*}(q, t) = 1.$$

These results are used to show

(9.40)
$$p_i^{L,\mu}(q, t) = \int d^3p \, p^\mu f_i^{(g)}(q, p, t)$$
$$= \frac{2\pi m_i}{b_k(t)}\nu_i^{(g)*}(q, t) \int_{-\infty}^{\infty} dp^\mu \, p^\mu e^{-b_k(t)[p^\mu - p_i^{*,\mu}(q,t)]^2/2m_i}$$
$$= \frac{1}{\sqrt{\pi}} \left[\frac{2\pi m_i}{b_k(t)}\right]^{\frac{3}{2}} \nu_i^{(g)*}(q, t) \int_{-\infty}^{\infty} dy \left[\sqrt{\frac{2m_i}{b_k(t)}}y + p_i^{*,\mu}(q, t)\right] e^{-y^2}$$
$$= \nu_i(q, t)p_i^{*,\mu}(q, t),$$

This leads by (9.32) to

(9.42a) $$g_i^\mu(q, t) = b_k^a(t)v_i^{*,\mu}(q, t) = b_k^a(t)[m_i\nu_i(q, t)]^{-1}p_i^{L,\mu}(q, t).$$

In the notation of Chapter 7, $p_i^{L,\mu}(q, t) = m_i v_i^{L,\mu}(q, t) = m_i \nu_i(q, t)\bar{v}_i^\mu(q, t)$, where $\bar{v}_i^\mu(q, t)$ is the local stream velocity at (q, t). By a similar calculation, it also follows that

(9.42b) $$b_k^a(t) = \frac{2}{3K_i^{L,a}(t)}.$$

The fact that the calculation leading to (9.35a) implies that $K_i^{L,a}(t)$ must be a quasilocal function if the other requirements on L_k are to be met was also used. Observe that $b_k^a(t) > 0$ if $K_i^{L,a}(t) < \infty$ for $i \in k$.

The relations (9.42) do not follow, however, when the integration over (Q_k, P_k) is restricted to the bounded set $\Omega_k(t)$ in phase space. The results for this case will be computed next. The TIS phase average for system k described by system distribution L can be written in the form

$$(9.43) \quad f_i(q,p,t) = \langle \delta(q - q_i)\delta(p - p_i) \rangle_{Lt} = e^{-b_k(t)K_i(p-p_i^*(q,t))}\nu_i^*(q,p,t),$$

where

(9.44)

$$\nu_i^*(q,p,t) = \frac{1}{Z_k^L(t)} \int dQ_k \int dP_k\, \delta(q - q_i)\delta(p - p_i)\chi_{\Omega_k(t)}(Q_k, P_k)$$

$$\times\, e^{-b_k(t)\sum_{j \in k - i} K_j(p_j - p_j^*(q_j, t)) - R_k(Q_k, t)}.$$

The partition function $Z_k^L(t)$ is obtained, as for the global case, by integrating both sides of (9.43) over $q \in \mathbf{R}^3$ and $p \in \mathbf{R}^3$ and setting the left side equal to 1.

Let us ignore the bounds $\mathcal{P}_k^u$ and $\mathcal{P}_k^l$ on the momentum and the bounds $\mathcal{J}_k^u$ and $\mathcal{J}_k^l$ on the angular momentum in (5.18) and focus on those on the energy. With this assumption, the representation (5.18) of $\chi_{\Omega_k(t)}(Q_k, P_k)$ and the symmetry of $\mathcal{H}_k(Q_k, P_k, t)$ in p_i show that $\nu_i^*(q,p,t)$ does not depend on $p_i^*(q_i, t)$. It follows immediately that $\nu_i^*(q, -p, t) = \nu_i^*(q,p,t)$.

To define B_1 and B_2, let us use $0 \leq K_k(P_k) \leq H_k^u - \Phi_k(Q_k, t)$ and $\Phi_k(Q_k, t) \geq H_k^l$ to show for any $i \in k$,

$$(9.45) \qquad 0 \leq [p_i]^2 \leq 2m_i K_k(P_k) \leq 2m_i[H_k^u - H_k^l] = 2m_i \mathcal{C}_{H,k},$$

which is obtained from the definitions of the upper and lower bounds of $\mathcal{H}_k$ and (2.30c). This result is used to define B as $B = \sqrt{2m_i\mathcal{C}_{H,k}}$. Then, by symmetry, it follows that $B_1 = -B$, $B_2 = B$. Using this with (9.43) and (9.44), yields

$$(9.46) \qquad \int d^3p\, f_i(q,p,t) = \int d^3p\, \theta(B - |p|)f_i(q,p,t) = \nu_i(q,t).$$

The phase average of p^μ is then given in this formalism by

(9.47)

$$p_i^{L,\mu}(q,t) = \int d^3p\, p^\mu f_i(q,p,t) = \int d^3y\, [y^\mu + p_i^{*,\mu}(q,t)]f_i(q, y + p_i^*(q,t), t)$$

$$= \nu_i(q,t)p_i^{*,\mu}(q,t) + \int d^3y\, y^\mu f_i(q, y + p_i^*(q,t), t).$$

Let us define for $|p_i^{L,\mu}(q,t)| > 0$ the quantity

$$(9.48) \qquad \lambda^*(\mathcal{C}_{\mathcal{H},k}) = \frac{1}{p_i^{L,\mu}(q,t)} \int d^3y \, y^\mu f_i(q, y + p_i^*(q,t), t).$$

Using this definition with (9.47) gives for $\nu_i(q,t) > 0$ the result
(9.49)
$$p_i^{*,\mu}(q,t) = [\nu_i(q,t)]^{-1} p_i^{L,\mu}(q,t)[1 - \lambda^*(\mathcal{C}_{\mathcal{H},k})] = m_i \bar{v}_i^\mu(q,t)[1 - \lambda^*(\mathcal{C}_{\mathcal{H},k})].$$

By (9.37) and (9.38), it follows from $\lim_{\mathcal{C}_{\mathcal{H},k}\to\infty} \chi_{\Omega_k(t)}(Q_k, P_k) = \chi_{D_k(t)}(Q_k)$ that $\lim_{\mathcal{C}_{\mathcal{H},k}\to\infty} f_i(q,p,t) = f_i^{(g)}(q,p,t)$. Using this with the symmetry of $f_i^{(g)}(q,p,t)$ yields

$$(9.50) \qquad \lim_{\mathcal{C}_{\mathcal{H},k}\to\infty} \lambda^*(\mathcal{C}_{\mathcal{H},k}) = 0.$$

This means that

$$(9.51) \qquad \lim_{\mathcal{C}_{\mathcal{H},k}\to\infty} p_i^{*,\mu}(q,t) = \nu_i(q,t) p_i^{L,\mu}(q,t) = \mu_i(q,t)\bar{v}_i^\mu(q,t)$$

is true in the thermodynamic limit as claimed above. Except for very small isolated systems, the value of the function $\lambda^*(\mathcal{C}_{\mathcal{H},k})$ will be small, so the difference between $p_i^*(q,t)$ and $[\nu_i(q,t)]^{-1} p_i^L(q,t)$ will usually be small.

To complete the calculation of Clausius' inequality, the relation (9.33) is used for L_k in (9.26) followed by use of the closure condition (5.29) to eliminate the term $R_k(Q_k, t)$ and other terms not dependent on p_i. For $q \in \partial d_k(t)$, it is also the case that $\bar{p}_i(q,t) = [\nu_i(q,t)]^{-1} p_i^L(q,t) = m_i[\nu_i(q,t)]^{-1} v_i^L(q,t) = m_i v_{\partial k}(q,t)$, so $\lim_{\mathcal{C}_{\mathcal{H},k}\to\infty} p_i^{*,\mu}(q,t) = m_i \nu_i(q,t) v_{\partial k}(q,t)$. The definitions and results (7.57)–(7.60) are used with this fact to obtain

(9.52)

$$\frac{dS_i(q,t)}{dt} \geq -k_B \left\langle \delta(q - q_i) \frac{\dot{\chi}_{d_k(t)}(q_i, v_i)}{\chi_{d_k(t)}(q_i)} \ln L_k \right\rangle_{kt}$$

$$= k_B \, b_k(q,t) \left\langle \delta(q - q_i) \frac{\dot{\chi}_{d_k(t)}(q_i, v_i)}{\chi_{d_k(t)}(q_i)} K_i(p_i - p_i^*) \right\rangle_{kt}$$

$$= k_B \, b_k(q,t)[Q_{\partial i}(q,t) + \lambda^*(\mathcal{C}_{\mathcal{H},k}) W_{\partial i}(q,t)],$$

where $K_i(p_i - p_i^*) = K_i(p_i - \bar{p}_i) + p_i \cdot (\bar{v}_i - v_i^*) - K_i(\bar{p}_i) + K_i(p_i^*)$ was used with (9.49) and (5.29). Summing both sides of (9.49) over $i \in \gamma_k^a$ and then over $a \in \pi(t)$ yields *Clausius' inequality*

$$(9.53) \qquad \frac{dS_k(q,t)}{dt} \geq k_B b_k(q,t) Q_{\partial k}^*(q,t)$$

where

$$(9.54) \qquad Q_{\partial k}^*(q,t) = Q_{\partial k}(q,t) + \lambda^*(\mathcal{C}_{\mathcal{H},k}) W_{\partial k}(q,t).$$

If the quasilocal phase function $b_k(q, t)$ can be identified with the local ther-mature, $\beta_k(q, t)$, the inequality (9.53) will confirm axiom 6.2. The question of this identification will be addressed in Chapter 12.

The expression of Clausius' inequality in (9.53) reduces to the standard form, $k_B b_k(q, t) Q_{\partial k}(q, t)$, in the following cases: (i) when the wall is at rest (i.e., $v_{\partial k}(q, t) = 0$ for $q \in d_k(t)$) in the fixed k system rest frame; (ii) by the remark after (9.31b), when the system is in the k particle rest frame; (iii) when boundary forces vanish so that $F_{\partial k}^\mu(q, t) = 0$; or (iv) in the thermodynamic limit (9.51).

CHAPTER 10

Volume

In this chapter, the volume descriptors are given a formal definition. In addition, a macroscopic operator, called the volume partial derivative operator, and a corresponding microscopic analog phase volume derivative operator are defined using the system projection operator as a basis.

10.1 The Definition of the System Volume

As macroscopic entities, the objects we deal with occupy space, have boundaries, and interact with other objects. However, when these objects are examined from a microscopic perspective, the concepts of object volume and object boundary break down. The choice of where to place a boundary around a collection of microscopic particles has no more significance than the choice of where to place the shoreline around an island. The placement of the system boundary therefore has a conventional aspect that is not connected with the physics of the particle system. In previous discussions, it was mentioned that this observation is supported by the fact that the volume of a system of particles does not play a role in its Hamiltonian dynamics. It follows that the volume is a purely macroscopic concept. It is essential to the thermodynamic description, but not to the microscopic description of a system.

The system volume, and other macroscopic parameters, may be adjusted by macroscopic means. These macroscopic parameters specify the boundary conditions for the microscopic particle dynamics internal to the system. The microscopic system dynamics in turn affects the forces needed to maintain the boundary conditions chosen. It follows that the separation in Chapter 5 of the partition function $Z_k(t)$, which depends on the macroscopic volume, from the system distribution F_k, which does not, is both meaningful and necessary. In this sense, the partition function helps mediate aspects of the relation between the macroscopic and microscopic descriptions of a system.

In Chapter 2, the volume set $d_k(t)$, with $k \in [t]$ or $k = t$, was introduced as an open, connected, bounded set in $\mathbf{R}^3$ with a piecewise smooth boundary that does not intersect itself. An examination of the particle coordinates in the system evolution equation shows that they always appear as differences such as $q_i - q_j$. Some quantities that are not written in this form, such as $Q_k \cdot \partial\Phi_k(Q_k,t)/\partial Q_k$, fall into this category because, as shown below, this quantity can be expressed in the form $\frac{1}{2}\sum_{i \in k}\sum_{j \in k-i}(q_i - q_j) \cdot (\partial\phi_{ii}(q_i - q_j)/\partial q_i)$. Quantities that depend on inner products of vectors representing coordinate differences are invariant under translations, rotations, and changes in the velocity frame. This includes the phase functions $\phi_{ij}(q_i - q_j)$, which depend on $|q_i - q_j|$, and the extended $3\mathcal{N}_k$-dimensional vector inner product $Q_k \cdot \partial\Phi_k(Q_k,t)/\partial Q_k$ among others.

The introduction of a boundary surface around a system volume introduces an apparently absolute location into thermodynamics that is absent in the dynamics of the underlying particle system that depends only on co-ordinate differences. The location of the system boundary surface and the volume set are fixed in the macroscopic reference frame. This means that a location q in a system boundary will represent an absolute location from the microscopic standpoint if the macroscopic and microscopic reference frames are transformed independently. This would disrupt the mechanics of the system because it would change the forces acting at the boundaries. The system projection operator $\chi_{\Omega_k(t)}(Q_k, P_k)$ used in system phase averages of thermo-dynamic quantities would also be affected. It follows that the macroscopic frame and the microscopic frames must be transformed in the same way at the same time. This means in effect that there is only one frame and that the macroscopic coordinates (q, t) are defined within the same frames as the microscopic particle locations and velocities.

The fact that the range of energies available in k, defined in (2.30c) as $C_{H,k}$, is bounded implies by (4.28) that there always exists a minimum distance between the particles, $r_c > 0$, which was called the distance of closest approach. Because two particles cannot occupy the same point, the boundary will never get "pinched" between two particles and the semi-norm, $\|\phi\|$, of ϕ_{ij} will always remain finite. In Chapter 2, the volume set $d_k(t)$ was defined by a rule, operating on the macroscopic level, for assigning particles to a system; an additional requirement was placed on the boundary of this set so that a closed 2-dimensional boundary set, $\partial d_k(t)$, can be specified that is consistent with the assignment of particles to the system. The volume set and its boundary are required to satisfy the conditions on the volume sets and their boundaries stated in Chapter 2.

Mathematically, a suitably chosen real valued function of the macro-scopic spacetime coordinates, which is designated by $B_k(q, t)$, can be used to define the boundary $\partial d_k(t)$. $B_k(q, t)$ is called the *boundary function*. It is assumed to be continuous, piecewise smooth, have bounded curvature, and a finite number of disjoint components. The function $B_k(q, t)$ is assumed to be $C^\infty(\mathbf{R}^4)$ almost everywhere. It defines the boundary set $\partial d_k(t)$ as the set of points q that satisfy the condition

$$(10.1) \qquad\qquad B_k(q, t) = 0.$$

The bounded curvature and smoothness conditions imply that the relation

$$(10.2) \qquad\qquad \alpha_k(t) = \int d^3q \, \chi_{\partial d_k(t)}(q) < \infty$$

for the k *system surface area* $\alpha_k(t)$ is valid. Because the particles do not come closer together than the distance $r_c > 0$, a suitable form of Urysohn's lemma shows that a function $B_k(q, t)$, which defines a boundary set that separates the particles located in system k from those outside system k, always exists. In light of the fact that the boundary is determined macroscopically, a boundary set will be useful for macroscopic purposes if it has bounded curvature on the microscopic level. Because of its role, it is important that the boundary is suitable for the purpose of separating macroscopic bodies rather than requiring that it meet the microscopic condition that a particular class of particles will always be found within it. A given boundary function $B_k(q, t)$ will be acceptable if the boundary it defines by the condition (10.1) meets the macroscopic criteria and satisfies (10.2) as well.

The formalism on surfaces in Clifford Truesdell and Richard Toupin [**1960**], pp. 498–500, or James Serrin [**1959**], p. 137, is the basis for this analysis. Let $\bar{v}_{Bk}(q,t)$ be a given velocity field, defined almost everywhere for q in a suitable bounded subset of $\mathbf{R}^3$ which contains $d_k(t)$. For points (q,t) such that $q \in \partial d_k(t)$, the function $\bar{v}_{Bk}(q,t)$ defines the velocity of the boundary $v_{\partial k}(q,t)$. There is also another vector field defined in a neighborhood of $\partial d_k(t)$ designated by $\hat{n}_{Bk}(q,t)$. At points $q \in \partial d_k(t)$, the vector function $\hat{n}_{Bk}(q,t)$ is required to have unit length and be normal to the plane tangent to the boundary surface. This vector function is defined by $\partial B_k(q,t)/\partial q$. At the system k surface it defines the inward pointing normal unit vector designated by $\hat{n}_{\partial k}(q,t)$.

In terms of these definitions, the function $B_k(q,t)$ is assumed to have the following properties:

1. $B_k(q,t)$ satisfies the conditions

(10.3)
$$
\begin{aligned}
B_k(q,t) &> 0, & q &\in d_k(t), \\
B_k(q,t) &= 0, & q &\in \partial d_k(t), \\
B_k(q,t) &< 0, & q &\notin d_k(t);
\end{aligned}
$$

2. $B_k(q,t)$ is the solution of

(10.4)
$$
\frac{\partial B_k(q,t)}{\partial t} + \bar{v}_{Bk}(q,t) \cdot \frac{\partial B_k(q,t)}{\partial q} = 0;
$$

3. The vector $\hat{n}_{Bk}(q,t)$ is defined by

(10.5)
$$
\hat{n}^{\mu}_{Bk}(q,t) = \frac{\partial B_k(q,t)}{\partial q_{\mu}}.
$$

For each $q \in \partial d_k(t)$, $\hat{n}_{Bk}(q,t)$ and $v_{Bk}(q,t)$ must satisfy the boundary conditions

(10.6a) $$|\hat{n}_{Bk}(q,t)| = 1,$$
(10.6b) $$\hat{n}_{Bk}(q,t) = \hat{n}_{\partial k}(q,t),$$
(10.6c) $$\bar{v}_{Bk}(q,t) = v_{\partial k}(q,t).$$

The conditions (10.3) are chosen so that $\hat{n}^{\mu}_{\partial k}(q,t)$ points into the k volume. Since (10.5) defines the gradient of $B_k(q,t)$ at the surface, $\hat{n}^{\mu}_{\partial k}(q,t)$ is normal to that surface. By (10.4), (10.5) and (10.6), it follows for $q \in \partial d_k(t)$ that

(10.7)
$$
\hat{n}^{\mu}_{\partial k}(q,t) \cdot v_{\partial k}(q,t) = -\frac{\partial B_k(q,t)}{\partial t}.
$$

The fact that the kl interface is shared is expressed for $q \in w_{kl}(t)$ in the relations

(10.8a) $$B_l(q,t) = B_k(q,t) = 0,$$

(10.8b) $$\hat{n}_{\partial l}(q,t) \cdot v_{\partial l}(q,t) = -\hat{n}_{\partial k}(q,t) \cdot v_{\partial k}(q,t).$$

Note that (10.8a) allows the existence of lines, e.g., $w_{kl}(t) \cap w_{ks}(t)$, which are located where the systems k, l, and s come together, on which equation (10.4) and, hence, equations (10.6) and (10.7) are not defined. These lines have Lebesgue measure zero in $\mathbf{R}^2$. Equation (10.8b) is in accord with (2.5). It is also easily seen that there is at least one point inside $d_k(t)$ at which the $\hat{n}_B(q,t)$ and $\bar{v}_{Bk}(q,t)$ vector fields vanish. Note also that condition (10.8b) implies that the k and l boundary surfaces do not separate from each other. They may slip past each other tangentially, however, unless (2.6) applies and $v_{\partial l}(q,t) = v_{\partial k}(q,t)$ for $q \in w_{kl}(t)$.

10.2 The Boundary Flux Operator

The indicator function of the open set $d_k(t)$ is defined by

(10.9) $$\chi_{d_k(t)}(q) = \theta(B_k(q,t)).$$

Using this, the expression (6.35) for the total time derivative of the system projection operator can be justified in terms of the formalism of this chapter by observing first that

(10.10a) $$\frac{\partial \chi_{d_k(t)}(q)}{\partial t} = \delta(B_k(q,t)) \frac{\partial B_k(q,t)}{\partial t},$$

(10.10b) $$\frac{\partial \chi_{d_k(t)}(q)}{\partial q} = \delta(B_k(q,t)) \frac{\partial B_k(q,t)}{\partial q}.$$

Next, the distribution $\delta(B_k(q,t))$ is written in the form

(10.11) $$\delta(B_k(q,t)) = \int d^3r \, \chi_{\partial d_k(t)}(r)\delta(q-r) = \int_{\partial k} d^2r \, \delta(q-r),$$

where $\partial k = \partial d_k(t)$ and the notation has been abused slightly to keep it simple in moving to the form on the far right. This formula can be justified at each point q on a system surface by using an appropriately chosen coordinate system in relation to the tangent plane to the surface at that point. Using (10.6) and (10.7) with (10.5) in (10.10) gives the useful formulas

(10.12a)
$$\frac{\partial \chi_{d_k(t)}(q)}{\partial t} = -\int_{\partial k} d^2r \, \delta(q-r)\hat{n}_{\partial k}(r,t) \cdot v_{\partial k}(r,t),$$

(10.12b)
$$\frac{\partial \chi_{d_k(t)}(q)}{\partial q_\mu} = \int_{\partial k} d^2r \, \delta(q-r)\hat{n}^\mu_{\partial k}(r,t).$$

Next, by (10.12) and the definition of the total phase time derivative, it follows for $q \in d_k(t)$ that

$$(10.13) \quad \dot{\chi}_{d_k(t)}(q, v) = \frac{d\chi_{d_k(t)}(q)}{dt} = \int_{\partial k} d^2 r \, \delta(q - r) \, \hat{n}_{\partial k}(r, t) \cdot (v - v_{\partial k}(r, t)).$$

The representation of the volume given in (2.10) is used with (10.12a) to compute the time derivative of the volume as

$$(10.14) \quad \frac{d\delta_k(t)}{dt} = - \int_{\partial k} d^2 r \, \hat{n}_{\partial k}(r, t) \cdot v_{\partial k}(r, t).$$

By (10.6c), (10.5), and the divergence theorem, (10.14) can be written

$$(10.15) \quad \frac{d\delta_k(t)}{dt} = \int_k d^3 q \, \frac{\partial \bar{v}_k^\mu(q, t)}{\partial q^\mu}.$$

In accord with Euler's formula, the quantity $\mathbf{div}(\bar{v}_k(q, t))$ is called the *local average volume dilation*.[1]

10.3 The Virial and Volume Phase Derivatives

The analysis of the pressure, as well as other quantities used in the formalism of thermodynamics, will require the definition of a volume partial derivative operator that acts on thermodynamic quantities. These quantities depend on the volume only implicitly through the projection operator $\chi_{\Omega_k(t)}(Q_k, P_k)$ and the partition function $Z_k(t)$. The projection operator and the partition function, in turn, depend on the volume set $d_k(t)$. This means that the problem must be approached indirectly by working with the volume set.

The first step is to examine the description of a volume dilation. A coordinate scale transformation of the form $q \to \lambda q$ is a type of conformal transformation that was introduced by Herbert S. Green [**1947**] for this purpose. For phase functions, this coordinate parameterization transformation has the effect of changing all interparticle distances, $|q_i - q_j|$, by a factor of λ. A new volume set, $d_k(t; \lambda)$, which increases its Lebesgue measure as λ increases, is defined by

$$(10.16a) \quad d_k(t; \lambda) = \{\, q \mid \lambda^{-1} q \in d_k(t) \,\}.$$

The volume of this set is then the Lebesgue measure

$$(10.16b) \quad \delta_k(t; \lambda) = \int d^3 q \, \chi_{d_k(t; \lambda)}(q) = \int d^3 q \, \chi_{d_k(t)}(\lambda^{-1} q) = \lambda^3 \delta_k(t).$$

---

[1]See Serrin [**1959**], pp. 130-131, or Truesdell and Toupin [**1960**], p. 342, for more on Euler's formula.

The derivative of this with respect to λ is

$$(10.16c) \qquad \frac{\partial \delta_k(t; \lambda)}{\partial \lambda} = 3\lambda^2 \delta_k(t).$$

As a check, it is easy to show

$$(10.17) \qquad \lim_{\lambda \to 1} \frac{d\chi_{d_k(t)}(\lambda^{-1}q)}{d\lambda} = -q \cdot \frac{\partial \chi_{d_k(t)}(q)}{\partial q} = -\delta(B_k(q,t))\hat{n}_{\partial k}(q,t) \cdot q.$$

Using (10.11) to replace $\delta(B_k(q,t))$ with its definition as an integral of $\delta(q-r)$ over the system surface and then integrating over $q \in d_k(t)$, the divergence theorem gives the result $3d_k(t)$ as required.

Coordinate parameterization was used to define a volume derivative in H. S. Green [**1947**], pp. 111, 115–117, Jack Irving and John Kirkwood [**1950**], P. Martin and Julian Schwinger [**1959**], pp. 1345–1347, and Dmitrii Zubarev [**1971**], pp. 49–52, 242–246, 256–257, and in the work of a number of other authors. These calculations take account of the effect of the changes in the system volume on the potential energy of a system. However, as calculations and experiments with a gas confined by a piston in a cylinder show, the motion of the piston in changing the volume will also change the momentum spectrum of the particles in the gas. It follows that preserving the Hamiltonian basis of the theory under a coordinate parameterization transformation means that the transformation must be extended to a canonical one. To make the coordinate dilation transformation canonical, it is required that the transformation $q_i \to \lambda_i q_i$ of the coordinates is accompanied by a simultaneous transformation $p_i \to \lambda_i^{-1} p_i$ of the momenta for each $i \in t$.[2] This idea will be used to extend Green's coordinate parameterization transformation to a canonical one. The first step is to define the transformation for all the particles at once by

$$(10.18a) \qquad\qquad \Lambda Q_k = \times_{i \in k} \lambda_i q_i,$$

$$(10.18b) \qquad\qquad \Lambda^{-1} P_k = \times_{i \in k} \lambda_i^{-1} p_i.$$

This transformation also preserves the quantity $Q_k \cdot P_k$, which will play an important role in subsequent work on pressure in both classical and quantum mechanics. The function $Q_k \cdot P_k$ is called the *phase action* because it has the same physical units as the *mechanical action* of classical mechanics, which is defined as the path integral $\int dt \, \dot{Q}_k(t) \cdot P_k(t)$ for a trajectory $(Q_k(t), P_k(t))$ in phase space. The integrand of the mechanical action, $\dot{Q}_k \cdot P_k$, plays a role

---

[2]It is not hard to show that this simultaneous transformation preserves the value of the Poisson bracket operator. See Chapter 15 for a further discussion and demonstration that the transformation $(q_i, p_i) \to (\lambda q_i, \lambda^{-1} p_i)$ is canonical.

in the Legendre transformations connecting the Lagrangian and Hamiltonian functions for the k system.[3]

For any phase function G_k defined in the phase volume $\Omega_k(t)$, the i-particle *dilation derivative* and the k *dilation derivative* are defined by

(10.19a)
$$\frac{dG_k(Q_k, P_k, t)}{dd_i} = \lim_{\lambda_i \to 1} \frac{dG_k(\Lambda Q_k, \Lambda^{-1} P_k, t)}{d\lambda_i}$$
$$= \frac{1}{3}\left[q_i \cdot \frac{\partial G_k(Q_k, P_k, t)}{\partial q_i} - p_i \cdot \frac{\partial G_k(Q_k, P_k, t)}{\partial p_i} \right],$$

(10.19b)
$$\frac{dG_k(Q_k, P_k, t)}{d\mathcal{D}_k} = \sum_{i \in k} \frac{dG_k(Q_k, P_k, t)}{dd_i}.$$

This dilation derivative operator is also used to define the k *volume phase derivative* of $G_k(Q_k, P_k, t)$, with $\lambda_i = \lambda$ for $i \in k$, by

(10.20)
$$\frac{dG_k(Q_k, P_k, t)}{d\Omega_k} = \lim_{\lambda \to 1} \left[\frac{\partial \delta_k(t; \lambda)}{\partial \lambda} \right]^{-1} \frac{dG_k(\Lambda Q_k, \Lambda^{-1} P_k, t)}{d\Lambda}$$
$$= \frac{1}{\delta_k(t)} \frac{\partial G_k(Q_k, P_k, t)}{\partial \mathcal{D}_k}$$
$$= \frac{1}{3\delta_k(t)} \left[Q_k \cdot \frac{\partial G_k(Q_k, P_k, t)}{\partial Q_k} - P_k \cdot \frac{\partial G_k(Q_k, P_k, t)}{\partial P_k} \right].$$

where the relation $\lim_{\lambda_i \to 1} d\delta_k(t; \lambda_i)/d\lambda_i = 3\delta_k(t)$ obtained from (10.16c) was used.

The dilation operators above are all special cases of a family of more general operators called the *virial tensor derivative* operators. These operators are defined by using the same pattern as was used for the dilation derivative, but with infinitesimal operators for translations and rotations of the particle coordinate and momentum vectors q_i and p_i. The most general infinitesimal transformation of either q_i or p_i depends on a dilation parameter λ_i, a general infinitesimal translation q_a^μ, a general infinitesimal velocity change v_a^μ, and an infinitesimal rotation transformation given by the matrix $a_r^{\mu\nu}$. For q_i this transformation takes the form $q_i^\mu \to \lambda_i q_i^\mu + q_a^\mu + a_r^\mu q_{i,\nu}$. The limits $\lambda_i \to 1$, $q_a^\mu \to 0$, and $a_r^{\mu\nu} \to 0$ are taken at the end of the calculation. Similarly, the infinitesimal transformation of p_i^μ is $p_i^\mu \to \lambda_i^{-1} p_i^\mu + p_{a,i}^\mu - a_r^{\mu\nu} p_{i,\nu}$, where $p_{a,i}^\mu = m_i v_a^\mu$. Because it represents an orthogonal rotation, $a_r^{\mu\nu}$ is an

---

[3]The mechanical action, the phase action $Q_k \cdot P_k$, and the Legendre transformations are discussed, for example, in Goldstein [**1950**]. The use of the phase action in the Legendre transformations of the system Lagrangian function is discussed in EIS. Further development of this approach from the TIS perspective, which includes historical references and a discussion of the principle of least action and associated Lagrangian methods, is pursued for both classical and quantum cases in QTS.

antisymmetric matrix with three independent components that meet the requirements for representing a proper orthogonal rotation about a given axis.

With this transformation, the derivatives of analog quantities with respect to λ_i, q_a^μ, $p_{a,i}^\mu$, and $a_r^{\mu\nu}$ all lead to versions of the general virial derivative or virial derivative tensor. The particular transformation and the derivative chosen with respect to λ_i, q_a^μ, $p_{a,i}^\mu$, or $a_r^{\mu\nu}$, is said to *induce* the transformation that represents the derivative operator that acts on phase functions. These operators are used to compute the response of a phase function to the type of change envisioned, whether dilation, translation, or rotation. The classical virial derivative operators take the form $q_i^\mu(\partial/\partial q_{i,\nu})$ and $p_i^\mu(\partial/\partial p_{i,\nu})$. The dilation derivative uses both of these forms.

Let us define the more general *i-particle virial tensor derivative* and *k virial tensor derivative* operators by

$$(10.21a) \qquad \frac{d}{dd_{i,\mu\nu}} = q_i^\mu \frac{\partial}{\partial q_{i,\nu}} - p_i^\mu \frac{\partial}{\partial p_{i,\nu}},$$

$$(10.21b) \qquad \frac{d}{dD_{k,\mu\nu}} = \sum_{i\in k} \frac{d}{dd_{i,\mu\nu}}.$$

As an example of this operator in action, the effect of the i-particle operator $d/dd_{i,\mu\nu}$ on the phase action $q_i \cdot p_i$ is

$$(10.22) \qquad \frac{d(q_i \cdot p_i)}{dd_{i,\mu\nu}} = [q_i^\mu p_i^\nu - q_i^\nu p_i^\mu] = 2q_i^{[\mu} p_i^{\nu]},$$

where antisymmetric alternation was used for the last equality. It is not hard to see that the quantity (10.22) is a generating function for an infinitesimal rotation.[4] It follows from (10.22) that

$$(10.23a) \qquad \delta_{\mu\nu} \frac{d(q_i \cdot p_i)}{dd_{i,\mu\nu}} = 0,$$

$$(10.23b) \qquad \epsilon_{\xi\mu\nu} \frac{d(q_i \cdot p_i)}{dd_{i,\mu\nu}} = e_{\xi\mu\nu} q_i^\mu p_i^\nu = j_{i,\xi}(q_i, p_i).$$

The effect of the virial tensor derivative $d/d_{i,\mu\nu}$ on the angular momentum $j_i^\xi(q_i, p_i)$ is

$$(10.24) \qquad \frac{dj_i^\xi(q_i, p_i)}{d_{i,\mu\nu}} = \epsilon^{\xi\kappa\sigma} [\delta_\kappa^\nu q_i^\mu p_{i,\sigma} - \delta_\sigma^\nu q_{i,\kappa} p_i^\mu].$$

---

[4]See, for example, the discussion of this quantity in Goldstein [**1950**], p. 262.

This leads, with the use of symmetric mixing, to

(10.25a) $$\delta_{\mu\nu}\frac{dj_i^\xi(q_i,p_i)}{dd_{i,\mu\nu}}=0,$$

(10.25b) $$\epsilon_{\rho\mu\nu}\frac{dj_i^\xi(q_i,p_i)}{dd_{i,\mu\nu}}=2[\delta_{\rho\lambda}q_i^{(\xi}p_i^{\lambda)}-\delta_\rho^\xi q_i\cdot p_i].$$

It follows immediately from (10.25b) that $\delta_\xi^\rho\epsilon_{\rho\mu\nu}(dj_i^\xi(q_i,p_i)/dd_{i,\mu\nu})=-4q_i\cdot p_i$.

The virial derivative components, $q_{i,\xi}(\partial/\partial q_{i,\nu})$ and $p_{i,\xi}(\partial/\partial p_{i,\nu})$, are a combination of a covariant vector $q_{i,\xi}$ and a contravariant vector $\partial/\partial q_{i,\nu}$ in the sense of differential geometry. The combined effect of these operators on the phase action shown in (10.23) and on the angular momentum shown in (10.25) indicates that the virial derivative operators play an important role in dynamics.

The *i-particle dilation derivative*, the *k dilation derivative tensor*, the *k dilation derivative*, and the *k volume phase derivative*, which act only on phase functions, are defined in terms of the virial tensor derivative operators by

(10.26a) $$\frac{d}{dd_i}=\frac{\delta_{\mu\nu}}{3}\frac{d}{dd_{i,\mu\nu}},\qquad \frac{d}{d\mathcal{D}_{k,\mu\nu}}=\sum_{i\in k}\frac{d}{dd_{i,\mu\nu}},$$

(10.26b) $$\frac{d}{d\mathcal{D}_k}=\frac{\delta_{\mu\nu}}{3}\frac{d}{d\mathcal{D}_{k,\mu\nu}},\qquad \frac{d}{d\Omega_k}=\frac{1}{\delta_k(t)}\frac{d}{d\mathcal{D}_k}.$$

There are several other operators that are based on the basic virial derivative operators of the form $q_{i,\mu}(\partial/\partial q_i^\nu)$ and $p_{i,\xi}(\partial/\partial p_i^\nu)$. The trace $\delta^{\mu\nu}q_{i,\mu}(\partial/\partial q_i^\nu)$ of the coordinate virial tensor derivative operator, for example, yields the scalar virial $\Xi_i(Q_k,t)$ itself when it is applied to the system potential or the system Hamiltonian. This operator is a measure of the force in the direction of q_i. Similarly, the vector operator $[q_i\times(\partial/\partial q_i)]^\mu$ represents the *i-particle rotation* operator $\tau_i^\mu=\epsilon^{\mu\nu\xi}q_{i,\nu}(\partial/\partial q_i^\xi)$ when it is applied to the system potential or system Hamiltonian functions. This operator is essentially a measure of the rotation of a function representing a physical quantity around the location of particle i.

For future reference, the effect of the dilation operator on various system quantities including the system projection operator will be computed. The dilation derivatives of the system phase energy, coordinate vector, momentum vector, total momentum, total angular momentum, and total phase action

$Q_k \cdot P_k$ are

(10.27a)
$$\frac{d\mathcal{H}_k(Q_k, P_k, t)}{d\mathcal{D}_k} = \frac{1}{3}\left[Q_k \cdot \frac{\partial \Phi_k(Q_k, t)}{\partial Q_k} - 2\mathcal{K}_k(P_k)\right],$$

(10.27b)
$$\frac{dQ_k}{d\mathcal{D}_k} = \tfrac{1}{3}Q_k, \qquad \frac{dP_k}{d\mathcal{D}_k} = -\tfrac{1}{3}P_k,$$

(10.27c)
$$\frac{d\mathcal{J}_k(Q_k, P_k)}{d\mathcal{D}_k} = 0, \qquad \frac{d(Q_k \cdot P_k)}{d\mathcal{D}_k} = 0.$$

It is not hard to show by (10.12b) that
(10.28)
$$Q_k \cdot \frac{\partial \chi_{D_k(t)}(Q_k)}{\partial Q_k} = \chi_{D_k(t)}(Q_k) \sum_{i \in k} \frac{1}{\chi_{d_k(t)}(q_i)} \int_{\partial k} d^2 r\, \delta(q_i - r)\hat{n}_{\partial k}(r, t) \cdot q_i.$$

The definition (5.18) is used next to obtain

(10.29)
$$\frac{d\chi_{\Omega_k(t)}(Q_k, P_k)}{d\mathcal{D}_k} = \chi_{\Omega_k(t)}(Q_k, P_k)\left\{Q_k \cdot \frac{d\chi_{D_k(t)}(Q_k)}{dQ_k}\right.$$
$$- \frac{d\mathcal{H}_k(Q_k, P_k, t)}{d\mathcal{D}_k}\left[\frac{\delta(\mathcal{H}_k^u - \mathcal{H}_k(Q_k, P_k, t))}{\theta(\mathcal{H}_k^u - \mathcal{H}_k(Q_k, P_k, t))} - \frac{\delta(\mathcal{H}_k(Q_k, P_k, t) - \mathcal{H}_k^l)}{\theta(\mathcal{H}_k(Q_k, P_k, t) - \mathcal{H}_k^l)}\right]$$
$$- \frac{dP_k}{d\mathcal{D}_k}\left[\frac{\delta(\mathcal{P}_k^u - \mathcal{P}_k)}{\theta(\mathcal{P}_k^u - \mathcal{P}_k)} - \frac{\delta(\mathcal{P}_k - \mathcal{P}_k^l)}{\theta(\mathcal{P}_k - \mathcal{P}_k^l)}\right]$$
$$\left. - \frac{d\mathcal{J}_k(Q_k, P_k)}{d\mathcal{D}_k}\left[\frac{\delta(\mathcal{J}_k^u - \mathcal{J}_k(Q_k, P_k))}{\theta(\mathcal{J}_k^u - \mathcal{J}_k(Q_k, P_k))} - \frac{\delta(\mathcal{J}_k(Q_k, P_k) - \mathcal{J}_k^l)}{\theta(\mathcal{J}_k(Q_k, P_k) - \mathcal{J}_k^l)}\right]\right\}.$$

The quantity $d\chi_{\Omega_k(t)}(Q_k, P_k)/d\mathcal{D}_k$ appears in phase average integrals as part of a volume derivative calculation, but will be used in the explicit form (10.29) only rarely if at all. Except for an exact state calculation, an integration by parts is almost always used in these cases to move the volume derivative from $\chi_{\Omega_k(t)}(Q_k, P_k)$ to the other factors, such as a phase analog function and the system distribution, in the integrand of the phase average integral.

For convenience, let us define the function $\mathrm{P}_k(Q_k, P_k, t)$ by

(10.30)
$$3\delta_k(t)\mathrm{P}_k(Q_k, P_k, t) = -\frac{d\mathcal{H}_k(Q_k, P_k, t)}{d\mathcal{D}_k}.$$

Using this in (10.29) with (10.27) and (10.28) gives the result

(10.31)
$$\frac{d\chi_{\Omega_k(t)}(Q_k, P_k)}{d\mathcal{D}_k} = \chi_{\Omega_k(t)}(Q_k, P_k)\left\{Q_k \cdot \frac{d\chi_{D_k(t)}(Q_k)}{dQ_k}\right.$$
$$+ 3\delta_k(t)\mathrm{P}_k(Q_k, P_k, t)\left[\frac{\delta(\mathcal{H}_k^u - \mathcal{H}_k(Q_k, P_k, t))}{\theta(\mathcal{H}_k^u - \mathcal{H}_k(Q_k, P_k, t))} - \frac{\delta(\mathcal{H}_k(Q_k, P_k, t) - \mathcal{H}_k^l)}{\theta(\mathcal{H}_k(Q_k, P_k, t) - \mathcal{H}_k^l)}\right]$$
$$\left. + P_k\left[\frac{\delta(\mathcal{P}_k^u - \mathcal{P}_k)}{\theta(\mathcal{P}_k^u - \mathcal{P}_k)} - \frac{\delta(\mathcal{P}_k - \mathcal{P}_k^l)}{\theta(\mathcal{P}_k - \mathcal{P}_k^l)}\right]\right\}.$$

Phase averaging the terms in (10.31) shows that the phase average of the dilation derivative of the system projection operator can be expressed in the form

$$
(10.32) \quad \left\langle \frac{d\chi_{\Omega_k(t)}(Q_k, P_k)}{d\mathcal{D}_k} \right\rangle_{kt} = \sum_{i \in k} \left[\int_{\partial k} d^2 r\, \hat{n}_{\partial k}(r, t) \cdot r \left\langle \frac{\delta(q_i - r)}{\chi_{d_k(t)}(q_i)} \right\rangle_{kt} \right]
$$

$$
+3\delta_k(t) \left\langle P_k(Q_k, P_k, t) \left[\frac{\delta(\mathcal{H}_k^u - \mathcal{H}_k(Q_k, P_k, t))}{\theta(H_k^u - \mathcal{H}_k(Q_k, P_k, t))} - \frac{\delta(\mathcal{H}_k(Q_k, P_k, t) - \mathcal{H}_k^l)}{\theta(H_k(Q_k, P_k, t) - H_k^l)} \right] \right\rangle_{kt}
$$

$$
+ \left\langle P_k \left[\frac{\delta(\mathcal{P}_k^u - \mathcal{P}_k)}{\theta(\mathcal{P}_k^u - \mathcal{P}_k)} - \frac{\delta(\mathcal{P}_k - \mathcal{P}_k^l)}{\theta(\mathcal{P}_k - \mathcal{P}_k^l)} \right] \right\rangle_{kt}.
$$

Let us use the assumptions, introduced in Chapter 5, that the points in phase space at which the system total energy is at its maximum or minimum are isolated and that processes cannot transfer matter, momentum, angular momentum, or energy, outside of $\Omega_k(t)$. This means that the set of points at which the system total energy is at an extremum is a set of Lebesgue measure zero. Let us assume similarly that the set of points on which the system total momentum is at its maximum or minimum is a set of measure zero. Finally, division by $\chi_{d_k(t)}(q_i)$ in the first term on the right in (10.32) removes the restriction on particle i to $d_k(t)$ and means that the boundary particle density, defined by

$$
(10.33) \qquad \nu_{\partial k}(r, t) = \sum_{i \in k} \left\langle \frac{\delta(q_i - r)}{\chi_{d_k(t)}(q_i)} \right\rangle_{kt},
$$

is defined for $r \in \partial d_k(t)$. These assumptions imply that (10.32) can be written as

$$
(10.34) \qquad \left\langle \frac{d\chi_{\Omega_k(t)}(Q_k, P_k)}{d\mathcal{D}_k} \right\rangle_{kt} = \frac{1}{3} \int_{\partial k} d^2 r\, \hat{n}_{\partial k}(r, t) \cdot r\, \nu_{\partial k}(r, t)
$$

and that

$$
(10.35) \qquad \left\langle \frac{d\chi_{\Omega_k(t)}(Q_k, P_k)}{d\Omega_k} \right\rangle_{kt} = \frac{1}{3\delta_k(t)} \int_{\partial k} d^2 r\, \hat{n}_{\partial k}(r, t) \cdot r\, \nu_{\partial k}(r, t).
$$

10.4 The Volume Partial Derivative Operator

The volume partial derivative operator plays an important role in many formulas of thermodynamics. A typical example from standard equilibrium thermodynamics is the formula $\partial \mathcal{H}_k^\epsilon / \partial V = -\mathrm{P}_k^\epsilon - \beta_k \partial \mathrm{P}_k^\epsilon / \partial \beta_k$ that connects the volume derivative of the k system equilibrium Hamiltonian to the equilibrium pressure and the equilibrium latent heat with respect to volume. In TIS, the volume derivative of a thermodynamic quantity is represented by an operator that maps that quantity, which must be the phase average of a thermodynamic analog phase function, to the thermodynamic quantity that represents its volume derivative. The TIS version of the *volume partial derivative* operator is written $\partial/\partial\delta_k$.

The TIS approach is based on the fact that system projection operator, $\chi_{\Omega_k(t)}(Q_k, P_k)$, defines the system boundaries in phase space in the phase averages of all thermodynamic quantities. The volume partial derivative is therefore defined in terms of its action on the system projection operator. The following relation represents this action in terms of the phase volume derivative as

$$(10.36) \qquad \frac{\partial\chi_{\Omega_k(t)}(Q_k, P_k)}{\partial\delta_k} = -\frac{d\chi_{\Omega_k(t)}(Q_k, P_k)}{d\Omega_k}.$$

A somewhat more general definition based on the k dilation tensor operator will also prove useful below. This generalization is called the *volume partial derivative tensor* operator. It is defined using the definition the dilation tensor phase derivative operator by its action on $\chi_{\Omega_k(t)}(Q_k, P_k)$ in parallel to the definition of the volume partial derivative operator in (10.36) by

$$(10.37) \qquad \frac{\partial\chi_{\Omega_k(t)}(Q_k, P_k)}{\partial\delta_{k,\mu\nu}} = -\frac{1}{\delta_k(t)}\frac{d\chi_{\Omega_k(t)}(Q_k, P_k)}{d\mathcal{D}_{k,\mu\nu}}.$$

It follows that $\partial/\partial\delta_k = (\delta_{\mu\nu}/3)(\partial/\partial\delta_{k,\mu\nu})$. The action of $\partial/\partial\delta_k$ on the thermodynamic quantity $G_k(t)$ is then expressed in terms of its action on the integral representing the phase average of the analog $G_k(Q_k, P_k, t)$ using (10.36) to replace $\partial\chi_{\Omega_k(t)}/\partial\delta_k$ under the integral sign by $-d\chi_{\Omega_k(t)}/d\Omega_k$. This step is followed by an integration by parts to move the phase derivative from $\chi_{\Omega_k(t)}$ to the other components in the integrand of the phase average.

The operator $\partial/\partial\delta_k$ plays two roles. In addition to being a macroscopic operator that is defined by its action on $\chi_{\Omega_k(t)}(Q_k, P_k)$ in (10.36), it is also used as an ordinary derivative with respect to the system volume δ_k when either the symbol δ_k or the symbol $\delta_k(t)$ appears explicitly in a phase analog function. Thus, the operators $\partial/\partial\delta_k$ and $\partial/\partial\delta_{k,\mu\nu}$ act under the integral sign of a phase average integration only on the system projection operator $\chi_{\Omega_k(t)}(Q_k, P_k)$ and on phase functions $G_k(Q_k, P_k, t)$ that depend explicitly on the volume δ_k. In general, phase operators should not depend on the system volume because this mixes macroscopic and microscopic notions, so an explicit dependence of a phase function on δ_k should be rare.

Consider now a system k represented by the system distribution density $F_k(Q_k, P_k, t)$ and the partition function $Z_k(t)$. The macroscopic volume partial derivative of the thermodynamic quantity $G_k(t) = \langle G_k(Q_k, P_k, t)\rangle_{kt}$ is given by $\partial G_k(t)/\partial\delta_k = \partial\langle G_k(Q_k, P_k, t)\rangle_{kt}/\partial\delta_k$. Using the definition (10.36) followed by integration by parts, the volume partial derivative of $G_k(t)$ is defined by

(10.38)
$$\frac{\partial G_k(t)}{\partial \delta_k} = \frac{1}{Z_k(t)} \frac{\partial}{\partial \delta_k} \int dQ_k \int dP_k \, F_k \chi_{\Omega_k(t)} G_k(Q_k, P_k, t) - \frac{\partial \ln Z_k(t)}{\partial \delta_k} G_k(t)$$
$$= \frac{1}{Z_k(t)} \int dQ_k \int dP_k \, \chi_{\Omega_k(t)}(Q_k, P_k) \left[\frac{\partial(F_k G_k)}{\partial \delta_k} + \frac{d(F_k G_k)}{d\Omega_k} \right]$$
$$- \frac{\partial \ln Z_k(t)}{\partial \delta_k} G_k(t),$$

where the partial derivative with respect to δ_k under the integrand is ordinary differentiation with respect to the parameter δ_k. The quantity $\partial \ln Z_k(t)/\partial \delta_k$ will be computed below and given a physical interpretation in Chapter 12.

The notation can be simplified by defining the *i-particle total volume phase derivative* and *k total volume phase derivative* operators by

(10.39a)
$$\frac{d}{d\delta_i} = \frac{\partial}{\partial \delta_k} + \frac{1}{\delta_k(t)} \frac{d}{dd_i},$$

(10.39b)
$$\frac{d}{d\delta_k} = \frac{\partial}{\partial \delta_k} + \frac{d}{d\Omega_k}.$$

These operators act only on phase quantities. The quantity that the operator $\partial/\partial \delta_k$ is acting on, whether a microscopic phase quantity $G_k(Q_k, P_k, t)$ or a macroscopic thermodynamic quantity $G_k(t)$, determines which version of the operator is being used. The total volume phase derivative of the phase analog $G_k(Q_k, P_k, t)$ is written in this formalism as

(10.40)
$$\frac{dG_k(Q_k, P_k, t)}{d\delta_k} = \frac{\partial G_k(Q_k, P_k, t)}{\partial \delta_k} + \frac{dG_k(Q_k, P_k, t)}{d\Omega_k}.$$

Exceptions to the rule that δ_k does not appear in phase analog functions are the phase analogs that are obtained as phase volume derivatives, such as the average system pressure analog defined in Chapter 11 as $-d\mathcal{H}_k/d\Omega_k$, because these operators explicitly introduce the symbol δ_k into the phase quantity that results from applying them.

The system distribution is a microscopic quantity and does not depend on the volume, so it is always the case that

(10.41)
$$\frac{\partial F_k(Q_k, P_k, t)}{\partial \delta_k} = 0.$$

Using (10.41) with the entropy analog implies immediately that

(10.42a)
$$\frac{\partial S_k(Q_k, P_k, t)}{\partial \delta_k} = 0$$

and

(10.42b)
$$\frac{dS_k(Q_k, P_k, t)}{d\delta_k} = \frac{dS_k(Q_k, P_k, t)}{d\Omega_k}$$

as well.

The entropy analog associated with the system probability distribution density $F_k^{(n)} = [Z_k(t)]^{-1} F_k$ is $S_k^{(n)}(Q_k, P_k, t) = -k_B \ln F_k^{(n)}(Q_k, P_k, t)$, so the time-dependent version of this entropy is $S_k^{(n)}(Q_k, P_k, t) = S_k(Q_k, P_k, t) + k_B \ln Z_k(t)$. The definitions (10.39) and (10.40) are used with (10.42) to show that the total phase volume derivative of this entropy analog is

(10.43)
$$\frac{dS_k^{(n)}(Q_k, P_k, t)}{d\delta_k} = \frac{dS_k(Q_k, P_k, t)}{d\Omega_k} + k_B \frac{\partial \ln Z_k(t)}{\partial \delta_k}.$$

By (10.38), (10.40), and (10.42), the action of the k *volume partial derivative* operator on any thermodynamic function $G_k(t)$ is

(10.44)
$$\frac{\partial G_k(t)}{\partial \delta_k} = \left\langle \frac{dG_k(Q_k, P_k, t)}{d\delta_k} \right\rangle_{kt} - \frac{1}{k_B} \left\langle G_k(Q_k, P_k, t) \frac{dS_k(Q_k, P_k, t)}{d\Omega_k} \right\rangle_{kt}$$
$$- \frac{\partial \ln Z_k(t)}{\partial \delta_k} G_k(t).$$

The relation

(10.45)
$$\frac{dF_k(Q_k, P_k, t)}{d\Omega_k} = -\frac{1}{k_B} F_k(Q_k, P_k, t) \frac{dS_k(Q_k, P_k, t)}{d\Omega_k}$$

was also employed to write the $dF_k/d\Omega_k$ term as a phase average in (10.44).

There is a consistency condition on the use of the volume derivative operator if it is to act as a true derivative. To compute this condition, let us set $G_k(Q_k, P_k, t) = 1$ so that $G_k(t) = \langle 1 \rangle_{kt} = 1$. The volume partial derivative of $\langle 1 \rangle_{kt}$ is computed by substituting 1 for $G_k(Q_k, P_k, t)$ in (10.44) with the result

(10.46)
$$0 = \frac{\partial \langle 1 \rangle_{kt}}{\partial \delta_k} = -\frac{1}{k_B} \left\langle \frac{dS_k(Q_k, P_k, t)}{d\Omega_k} \right\rangle_{kt} - \frac{\partial \ln Z_k(t)}{\partial \delta_k}.$$

This implies

(10.47)
$$\frac{\partial \ln Z_k(t)}{\partial \delta_k} = -\frac{1}{k_B} \left\langle \frac{dS_k(Q_k, P_k, t)}{d\Omega_k} \right\rangle_{kt}.$$

The result (10.47) expresses the volume partial derivative of $\ln Z_k(t)$ as the phase average of the volume phase derivative of the entropy. This relation can also be computed directly using the definition (5.3) of the partition function and the definition of the volume partial derivative operator. It follows that although the partition function itself is not a thermodynamic observable, its derivative with respect to volume is. The result (10.47) is used with (10.44) to reexpress volume partial derivative of $G_k(t)$ in the form

(10.48)
$$\frac{\partial G_k(t)}{\partial \delta_k} = \left\langle \frac{dG_k(Q_k, P_k, t)}{d\delta_k} - \frac{1}{k_B}[G_k(Q_k, P_k, t) - G_k(t)]\frac{dS_k(Q_k, P_k, t)}{d\Omega_k} \right\rangle_{kt}.$$

The volume derivative of the entropy is computed using (10.48) with (10.42). The resulting formula is

(10.49) $\quad \dfrac{\partial S_k(t)}{\partial \delta_k} = -\dfrac{1}{k_B} \left\langle [S_k(Q_k, P_k, t) - S_k(t) - k_B]\dfrac{dS_k(Q_k, P_k, t)}{d\Omega_k} \right\rangle_{kt}.$

Looking ahead, it will be shown in Chapter 18 that the result (10.49) is consistent with equation (18.33b) obtained for the volume partial derivative of the equilibrium entropy.

For future reference, the action of the volume partial derivative operator on $G_k(t)$ is separated into an entropy fixed term and a term that represents the contribution of the change in the entropy

(10.50) $\quad \dfrac{\partial G_k(t)}{\partial \delta_k} = \dfrac{\partial}{\partial \delta_k} \langle G_k(Q_k, P_k, t)\rangle_{kt}$

$$= \frac{\partial G_k(t)}{\partial \delta_k}\bigg|_S - \frac{1}{k_B} \left\langle [G_k(Q_k, P_k, t) - G_k(t)]\frac{dS_k(Q_k, P_k, t)}{d\Omega_k} \right\rangle_{kt}.$$

The term with the entropy fixed is written as a phase average in the form

(10.51) $$\frac{\partial G_k(t)}{\partial \delta_k}\bigg|_S = \left\langle \frac{dG_k(Q_k, P_k, t)}{d\delta_k} \right\rangle_{kt}.$$

This separation reflects a similar distinction, which was introduced in Boltzmann [1871b], between the mechanical terms in a variation calculation, which are obtained by varying the mechanical analog, and the heat term that is obtained by varying the system distribution.[5]

A second consistency condition is the requirement that the closure conditions, expressed in (5.29), are preserved under the volume derivative map. The volume partial derivatives of $\dot{\chi}_{\Omega_k(t)}$ and other quantities obtained from $\chi_{\Omega_k(t)}(Q_k, P_k)$ are treated in the same way as the volume partial derivative of $\chi_{\Omega_k(t)}(Q_k, P_k)$ in (10.36) because a volume derivative involves moving the limits of the system integration in phase averages. Computing the volume derivative of the quantity $\langle \dot{\chi}_{\Omega_k(t)} \rangle_{kt}$ with this rule gives

(10.52) $\quad \dfrac{\partial \langle \dot{\chi}_{\Omega_k(t)}(Q_k, P_k) \rangle_{kt}}{\partial \delta_k} = \left\langle \dfrac{\partial \dot{\chi}_{\Omega_k(t)}(Q_k, P_k)}{\partial \delta_k} \right\rangle_{kt}$

$$= -\left\langle \frac{d\dot{\chi}_{\Omega_k(t)}(Q_k, P_k)}{d\Omega_k} \right\rangle_{kt} = -\frac{1}{k_B} \left\langle \dot{\chi}_{\Omega_k(t)}(Q_k, P_k)\frac{dS_k(Q_k, P_k, t)}{d\Omega_k} \right\rangle_{kt}.$$

---

[5]This is discussed in Chapter 22.

Because the systems are closed, the relation $\left\langle \dot{\chi}_{\Omega_k(t)} \right\rangle_{kt} = 0$ must be preserved. This imposes a condition on the system distribution that takes the form

$$(10.53) \quad \left\langle \frac{d\dot{\chi}_{\Omega_k(t)}(Q_k, P_k)}{d\Omega_k} \right\rangle_{kt} = \frac{1}{k_B} \left\langle \dot{\chi}_{\Omega_k(t)}(Q_k, P_k) \frac{dS_k(Q_k, P_k, t)}{d\Omega_k} \right\rangle_{kt} = 0.$$

As a further check on the formalism, it is interesting to compute the volume partial derivative of the volume $\delta_k(t)$ defined in (2.10). It is easy to show using the transformation (10.18) that

$$(10.54) \quad \frac{\partial \delta_k(t)}{\partial \delta_k} = \int d^3q \, \frac{\partial \chi_{d_k(t)}(q)}{\partial \delta_k} = -\int d^3q \, \frac{\partial \chi_{d_k(t)}(q)}{d\Omega_k}$$

$$= -\frac{1}{3\delta_k(t)} \int d^3q \, q \cdot \frac{\partial \chi_{d_k(t)}(q)}{\partial q} = 1.$$

This result shows that the volume and volume derivative formalisms are compatible with each other.

If $G_k(Q_k, P_k, t)$ is a particle sum function, its phase average can be defined locally and the formalism for the volume partial derivative operator extended to it. By convention, the volume partial tensor derivative and the volume partial derivative of a local particle function such as $G_k(q, t)$ are interpreted as

$$(10.55a) \quad \frac{\partial G_k(q, t)}{\partial \delta_{k,\mu\nu}} = \sum_{i \in k} \frac{\partial G_i(q, t)}{\partial \delta_{i,\mu\nu}},$$

$$(10.55b) \quad \frac{\partial G_k(q, t)}{\partial \delta_k} = \sum_{i \in k} \frac{\partial G_i(q, t)}{\partial \delta_i}.$$

The reason for using these forms is that the local partial derivative of an analog is not divided by the volume as is the case for $\partial/\partial\delta_k$ acting on a global function $G_k(t)$. For the volume partial derivative of $G_i(q, t)$, the definition (5.10a) is used and the pattern of the derivation of (10.48) is followed to obtain

$$(10.56) \quad \frac{\partial G_i(q, t)}{\partial \delta_i} = \left\langle \frac{d(\delta(q - q_i)G_i(Q_k, p_i, t))}{dd_i} \right\rangle_{kt}$$

$$- \frac{1}{k_B} \left\langle \delta(q - q_i)G_i(Q_k, p_i, t) \frac{dS_k(Q_k, P_k, t)}{dd_i} \right\rangle_{kt} - \frac{\partial \ln Z_k(t)}{\partial \delta_i} G_i(q, t).$$

The first term on the right can be computed using the phase coordinate parameterization transformation with the result
$$(10.57)$$
$$\left\langle \frac{d(\delta(q - q_i)G_i(Q_k, p_i, t))}{dd_i} \right\rangle_{kt} = \left\langle \delta(q - q_i) \frac{dG_i(Q_k, p_i, t)}{dd_i} \right\rangle_{kt} - \frac{q}{3} \cdot \frac{\partial G_i(q, t)}{\partial q}.$$

The term $\partial \ln Z_k(t)/\partial \delta_i$ can be computed in the same way as was done in (10.46) by setting $G_i(Q_k, p_i, t) = 1$ and $G_i(q, t) = 1$ in (10.56) and obtaining

$$(10.58) \qquad \frac{\partial \ln Z_k(t)}{\partial \delta_i} = -\frac{1}{k_B} \left\langle \frac{dS_k(Q_k, P_k, t)}{dd_i} \right\rangle_{kt}$$

In parallel with the total phase volume derivatives defined in (10.39), let us define the *i-particle total volume derivative* operator, which acts on thermodynamic functions, by

$$(10.59) \qquad \frac{d}{d\delta_i} = \frac{\partial}{\partial \delta_i} + \frac{q}{3} \cdot \frac{\partial}{\partial q}.$$

Using the above results and definition (10.59) with (10.56) yields

$$(10.60)$$
$$\frac{dG_i(q,t)}{d\delta_i} = \left\langle \delta(q - q_i) \frac{dG_i(Q_k, p_i, t)}{dd_i} \right\rangle_{kt}$$
$$- \frac{1}{k_B} \left\langle \delta(q - q_i)[G_i(Q_k, p_i, t) - G_i(q, t)] \frac{dS_k(Q_k, P_k, t)}{dd_i} \right\rangle_{kt}.$$

The thermodynamic volume partial derivative of $G_k(q, t)$ is obtained by summing $\partial G_i(q, t)/\partial \delta_i$ over $i \in k$. The volume partial derivative of $G_k(t)$ is obtained by averaging $\partial G_k(q, t)/\partial \delta_k$ over the volume, which means integrating it over $q \in d_k(t)$ and dividing the result by $\delta_k(t)$. It is not hard to show that these steps yield the same quantity as that expressed in relation (10.48).

Because the entropy is not defined for the singular exact state system distribution, formula (10.48) cannot be used to compute the volume derivative of exact state thermodynamic functions. The volume derivative will be calculated directly for this case using (10.36). This yields

$$(10.61)$$
$$\frac{\partial G_k^\delta(t)}{\partial \delta_k} = -\int dQ_k \int dP_k \frac{d\chi_{\Omega_k(t)}(Q_k, P_k)}{d\Omega_k} G_k(Q_k, P_k, t) F_k^\delta(Q_k, P_k, t)$$
$$= -\frac{d\chi_{\Omega_k(t)}(Q_k, P_k)}{d\Omega_k} \bigg|_{(Q_k, P_k) = (Q_k(t), P_k(t))} G_k(Q_k(t), P_k(t), t).$$

Because $d\chi_{\Omega_k(t)}(Q_k, P_k)/d\Omega_k$ is singular and non-zero only at the boundary of the k system, the result (10.61) indicates that the k volume partial derivative of a thermodynamic quantity defined by the exact state is singular and nonzero only at the boundary of the k system. It follows that volume derivative of the exact state is either 0 or undefined. This is to be expected in that the trajectories representing the exact state are based only on the microscopic Hamiltonian function and the macroscopic system volume is not part of its definition.

10.5 The Accessible Volume

There is an element of arbitrariness in the definition of the system volume because the choice of where to place the boundaries has latitude as long as all particles belonging to the system fall within it. The system particles need not fill the volume assigned to the system and parts of the assigned volume may not be accessible to the particles. When this is the case, moving the system boundary in regions not occupied by particles or not accessible to them should have little or no effect on the particle physics or thermodynamics of the system. Normally, this is of no consequence because the boundary is simply there as a macroscopic bookkeeping device. However, because the volume derivative of the system will be used to define the system pressure, this arbitrariness can matter. The aspect of the volume important to the pressure of a system is the volume accessible to each of its particles and how they are confined. If this accessible volume changes, through expansion, compression, or some other means, there will be thermodynamic consequences.

The volume accessible to a particle within a system k is governed by its energy, momentum, angular momentum, and the system projection operator $\chi_{\Omega_k(t)}(Q_k, P_k)$. The system projection operator plays a larger role than just maintaining the system boundaries. Its energy, momentum, and angular momentum range limitations imply that particles cannot come too close to each other and that there is a forbidden volume around each particle inaccessible to other particles. Particles that have an interaction potential that rises very sharply as they get closer have well defined volumes from this perspective.

To make the idea of the volume accessible to a particle i more precise, let s be a non-negative real number called the *support threshold*. This is used to define the *i-particle accessible volume set at threshold s* by

$$(10.62) \qquad a_i^{(s)}(t) = \{\, q \in d_k(t) \mid \nu_i(q,t) > s \,\}.$$

The set $a_i^{(0)}(t)$ is the *support* of the i-particle local density at time t, which is defined as the complement of the largest open set on which $\nu_i(q,t)$ vanishes. The *i-particle accessible volume at threshold s* is then defined by

$$(10.63) \qquad \delta_i^{(s)}(t) = \int d^3q \, \chi_{a_i^{(s)}(t)}(q).$$

Because $q_i \notin a_i^{(0)}(t)$ if particle i cannot access the location $q_i \in d_k(t)$, the Lebesgue measure of $a_i^{(0)}(t)$ is the volume accessible to particle i. The k *average accessible volume at threshold s* for the particles in k is then given by

$$(10.64) \qquad \delta_k^{(s)}(t) = \frac{1}{\mathcal{N}_k} \sum_{i \in k} \delta_i^{(s)}(t).$$

The fact that $a_i^{(0)}(t) \subset d_k(t)$ for $i \in k$ means that

$$(10.65a) \qquad\qquad \delta_i^{(s)}(t) \leq \delta_k(t),$$

$$(10.65b) \qquad\qquad \delta_k^{(s)}(t) \leq \delta_k(t).$$

The particle accessible volume and system average accessible volume are used next to define the *i-particle inaccessible volume at threshold s* and the *k inaccessible volume at threshold s* by

$$(10.66a) \qquad\qquad b_i^{(s)}(t) = \delta_k(t) - \delta_i^{(s)}(t),$$

$$(10.66b) \qquad\qquad b_k^{(s)}(t) = \delta_k(t) - \delta_k^{(s)}(t).$$

For most calculations, the threshold s will be set to 0.

As hard-sphere calculations, van der Waals' equation, and similar approximations suggest, it is the accessible volume rather than the nominal system volume that is physically significant in the kinetic component of pressure calculations.[6] The need to use the accessible volume in pressure calculations can be substantiated by kinetic theory calculations. The accessible volume will be discussed further in EIS where it will be shown that it emerges naturally from the TIS formalism.

---

[6]Attempts to approximate the accessible volume were made by Boltzmann [**1896a**], Ursell [**1927**] and then others afterward.

Pressure

The usual equilibrium relation that connects working, pressure, and the change in the volume, is $W = -PdV$. This formula is not compatible with a general particle-based theory. The TIS formalism for pressure and its impact on the relation between work and pressure and on the Euler-Cauchy and Navier-Stokes equations are discussed in this chapter.

11.1 Historical Notes

The origins of kinetic theory and its use in explaining pressure can be traced to the seventeenth century.[1] From the standpoint of kinetic theory, the concept of pressure was given an important treatment by David Bernoulli [**1738**].[2] He computed the pressure of an ideal gas in a cubic box with sides of length l in terms of the momentum change at a surface of the box. A particle i with mass m_i and velocity $v_{i,x} = v_x$ moving in the x-direction in a frame in which the box is at rest will strike a particular wall perpendicular to the x-axis $v_{i,x}/2l$ times a second. The momentum transferred to this wall by a collision is $2p_{i,x}$. The momentum transferred per second to this wall by these collisions is then $2p_{i,x}v_{i,x}/2l = m_i v_{i,x}^2/l$. Summing $p_{i,x}v_{i,x}/l$ over x, y, and z, and multiplying by 2 to account for the walls in each direction, it follows that the total force, in terms of the transfer of momentum per second to all 6 walls, is $2p_i \cdot v_i/l = 4K_i(p_i)/l$. Summing over $i \in k$ and setting the resulting total force equal to $P_k 6l^2$, which is the pressure times the area of the 6 walls, yields the final result

$$(11.1) \qquad P_k \delta_k = \tfrac{2}{3}\mathcal{K}_k.$$

When the temperature is held constant, which means that the average kinetic energy is constant, this result yields the Boyle-Towneley law

$$(11.2) \qquad P_k \delta_k = \text{const.}$$

In later work, the average kinetic energy was associated with the temperature in the formulas $K_i = \tfrac{3}{2}k_B T_k$, where T_k is the absolute temperature, so that $\mathcal{K}_k = \tfrac{3}{2}\mathcal{N}_k k_B T_k$. Using this with (11.1) leads to the law of Gay-Lussac (which is sometimes also attributed to Mariotte)

$$(11.3) \qquad P_k \delta_k = \mathcal{N}_k k_B T_k.$$

A significant step beyond this was made by Clausius [**1870**] in his paper introducing the virial. Clausius' virial theorem is often used in current statistical mechanical treatments to introduce the pressure. Clausius began the proof of his theorem with the sequence of calculations[3]

$$(11.4) \qquad \frac{d}{dt}\left(\sum_{i \in k} m_i v_i \cdot q_i\right) = \sum_{i \in k}[m_i v_i^2 + m_i \dot{v}_i \cdot q_i] = \sum_{i \in k}[2K_i + q_i \cdot F_i].$$

---

[1]See the brief references in James Jeans [**1925**], p. 12, for example, to the work of Pierre Gassendi and Robert Hooke. For a more detailed treatment, see Bailyn [**1994**], Chapters 1 and 2.

[2]Peter Guthrie Tait [**1885**], pp. 275–276, observed that Robert Hooke published essentially same ideas previously in a pamphlet published in 1678 called "De Potentiâ Restitutivâ".

[3]In TIS notation, this equation is written $d(Q_k \cdot P_k)/dt = 2\mathcal{K}_k + Q_k \cdot \mathcal{F}_k$.

The α time average of a thermodynamic mechanical quantity such as $K_i(t)$ is defined as

$$(11.5) \qquad\qquad \overline{K_i}(\alpha) = \frac{1}{\alpha} \int_0^\alpha dt\, K_i(t).$$

Consider next an isolated system with a fixed total energy in a bounded volume, such as a gas in a box. Assuming that the quantity $\sum_{i \in k} m_i v_i \cdot q_i$, which is called the *phase action* in TIS, is bounded by the constant M at each point in time, the time average of equation (11.4) is then

$$(11.6) \qquad\qquad |2\overline{K_k} + \overline{\sum_{i \in k} q_i \cdot F_i}| \le \frac{M}{\alpha}.$$

Taking the $\alpha \to \infty$ limit gives the result

$$(11.7) \qquad\qquad -2\overline{K_k} = \overline{\sum_{i \in k} q_i \cdot F_i},$$

where the quantity $\overline{\sum_{i \in k} q_i \cdot F_i}$ is called the *virial*.

In order to associate this calculation with the pressure, the force F_i on a surface element dA is expressed in terms of the i-particle pressure by $F_i = -P_i dA$.[4] The i-particle virial is next expressed in terms of the i-particle pressure as $-P_i q_i \cdot \hat{n}_{\partial k}(q_i) dA$, where $\hat{n}_{\partial k}(q_i)$ is the inward pointing unit vector at the surface and dA is a surface element. This quantity is integrated over the surface and averaged over time to obtain the quantity $\overline{q_i \cdot F_i}$ for each particle $i \in k$.

As a simple example of a pressure calculation using the virial, consider a spherical volume centered at $r = 0$. In this special case, it follows that $dA\, q_i \cdot \hat{n}_{\partial k}(q_i) = r^3 d\omega$, where $d\omega$ is a solid angle differential. Integrating over the solid angle and setting $\delta_k = 4\pi r^3/3$ yields immediately $-3P_i \delta_k = \overline{q_i \cdot F_i}$, where the average is now over both the time and the surface of the volume and P_i now represents the average i-particle pressure over the surface. Redoing this calculation with the total average pressure P_k of all the particles over the whole surface gives

$$(11.8) \qquad\qquad -3P_k \delta_k = \overline{\sum_{i \in k} q_i \cdot F_i}.$$

By (11.7), this yields the result (11.1) again. The spherical shape of the container was introduced for simplicity, but it is not a requirement for obtaining the result (11.8). This calculation extends Bernoulli's computation with ideal gases to include interparticle interactions in the calculation of the pressure,

---

[4]See the discussion of a similar calculation in Watson [**1876**], pp. 41–45, or Jeans [**1925**], pp. 129–134, 169.

but raises the question of why the inclusion of the interparticle forces does not change the result.

This question was addressed by Pascual Jordan [**1933**], who expressed the force in terms of the interparticle potential by (TIS notation) $F_i = -\partial\phi_{ij}(q_i - q_j)/\partial q_i$. It follows easily that $q_i \cdot F_i = -\frac{1}{2}\sum_{i\in k-j}(q_i - q_j) \cdot \partial\phi_{ij}(q_i - q_j)/\partial q_i$. Jordan, p. 4, used this last relation to explain the coincidence of Bernoulli's result with Clausius' as due to the fact that $\partial\phi_{ij}(q_i - q_j)/\partial q_i \approx 0$ unless $q_i \approx q_j$. When $q_i \approx q_j$, the $(q_i - q_j)$ factor is small, so Jordan concluded that the contribution of the virial is small and the kinetic term strongly dominates. However, the relation (11.7) indicates that the contribution of the virial is twice that of the average kinetic energy, so Jordan's answer to this question does not work. This issue will be examined again from the TIS perspective below.

There are a number of problems with the above strategy for computing the pressure using the virial. A survey of the attempts to do this shows that each of the approaches using the virial to compute the pressure makes unrealistic assumptions about the constancy of the pressure on the boundary of the system. This means that the calculation requires a quiescent system and cannot be used in nonequilibrium situations. Moreover, many derivations of the pressure using the virial require that the system be at equilibrium so that it can be represented by a classical or quantum canonical distribution. In addition to these problems, viewing the pressure as the force on the boundary is not appropriate in TIS because forces are always between particles and the boundary in TIS is a conventional construct that may or may not contain particles at any given time.

For quite some time, there was little change in the formalism related to the pressure. New work on this formalism began with Green's introduction of coordinate parameterization.[5] The TIS approach to the pressure is based on the volume derivative operator that extends the coordinate parameterization of Q_k to the phase coordinate parameterization of (Q_k, P_k).

11.2 The Definition of Pressure

Pressure is a bridging concept that connects the macroscopic and the microscopic worlds. Just as a macroscopic body must maintain its boundaries against external pressure, so too must a particle resist the pressure of those particles surrounding it. Pressure is usually expressed as "force per unit area" measured on a bounding surface. The work dW done by a force F over a distance dq is $dW = F \cdot dq$. This relation can be used to connect the pressure at a system boundary with the work done on the body by its moving walls. In standard equilibrium thermodynamics, this is expressed as $dW = -PdV$.

---

[5]The use of coordinate parameterization to define a volume derivative was discussed in the last chapter.

There are four definitions of pressure that are used in thermodynamics. These are (1) the definition in terms of the virial theorem presented above, (2) minus the derivative of the system energy with respect to the volume with the entropy held constant, $P = -(\partial E/\partial V)_S$, (3) minus the derivative of the work with respect to the volume, $P = -dW/dV$, and (4) minus one third the trace of the Euler-Cauchy stress tensor, $P = -\frac{1}{3}\delta_{\mu\nu}R^{\mu\nu}$. However, these four definitions are not mutually compatible when one attempts to express them in terms of phase analog functions based on an underlying particle model.

The third definition is used in equilibrium theory to relate the pressure and the change in volume to the rate of doing work in a quasistatic process.[6] This relation between the working and pressure does not hold for real processes, which means that it is not a suitable choice on which to base the definition of pressure.

The fourth definition does not fare better. The problems with the Euler-Cauchy equations vis à vis the underlying particle mechanics will be discussed in Section 11.8 below. For now, this option is simply rejected. Similarly, the identification of pressure with the force on the boundary of a system in the first definition runs into difficulties because the system boundary is not a mechanically defined quantity on which forces act as they do with particles. This means that the standard definition of pressure as force per unit area on a boundary is only an approximation.

The definition of the pressure adopted here is based on option (2). Since a local pressure function is desired, the phase analog for the pressure will be a particle sum function. While this is not a necessary requirement, it is in accord with Dalton's law of partial pressures and allows us to track the contributions of the individual particles to the system pressure.

In the last chapter, the accessible volume was identified as the important volume in pressure calculations. In connection with this, Johannes van der Waals [**1881**] realized that the volume occupied by the particles themselves in a system has an effect on the pressure. For this reason he replaced the system volume δ_k with $\delta_k - \mathcal{N}_k b$, where b is a volume occupied by one of the particles that is inaccessible to the others. There may also be external potential fields that render a portion of the system volume inaccessible to the particles in the system as well.

In the TIS treatment, the interparticle forces make a region near one particle inaccessible to other particles. This is reflected in the limit r_c for the closest approach between particles. Because the Hamiltonian plays a role in the definition of $\Omega_k(t)$, these forces are taken account of by the system projection operator, which takes the value zero for particle configurations that violate this energy bound. This is most easily seen by transforming the $6\mathcal{N}_k$ phase coordinates (Q_k, P_k) into the $6\mathcal{N}_k$ energy coordinates (E_k, Υ_k),

---

[6]For more on the analysis of pressure and work at the system boundary, see Section 11.7. Quasistatic processes are discussed in Chapter 17.

where $E_k = \times_{i \in k} E_i$ is a $\mathcal{N}_k$-dimensional particle energy variable and Υ_k is a $5\mathcal{N}_k$-dimensional set of coordinates orthogonal to these energies. For each E_i in the range of possible i-particle energies, the system projection operator makes certain parts of the volume inaccessible to particle i with the energy E_i.

11.3 The Pressure Analog

The symbolic formula $(\partial H / \partial V)_S$ suggests a definition of the pressure as the volume derivative of the system global Hamiltonian energy with the entropy fixed. The Hamiltonian energy function $\mathcal{H}_k(t)$ is the appropriate choice for these calculations instead of the system energy function $\mathcal{E}_k(t)$ because the Hamiltonian energy is used to compute forces. Applying the volume partial derivative operator defined in equation (10.48) to the k system Hamiltonian gives rise to terms related to the mechanical changes associated with pressure and to terms connected with changes in the heat contained in the system that result from the changes in the particle distribution as the volume changes.

Before proceeding, let us examine the dependence of the pressure on the coordinate and momentum frame chosen for the calculation. David Bernoulli's simple ideal gas calculation will be reconsidered for this purpose. On the face of it, neither the frequency at which the particle strikes the walls of the box nor the momentum transferred to the box on each impact should depend on the frame of reference within which it is observed.

In the original frame the walls are at rest so $v_{w,x} = 0$, where $v_{w,x}$ is the velocity of a wall in the x direction. Suppose a Galilean transformation is used to transform the original frame to a new frame in which the box is moving in the $-x$ direction with a velocity $-v_a$ so that the walls have a velocity $v_{w,x} = -v_a$ in the new frame. Let us assume that $|v_x| > |v_a|$. When the particle is moving in the positive x direction with velocity v_x in the original frame, it has the velocity $v_x + v_a$ in the new frame. Let us label the wall on the left as 0 and the wall on the right as 1. The time required for the particle to move from wall 0 to 1 is determined in the new frame by the motion of both the particle and the wall as $(v_x + v_a)t_{01} - v_a t_{01} = l$. Solving this gives $t_{01} = l/v_x$, which is the same result as before.

In the original frame, the particle approaches wall 1 with a relative velocity v_x, so it is reflected with a relative velocity $-v_x$ and the momentum change at wall 1 is $2p_x = 2m_i v_x$. In the new frame, it approaches the wall with the relative velocity $(v_x + v_a) - v_a = v_x$ and recedes from it with a relative velocity $-(v_x + v_a) + v_a = -v_x$. The momentum change at wall 1 is $2p_x$ as before.

The time required for the particle to move from wall 1 to wall 0 with the velocity $-(v_x + v_a)$ is determined by $-(v_x + v_a)t_{10} + v_a t_{01} - -l$. Solving this also yields $t_{10} = l/v_x$. It follows that the total time is $2l/v_x$ as was found in the original frame. At wall 0, the relative velocity is $-v_x$ and the momentum change at wall 0 is $2p_x$ as before.

Computing the pressure on all 6 walls as before gives the same result $P\delta_k = (2/3)\mathcal{K}(p)$, where $\mathcal{K}(p) = (m/2)[v_x^2 + v_y^2 + v_z^2]$. This ideal gas example shows that the kinetic energy term in the virial equation should be evaluated in the original rest frame of the box. In TIS terminology, it is the system rest frame, which is the rest frame of the walls defined in Chapter 2, that is the appropriate Galilean frame for the calculation. For fixed walls, the value $2\mathcal{K}_k(P_k)$ of the kinetic component of the k pressure phase analog is always calculated in this special frame and therefore is independent of the reference frame of the observer.

The k system pressure will be defined below as the expectation function of the volume derivative of the system Hamiltonian. In order to obtain a definitive result for the pressure and to facilitate the comparison of the pressure in different systems, this calculation must be made in the same coordinate and momentum frame for all systems involved in a given calculation. The discussion above indicates that the k system rest frame is the natural frame in which to compute the average system pressure. For a collection of systems, the system rest frame of the collection as a whole must be used. In certain cases, such as the measurement of the pressure in a system by an instrument that is not attached to a wall, the local rest frame of the instrument is the important frame for computing the pressure it will measure.

For simplicity, it will be assumed that none of the phase quantities used in this chapter depends explicitly on the volume $\delta_k(t)$ as a parameter.

Using (10.50), the thermodynamic volume partial derivative of the Hamiltonian is[7]

$$(11.9) \qquad \frac{\partial \mathcal{H}_k(t)}{\partial \delta_k} = -P_k(t) + \Lambda_{k,V}(t),$$

where the k *system pressure* is given by a definition based on (10.51) as

$$(11.10) \qquad P_k(t) = -\left. \frac{\partial \mathcal{H}_k(t)}{\partial \delta_k} \right|_S ,$$

---

[7]Equation (11.9) is the nonequilibrium generalization of the equilibrium equation that is written in the standard notation of equilibrium theory in the form

$$\left(\frac{\partial H}{\partial V} \right)_T = -P + T \left(\frac{\partial P}{\partial T} \right)_V ,$$

where V is the volume, T is the absolute temperature and P is the pressure. William Thomson obtained a result equivalent to this in 1851 (see Thomson [**1882**], Vol. I, p. 227) as part of his research on the dynamical theory of heat. For the equilibrium version of this formula in TIS notation and its derivation in the equilibrium setting, see equation (18.28). See also Chapter 17 for a discussion of the latent heat with respect to volume in the context of classical thermodynamics.

and the k *latent heat with respect to volume* is

(11.11)
$$\Lambda_{k,V}(t) = -\frac{1}{k_B}\left\langle \mathcal{H}_k(Q_k, P_k, t)\frac{dS_k(Q_k, P_k, t)}{d\Omega_k}\right\rangle_{kt} - \frac{\partial \ln Z_k(t)}{\partial \delta_k}\mathcal{H}_k(t)$$
$$= -\frac{1}{k_B}\left\langle [\mathcal{H}_k(Q_k, P_k, t) - H_k(t)]\frac{dS_k(Q_k, P_k, t)}{d\Omega_k}\right\rangle_{kt}.$$

The fact that $S_k(Q_k, P_k, t)$ does not depend explicitly on δ_k was also used to set $dS_k(Q_k, P_k, t)/d\delta_k = dS_k(Q_k, P_k, t)/d\Omega_k$ in (11.11).

The term $(\partial \mathcal{H}_k(t)/\partial \delta_k)_S$ is computed next. Since $\mathcal{H}_k(Q_k, P_k, t)$ does not depend on the volume $\delta_k(t)$, using the definition (10.51) with (11.10) gives

(11.12) $$P_k(t) = \langle P_k(Q_k, P_k, t)\rangle_{kt} = -\left\langle \frac{d\mathcal{H}_k(Q_k, P_k, t)}{d\Omega_k}\right\rangle_{kt}.$$

It follows that the analog for the pressure is

(11.13)
$$P_k(Q_k, P_k, t) = -\frac{d\mathcal{H}_k(Q_k, P_k, t)}{d\Omega_k}$$
$$= -\left[\frac{d\mathcal{K}_k(P_k)}{d\Omega_k} + \frac{d\Phi_{k,n}(Q_k)}{d\Omega_k} + \frac{d\Phi_{k,x}(Q_k, t)}{d\Omega_k}\right].$$

The first term on the right in equation (11.13) is called the *kinetic virial*, the second term the *internal virial* and the third term is the *external virial*. The k *phase latent heat with respect to volume* analog is

(11.14) $$\Lambda_{k,V}(Q_k, P_k, t) = -\frac{1}{k_B}[\mathcal{H}_k(Q_k, P_k, t) - H_k(t)]\frac{dS_k(Q_k, P_k, t)}{d\Omega_k}.$$

Calculating the volume derivatives in (11.13) leads to the k *phase pressure analog*
(11.15)
$$P_k(Q_k, P_k, t) = \frac{1}{3\delta_k(t)}\left[2\mathcal{K}_k(P_k) - Q_k \cdot \frac{\partial \Phi_{k,n}(Q_k)}{\partial Q_k} - Q_k \cdot \frac{\partial \Phi_{k,x}(Q_k, t)}{\partial Q_k}\right].$$

It is not hard to see that the definition of the total time derivative can be used to write this in the form

(11.16) $$P_k(Q_k, P_k, t) = \frac{1}{3\delta_k(t)}\frac{d(Q_k \cdot P_k)}{dt}.$$

This result illustrates the deep connection between the time derivative of the phase action $Q_k \cdot P_k$, the virial, and the pressure that was first noticed by Clausius.[8]

---

[8]This form for the pressure analog is not new. It has been derived in various settings, using time derivatives of the Jacobi function, $J_k^{(J)}(Q_k) = (1/2)\sum_i m_i q_i^2$, in both classical

The next step is to generalize these quantities to the case of the i-particle pressure tensor analog. This will be done in the next section after the discussion of the i-particle virial phase tensors.

From a historical perspective, the division of the volume derivative into the pressure and the latent heat with respect to volume reflects a distinction made by Boltzmann identifying the changes in a thermodynamic quantity as a combination of changes in the analog of the quantity itself and changes in the particle distribution it is averaged over.[9] The definition of the pressure as the phase average of the volume phase derivative of the Hamiltonian in (11.12) and its expression in terms of other functions in (11.13) represent mechanical changes due to a change in the system volume. The presence of the volume phase derivative of the entropy in the analog of the latent heat with respect to volume in (11.14) indicates that this is a heat-related function connected with changes in the system distribution.

11.4 The Virial Tensors

The virial tensors associated with a system Hamiltonian are defined using the *virial tensor derivative* operator introduced above in (10.21). It will be used to define a generalization of Clausius' virial quantities. Beginning with the kinetic energy component of the Hamiltonian, the *i-particle phase kinetic virial tensor* is

$$(11.17) \qquad \frac{d\mathcal{K}_k(p_i)}{dd_{i,\mu\nu}} = -p_i^{\mu\nu}(p_i),$$

where the i-particle momentum current $p_i^{\mu\nu}$ is defined in (7.11). The *i-particle phase internal virial tensor, k phase internal virial tensor, i-particle phase internal virial* and the *k phase internal virial* are defined similarly by

$$(11.18a) \qquad \Xi_{i,n}^{\mu\nu}(Q_k) = \frac{d\Phi_{k,n}(Q_k)}{dd_{i,\mu\nu}} = \frac{1}{2} \sum_{j \in k-i} q_i^{(\mu} \frac{\partial \phi_{ij}(q_i - q_j)}{\partial q_{i,\nu)}},$$

$$(11.18b) \qquad \Xi_{k,n}^{\mu\nu}(Q_k) = \sum_{i \in k} \Xi_{i,n}^{\mu\nu}(Q_k),$$

$$(11.18c) \quad \Xi_{i,n}(Q_k) = \tfrac{1}{3}\delta_{\mu\nu}\Xi_{i,n}^{\mu\nu}(Q_k), \qquad \Xi_{k,n}(Q_k) = \tfrac{1}{3}\delta_{\mu\nu}\Xi_{k,n}^{\mu\nu}(Q_k).$$

It is easy to show using (4.2e) that $\Xi_{k,n}(Q_k)$ can be written in the form

$$(11.19) \ \Xi_{k,n}(Q_k) = \tfrac{1}{6} \sum_{i \in k} \sum_{j \in k-i} (q_i - q_j) \cdot \frac{\partial \phi_{ij}(q_i - q_j)}{\partial q_i} = \tfrac{1}{3} Q_k \cdot \frac{\partial \Phi_{k,n}(Q_k)}{\partial Q_k}.$$

---

and quantum virial approaches. The role of the Jacobi function in mechanics is discussed in QTS.

[9]Boltzmann's [**1871b**] interpretation of thermodynamic changes in these terms is discussed in Chapter 22 and in Bailyn [**1994**], pp. 450–451.

The *i-particle phase external virial tensor*, *k phase external virial tensor*, *i-particle phase external virial*, and the *k phase external virial* are defined similarly by

$$\Xi_{i,x}^{\mu\nu}(q_i,t) = \frac{d\Phi_{k,x}(Q_k,t)}{dd_{i,\mu\nu}} = \sum_{l\in[t]-k}\sum_{j\in l}\left\langle q_i^{(\mu}\frac{\partial\phi_{ij}(q_i-q_j)}{\partial q_{i,\nu)}}\right\rangle_{lt}$$

(11.20a)

$$= q_i^{(\mu}\frac{\partial\Phi_{i,x}(q_i,t)}{\partial q_{i,\nu)}},$$

(11.20b)

$$\Xi_{k,x}^{\mu\nu}(Q_k,t) = \sum_{i\in k}\Xi_{i,x}^{\mu\nu}(q_i,t),$$

(11.20c)

$$\Xi_{i,x}(q_i,t) = \tfrac{1}{3}\delta_{\mu\nu}\Xi_{i,x}^{\mu\nu}(q_i,t), \qquad \Xi_{k,x}(Q_k,t) = \tfrac{1}{3}\delta_{\mu\nu}\Xi_{k,x}^{\mu\nu}(Q_k,t).$$

The *i-particle total virial tensor* and *k total virial tensor* are expressed in terms of the internal and external virials by

(11.21a) $$\qquad\qquad \Xi_i^{\mu\nu}(Q_k,t) = \Xi_{i,n}^{\mu\nu}(Q_k) + \Xi_{i,x}^{\mu\nu}(q_i,t),$$

(11.21b) $$\qquad\qquad \Xi_k^{\mu\nu}(Q_k,t) = \Xi_{k,n}^{\mu\nu}(Q_k) + \Xi_{k,x}^{\mu\nu}(Q_k,t),$$

It is a consequence of these definitions that one-third the trace of $\Xi_k^{\mu\nu}(Q_k,t)$ is

(11.21c) $$\tfrac{1}{3}\delta_{\mu\nu}\Xi_k^{\mu\nu}(Q_k,t) = \tfrac{1}{3}Q_k\cdot\frac{\partial\Phi_k(Q_k,t)}{\partial Q_k} = \Xi_{k,n}(Q_k) + \Xi_{k,x}(Q_k,t).$$

Clausius [1870], (Kestin [1976], p. 176), called both the quantity designated here by $3\Xi_{k,n}(Q_k)$ and its time average for an exact state designated by $3\overline{\Xi}_{k,n}(\alpha) = 3\overline{\Xi_{k,n}(Q_k)}$ the *virial* of the system. The time averaged version was used in the virial theorem (11.7). The pressure was obtained from this result in (11.8) using the divergence theorem and some implicit assumptions.[10] For reasons similar to those discussed below in Section 11.8, this method cannot be adapted to a particle oriented analysis so it will not be used to compute the pressure here.

The terminology of Clausius [1870], p. 176 was followed in adopting the names internal virial and external virial for $\Xi_{k,n}$ and $\Xi_{k,x}$. The term kinetic virial was chosen for $\sum_{i\in k}d\mathcal{K}_k(p_i)/dd_i$ by extension of this terminology to characterize the result of using the virial derivative on the kinetic energy. All of these virial quantities play a role in the expression of the pressure and its analog. In opposition to the approach of Clausius, however, time averaging is not used in the TIS computation of the pressure.

---

[10]Clausius' work and this method of defining the pressure will be discussed further in EIS for the classical case and the "virial controversy" concerning the virial theorem for both the classical and quantum cases is analyzed in QTS.

Let us now add some definitions to those given above. The local and global phase averages of the i-particle and k internal and external virial tensors yields the *i-particle* and *k local and global kinetic, internal, and external virial density tensors*

(11.22a)
$$p_i^{\mu\nu}(q,t) = \langle \delta(q-q_i) p_i^{\mu\nu}(p_i) \rangle_{kt}, \qquad p_i^{\mu\nu}(t) = \int d^3q \, p_i^{\mu\nu}(q,t),$$

(11.22b) $\quad \mathcal{P}_k^{\mu\nu}(q,t) = \sum_{i\in k} p_i^{\mu\nu}(q,t), \qquad \mathcal{P}_k^{\mu\nu}(t) = \int d^3q \, \mathcal{P}_k^{\mu\nu}(q,t),$

(11.22c)
$$\Xi_{i,n}^{\mu\nu}(q,t) = \langle \delta(q-q_i) \Xi_{i,n}^{\mu\nu}(Q_k) \rangle_{kt}, \qquad \Xi_{i,n}^{\mu\nu}(t) = \int d^3q \, \Xi_{i,n}^{\mu\nu}(q,t),$$

(11.22d) $\quad \Xi_{k,n}^{\mu\nu}(q,t) = \sum_{i\in k} \Xi_{i,n}^{\mu\nu}(q,t), \qquad \Xi_{k,n}^{\mu\nu}(t) = \int d^3q \, \Xi_{k,n}^{\mu\nu}(q,t),$

(11.22e)
$$\Xi_{i,x}^{\mu\nu}(q,t) = \langle \delta(q-q_i) \Xi_{i,x}^{\mu\nu}(q_i,t) \rangle_{kt}, \qquad \Xi_{i,x}^{\mu\nu}(t) = \int d^3q \, \Xi_{i,x}^{\mu\nu}(q,t),$$

(11.22f) $\quad \Xi_{k,x}^{\mu\nu}(q,t) = \sum_{i\in k} \Xi_{i,x}^{\mu\nu}(q,t), \qquad \Xi_{k,x}^{\mu\nu}(t) = \int d^3q \, \Xi_{k,x}^{\mu\nu}(q,t).$

The traces of the local and global phase averages of the i-particle and k internal and external virial tensors yields the *i-particle* and *k local and global kinetic, internal, and external virial densities* given by

(11.23a) $\qquad K_i(q,t) = \frac{1}{2}\delta_{\mu\nu} p_i^{\mu\nu}(q,t), \qquad K_i(t) = \frac{1}{2}\delta_{\mu\nu} p_i^{\mu\nu}(t),$

(11.23b) $\qquad \mathcal{K}_k(q,t) = \sum_{i\in k} K_i(q,t), \qquad \mathcal{K}_k(t) = \sum_{i\in k} K_i(t),$

(11.23c) $\qquad \Xi_{i,n}(q,t) = \frac{1}{3}\delta_{\mu\nu} \Xi_{i,n}^{\mu\nu}(q,t), \qquad \Xi_{i,n}(t) = \frac{1}{3}\delta_{\mu\nu} \Xi_{i,n}^{\mu\nu}(t),$

(11.23d) $\qquad \Xi_{k,n}(q,t) = \sum_{i\in k} \Xi_{i,n}(q,t), \qquad \Xi_{k,n}(t) = \sum_{i\in k} \Xi_{i,n}(t),$

(11.23e) $\qquad \Xi_{i,x}(q,t) = \frac{1}{3}\delta_{\mu\nu} \Xi_{i,x}^{\mu\nu}(q,t), \qquad \Xi_{i,x}(t) = \frac{1}{3}\delta_{\mu\nu} \Xi_{k,x}^{\mu\nu}(t),$

(11.23f) $\qquad \Xi_{k,x}(q,t) = \sum_{i\in k} \Xi_{i,x}(q,t), \qquad \Xi_{k,x}(t) = \Xi_{i,x}(t).$

The purpose of this blizzard of definitions and relations is to allow easy reference to the forms of these quantities and their connections that will be needed below.

11.5 The Local Pressure Tensor and Pressure

It is time now to put together the components presented in the last few sections. Let us define the *i-particle phase pressure tensor* and the *k phase pressure tensor* by

(11.24a)
$$\mathrm{P}_i^{\mu\nu}(Q_k, p_i, t) = p_i^{\mu\nu}(p_i) - \Xi_{i,n}^{\mu\nu}(Q_k) - \Xi_{i,x}^{\mu\nu}(q_i, t),$$

(11.24b)
$$\mathrm{P}_k^{\mu\nu}(Q_k, P_k, t) = \sum_{i\in k} \mathrm{P}_i^{\mu\nu}(Q_k, p_i, t).$$

The phase pressure analog is defined as one-third of the trace of the pressure tensor. With this definition, (11.24), and the definition (11.15), the *i-particle phase pressure* and *k phase pressure* are

(11.25a)
$$P_i(Q_k, p_i, t) = \tfrac{1}{3}\delta_{\mu\nu}\mathrm{P}_i^{\mu\nu}(Q_k, p_i, t),$$

(11.25b)
$$P_k(Q_k, P_k, t) = \sum_{i\in k} P_i(Q_k, p_i, t).$$

The *i particle local pressure* is the local phase average of the *i*-particle phase pressure. It is defined by

(11.26)
$$P_i(q, t) = \langle \delta(q - q_i)P_i(Q_k, p_i, t)\rangle_{kt},$$

Phase averaging each side of equation (11.24) gives the *i-particle local pressure tensor* and the *k local pressure tensor*

(11.27a)
$$\mathrm{P}_i^{\mu\nu}(q, t) = p_i^{\mu\nu}(q, t) - \Xi_{i,n}^{\mu\nu}(q, t) - \Xi_{i,x}^{\mu\nu}(q, t),$$

(11.27b)
$$\mathrm{P}_k^{\mu\nu}(q, t) = \mathcal{P}_k^{\mu\nu}(q, t) - \Xi_{k,n}^{\mu\nu}(q, t) - \Xi_{k,x}^{\mu\nu}(q, t).$$

These results are in accord with Axiom 8.1. One-third of the trace of (11.27) gives the *i-particle local pressure* and the *k local pressure* in the form

(11.28a)
$$P_i(q, t) = \tfrac{1}{3}\left[2K_i(q, t) - \Xi_{i,n}(q, t) - \Xi_{i,x}(q, t)\right],$$

(11.28b)
$$P_k(q, t) = \tfrac{1}{3}\left[2\mathcal{K}_k(q, t) - \Xi_{k,n}(q, t) - \Xi_{k,x}(q, t)\right].$$

For an isolated system t, it is easy to show that $\Xi_{i,x}^{\mu\nu}(q, t) = \Xi_{t,x}^{\mu\nu}(q, t) = 0$ in accord with (2.21b) and that the pressure is defined in terms of $\mathcal{K}_t$ and $\Xi_{t,n}$.

11.6 The System Pressure Tensor and Pressure

The system pressure tensor and pressure are next defined in terms of the local pressure tensor and local pressure. By construction, the i-particle system pressure tensor is the average over the volume of the i-particle local pressure tensor and the i-particle system pressure is the average over the volume of the i-particle local pressure:

$$(11.29a) \qquad P_i^{\mu\nu}(t) = \frac{1}{\delta_k(t)} \int d^3q \, P_i^{\mu\nu}(q,t).$$

$$(11.29b) \qquad P_i(t) = \frac{1}{\delta_k(t)} \int d^3q \, P_i(q,t).$$

Summing (11.29b) over $i \in k$ yields the formula (3.20c). The definitions (11.27) and (11.28) are used to give us the *i particle pressure tensor, k pressure tensor, i particle pressure,* and *k pressure* as

$$(11.30a) \qquad P_i^{\mu\nu}(t) = \frac{1}{\delta_k(t)} \left[2p_i^{\mu\nu}(t) - \Xi_{i,n}^{\mu\nu}(t) - \Xi_{i,x}^{\mu\nu}(t) \right],$$

$(11.30b)$

$$P_k^{\mu\nu}(t) = \sum_{i \in k} P_i^{\mu\nu}(t),$$

$$(11.30c) \qquad P_i(t) = \frac{1}{3\delta_k(t)} \left[2K_i(t) - \Xi_{i,n}(t) - \Xi_{i,x}(t) \right],$$

$$(11.30d) \qquad P_k(t) = \sum_{i \in k} P_i(t).$$

The derivatives of the macroscopic pressure $P_k(t)$ with respect to time, thermature, and volume, are readily computed using the TIS formalism. The thermature derivative of the pressure will be investigated in Part II and the first and second volume derivatives of the pressure will be examined in EIS.

The $[t]$ system pressure is defined as the weighted average over the subsystem pressures:

$$(11.31) \qquad P_{[t]}(t) = \frac{1}{\delta_t(t)} \sum_{k \in [t]} \delta_k(t) P_k(t).$$

This choice is adopted because it yields the systems level version of Dalton's law of partial pressures. The definition (11.31) is in accord with Axiom 8.2.

It should be noted that the additivity of the pressure components used in the above definitions depends on the additivity of the particle-particle interaction potential energy terms and may be false if the latter additivity does not hold. The definition (11.9) will be used in any case.

11.7 Pressure and Work

Bailyn [1994], p. 297, pointed out that Gibbs found that the "hydrostatic work", $dW = -PdV$, is not adequate to account for work done in an elastic solid. This is not surprising because this definition of the work done in terms of the pressure and the change in volume is not particle based. However, in many circumstances it is an adequate approximation within experimental accuracy. It therefore retains its importance and usefulness in equilibrium calculations even though it is not part of the general theory.

To explore the relation between the equilibrium definition and the TIS definition, the approximations necessary to express the total working at the boundary in the standard form $W_k^{(\mathrm{std})}(t) = -\mathrm{P}_k(t)\dot{\delta}_k(t)$ will be discussed. The usual derivation of the thermodynamic working in terms of the pressure and change of volume begins with the definition of pressure as 'force per unit area'. The TIS representation of the changing volume uses (10.14) to obtain $d\delta_k = -dt \int_{\partial k} d^2r\, \delta(q-r)\hat{n}_{\partial k}(r,t) \cdot v_{\partial k}(r,t)$. This relation expresses the fact that it is motion normal to the boundary that contributes to the change in volume. Proceeding informally, the local approximate "wall pressure" on a small area A centered at $q \in \partial d_k(t)$ can be defined in terms of a wall force density by

$$(11.32) \qquad \mathcal{F}_{\partial k}^{(\mathrm{wall})\mu}(q,t) = -\hat{n}_{\partial k}^{\mu}(q,t)\, \mathrm{P}_k^{(\mathrm{wall})}(q,t)A.$$

The presence of the arbitrary area A in this formula indicates that this is an approximation. The wall pressure defined in (11.32) represents the pressure at the boundary due to the effects of the particles in the k system and may be positive or negative depending on the sign of $\mathcal{F}_{\partial k}^{(\mathrm{wall})\mu}(q,t)$. The direction and speed of the motion of the boundary, $v_{\partial k}(q,t)$, depends on both this pressure and that due to any other system sharing this boundary at the point q.

For a boundary moving at $v_{\partial k}(q,t)$, the rate of change of the standard local working on this area for the k system is then $\dot{W}_k^{(\mathrm{std})}(q,t) = v_{\partial k}(q,t) \cdot F_{\partial k}^{(\mathrm{wall})}(q,t)$. The relation (11.32) is used to express this in the form $\dot{W}_k^{(\mathrm{std})}(q,t) = -\hat{n}_{\partial k}(q,t) \cdot v_{\partial k}(q,t)\mathrm{P}_k^{(\mathrm{wall})}(q,t)A$. The singular operator $(1/A) \int_{\partial k} d^2r\, \delta(q-r)$ is applied next to both sides of this equation followed by an integration of both sides over $q \in d_k(t)$. These steps lead to the expression

$$(11.33) \qquad dW_k^{(\mathrm{std})}(t) = -dt \int_{\partial k} d^2r\, \hat{n}_{\partial k}(r,t) \cdot v_{\partial k}(r,t)\mathrm{P}_k^{(\mathrm{wall})}(r,t),$$

where the differential of the standard thermodynamic work is defined as $dW_k^{(\mathrm{std})}(t) = dt \int_{\partial k} d^2r\, \dot{W}_k^{(\mathrm{std})}(r,t)$. If the wall pressure at the boundary is uniform and equal to the system wall pressure, so that $\mathrm{P}_k^{(\mathrm{wall})}(r,t) = \mathrm{P}_k^{(\mathrm{wall})}(t)$, then (10.14) allows us to write (11.33) as

$$(11.34) \qquad dW_k^{(\mathrm{std})}(t) = -d\delta_k(t)\mathrm{P}_k^{(\mathrm{wall})}(t).$$

The major problems with this derivation of the standard work differential from the approximate wall pressure are

1. the force on the boundary defined in (11.32) is not connected with the particles exerting it and the particles experiencing it;
2. the work done on the particles by shearing forces is neglected;
3. the local wall pressure is required to be uniform on the boundary and equal to the system wall pressure.

These problems show that the work differential in (11.34) can only be used in very special cases, such as equilibrium, and cannot be used to express the working within a general thermodynamics that is consistent with the underlying particle mechanics.

Let us next see how (11.33) can be approximated within the TIS formalism. To define the pressure, the rate of change of the energy and the momentum at a boundary point $q \in \partial d_k(t)$ of system k will be evaluated in the reference frame in which the k system is at rest. The first step is to consider the force due to the momentum flow into the surface. By (7.18) and (2.12), the total force at the boundary in this frame is given by the rate of change of the local i-particle momentum density at the boundary as $F_{\partial k}^{(\text{wall})\mu}(q,t) = \partial p_i^\mu(q,t)/\partial t = -\partial p_i^{\mu\nu}(q,t)/\partial q^\nu + F_{b,i}^\mu(q,t) + F_{\partial i}^\mu(q,t)$ for $q \in \partial d_k(t)$. Observe that in accord with TIS rules, this wall force density at the boundary is due only to particles that have a non-zero probability of being at the location $q_i = q \in \partial d_k(t)$ at time t. For the wall force density to be useful in computing an approximate pressure at the wall, it is necessary that the long-range external forces that pass through the system boundary are negligible so that $\mathcal{F}_{k,x}(q,t) \approx 0$.

The local surface working density is expressed in terms of the wall force density and the boundary velocity as $\mathcal{W}_{\partial k}^{(\text{wall})}(q,t) = -\mathcal{F}_{\partial k}^{(\text{wall})}(q,t) \cdot v_{\partial k}(q,t)$ in (2.19). The minus sign is included so that work will be done on the system, increasing the energy $W_{\partial k}^{(\text{wall})}$, when $\mathcal{F}_{\partial k}^{(\text{wall})}(q,t) \cdot v_{\partial k}(q,t) < 0$.

In the first step, shearing forces at the boundary are neglected and the local wall force density is replaced by its component normal to the boundary: $\mathcal{F}_{\partial k}^{(\text{wall})}(q,t) \to -\hat{n}_{\partial k}(q,t) \cdot \mathcal{F}_{\partial k}^{(\text{wall})}(q,t)$. The fact that $\hat{n}_{\partial k}(q,t)$ is inward pointing accounts for the minus sign. Similarly, the boundary velocity is replaced by its component normal to the boundary: $v_{\partial k}(q,t) \to -\hat{n}_{\partial k}(q,t) \cdot v_{\partial k}(q,t)$. These steps yield

$$(11.35) \quad \dot{W}_k(q,t) \approx \mathcal{W}_{\partial k}^{(\text{wall})}(q,t) = -\hat{n}_{\partial k}(q,t) \cdot \mathcal{F}_{\partial k}^{(\text{wall})}(q,t)\hat{n}_{\partial k}(q,t) \cdot v_{\partial k}(q,t).$$

Integrating this over $q \in \partial d_k(t)$, gives the result
(11.36)
$$\dot{W}_k(t) \approx \mathcal{W}_{\partial k}^{(\text{wall})}(t) = -\int d^3q\, \hat{n}_{\partial k}(q,t) \cdot \mathcal{F}_{\partial k}^{(\text{wall})}(q,t)\hat{n}_{\partial k}(q,t) \cdot v_{\partial k}(q,t).$$

Using (10.14), this is approximated by

(11.37)

$$\dot{W}_k(t) \approx \mathcal{W}_{\partial k}^{(\text{wall})}(t) \approx - \int d^3q\,\hat{n}_{\partial k}(q,t)\cdot\mathcal{F}_{\partial k}^{(\text{wall})}(q,t) \int d^3q\,\hat{n}_{\partial k}(q,t)\cdot v_{\partial k}(q,t)$$

$$= -\mathrm{P}_k^{(\text{wall})}(t)\frac{d\delta_k(t)}{dt},$$

where $\mathrm{P}_k^{(\text{wall})}(t) = -\int_{\partial k} d^2r\,\hat{n}_{\partial k}(r,t)\cdot\mathcal{F}_{\partial k}^{(\text{wall})}(r,t)$ is the approximate pressure obtained by integrating the outward normal surface force density over the surface $\partial d_k(t)$.

The assumptions needed to make equation (11.37) valid are clearly too restrictive for any sort of general theory, but they are valid for the very important equilibrium case. The use of the form (11.37) in equilibrium theory will be examined and the approximation (11.37) will be used in the TIS version of the theory of the phase boundary in EIS, Part IV.

11.8 The Euler-Cauchy and Navier-Stokes Equations

The Euler-Cauchy and Navier-Stokes equations are two sets of macroscopic equations describing the behavior of fluid matter. These macroscopic equations are obtained using the methods of continuum mechanics and employ specific approximations in their derivations. In these derivations, the macroscopic equations for the transfer of conserved quantities are obtained by considering small volumes within a body. These volumes are assumed to be small enough to represented by an infinitesimal volume element $d\delta_k$ in the calculus, but large enough so that they contain many particles. A force vector, representing the net force on the volume by particles outside it, is assigned to each of these small volumes. The volume also is assigned a momentum vector and the rates of change of various quantities are computed using the laws of motion with these volumes. The Navier-Stokes and Euler-Cauchy equations are two sets of equations of motion for physical quantities obtained using this procedure.

The equations of Leonhard Euler for a perfect fluid, which is a fluid in which the tangential or shearing stresses can be neglected, were developed in the seventeenth century. For an irrotational, incompressible fluid, for example, these equations led to Daniel Bernoulli's equation. In TIS notation, Bernoulli's equation expresses the system pressure in terms of a constant $\mathcal{C}_k$, the potential energy $\Phi_k(t)$, the mass density $\mu_k(t)$, and the average velocity $\bar{v}_k(t)$ in the form $\mathrm{P}_k(t) = \mathcal{C}_k - \Phi_k(t) + (1/2)\mu_k(t)\bar{v}_k^2(t)$.[11] Euler's equations were extended in the nineteenth century by Augustin Cauchy to include shearing stresses expressed in the form of a stress tensor, which will be designated

---

[11] For a mathematical discussion of the equations of motion for fluids and Bernoulli's equation, see Morse and Feshbach [**1953**], Vol. 1, pp. 151–171. A summary of the associated global balance equations and their relations appears in Bailyn [**1994**], Chapter 5.

here by $R_k^{\mu\nu}$. Claude-Louis-Marie-Henri Navier presented a similar set of equations to the French Academy in 1822. These were soon extended by George Gabriel Stokes and others.

While the approach of continuum mechanics and the approximating functions it employs, such as the coefficients of viscosity and the coefficient of expansive friction, offer a plausible description of the mechanics of fluids, the assumptions underlying this approach are not compatible with a particle based viewpoint. In this section, the problems associated with reconciling the assumptions and approximations of continuum mechanics with the particle approach will be investigated.

Some time after the development of the Euler-Cauchy and Navier-Stokes equations, both Maxwell and Boltzmann attempted to justify macroscopic hydrodynamic equations in terms of the kinetic theory of microscopic particles. Maxwell was aware that his original derivations of the transport equations of hydrodynamics from kinetic theory in Maxwell [1860], [1866], required too many approximations and neglected too many terms, so he remarked in Maxwell [1879a] and that he had made new calculations in this area and had new results. After Maxwell's death, Boltzmann [1894] published a plea in the Report of the British Association that Maxwell's papers be examined to see if a manuscript could be found. Boltzmann raised the question of whether the problems in the previous derivations were due to defects in the theory of gases or in hydrodynamics. He stated that he had been making calculations in this area himself, but was holding back publication of his work pending the discovery of Maxwell's work because his results did not agree with the remarks published by Maxwell and the "danger of falling into errors on this subject is very great."

A detailed comparison of the methods of continuum mechanics with those of TIS will illustrate the differences and similarities of these ways of approaching the subject. To facilitate comparing the TIS formalism with the formalism of continuum mechanics, certain quantities and equations of continuum mechanics will be expressed in terms of the TIS quantities and equations defined above.

The *material stream velocity* is expressed in terms of the local momentum density and the mass density by

$$(11.38) \qquad \bar{u}_k^\mu(q,t) = [\mu_k(q,t)]^{-1} p_k^\mu(q,t).$$

The *material time derivative*, also called the "convective derivative" in Bailyn [1994], p. 160, is the time derivative in a frame moving with the local stream velocity. The material time derivative of any local thermodynamic quantity $G_k(q,t)$, which may have tensor indices, is defined by

$$(11.39a) \qquad \frac{d_m G_k(q,t)}{dt} = \frac{\partial G_k(q,t)}{\partial t} + \bar{u}_k^\mu(q,t) \frac{\partial G_k(q,t)}{\partial q^\mu}.$$

This can be expressed in terms of quantities defined in the Theory of Interacting Systems by

$$(11.39\text{b}) \qquad \frac{d_m G_k(q,t)}{dt} = \frac{dG_k(q,t)}{dt} + \bar{u}_k(q,t) \cdot \frac{\partial G_k(q,t)}{\partial q} - \frac{\partial G_k^\mu(q,t)}{\partial q^\mu},$$

where $dG_k(q,t)/dt$ is the macroscopic total time derivative in TIS defined in (2.12). This equation illustrates clearly the differences in the two approaches. The material derivative views the rate of change of a macroscopic quantity in terms of small elements of substance and attributes a velocity $\bar{u}_k$ to the whole element. This approach gives rise to the flow term $\bar{u}_k \cdot (\partial G_k(q,t)/\partial q)$ in $d_m G_k(q,t)/dt$. In the TIS version, the current $G_k^\mu(q,t)$ is the phase average of $\sum_{i \in k} v_i^\mu G_i(Q_k, p_i, t)$ at the spacetime point (q,t) and its divergence is the flow term.

The equation of continuity for the mass density is

$$(11.40) \qquad \frac{\partial \mu_k(q,t)}{\partial t} + \frac{\partial}{\partial q^\mu} [\mu_k(q,t) \bar{u}_k^\mu(q,t)] = 0.$$

Using this with (11.39a) for the material time derivative of the mass density gives

$$(11.41) \qquad \frac{d_m \mu_k(q,t)}{dt} = -\mu_k(q,t) \frac{\partial \bar{u}_k^\mu(q,t)}{\partial q^\mu}.$$

The k *local vorticity vector* is related to the rotational component of the velocity and defined by

$$(11.42) \qquad \bar{w}_k^\mu(q,t) = \tfrac{1}{2} \epsilon^{\mu\nu\sigma} \frac{\partial \bar{u}_{k,\nu}(q,t)}{\partial q^\sigma}.$$

The definition (11.38) can be used to state this in terms of the quantities $\mu_k(q,t)$ and $p_k^\mu(q,t)$. The material time derivative of the stream velocity may be written in terms of the vorticity as

$$(11.43) \qquad \frac{d_m \bar{u}_k^\mu(q,t)}{dt} = \frac{\partial \bar{u}_k^\mu(q,t)}{\partial t} + \frac{1}{2} \frac{\partial \bar{u}_k^2(q,t)}{\partial q_\mu} + 2\epsilon^{\mu\nu\sigma} \bar{u}_{k,\nu}(q,t) \bar{w}_{k,\sigma}(q,t).$$

Assume next that the k system occupies a small volume element $\Delta \delta_k(t)$ and let the quantity $R_k^{\mu\nu}(q,t; \bar{u}_k(q,t))$ be the k *local stress tensor*. It is defined in such a way that the body force on the volume element $\Delta \delta_k$, centered at the spacetime point (q,t) and moving as a whole with local stream velocity $\bar{u}_k(q,t)$, is expressed in terms of the divergence of the stress tensor by $\mathcal{T}_{k,b}^\mu(q,t) = \partial R_k^{\mu\nu}(q,t; \bar{u}_k(q,t))/\partial q^\nu$. The total force $\mathcal{F}_k^\mu(q,t)$ is the sum of the body force and the external force $\mathcal{F}_{k,x}^\mu(q,t)$ defined in equation (7.15). This is written in terms of Newton's second law in a frame in which the volume element is moving with velocity $\bar{u}_k(q,t)$ as[12]

(11.44a) $$\mu_k(q,t)\frac{d_m\bar{u}_k^\mu(q,t)}{dt} = \frac{\partial R_k^{\mu\nu}(q,t;\bar{u}_k(q,t))}{\partial q^\nu} + \mathcal{F}_{k,x}^\mu(q,t).$$

Compare this to the version of the *Euler-Cauchy balance of linear momentum* equation, presented as Cauchy's first law of motion, in C. Truesdell and R. Toupin [1960], p. 545 (TIS notation):

(11.44b) $$\frac{\partial p_k^\mu(q,t)}{\partial t} = \frac{\partial R_k^{\mu\nu}(q,t;\bar{u}_k(q,t))}{\partial q^\nu} + \mathcal{F}_{k,x}^\mu(q,t).$$

In spite of apparent differences, these forms are the same. Newton's law for the volume element $\Delta\delta_k$, which expresses the total force in the volume element in terms of the rate of change of the momentum of the volume element, is written as $\mathcal{F}_k^\mu(q,t)\Delta\delta_k = \mu_k(q,t)(d_m\bar{u}_k^\mu(q,t)/dt)\Delta\delta_k$. This relation was used by Morse and Feshbach [1953] to obtain (11.44a). The rate of change of $p_k^\mu(q,t)$ in (11.44b), on the other hand, is based on continuum mechanics. To compare them, equation (11.38) is used with (11.41) to show

(11.45) $$\frac{d_m p_k^\mu(q,t)}{dt} = \mu_k(q,t)\left\{\frac{d_m\bar{u}_k^\mu(q,t)}{dt} - \frac{\partial\bar{u}_k^\sigma(q,t)}{\partial q^\sigma}\bar{u}_k^\mu(q,t)\right\}.$$

The relation (11.39a) is then used to obtain from this

(11.46)
$$\frac{\partial p_k^\mu(q,t)}{\partial t} = \mu_k(q,t)\left\{\frac{d_m\bar{u}_k^\mu(q,t)}{dt} - \frac{\partial\bar{u}_k^\sigma(q,t)}{\partial q^\sigma}\bar{u}_k^\mu(q,t)\right\}$$
$$- \bar{u}_k^\sigma(q,t)\frac{\partial p_k^\mu(q,t)}{\partial q^\sigma}.$$

The definition $p_k^\mu(q,t) = \mu_k(q,t)\bar{u}_k^\mu(q,t)$ can be used to eliminate $p_k^\mu(q,t)$ from equation (11.46) and show by (11.41) that (11.44a) and (11.44b) are equivalent.

The interpretation of the stress tensor usually proceeds along lines similar to the following: A fluid in motion will experience internal "frictional stress" due to the rate of change of the strain in the fluid. The symmetric k *local rate of change of strain tensor* in a fluid in motion is expressed by[13]

(11.47) $$\varpi_k^{\mu\nu}(q,t) = \frac{\partial\bar{u}_k^{(\mu}(q,t)}{\partial q_{\nu)}}.$$

---

[12]This equation is obtained from Philip Morse and Herman Feshbach [1953], pp. 154–160. See also Bailyn [1994], pp. 166–168.

[13]This treatment follows the analysis in Morse and Feshbach [1953], pp. 158–161, here. See also Truesdell and Toupin [1960], pp. 230–231.

For a fluid at rest, the symmetric total stress tensor is the negative of the local pressure, which is written as $R_k^{\mu\nu}(q,t;0) = -P_k(q,t)\delta^{\mu\nu}$. For purely expansive motion, this is expressed as $R_k^{\mu\nu}(q,t;\bar{u}_k(q,t)) = (-P_k(q,t)+\lambda \mathbf{div}\bar{u}_k(q,t))\delta^{\mu\nu}$, where λ is the *coefficient of expansive friction*. Another quantity often used is the rate of change of the strain expressed in terms of a traceless pure shearing rate by

$$(11.48) \qquad \varpi_{k(s)}^{\mu\nu}(q,t) = \varpi_k^{\mu\nu}(q,t) - \frac{1}{3}\frac{\partial\bar{u}_k^\sigma(q,t)}{\partial q^\sigma}\delta^{\mu\nu}.$$

The symmetric k *local total stress tensor* can then be written in terms of these components as

(11.49)

$$R_k^{\mu\nu}(q,t;\bar{u}_k(q,t)) = -\left(P_k(q,t) - \lambda\frac{\partial\bar{u}_k^\sigma(q,t)}{\partial q^\sigma}\right)\delta^{\mu\nu} + 2\eta\varpi_{k(s)}^{\mu\nu}(q,t)$$

$$= -\left(P_k(q,t) + \gamma\frac{\partial\bar{u}_k^\sigma(q,t)}{\partial q^\sigma}\right)\delta^{\mu\nu} + 2\eta\varpi_k^{\mu\nu}(q,t),$$

where $\gamma = (\frac{2}{3}\eta - \lambda)$ is the *second viscosity coefficient*. If the form (11.49) for $R_k^{\mu\nu}(q,t;\bar{u}_k(q,t))$ is used in (11.44a), a little manipulation yields one form of the *Navier-Stokes equation*

$$(11.50) \quad \mu_k(q,t)\frac{\partial\bar{u}_k^\mu(q,t)}{\partial t} + \mu_k(q,t)\bar{u}_k^\nu(q,t)\frac{\partial\bar{u}_k^\mu(q,t)}{\partial q^\nu} + \frac{\partial P_k(q,t)}{\partial q_\mu}$$

$$- (\tfrac{1}{3}\eta + \lambda)\frac{\partial^2\bar{u}_k^\sigma(q,t)}{\partial q_\mu\partial q^\sigma} - \eta\frac{\partial^2\bar{u}_k^\mu(q,t)}{\partial q^\sigma\partial q_\sigma} = \mathcal{F}_{xk}^\mu(q,t).$$

Let us turn now to the question of the compatibility of the Euler-Cauchy balance of linear momentum in (11.44b) with the rate of change of the momentum stated in equation (3.5). The analysis of (11.44a) will yield a similar result. Comparing (11.44b) with (3.5) and using (2.17), it follows that the relation

$$(11.51) \qquad \frac{\partial R_k^{\mu\nu}(q,t;\bar{u}_k(q,t))}{\partial q^\nu} = -\frac{\partial\mathcal{P}_k^{\mu\nu}(q,t)}{\partial q^\nu} + \mathcal{F}_{nk}^\mu(q,t) + \mathcal{F}_{\partial k}^\mu(q,t)$$

must hold if $R_k^{\mu\nu}$ is to find an interpretation within the Theory of Interacting Systems particle-based approach. In order that a local tensor density, $R_k^{\mu\nu}$, satisfying (11.51) exist, however, it is necessary to write $\mathcal{F}_{k,n}^\mu(q,t)$ and $\mathcal{F}_{\partial k}^\mu(q,t)$ as divergences of second order tensors. In considering this possibility, the discussion will be limited to $\mathcal{F}_{k,n}^\mu(q,t)$. Similar remarks apply to $\mathcal{F}_{\partial k}^\mu(q,t)$.

Various methods have been used to write $\mathcal{F}_{k,n}^\mu(q,t)$ as the divergence of a tensor. Green [**1947**], pp. 115–117, used an approximation scheme as did Zubarev [**1971**], pp. 245–246. Irving and Kirkwood [**1950**], pp. 822, used a truncated Taylor expansion of the δ measure to obtain their result.

Daniel Massignon [**1957**], pp. 32–34, used the full Taylor expansion of the δ measure to the same end. Walter Noll [**1955**] used an integration coupled with a differentiation for an equivalent result. Another method was used by A. Kugler [**1967**], p. 238, who employed the relation

$$(11.52) \qquad \delta(q - q_i) = -\frac{1}{4\pi} \frac{\partial}{\partial q} \cdot \frac{\partial}{\partial q} |q - q_i|^{-1}$$

to replace the δ measure with a term expressed as a divergence.

The truncated series approximations to $\delta(q - q_i)$ are of dubious value and add an unwanted element of "irreversibility" to the equations. These will not be considered further. For the full Taylor expansion of $\delta(q-q_i)$ to be meaningful, integration by parts to all orders is required. This leads to the stringent requirement that $F_k(Q_k, P_k, t) \in C^\infty(\Omega_k(t))$ at each $t \in [0, \alpha)$—with similar requirements for other quantities. Even if these requirements are met, the consequence of using the Taylor expansion the δ measure is that the particle coordinate q_i is no longer fixed at the observation point q. This is also true when the representation (11.52) of the δ measure is used. These approaches are rejected because they violate the fundamental requirement in TIS that the forces at the point (q, t) must be connected with particles that experience them there.

Approaching the problem directly, the divergence of a local tensor density $G_i^{\mu\nu}(q, t)$, which is the phase average of the phase function $G_i^{\mu\nu}(Q_k, p_i, t)$, is computed. The result is

(11.53)
$$\frac{\partial G_i^{\mu\nu}(q, t)}{\partial q^\nu} = \left\langle \delta(q - q_i) \frac{\partial G_i^{\mu\nu}(Q_k, p_i, t)}{\partial q_i^\nu} \right\rangle_{kt}$$
$$+ \left\langle \delta(q - q_i) G_i^{\mu\nu}(Q_k, p_i, t) \frac{\partial \chi_{\Omega_k(t)}(Q_k, P_k)}{\partial q_i^\nu} \right\rangle_{kt}$$
$$- \frac{1}{k_B} \left\langle \delta(q - q_i) G_i^{\mu\nu}(Q_k, p_i, t) \frac{\partial S_k(Q_k, P_k, t)}{\partial q_{i,\nu}} \right\rangle_{kt}.$$

The second and third terms on the right-hand side of this equation are not zero, in general. Comparing the definition of $\mathcal{F}_{i,n}^\mu(q, t)$ based on (7.14) to the form (11.53) shows that a local tensor density $G_i^{\mu\nu}(q, t)$, with $\mathcal{F}_{i,n}^\mu(q, t) = \partial G_i^{\mu\nu}(q, t)/\partial q^\nu$, does not exist in general.

This conclusion makes clear the tension between the continuum approach and the particle approach. In the Theory of Interacting Systems, averages and approximations are introduced only with respect to the independence of systems and the range of microscopic forces. The hydrodynamic equations of motion for matter within one system, as stated in the TIS axioms, do not contain the viscosity because the interparticle forces within the system are treated exactly. In other words, viscosity is a macroscopic quantity.

This does not make it less important, but it does indicate that an appropriate approximation method must be introduced to obtain the viscosity from the underlying particle model in a way that is consistent with the particle dynamics. If such an approximation scheme exists only for certain system distributions, the use of a quantity representing the viscosity in an approximate macroscopic equation of motion will only be justified when the system is described by one of these distributions.

Physically, the problem of reconciling these approaches is intrinsic to the assumptions of the particle model, where particles are point "seats of interaction," compared to the assumptions underlying a continuum model. Thus, what Serrin [1959], p. 134, footnote 1, called a "plausible analog" for the case of particle based mechanics expressed in the formalism of continuum mechanics is in fact false in the theory developed here. We are dealing with the case of "concentrated loads" in the terminology of Truesdell and Toupin [1960], p. 538, which are "hardly in the spirit of continuum mechanics." In short, a thermodynamics/hydrodynamics based on continuum mechanics is not reducible to the particle-based thermodynamics developed in this book.

CHAPTER 12

Temperature

Temperature is a purely thermodynamic concept that has no counterpart in mechanics. It has a macroscopic aspect that is associated with heat and heat flow in thermodynamic systems. Microscopically, it is connected with the average energy of individual particles in microscopic systems and the distribution of these particles among energy levels. Meshing these different aspects into one concept proves to be difficult. In this discussion of the concept of temperature and its role in the theory, reference will be made either to the thermature or the temperature as seems fit.

12.1 Historical Backdrop

The notion of temperature has long been associated with that of heat and, until the development of the calorimeter in the late eighteenth century, the change in temperature of a fixed amount of a given substance was used as a measure of the heat exchanged. After the concept of temperature was separated from that of heat, a temperature difference was used to predict the direction of the flow of heat. Newton's law of cooling and Fourier's heat equation went further and made quantitative predictions of the rate of flow of heat as a function of the difference in temperature of two bodies. Various axiomatic approaches to thermodynamics use heat relationships to determine a temperature scale.

The distinction between the "quantity of heat" and the "intensity of heat" grew slowly during the eighteenth century. It was given an important stimulus in the early part of that century when a scale for the thermometer was devised by Daniel G. Fahrenheit. One of the early questions to be investigated experimentally was how to compute the final temperature of a mixture of two quantities of two substances at particular initial temperatures. For a single substance that is not undergoing phase changes, the most successful theoretical formula for computing the temperature of a mixture from the temperatures of its components was introduced by G. W. Richmann in 1747. He proposed the law that the final temperature, T_f, of a mixture of two volumes of the same substance, with masses and temperatures (m_1, T_1) and (m_2, T_2), respectively, is given by

$$(12.1) \qquad T_f = \frac{m_1 T_1 + m_2 T_2}{m_1 + m_2}.$$

Richmann also generalized this formula to the case of many components of the same substance. This law usually fails, however, when two different substances are used.[1]

Experiments and analysis by Joseph Black [**1803**] in the mid to late 18th century established clearly the distinction between the quantity of heat and the intensity of heat and led to his discovery of latent and specific heats. As

---

[1]The modern version of this law is based on the heat capacities, C_V^1 and C_V^2, of the two quantities of matter to be combined. Assume for simplicity two systems with equal mass. System 1 is at temperature T_1, system 2 is at temperature T_2, and the final temperature is T_f. Then, if the heat capacity of 1 is constant in the temperature range (T_1, T_f) and the heat capacity of 2 is constant in the range (T_2, T_f), the fact that the heat emitted by 1 is absorbed by 2 and vice versa is used to show that the final temperature is

$$T_f = \frac{C_V^1 T_1 + C_V^2 T_2}{C_V^1 + C_V^2}.$$

Black pointed out, these latter concepts explained why mixtures of water and snow or of water and mercury did not obey laws like that of Richmann.[2]

The *caloric theory of heat,* which viewed heat as a special kind of particle, had a brief ascendancy in the early to mid nineteenth century. There was also a competing theory of heat, based on the work of Thomas Young and others on radiant energy, called the *wave theory of heat.*[3] It was not long before both of these theories were eclipsed by the developing kinetic theory.

The rise of kinetic theory and the mechanical theory of heat in the mid nineteenth century led to the search for a mechanical analog for temperature. In an investigation of the early kinetic theories and how temperature was treated in them, G. R. Talbot and A. J. Pacey [**1966**], p. 145, discussed an analogy between the momentum of particles and the temperature that was proposed by John Herapath in 1821. A relation between the kinetic energy of particles and the temperature was also proposed by John J. Waterston [**1846**], but this work was not generally known or published until 1893.[4] An early discussion of such an analog can also be found in Clausius [**1850**].[5]

Boltzmann [**1866**], p. 14, proposed that the system temperature be defined as the time average of the kinetic energy of the system.[6] Clausius and others subsequently used Boltzmann's definition of the system temperature in terms of the average kinetic energy for the system. Another analog based in kinetic theory was proposed by Boltzmann [**1871a**]. This analog is included in the discussion of the Maxwell-Boltzmann theory in Part III. Yet another of definition of temperature by Boltzmann is presented below in equation (12.9) and discussed in the next section. It was used as the definition of the temperature by Khinchin [**1943**].

Gibbs [**1902**] proposed several analogs for the temperature. He, p. 93, refined his extension-in-phase function $\delta_k^{(G)\epsilon}(E_k)$, defined in (9.5), into the 'kinetic extension in phase' by using the kinetic energy $\mathcal{K}_k(P_k)$ in place of the equilibrium Hamiltonian energy $\mathcal{H}_k^\epsilon(Q_k, P_k)$ in (9.5). The *kinetic extension in phase* is represented by a Dirichlet integral and its calculation for the unbounded momentum case takes the form (TIS notation)

$$(12.2) \qquad \delta_k^{(Gp)}(E_k^p) = \int dP_k\, \theta(E_k^p - K_k(P_k)) = \frac{(2\pi E_k^p)^{\frac{N_k}{2}}}{\Gamma\left(\frac{N_k}{2} + 1\right)}.$$

---

[2] For further information on these historical matters, including Black's work, see Douglas McKie and Niels Heathcote [**1935**]. The work of Richmann is discussed primarily on pp. 63–76.

[3] See Brush [**1970**].

[4] See Brush [**1965**], Vol. 1, pp. 10–42, for more on Waterston.

[5] Work in the late nineteenth century is reviewed in the summary discussion in George H. Bryan [**1891**].

[6] For a discussion of Boltzmann's effort at a kinetic definition of the temperature, see G. Bierhalter [**1992**], pp. 30–35.

It follows immediately from (12.2) that

$$(12.3) \qquad \frac{d\delta_k^{(\text{Gp})}(E_k^p)}{dE_k^p} = \frac{[2\pi]^{\frac{N_k}{2}} [E_k^p]^{\frac{N_k}{2}-1}}{\Gamma\left(\frac{N_k}{2}\right)}.$$

This quantity is actually the Jacobian of the transformation from the P_k coordinates to the E_k^p coordinates. The entropies that Gibbs associated with the quantities (12.2) and (12.3) are (TIS notation)

$$(12.4a) \qquad S_k^{(\text{Gp2})}(E_k^p) = -k_B \ln(\delta_k^{(\text{Gp})}(E_k^p)),$$

$$(12.4b) \qquad S_k^{(\text{Gp3})}(E_k^p) = -k_B \ln\left(\frac{d\delta_k^{(\text{Gp})}(E_k^p)}{dE_k^p}\right).$$

Gibbs, p. 94, used the result (12.3) to show that the phase average of any function $U_k(E_k^p)$ of the kinetic energy is given by

$$(12.5) \qquad \bar{U}_k = \frac{[2\pi\beta_k]^{\frac{N_k}{2}}}{\Gamma\left(\frac{N_k}{2}\right)} \int_0^\infty dE_k^p\, e^{-\beta_k E_k^p} [E_k^p]^{\frac{N_k}{2}-1} U_k(E_k^p).$$

Gibbs subsequently inverted the thermodynamic relation $dE/dS = T$ and used the definitions (12.4) to show that the phase average of the entropy analogs over the canonical state is (TIS notation)[7]

$$(12.6a) \qquad \beta_k^{(\text{G1})} = \frac{1}{k_B} \overline{\frac{dS_k^{(\text{Gp2})}(E_k^p)}{dE_k^p}}, \qquad \text{for } N_k > 2,$$

$$(12.6b) \qquad \beta_k^{(\text{G2})} = \frac{1}{k_B} \overline{\frac{dS_k^{(\text{Gp3})}(E_k^p)}{dE_k^p}}, \qquad \text{for } N_k > 2.$$

Because the entropies used in these definitions do not generalize to the non-equilibrium case, the thermature definitions based on them will not be pursued further.[8]

Another calculation of the thermature was presented by Gibbs [1902], p. 120, formula (384), and took the form

$$(12.7) \qquad \beta_k^{(\text{G3})}(t) = (3N_k - 2) \left\langle \frac{1}{2K_k(P_k)} \right\rangle_{kt}.$$

___

[7]The suitability of these quantities as analogs for the temperature is discussed in Gibbs [1902], Chapter XIV. Einstein [1902], pp. 423, 427–428, [1903], also proposed a similar differential analog for temperature. See Mandelbrot [1962], pp. 1031–1036, for a modern discussion of the statistical aspects of these analogs and Mandelbrot [1964] for a physical discussion of Gibbs' analogs.

[8]Gibbs' [1902] definitions of $\delta_k^{(\text{G})e}(E_k)$, where E_k is the total energy, and $\delta_k^{(\text{Gp})}(E_k^p)$ will be investigated further in EIS in a discussion of approximating the partition function.

A similar formula is obtained if the thermature analog is defined by

$$(12.8) \qquad \beta_k^{(\star)}(Q_k, P_k, t) = \frac{P_k}{2k_B \mathcal{K}_k(P_k)} \cdot \frac{\partial S_k(Q_k, P_k, t)}{\partial P_k}.$$

The relation between this analog and the analog $(3\mathcal{N}_k - 2)/2\mathcal{K}_k(P_k)$ stated in (12.7) is based on an integration by parts in a setting in which the momentum volume set for the integration is unbounded.

The analog proposed by Gibbs in (12.7) has much to recommend it. It is not hard to show that $\beta_k^{(G3)}(t) > 0$. This version also captures our feeling that the kinetic energy is the determining factor in the transmission of energy from one body to another. Using the representation of $e^{-\beta_k \mathcal{K}_k(P_k)}/2\mathcal{K}_k(P_k) = \frac{1}{2}\int_{\beta_k}^{\infty} d\gamma_k \, e^{-\gamma_k \mathcal{K}_k(P_k)}$, making the transformation $p_i \to (\sqrt{\gamma_k/2m_i})p_i$, followed by extending the P_k integrations to ∞ and performing them, the equilibrium formalism leads to the result $\beta_k^{(G3)} = \frac{3\mathcal{N}_k - 2}{2}\beta_k^{\frac{3\mathcal{N}_k}{2}}\int_{\beta_k}^{\infty} d\gamma_k \, [\gamma_k]^{-\frac{3\mathcal{N}_k}{2}} = \beta_k$. However, this analog is too closely tied to the canonical distribution and gives a quantity with the wrong sign when its derivative with respect to the thermature derivative operator defined below is computed.

Khinchin [1943], pp. 76–77, 88, used the logarithm of the system partition function as the basis for his work, which he acknowledged was based on Boltzmann's conceptions.[9] Khinchin used the convexity of the logarithm function, along with several general and physically plausible assumptions, to show that the equation

$$(12.9) \qquad \frac{\partial \ln Z_k^\epsilon(\alpha_k)}{\partial \alpha_k}\bigg|_{\alpha_k = \beta_k} + \bar{\mathcal{H}}_k^\epsilon = 0,$$

where $\bar{\mathcal{H}}_k^\epsilon$ is the average equilibrium energy of the system, has a unique solution $\alpha_k = \beta_k$ for some thermature $\beta_k > 0$.

To interpret this definition of the temperature in the nonequilibrium case, Khinchin considered matching the general probability distribution density $F_k^{(n)}(Q_k, P_k, t)$ with the closest canonical distribution $[Z_k(\beta_k)]^{-1}e^{-\beta_k \mathcal{H}_k}$ for some β_k. This requires a measure of the distance between these distribution densities and interprets 'closest' as the minimum of this measure for some value of β_k. The function $\sup_{0 \le E_k \le \infty}\{|F_{k,\text{true}}(E_k, t) - F_{k,\text{approx}}(E_k, t)|e^{\beta_k E_k}\}$ for example, was used by Khinchin [1951], Supplement 6, to show the self-consistency of the temperature estimation. The $\chi_k^2(F_k^{(n)}, F_k^\epsilon(\beta_k))$ function defined in (26.15) is another example of a measure that could be used to select the canonical distribution that is closest to the F_k distribution. However, an investigation of this approach to defining the temperature indicates that there is generally not a unique minimum for this equation as a function of β_k when $F_k^{(n)}$ is a highly skewed or multimodal distribution.

---

[9] The definition of the thermature presented in equation (12.9) was called the *Boltzmann analog* by Mandelbrot [1962] in his discussion of temperature analogs.

Other work appears in John Kirkwood [1946], p. 185, Green [1947], and Irving and Kirkwood [1950], p. 824. Irving and Kirkwood approached the temperature from a statistical perspective and defined T_k as a quantity related to the variance of the system momentum. This variance is called in TIS the *heat function* and defined by $Q_k(t) = \langle \mathcal{K}_k(P_k - \bar{P}_k) \rangle_{kt}$. The definition by Irving and Kirkwood has the advantage that it does not change when the origin of the momentum frame is changed.

The definition of Irving and Kirkwood is a more sophisticated version of the definition proposed by Boltzmann in 1871. The $3\mathcal{N}_k$-dimensional average momentum $\bar{P}_{k,0}$ is defined by $\bar{P}_{k,0} = \times_{i\in k}\bar{p}_0$, where $\bar{p}_0 = [\mathcal{N}_k]^{-1} \sum_{i\in k} p_i = [\mathcal{N}_k]^{-1}\mathcal{P}_k$. In the particle rest frame of the k system, the total momentum vanishes, so $\bar{p}_0 = 0$ and $\bar{P}_{k,0} = 0$, so the resulting definition of the temperature takes the form

$$(12.10) \qquad T_k^{(IK)}(t) = \frac{2}{3k_B\mathcal{N}_k} \langle \mathcal{K}_k(P_k) \rangle_{kt} = \frac{2\mathcal{K}_k(t)}{3k_B\mathcal{N}_k}.$$

The corresponding definition of the thermature in terms of this relation is

$$(12.11) \qquad \beta_k^{(IK)}(t) = \frac{3\mathcal{N}_k}{2\mathcal{K}_k(t)},$$

which is close to Gibbs' $\beta_k^{(G3)}$.

These forms, which base the temperature on the average kinetic energy, were considered recently by Byung Chan Eu and L. S. García-Colin [1996]. They, p. 2506, rejected a connection of the form (12.10) on the grounds that it is not valid for quantum and relativistic theories and is therefore not universal. From a more general perspective, Eu and García-Colin investigated other definitions of the temperature based on the relation $T_k^{-1} = dS/d\mathcal{E}$ using Boltzmann's equation, Extended Irreversible Thermodynamics (EIT), and other theories. They concluded that "the temperature of the system, whether equilibrium or nonequilibrium, is a phenomenological attribute of the system, even in the statistical theory of molecular systems." In other words, temperature is a macroscopic phenomenon. Eu and García-Colin, p. 2509, went on to say "the misconception about this subtle but important point has been the cause of numerous confusing results and propositions in EIT." They concluded that the definition of the temperature in nonequilibrium cases depends essentially on the zeroth law of thermodynamics, which is a form of the theory of exchanges, and the phenomenological relation obtained from Clausius' inequality in conjunction with Gibbs' equilibrium canonical distribution.[10]

---

[10]For more information on other attempts to define the temperature, and on Extended Irreversible Thermodynamics in particular, see the extensive references in Eu and García-Colin [1996].

12.2 The Properties of Temperature

Many authors have maintained in articles written over the past 130 years that the temperature of a system can only be defined at equilibrium. These authors argued that the equilibrium version of the second law, written $T^{-1} = dS/dQ$, is the only way the temperature can be defined in terms of thermodynamics. This viewpoint was challenged by Planck [**1904**], p. 79, who felt that the concept of temperature can only be understood with respect to its "complete meaning" from the standpoint of irreversibility. Planck observed that other physical quantities are meaningful for systems not at equilibrium, such as the entropy, and raised the question of why the temperature cannot also be defined for these states.

From the standpoint of heat exchange, two adjacent bodies are said to have the same temperature if the net heat transferred between them is zero. In continuum versions of thermodynamics, this became an axiom defining the notion of the same temperature in two bodies. In the version of thermodynamics based on kinetic theory, the theory of exchanges held that if the net exchange of heat between two volumes is zero, their average kinetic energies are the same. This was the basis for the suggestions in the mid nineteenth century, mentioned above, that the particle kinetic energy be used as an analog for temperature. Subsequent thinking by kinetic theorists associated temperature with the variance in the particle momentum of a body. This variance is closely related to the function $\mathcal{K}_k(P_k - \bar{P}_k)$.

The problem with associating temperature with the direction of heat flow in macroscopic bodies and then connecting these concepts with microscopic concepts is that there is no guarantee provided by mechanics that a stream of particles with a low temperature will not carry heat energy into a region in which the temperature is higher. Ordinary experience indicates that this is an unusual situation, but it is not forbidden. It can even be set up using macroscopic means under circumstances in which a flow of cool particles is directed into a region containing hot particles and deposit their kinetic energy there.

Because of the problems connected with the role of temperature in heat flow and the difficulty in defining a suitable analog mentioned above, it is the most difficult of the thermodynamic concepts to deal with formally. Temperature is an intensive quantity that depends on the number of degrees of freedom in a system and their efficacy in absorbing and releasing energy between various potential and kinetic modes. It plays one role as a parameter conditioning the energy in the particle equilibrium probability distribution density. It plays another role as one of the determining factors in the direction heat will flow. It is not a property attributed to particles in microscopic Hamiltonian mechanics, but it is often used as one of the macroscopic system boundary conditions.

To illustrate the significance of the system degrees of freedom with regard to the concept of temperature, let us anticipate the equilibrium results of Chapter 18 and observe that the equilibrium average energy for a system at the thermature β_k can be written as $(3\mathcal{N}_k/2\beta_k) + \bar{\bar{\Phi}}_k^\epsilon(\beta_k)$, where the first term is the kinetic term and $\bar{\bar{\Phi}}_k^\epsilon(\beta_k)$ is the average equilibrium potential. The importance of the number of degrees of freedom is manifest in the kinetic term. For the coordinate component, the temperature is tied up with the specific structure of the interparticle potentials. A model of these potentials is required to estimate their role in determining the temperature. The dependence of the thermature on the number of degrees of freedom and the structure of particle interactions makes finding a suitable analog for computing β_k and finding a corresponding thermature derivative operator harder.

From the standpoint of heat exchange, the closure conditions (5.29) allow kinetic energy, but not potential energy, to be transmitted through a boundary. The transmission of kinetic energy is also associated with the flow of heat as shown in equation (7.58). Internally, a system exchanges energy between its potential and kinetic energy components. Because the heat capacity of a system is related to this exchange, it is clear that both the kinetic and potential energies play a role in determining the temperature of a system. Furthermore, the heat and work passing through the boundaries of the k system are frame dependent, so the division of the flow of energy through the moving boundaries of the k system into heat and work depends on the velocity of the boundary relative to the system rest frame. This means that there is no invariant or conserved physical quantity whose transmission is associated with changes in the temperature.

Some recent definitions of temperature in thermodynamics use the equilibrium relation $T = (\partial E/\partial S)_{\text{vol}}$ to express it as the derivative of the energy with respect to the entropy at constant volume. In the TIS treatment of equilibrium, the entropy analog is $S_k^\epsilon = k_B \beta_k \mathcal{H}_k^\epsilon$. This formula can be used to interpret the derivative $1/T_k = k_B \beta_k = \partial S_k^\epsilon / \partial \mathcal{H}_k^\epsilon$ at equilibrium. For systems not at equilibrium, there is no special relation between the entropy of a distribution and its energy of this form, so such a definition does not have general significance.

The view that the only way to define the temperature is by reference to the second law of thermodynamics and Planck's response to it were mentioned above. R. H. Fowler [1936], p. 188, nevertheless, felt that the usual appeal to the properties of an ideal substance to define the temperature scale were "illogical" and then stated: "In thermodynamical theory the *absolute temperature* is defined in connection with the second law, and can only be defined in this way." Let us examine this viewpoint in more detail now in connection with the relation of heat flow, entropy, and temperature.

The change in the entropy is related to the temperature and the heat flow across the boundaries of a system in its system or particle rest frame by

Clausius' inequality (3.13b), (3.14a), where it is expressed in the form $dS_k \geq k_B \beta_k d\mathcal{Q}_{\partial k}^*$. Because there is equality in this relation only at equilibrium and $\mathcal{Q}_{\partial k}^* = \mathcal{Q}_{\partial k}$ only in the system or particle rest frame or for an isolated system, the change in entropy equals the heat flow across the system boundary divided by the temperature at the boundary only at equilibrium—where, strictly speaking, there is no heat flow. Clausius' inequality, and by implication the second law, are therefore not suitable as the basis for a general definition of temperature.

While the candidates proposed above all reduce to the proper thermature at equilibrium, each of them founders on one difficulty or another in the general setting. Moreover, when the possibility of coordinate and momentum transformations is considered as well, the question of the definition of the temperature in other reference frames becomes an issue. To help sort through these issues and possibilities, the way in which the temperature of a system can be measured in different reference frames is examined next.

Consider, for example, a beam of rapidly moving particles with a distribution of momenta about the mean momentum of the beam. In discussions of these particle beams in the literature, the process of reducing the variance of the momentum of the particles in the beam toward zero, thereby making the beam more monochromatic, is referred to as "cooling the beam" towards absolute zero. On the other hand, the beam carries kinetic energy and will deposit heat in a barrier placed in its way in spite of the low temperature assigned to it. The general question of the velocity dependence of temperature has been the source of significant controversy in the literature—especially in connection with the versions of thermodynamics in special relativity.

Because the temperature has often been interpreted in terms of the theory of exchanges mentioned above, in which equal temperatures for two or more bodies means that the net heat flow or net kinetic energy flow between them vanishes, it is viewed as a predictor for the direction of heat flow when two adjacent systems are at rest with respect to each other. Most axiomatic approaches to macroscopic thermodynamics adopt some statement concerning the flow of heat from hot regions to cold ones as an axiom. However, if the reference frame is changed to one moving uniformly with respect to the original one, a different heat flow will be seen.

For the statistical point of view, an important paper on the foundations of thermodynamics by Benoit Mandelbrot [1962] applied modern statistical methods to the analysis of equilibrium phenomenological thermodynamics. His first purpose was to apply the statistical concept of sufficiency to the study of Gibbs' canonical distribution, which in its energy form is characterized mathematically as an exponential Γ distribution. Mandelbrot pointed out that Szilard [1925a] had introduced a concept into phenomenological thermodynamics that is essentially equivalent to the statistical concept of *sufficiency* introduced soon afterward by Ronald A. Fisher. Szilard showed (in modern terminology) that under certain regularity conditions the canonical distribution of Gibbs is the only probability distribution with a single scalar sufficient statistic—the temperature. Mandelbrot, p. 1025, used the fact that the energy is an invariant in an isolated system as basis for a *criterion of sufficiency* for an isolated system: *"The nature of thermal equilibrium is such that, if a system is withdrawn from contact with a heat reservoir, the energy of that system is a necessary and sufficient (or minimal sufficient) statistic for the temperature of the heat reservoir."* Mandelbrot's work on phenomenological thermodynamics dovetails with the analysis of Tisza [1961] and Tisza and Quay [1963].

A second purpose of Mandelbrot's [1962] paper was to generalize the notion of temperature to isolated systems. In accord with Gibbs' views and the more precise analysis of Khinchin [1943], the temperature associated with a canonical distribution is defined for a small system in contact with a heat reservoir by the temperature of the reservoir. This definition cannot be used, however, to define the temperature of an isolated system with a fixed energy. For this case, Mandelbrot stated that *"... the temperature for systems-in-isolation should be viewed as a statistical estimate of the parameter of a conjectural canonical distribution."* He defined a thermometer as *"a physical system such that the value of its energy can be ascertained by direct observation."* Its readings are expressed in units of temperature rather than energy for traditional reasons and convenience. The assignment of a temperature to an isolated system is an issue of statistical estimation from this point of view. Mandelbrot referred to the temperature analog displayed in (12.6a) as *Gibbs integral analog,* the analog (12.6b) as *Gibbs differential analog,* and (12.9) as *Boltzmann's analog.* He discussed applying the notions of a self-consistent temperature or a self-unbiased temperature to estimating the temperature of an isolated system. Mandelbrot also pointed out that Boltzmann's analog (12.9) is an example of a *maximum likelihood estimator* when it is applied to the estimation of the temperature of an isolated system.

Mandelbrot's work made more precise the statistical aspects of the attribution of a temperature to a system described by Gibbs' equilibrium canonical distribution. However, the role of the temperature as a parameter conditioning a statistical distribution as presented by Mandelbrot is at odds with its physical role. As a physical parameter, we want it to be defined in all physically reasonable circumstances and have an analog. As a parameter conditioning a canonical distribution, it is well defined only for particle distributions of the exponential type. It is also the case that only for distributions of this form, or ones that are similar to it, which we can expect to be able to define a maximum likelihood function. As was pointed out following (12.9), there is no single parameter for most distributions that can be defined by an analog that could reasonably be called the thermature.

Another important aspect of Mandelbrot's discussion was that the isolated system had previously been attached to a reservoir. The "conjectural canonical distribution" associated with the isolated system after it is removed from contact with the reservoir actually refers to the likelihood that it is left with the energy E_k in the partition of the energy between itself and the reservoir. The temperature in this case is a "statistical estimate of the parameter" associated with this "partition of energy distribution" and does not refer to the isolated system itself. For a small system that remains in contact with a reservoir, the canonical distribution will properly reflect this division over time and can be used to define the temperature. When the system is isolated by removing it from the reservoir, its temperature can be estimated by this method, but the state attributed to the newly isolated system can no longer be a canonical distribution. The upshot of this tension between the statistical and physical roles of the temperature is that any definition of a phase analog function will be a compromise and perhaps not completely satisfactory for either role.

The viewpoints that temperature can only be defined at equilibrium or with reference to the second law run counter to our experience that measurements using thermometers, which do not depend on the second law, give useful results even for systems not at equilibrium. The statistical approach does offer a way to extend our conception of temperature, but it cannot be tied to Gibbs' canonical formalism.

12.3 The TIS Approach to Temperature

The TIS standpoint is that the temperature can be defined in any system, whether at equilibrium or not, if there is a suitable thermodynamic analog to serve as the basis for the calculation. If this analog allows localization, the local temperature in a system can be computed as well. While mindful of the point of view that sees temperature as an estimated statistic of a conjectural canonical distribution, it will be given a physical definition here in terms of the current microscopic state of a system. The temperature of the k system will be defined in the system rest frame.

There is a significant difference between the TIS approach to defining the temperature and that in standard equilibrium thermodynamics. In standard equilibrium thermodynamics, the entropy can serve as a state variable and derivatives with respect to it can be taken. The standard system Hamiltonian energy is expressed in terms of the entropy and volume as independent variables by $\mathcal{H}_k(S_k, \delta_k)$. It is natural that the inverse absolute temperature is usually defined in equilibrium thermodynamics as the partial derivative of the Hamiltonian energy with respect to the entropy. The extensive entropy and intensive temperature are conjugate variables in this setting. In the general thermodynamics of TIS, the Hamiltonian energy cannot usually be written as a function of the entropy, so another approach must be used.

The definition for the thermature chosen here is based on the equilibrium formula $\beta_k = [\mathrm{P}_k^\epsilon]^{-1} \partial \ln Z_k^\epsilon / \partial \delta_k$, where Z_k^ϵ is the equilibrium partition function and P_k^ϵ is the equilibrium pressure. Because both the pressure $\mathrm{P}_k(t)$ and the partition function $Z_k(t)$ are defined in the general theory in TIS, the k *thermature* is defined in the system rest frame by

$$(12.12\mathrm{a}) \qquad \beta_k(t) = \frac{1}{\mathrm{P}_k(t)} \frac{\partial \ln Z_k(t)}{\partial \delta_k} = -\frac{1}{k_B \mathrm{P}_k(t)} \left\langle \frac{dS_k(Q_k, P_k, t)}{d\Omega_k} \right\rangle_{kt}.$$

The second equality in (12.12a) follows from (10.47). The k *phase thermature* analog used in this definition is

$$(12.12\mathrm{b}) \qquad \beta_k(Q_k, P_k, t) = -\frac{1}{k_B \mathrm{P}_k(t)} \frac{dS_k(Q_k, P_k, t)}{d\Omega_k}.$$

The thermature and the temperature are defined in the system rest frame, as is the pressure, for the reasons given in previous chapters. If the pressure vanishes, so that $\mathrm{P}_k(t) = 0$, the thermature cannot be computed using the analog (12.12b), so a modified form of Gibbs' definition (12.7) is adopted to compute the thermature. This modified form is used in Chapter 14 in the computation of the thermature of a special harmonic oscillator distribution.

The physical justification for the definition (12.12) is that it is proportional to the ratio of the volume derivative of the entropy to the volume derivative of the Hamiltonian energy with the entropy fixed. Symbolically, this is expressed as $\beta_k \sim k_B^{-1}(\partial S_k / \partial \delta_k)(\partial \delta_k / \partial \mathcal{H}_k)_S$. It should be emphasized that this definition, stated as the ratio of two phase averages, is not a natural consequence of applying the TIS formalism to a dynamic quantity in the way that other quantities in the theory have been defined. It was chosen to produce as good a measure of the nonequilibrium thermature as possible and to become an exact measure when the system is at equilibrium.

This definition of the thermature in terms of the pressure and the phase average of the volume derivative of the entropy fits with the TIS requirement that the quantities defined in the theory have analogs. Note that this analog is an exception to the general rule in TIS that macroscopic and microscopic aspects are not mixed. This is because it is the ratio of two macroscopic quantities and not an exact definition.

The suitability of the mapping of $\ln Z_k(t)$ to $\partial \ln Z_k(t) / \partial \delta_k$ by the volume partial derivative operator follows from the fact that it is a canonical mapping of thermodynamic objects so the resulting formula is consistent with the Hamiltonian basis of the theory. Moreover, in spite of the fact that the entropy plays a role in the definition of the thermature, the second law does not. Finally, this definition has the major advantage of extending to general thermodynamics the equilibrium connection between the thermature, pressure, and the volume derivative of the partition function.

The thermature definition (12.12) can be generalized into a thermature tensor by using the definition (10.37) of the volume tensor partial derivative operator. This gives the k *thermature tensor* in the form

$$(12.13) \qquad \beta_k^{\mu\nu}(t) = -\frac{1}{P_k(t)} \frac{\partial \ln Z_k(t)}{\partial \delta_{k,\mu\nu}}.$$

The action of the volume tensor partial derivative operator $\partial/\partial \delta_{k,\mu\nu}$ on thermodynamic objects such as $Z_k(t)$ is computed using (10.37) in the integral defining $Z_k(t)$ and integrating by parts to obtain the result

$$(12.14)$$

$$\frac{\partial \ln Z_k(t)}{\partial \delta_{k,\mu\nu}} = \frac{1}{\delta_k(t)} \int dQ_k \int dP_k \, \chi_{\Omega_k(t)}(Q_k, P_k) \frac{dF_k(Q_k, P_k, t)}{dD_{k,\mu\nu}}$$

$$= -\frac{1}{k_B \delta_k(t)} \left\langle \frac{dS_k(Q_k, P_k, t)}{dD_{k,\mu\nu}} \right\rangle_{kt},$$

where $d/dD_{k,\mu\nu}$ is defined in (10.26a). The thermature itself is one-third the trace of the thermature tensor

$$(12.15) \qquad \beta_k(t) = \tfrac{1}{3} \delta_{\mu\nu} \beta_k^{\mu\nu}(t).$$

This formalism can be extended to the single particle level. The *i-particle thermature tensor analog* is defined in parallel with (12.12b) as

$$(12.16a) \qquad \beta_i^{\mu\nu}(Q_k, P_k, t) = -\frac{1}{k_B \delta_k(t) P_i(t)} \frac{dS_k(Q_k, P_k, t)}{dd_{i,\mu\nu}},$$

where $P_i(t)$ is the i-particle pressure. As in the case of the local i-particle time derivative of the entropy, this quantity is not a particle sum function because $S_k(Q_k, P_k, t)$ is not one. Using the i-particle dilation tensor derivative in place of the system version gives the *i-particle thermature tensor*

$$(12.16b) \quad \beta_i^{\mu\nu}(t) = \frac{1}{P_i(t)} \frac{d \ln Z_k(t)}{d\delta_{i,\mu\nu}} = -\frac{1}{k_B \delta_k(t) P_i(t)} \left\langle \frac{dS_k(Q_k, P_k, t)}{dd_{i,\mu\nu}} \right\rangle_{kt}.$$

The relation between the i-particle thermature tensor and the k system thermature tensor is expressed as the weighted average

$$(12.17) \qquad \beta_k^{\mu\nu}(t) = \frac{1}{P_k(t)} \sum_{i \in k} P_i(t) \beta_i^{\mu\nu}(t).$$

This law of combination reflects the fact that the thermature is an intensive quantity.

It is clear that the thermature tensor is symmetric by the definition of the operators that it is based on. This confirms Axiom 7 of Chapter 3. Similarly, the relations for the thermature and the definition of the absolute temperature in Chapter 3 are consistent with those of this chapter.

Turning to the local thermature, let us use (13.49a) or (18.12) to antici-
pate the equilibrium entropy analog written $S_k^\epsilon(Q_k, P_k) = k_B \beta_k \mathcal{H}_k^\epsilon(Q_k, P_k)$.
The local equilibrium phase average of the volume derivative of the equilib-
rium entropy is then given by

(12.18)
$$\sum_{i\in k} \left\langle \delta(q - q_i) \frac{dS_k^\epsilon(Q_k, P_k)}{dd_i} \right\rangle_{k\epsilon} = -k_B \beta_k \sum_{i\in k} \langle \delta(q - q_i) \mathrm{P}_i^\epsilon(Q_k, p_i) \rangle_{k\epsilon}$$
$$= -k_B \beta_k \mathrm{P}_k^\epsilon(q),$$

where $\langle \cdot \rangle_{k\epsilon}$ is the phase average over the equilibrium state. This means that
the local thermature in the equilibrium case should be written

(12.19) $$\beta_k^\epsilon(q) = -\frac{1}{k_B \mathrm{P}_k^\epsilon(q)} \sum_{i\in k} \left\langle \delta(q - q_i) \frac{dS_k^\epsilon(Q_k, P_k)}{dd_i} \right\rangle_{k\epsilon} = \beta_k.$$

For the general case, this pattern is used to define the *i-particle local thermature* and the *k local thermature* by

(12.20a)
$$\beta_i(q, t) = -\frac{1}{k_B \mathrm{P}_i(q, t)} \left\langle \delta(q - q_i) \frac{dS_k(Q_k, P_k, t)}{dd_i} \right\rangle_{kt},$$

(12.20b)
$$\beta_k(q, t) = \frac{1}{\mathrm{P}_k(q, t)} \sum_{i\in k} \mathrm{P}_i(q, t) \beta_i(q, t).$$

As before, equation (12.20b) shows that the local thermature is an intensive
quantity and not a true particle sum function. Comparing (12.20b) with
(12.12) shows that the law of composition for obtaining the global thermature
from the local thermature is

(12.21) $$\beta_k(t) = \frac{1}{\mathrm{P}_k(t)} \int d^3q \, \mathrm{P}_k(q, t) \beta_k(q, t).$$

The tensor version of this relation confirms (3.17b).

The local *i*-particle thermature can also be expressed in terms of the
volume derivative of the local density. The *i*-particle volume partial derivative
of the *i*-particle local density is

(12.22)
$$\frac{\partial \nu_i(q, t)}{\partial \delta_i} = -\frac{1}{k_B \delta_k(t)} \left\langle \delta(q - q_i) \frac{dS_k(Q_k, P_k, t)}{dd_i} \right\rangle_{kt} - \frac{q}{3\delta_k(t)} \cdot \frac{\partial \nu_i(q, t)}{\partial q}.$$

Using this with (12.20a) and the definition (10.39a) gives

(12.23) $$\beta_i(q, t) = \frac{\delta_k(t)}{\mathrm{P}_i(q, t)} \frac{d\nu_i(q, t)}{d\delta_i}.$$

It is also possible to define the local i-particle and system thermature tensor $\beta_i^{\mu\nu}(q,t)$ using $d/d\delta_{i,\mu\nu}$ in place of $d/d\delta_i$ in (12.23).

As with the pressure, the thermature is always computed in the system rest frame. This means that the thermature is invariant under Galilean transformations of the reference frame of the observer. The implications of this choice and its connection with mechanics are discussed in Chapter 15.

In addition to the failure of the thermature analog (12.12b) to be useful when the system pressure is zero, there is a second problem. It cannot be used in this form with an exact system state $F_k^\delta(Q_k, P_k, t)$ to define the thermature of a system because the system entropy is not defined for that state. To compute the thermature for an exact state, the relation between the thermature and the volume partial derivative of the partition function is retained. The exact state thermature is defined as the volume partial derivative of the exact state partition function $Z_k^\delta(t)$ divided by the exact state pressure $P_k^\delta(t)$. Computing the volume partial derivative of the exact state partition function using the TIS formalism yields

$$(12.24) \qquad \frac{\partial Z_k^\delta(t)}{\partial \delta_k} = -\int dQ_k \int dP_k\, F_k^\delta(Q_k, P_k, t) \frac{d\chi_{\Omega_k(t)}(Q_k, P_k)}{d\Omega_k}.$$

The volume derivative of the system projection operator is given in (10.35). Using this with the definition of $F_k^\delta(Q_k, P_k, t)$ in (12.24), and computing the result, gives

$$(12.25) \qquad \frac{\partial Z_k^\delta(t)}{\partial \delta_k} = -\frac{1}{3\delta_k(t)} \int_{\partial k} d^2 r\, \hat{n}_{\partial k}(r, t) \cdot r\, \nu_{\partial k}^\delta(r, t),$$

where

$$(12.26) \qquad \nu_{\partial k}^\delta(r, t) = \sum_{i \in k} \left\langle \frac{\delta(q_i - r)}{\chi_{d_k(t)}(q_i)} \right\rangle_{k\delta t} = \sum_{i \in k} \delta(q_i^{(s)}(t) - r)$$

is the macroscopic i-particle exact state boundary density. The exact state pressure is obtained similarly with the result

$$(12.27) \qquad P_k^\delta(t) = \frac{1}{3\delta_k(t)} \left[2\mathcal{K}_k(P_k^{(s)}(t)) - Q_k^{(s)}(t) \cdot \frac{\partial \Phi_k(Q_k^{(s)}(t), t)}{\partial Q_k} \right].$$

The exact state thermature is expressed in terms of (12.25), (12.26), and (12.27) as

$$(12.28) \qquad \beta_k^\delta(t) = -\frac{1}{3\delta_k(t) P_k^\delta(t)} \int_{\partial k} d^2 r\, \hat{n}_{\partial k}(r, t) \cdot r\, \nu_{\partial k}^\delta(r, t).$$

Because the integration on the right side of (12.28) is two-dimensional and the exact state boundary density $\nu_{\partial k}^\delta(q, t)$ is a three-dimensional δ measure, $\beta_k^\delta(t)$ is singular. This means that it requires averaging over a small domain using the formalism of Chapter 3.

With the special form for the exact state thermature stated in (12.28), the definition of the thermature in (12.12) in terms of a particle analog function meets the TIS requirements for an acceptable thermodynamic definition. In addition, when there is no matter or energy in a coordinate volume, the pressure is zero there and the effective thermature is infinite, which corresponds to a temperature of absolute zero. All energy radiated into this volume will pass through and none will be reflected back. There is no energy emitted from this volume either. From the standpoint of the theory of exchanges, this is equivalent to a temperature of absolute zero.

12.4 The Thermature at Equilibrium

In Chapter 9, a solution (9.31) of the variational equation for the reference distribution L_k was expressed in terms of four undetermined multipliers. The analog for the thermature will be used with this distribution to make further progress in defining the as yet underdetermined multiplier b_k in L_k associated with the thermature. In (9.33), L_k was established in the form

$$(12.29) \qquad L_k(Q_k, P_k, t) = e^{-\left[\sum_{i \in k} (b_i(q_i,t) K_i(p_i - p_i^*)) + R_k(Q_k,t)\right]}.$$

In this equation, $b_i(q_i, t)$ and $p_i^* = p_i^*(q, t)$ are quasilocal step functions.

In (9.35a) it was shown that $b_i(q_i, t)$ can be written $b^a(t)$ because it depends only on the quasilocal domain and does not depend on i, k, or l. It was also shown in (9.35c) that $v_i^*(q_i, t)$ also does not depend on k and l and depends on i only through the q_i coordinate. It can be written $v^{*,a}(q_i, t)$. The associated momentum is written $p_i^{*,a}(q_i, t) = m_i v^{*,a}(q_i, t)$. In the case in which the quasilocal domain is the whole k volume set, these quantities are expressed in terms of system values as $b^a(t) = b_k(t)$ and $v^{*,a}(q_i, t) = v_k^*(q_i, t)$.

In contrast to the work above on the pressure and temperature, the *particle rest frame* rather than the system rest frame is the important frame for these calculations. In this frame, the angular momentum implicit in the definition of $v_k^*(q_i, t)$ vanishes, which means that the function $s_k(t)$ vanishes in the definition (9.35c) of $v_i^*(q_i, t)$ as well. This implies that $v_k^*(q_i, t)$ does not depend on q_i in the k particle rest frame. It follows from (9.35c) that $v_k^*(q_i, t) = v_k^*(t) = c_k(t)$ and $p_i^*(q_i, t) = p_i^*(t) = m_i c_k(t)$. It is also true that $c_k(t) = 0$ in this frame, so it follows finally that $v_k^*(t) = 0$ and $p_k^*(t) = 0$ in the k particle rest frame.

More stringent conditions are obtained by solving the equation $dL_k/dt = 0$ to find out what is required for L_k to be a state of the k system. Let us anticipate the results (13.32) that state that $b_k(t) = b_k$ is a constant, that $R_k(Q_k, t) = b_k \Phi_k(Q_k, t)$, and $\Phi_k(Q_k, t) = \Phi_k^\epsilon(Q_k, t)$, where $\Phi_k^\epsilon(Q_k, t)$ is a special form determined by the conditions of equilibrium. It follows that L_k can be written in the particle rest frame of the k system in the form

$$(12.30) \qquad L_k(Q_k, P_k, t) = e^{-b_k \mathcal{H}_k(Q_k, P_k, t)}.$$

The entropy of the L_k distribution is $S_k^L = -k_B \ln L_k$, and using this with the definition (12.12) gives the thermature analog associated with L_k as

$$(12.31) \qquad \beta_k^L(Q_k, P_k, t) = -\frac{1}{k_B \mathrm{P}_k^L(t)} \frac{dS_k^L(Q_k, P_k, t)}{d\Omega_k} = \frac{b_k \mathrm{P}_k^L(Q_k, P_k, t)}{\mathrm{P}_k^L(t)},$$

where $\mathrm{P}_k^L(Q_k, P_k, t) = -d\mathcal{H}_k(Q_k, P_k, t)/d\Omega_k$ and $\mathrm{P}_k^L(t) = \langle \mathrm{P}_k^L(Q_k, P_k, t) \rangle_{Lt}$. Taking the L phase average of (12.31) and using the definition (12.12) yields

$$(12.32) \qquad\qquad \beta_k(t) = b_k = \beta_k^L.$$

This is the identification of b_k with β_k mentioned in Chapter 9. This identification means that the equilibrium state of this theory and calculations made with it correspond to those of Gibbs' theory.

As mentioned, the distribution density L_k has not yet been shown to be a system state. This question is examined in Chapter 13 where it is shown that for the time derivative of L_k to satisfy (5.16), required for L_k to be a system state, the additional constraints mentioned must be imposed on $b_k(q, t)$.

12.5 The Thermature Derivative Operator

The next concern is the definition of an operator that will be used to compute the derivative of thermodynamic functions with respect to thermature. Since the thermature plays a direct role as a parameter in thermodynamic formulas only at equilibrium, this operator must be defined indirectly as a map from thermodynamic functions to thermodynamic functions. As usual, the application of this thermature derivative operator to an equilibrium thermodynamic quantity must give the same result as the thermature derivative of this quantity obtained within standard equilibrium theory.

The work in this section can easily be rewritten to express it as a derivative with respect to the absolute temperature. This conversion takes the form

$$(12.33) \qquad\qquad T_k \frac{\partial}{\partial T_k} = -\beta_k \frac{\partial}{\partial \beta_k}.$$

Let us begin with the fact that the thermature is a property of the system distribution and not of the individual particles in it. Next, an examination of the definition of the equilibrium state (see (13.49) shows that the equilibrium relation $\partial F_k^\epsilon / \partial \beta_k = -\mathcal{H}_k^\epsilon(Q_k, P_k)F_k^\epsilon$ follows immediately.[11] This relation can be extended to the nonequilibrium case by defining the *k phase thermature derivative* operator for the unnormalized system distribution F_k by

$$(12.34) \qquad \frac{\partial F_k(Q_k, P_k, t)}{\partial \beta_k} = -\mathcal{H}_k(Q_k, P_k, t)F_k(Q_k, P_k, t).$$

---

[11]See the definition (13.49) of the equilibrium state and the calculation in equation (18.18) below.

When applying the thermature phase derivative operator to the system distribution, the convention will be used that if the system distribution depends on β_k as a parameter, as it does in the equilibrium case, the ordinary derivative with respect to β_k will be used. Otherwise, the definition (12.34) will be used. In either case, the result is the same as that in (12.34).

Applying (12.34) to the partition function $Z_k(t)$ associated with F_k gives the result

(12.35)
$$
\begin{aligned}
\frac{\partial Z_k(t)}{\partial \beta_k} &= \int dQ_k \int dP_k \, \frac{\partial[\chi_{\Omega_k(t)}(Q_k, P_k)F_k(Q_k, P_k, t)]}{\partial \beta_k} \\
&= \int dQ_k \int dP_k \left[\frac{\partial \chi_{\Omega_k(t)}(Q_k, P_k)}{\partial \beta_k} - \chi_{\Omega_k(t)}(Q_k, P_k)\mathcal{H}_k(Q_k, P_k, t) \right] \\
&\quad \times F_k(Q_k, P_k, t).
\end{aligned}
$$

In order that this calculation give the proper result $-\mathcal{H}_k^\epsilon$ at equilibrium, it is necessary to define the thermature derivative of the system projection operator by

(12.36)
$$
\frac{\partial \chi_{\Omega_k(t)}(Q_k, P_k)}{\partial \beta_k} = 0.
$$

Using this definition with (12.35) gives the result

(12.37)
$$
\frac{\partial \ln Z_k(t)}{\partial \beta_k} = - \langle \mathcal{H}_k(Q_k, P_k, t) \rangle_{kt} = -\mathcal{H}_k(t),
$$

which yields the correct formula when the system is at equilibrium as required. An additional consequence of (12.34) is

(12.38)
$$
\frac{\partial S_k(Q_k, P_k, t)}{\partial \beta_k} = k_B \mathcal{H}_k(Q_k, P_k, t).
$$

When acting on any phase function other than the system distribution, the phase thermature derivative is the ordinary partial derivative with respect to β_k. While it would be unusual for a phase analog particle function to depend on the macroscopic thermature, it is not excluded. For this reason, the *total thermature derivative* acting on analog functions $G_k(Q_k, P_k, t)$ is defined by

(12.39)
$$
\frac{dG_k(Q_k, P_k, t)}{d\beta_k} = \frac{\partial G_k(Q_k, P_k, t)}{\partial \beta_k} - \mathcal{H}_k(Q_k, P_k, t)G_k(Q_k, P_k, t).
$$

When acting on any system distribution density or its entropy, even at equilibrium, it is required that $dF_k/d\beta_k = \partial F_k/\partial \beta_k = -\mathcal{H}_k F_k$ for consistency.

With these definitions, (12.36) is used to show that the *i-particle local therma-ture derivative* and the k *local thermature derivative* of any thermodynamic function $G_k(t)$ is

(12.40)
$$\frac{\partial G_k(t)}{\partial \beta_k} = \frac{1}{Z_k(t)} \int dQ_k \int dP_k \, \chi_{\Omega_k(t)}(Q_k, P_k) \frac{\partial [F_k(Q_k, P_k, t) G_k(Q_k, P_k, t)]}{\partial \beta_k}$$

$$- \frac{\partial \ln Z_k(t)}{\partial \beta_k} G_k(t)$$

$$= \left\langle \frac{\partial G_k(Q_k, P_k, t)}{\partial \beta_k} - \mathcal{H}_k(Q_k, P_k, t) G_k(Q_k, P_k, t) \right\rangle_{kt} + \mathcal{H}_k(t) G_k(t)$$

$$= \left\langle \frac{dG_k(Q_k, P_k, t)}{d\beta_k} \right\rangle_{kt} + \mathcal{H}_k(t) G_k(t).$$

The term $\mathcal{H}_k(t) G_k(t)$ in the last line of (12.40) is obtained from (12.37) and stems from the requirement that the expectation function calculation remain normalized while the system thermature is changing. As noted above, the thermature is a macroscopic parameter so microscopic analog operators do not depend on it in the usual case and the relation $\partial G_k(Q_k, P_k, t)/\partial \beta_k = 0$ is generally valid. When this is true, (12.40) is written

(12.41) $$\frac{\partial G_k(t)}{\partial \beta_k} = - \left\langle [H_k(Q_k, P_k, t) - H_k(t)] G_k(Q_k, P_k, t) \right\rangle_{kt}.$$

The equilibrium version of (12.41) is presented in Chapter 18. As a con-sistency condition on $\partial/\partial \beta_k$ as a derivative operator, it is necessary that $\partial \langle 1 \rangle_{kt} / \partial \beta_k = 0$. To demonstrate that this is the case, set $G_k(Q_k, P_k, t) = 1$ and $G_k(t) = 1$ in (12.41). This leads to

(12.42) $$0 = \frac{\partial \langle 1 \rangle_{kt}}{\partial \beta_k} = - \left\langle \mathcal{H}_k(Q_k, P_k, t) - \mathcal{H}_k(t) \right\rangle_{kt},$$

which follows from the fact that $\mathcal{H}_k(t) = \langle \mathcal{H}_k(Q_k, P_k, t) \rangle_{kt}$.

For the local thermature derivative, the procedure above is applied to the definition of $G_i(q, t)$ to obtain the *local thermature derivative* of a local *i*-particle thermodynamic function $G_i(q, t)$ in the form

(12.43) $$\frac{\partial G_i(q, t)}{\partial \beta_k} = \left\langle \delta(q - q_i) \frac{dG_i(Q_k, p_i, t)}{d\beta_k} \right\rangle_{kt} + \mathcal{H}_k(t) G_i(q, t).$$

The other major consistency condition is the requirement that any trans-formation preserves the closure conditions (5.29) and (5.31). Using the func-tions $G_k(Q_k, t) = 1$ or $G_k(Q_k, P_k, t) = 1$ in either of these conditions yields $\left\langle \dot{\chi}_{\Omega_k(t)}(Q_k, P_k) \right\rangle_{kt} = 0$. Since changes in the thermature should preserve these closure conditions, (12.40) is used to define the action of the thermature derivative on $\dot{\chi}_{\Omega_k(t)}(Q_k, P_k)$ as

$$(12.44a) \qquad 0 = \frac{\partial \left\langle \dot{\chi}_{\Omega_k(t)}(Q_k, P_k) \right\rangle_{kt}}{\partial \beta_k} = \left\langle \frac{d\dot{\chi}_{\Omega_k(t)}(Q_k, P_k)}{d\beta_k} \right\rangle_{kt},$$

which has the consequence

$$(12.44b) \qquad \frac{d\dot{\chi}_{\Omega_k(t)}(Q_k, P_k)}{d\beta_k} = 0.$$

In accord with the definition (12.39), the result (12.44) can be guaranteed by defining the thermature derivative of $\dot{\chi}_{\Omega_k(t)}(Q_k, P_k)$ by

$$(12.45) \qquad \frac{\partial \dot{\chi}_{\Omega_k(t)}(Q_k, P_k)}{\partial \beta_k} = \dot{\chi}_{\Omega_k(t)}(Q_k, P_k)\mathcal{H}_k(Q_k, P_k, t).$$

It is proper to include the Hamiltonian energy $\mathcal{H}_k(Q_k, P_k, t)$ in (12.45) because the Hamiltonian energy is used in the definition of $\chi_{\Omega_k(t)}(Q_k, P_k)$. A relation of this form between $\dot{\chi}_{\Omega_k(t)}(Q_k, P_k)$ and β_k is expected because the change in thermature depends on the energy flux through the boundaries.

When acting on phase functions $G_k(Q_k, P_k, t)$, the operators $\partial/\partial\beta_k$ and $d/d\Omega_k$ commute because they are derivatives with respect to ordinary variables. It is only with respect to the system distribution and its entropy that there is a question about their commutativity. At equilibrium, it is not hard to show that these derivatives commute because the variables are independent. This question will be dealt with in general in Chapter 19.

It was mentioned that the adoption of the definition (12.12) for the thermature has the significant advantage of extending the important equilibrium thermodynamic relation $\beta_k P_k^\epsilon = \partial \ln Z_k^\epsilon / \partial \delta_k$ to the nonequilibrium case. The definition (12.12a) and the result (12.37) mean that both $\partial \ln Z_k(t)/\partial \delta_k = \beta_k(t)P_k(t)$ and $\partial \ln Z_k(t)/\partial \beta_k = -\mathcal{H}_k(t)$ play a role in general TIS thermodynamics. It will be shown next that these definitions are consistent with the normal behavior of derivative operators and they mesh smoothly and consistently with the definitions of other TIS quantities.

Consider the expression $\partial \mathcal{H}_k(t)/\partial \delta_k = -P_k(t) + \Lambda_{k,V}(t)$ obtained in (11.9). By (12.37), this can be written in the form

$$(12.46) \qquad \frac{\partial^2 \ln Z_k(t)}{\partial \delta_k \partial \beta_k} = P_k(t) - \Lambda_{k,V}(t).$$

On the other hand, it is easy to show

$$(12.47) \qquad \frac{\partial^2 \ln Z_k(t)}{\partial \beta_k \partial \delta_k} = \frac{\partial \beta_k(t)P_k(t)}{\partial \beta_k} = \frac{\partial \beta_k(t)}{\partial \beta_k}P_k(t) + \beta_k(t)\frac{\partial P_k(t)}{\partial \beta_k}.$$

If it is assumed that the differentiation operators $\partial/\partial\beta_k$ and $\partial/\partial\delta_k$ commute when acting on thermodynamic objects, as will be shown in Chapter 19, (12.46) and (12.47) can be set equal and it follows that

(12.48) $$\frac{\partial\beta_k(t)}{\partial\beta_k} = 1 - \frac{1}{\mathrm{P}_k(t)}\left[\Lambda_{k,V}(t) + \beta_k(t)\frac{\partial\mathrm{P}_k(t)}{\partial\beta_k}\right].$$

Using the fact that $d/d\Omega_k$ commutes with $\partial/\partial\beta_k$ (see Chapter 19), a direct calculation of $\partial\beta_k(t)/\partial\beta_k$ easily yields this result as well, so the formalism is consistent. Moreover, it is also the case that $\partial\beta_k^\epsilon/\partial\beta_k = 1$ at equilibrium because, as will be shown in Chapter 17, $\Lambda_{k,V}^\epsilon = -\beta_k\partial P_k^\epsilon/\partial\beta_k$. The failure of $\partial\beta_k(t)/\partial\beta_k = 1$ to be valid in general is due to the conflict, discussed above, between its role as the thermature parameter in a putative equilibrium distribution and its role as the negative of the energy operator in the general case.[12]

Let us now compute the thermature derivatives of quantities defined for the exact state distribution. For exact state phase averages $G_k^\delta(t)$, the thermature derivative is

(12.49)
$$\frac{\partial G_k^\delta(t)}{\partial\beta_k} = \frac{1}{Z_k^\delta(t)}\int dQ_k\int dP_k\,\chi_{\Omega_k(t)}(Q_k,P_k)\frac{\partial F_k^\delta(Q_k,P_k,t)}{\partial\beta_k}G_k(Q_k,P_k,t)$$
$$- \frac{\partial\ln Z_k^\delta(t)}{\partial\beta_k}G_k^\delta(t)$$
$$= \frac{1}{Z_k^\delta(t)}\int dQ_k\int dP_k\,\chi_{\Omega_k(t)}F_k^\delta(Q_k,P_k,t)G_k(Q_k,P_k,t)$$
$$\times [\mathcal{H}_k(Q_k,P_k,t) - \mathcal{H}_k^\delta(t)]$$
$$= [\mathcal{H}_k(Q_k(t),P_k(t),t) - \mathcal{H}_k^\delta(t)]G_k^\delta(t).$$

The relation

(12.50) $$\mathcal{H}_k^\delta(t) = \langle\mathcal{H}_k(Q_k,P_k,t)\rangle_{k\delta} = \mathcal{H}_k(Q_k(t),P_k(t),t)$$

is used to show that the thermature derivative of any exact state phase average is zero:

(12.51) $$\frac{\partial G_k^\delta(t)}{\partial\beta_k} = 0.$$

This is to be expected because the temperature and entropy do not play a role in the exact state representation.

As an application of the thermature derivative operator, the heat capacity will be computed. It is the derivative of the Hamiltonian energy with respect

---

[12]It is interesting to note that Gibbs' [**1902**], pp. 170–176, also found that his temperature analogs were not perfect.

to the system thermature. By (12.42), the *i-particle local heat capacity density* is

(12.52a)

$$K_{i,V}(q,t) = \frac{\partial H_i(q,t)}{\partial \beta_k} = H_i(q,t)\mathcal{H}_k(t) - \langle \delta(q - q_i)H_i(Q_k,P_k,t)\mathcal{H}_k(Q_k,P_k,t)\rangle_{kt}$$

and the k system *local heat capacity density* is

(12.52b)
$$K_{k,V}(q,t) = \sum_{i \in k} K_{i,V}(q,t).$$

Integrating the local heat capacities, $K_{i,V}(q,t)$ and $K_{k,V}(q,t)$, over $q \in d_k(t)$ yields $K_{i,V}(t)$ and $K_{k,V}(t)$, respectively. Computing the k *heat capacity*, $K_{k,V}(t)$ directly, it is easy to see that it is the negative of the variance of the energy for the state F_k:

(12.53)

$$K_{k,V}(t) = \frac{\partial \mathcal{H}_k(t)}{\partial \beta_k} = -\langle [\mathcal{H}_k(Q_k,P_k,t) - \mathcal{H}_k(t)]^2 \rangle_{kt}$$

$$= [\mathcal{H}_k(t)]^2 - \langle [\mathcal{H}_k(Q_k,P_k,t)]^2 \rangle_{kt} \leq 0.$$

The relation (12.33) implies that the relation between this heat capacity and the k *conventional heat capacity*, $C_{k,V}(t)$, is

(12.54)
$$C_{k,V}(t) = -k_B \beta_k^2 K_{k,V}(t),$$

The heat capacity of a mole of a substance is called the *molar heat capacity* and the heat capacity per unit mass is the *specific heat*. The calculation and measurement of the specific heat of a substance in the k system must be made in the particle rest frame of the k system. In Chapter 14, it is shown that the relation (12.54) is not valid in the $\beta_k \to \infty$ limit for systems at absolute zero. The equilibrium versions of the heat capacities will be discussed further in Chapters 17 and 18.

12.6 The Heat Conduction Equation

The heat conduction equation is a phenomenological relation between the macroscopic heat flow and the gradient of the temperature. In standard notation, this is expressed as $\mathcal{Q}_k^\mu(q,t) = -\kappa_Q \nabla_k^\mu T_k(q,t)$, where $\mathcal{Q}_k^\mu(q,t)$ is the local heat flux vector, κ_Q is the coefficient of heat conduction, and $T_k(q,t)$ is the local temperature. To examine this equation in the TIS context, let us define the k *local heat function* by

(12.55)
$$\mathcal{Q}_k(q,t) = \sum_{i \in k} \langle \delta(q - q_i)K_i(p_i - \bar{p}_i(q,t))\rangle_{kt},$$

where $\bar{p}_i(q, t)$ is the average i-particle local momentum at (q, t). Next, a k *local heat flux* vector is defined by

$$(12.56) \qquad \mathcal{Q}_k^\mu(q, t) = \sum_{i \in k} \langle \delta(q - q_i) K_i(p_i - \bar{p}_i(q, t)) v_i^\mu \rangle_{kt}.$$

In order to define a heat condition equation in the form given above, it is necessary to find a local temperature function $T_k(q, t)$ such that

$$(12.57) \qquad \mathcal{Q}_k^\mu(q, t) = \sum_{i \in k} \langle \delta(q - q_i) K_i(p_i - \bar{p}_i(q, t)) v_i^\mu \rangle_{kt} = -\kappa_Q \frac{\partial T_k(q, t)}{\partial q_\mu}.$$

An interpretation of this equation that is useful in the TIS context requires a particle-based definition of the coefficient of heat condition κ_Q such that this equation is valid. The key to a successful interpretation is an appropriate approximation of κ_Q that meets TIS requirements.[13]

Most "derivations" of the heat conduction equation make use of systems "near equilibrium" and make several additional approximations in passing. But these approximations are usually not given an adequate treatment that would (1) allow for the evaluation of their validity, (2) indicate the domain of their proper application, and (3) allow for the calculation of the "constant" κ_Q in terms of particle properties. These objections do not to mean that the heat conduction equation is not useful. It is just that its domain of validity is too limited to make it a general equation of the theory.

It is worth mentioning that the gradient of the local thermature function $\beta_i(q, t)$ or the local thermature tensor $\beta_i^{\mu\nu}(q, t)$ can be computed using TIS methods and can be related to some aspects of the conduction of heat.

Let us return to the results Maxwell [**1860**] obtained using his transport equations and consider the transport of heat. Maxwell used the heat function $\mathcal{Q}_k(q) = (1/2)m\overline{|v_k(q)|}$ in this calculation.[14] Using the pattern (7.42) for the normal current into a plane, the heat current at the point q_0 in this plane is obtained with the help of the relation $\partial \mathcal{Q}_k(q_0)/\partial q_\mu = (\partial \mathcal{Q}_k(q_0)/\partial T_k)(\partial T_k(q_0)/\partial q_\mu)$ in the form

$$(12.58) \qquad \mathcal{Q}_k^{(n)}(q_0) = -b_\varrho \lambda \nu_k(q_0) \frac{d\mathcal{Q}_k(q_0)}{dT_k} \hat{n}_\nu \left. \frac{dT_k(q)}{dq_\nu} \right|_{q=q_0}.$$

Comparing this with (12.57) yields the result

$$(12.59) \qquad \kappa_k^{(Q)}(q_0) = \tfrac{1}{3} \lambda \nu_k(q_0) \mathcal{C}_{k,V}^{(1)}(q_0),$$

for the k *coefficient of heat conductivity*, where $\mathcal{C}_{k,V}^{(1)}(q_0)$ is the k *heat capacity per particle*. Because the geometry is the same as for equation (7.43), the

---

[13]The derivation of such an equation in Kerson Huang [**1963**], pp. 95–107, for example, is not possible in the TIS formalism.

[14]This is taken from Bailyn [**1994**], p. 435.

constant satisfies $b_Q = 1/3$ here as well. The heat capacity per particle at the point q_0 is defined in Maxwell's terms by

$$(12.60) \qquad \mathcal{C}_{k,V}^{(1)}(q_0) = \frac{d\mathcal{Q}_k(q_0)}{dT_k} = \frac{m}{2} \frac{d\overline{|v_k(q_0)|}}{dT_k}.$$

Because $\lambda \sim 1/\nu_k(q_0)$ the coefficient of heat conduction is independent of the density just as the coefficients of viscosity and diffusion are.

Maxwell's result on heat conduction is clearly too specialized to be part of the general TIS theory.

12.7 Temperature "Fluctuations"

In many discussions of thermodynamics, "fluctuations" or a "dispersion" is attributed to the temperature of an isolated system when this system is described by the canonical state.[15] The idea is that if the temperature is fixed in a system described by the canonical state, the energy will have a positive variance. Similarly, since the energy is fixed in an isolated system, a positive variance is attributed to the temperature. This point of view is misleading at best. First, it is improper to describe an isolated system by the canonical state. Briefly, this is because the total energy, momentum and angular momentum for an isolated system are fixed on the manifold $\Omega_t(t)$. Because the total t system Hamiltonian energy is constant on t, i.e., $\mathcal{H}_t(Q_t, P_t, t) = I_{t,H}$, it follows that $F_t^{(n)\epsilon}(Q_k, P_k)$ is constant on this manifold. The $F_t^{\gamma}(Q_k, P_k)$ state is also constant on this manifold and, since $F_t^{(n)\epsilon}$ and F_t^{γ} are both are normalized, it follows that $F_t^{(n)\epsilon} = F_t^{\gamma}$.[16]

The second objection concerns the statistical ideas behind this point of view. As stated before, the variance of a thermodynamic quantity in any system is the statistical variance of that quantity considered as a random variable in the system described. This is a legitimate procedure when observing small samples of a large system.[17] However, this requirement is violated when fluctuation computations are made using the canonical distribution with an isolated system. Finally, the temperature in the canonical distribution is not a random variable in the way that the energy is, so statements about its variance or fluctuation do not fit with its statistical meaning. The correct viewpoint, according to the theory presented here, is that the computation of the temperature of an isolated system should be made using the formula (12.12). The temperature may change in time or vary locally but is not fluctuating in the above sense.

---

[15]See Gibbs [**1902**], p. 181, or Mandelbrot [**1962**], pp. 1031–1036, for example, on this.

[16]See the discussion in Gibbs [**1902**], Chapter X.

[17]See Khinchin [**1943**], pp. 118–122, 147–151. This topic is treated in detail in the work on asymptotic thermodynamics in EIS, Chapter 18.

CHAPTER 13

Reversible Processes and the Equilibrium State

The conditions stated in Chapter 9 that the wall of a system must meet if the reference L_k distribution is to be reversible are taken up in this chapter. An investigation of the additional requirements it must meet to be a system distribution leads to the definition of the equilibrium state.

13.1 Deterministic Walls

In Chapter 9, criteria for equality in Clausius' relation, which are the criteria for reversible states, were stated in equations (9.22). The requirement in condition (9.22a) that the transformations $P_k^\star(P_k, P_l)$ and $P_l^\star(P_k, P_l)$ be invertible is a strong, but necessary, one. This requirement reflects the point that a transition in the microscopic description of a reversible process must link the before and after microscopic states in a time reversible, and therefore invertible way. The strategy for demonstrating this is to choose U_i equal to the delta measure stated in (9.22a) and solve (9.19) for the interface transition probability, $W_{kl}^{(\star)}$, that corresponds to this. Following these steps yields

$$(13.1) \qquad W_{kl}^{(\star)} = \frac{|\hat{n}_{\partial k} \cdot u_i^+| L_k L_l \delta(P_k' - P_k^\star(P_k, P_l)) \delta(P_l' - P_l^\star(P_k, P_l))}{|\hat{n}_{\partial k} \cdot u_i^{-\prime}| L_k' L_l'}.$$

The fact that the map $P^\star$ is a proper, unimodular, map, which preserves the total momentum, angular momentum, and energy, means that $(P_k', P_l') = (P_k^\star(P_k, P_l), P_l^\star(P_k, P_l))$ if and only if the inverse relation, written as $(P_k, P_l) = ([P_k^\star]^{-1}(P_k', P_l'), [P_l^\star]^{-1}(P_k', P_l'))$, also has these properties. Let us also define v_i by the inverse of the transformation $P_i^\star(P_k', P_l')$ as $v_i = [m_i]^{-1}[P_i^\star]^{-1}(P_k', P_l')$. These facts, along with the fact that the Jacobian of the transformation is 1, are then used to show that

$$(13.2a) \qquad \int dP_k \int dP_l\, \delta(P_k' - P_k^\star(P_k, P_l)) \delta(P_l' - P_l^\star(P_k, P_l)) = 1$$

and

$$(13.2b) \qquad \int dP_k \int dP_l\, \delta(P_k' - P_k^\star(P_k, P_l)) \delta(P_l' - P_l^\star(P_k, P_l)) G(P_k, P_l)$$

$$= G([P_k^\star]^{-1}(P_k', P_l'), [P_l^\star]^{-1}(P_k', P_l')).$$

This information allows us to integrate (13.1) over (P_k, P_l) and set the result to 1 in order to satisfy (6.36b) for each $i \in k$ or $i \in l$. The result is
(13.3)

$$1 = \frac{|\hat{n}_{\partial k} \cdot u_i^+(q_i, v_i)| L_k(Q_k, [P_k^\star]^{-1}(P_k', P_l'), t) L_l(Q_l, [P_l^\star]^{-1}(P_k', P_l'), t)}{|\hat{n}_{\partial k} \cdot u_i^-(q_i, v_i')| L_k(Q_k, P_k', t) L_l(Q_l, P_l', t)},$$

where, by (6.30), $u_i^\pm(q_i, v_i) = \theta^\pm(q_i, v_i)(v_i^\mu - v_{\partial k}^\mu(q_i, t))$. The relation (13.3) is to be true for all (P_k', P_l').

It is assumed that the reference distributions take the form (9.33), so the function $P^\star$ is also required to be compatible with the quasilocal domain analysis of Section 6.2. Then, by the form (9.33) of L_k, L_l, and the conservation properties of $P^\star$, the relation
(13.4a)

$$L_k(Q_k, [P_k^\star]^{-1}(P_k', P_l'), t) L_l(Q_l, [P_l^\star]^{-1}(P_k', P_l'), t) = L_k(Q_k, P_k', t) L_l(Q_l, P_l', t)$$

is valid for all (P'_k, P'_l). This implies by (13.3) that

$$(13.4b) \qquad \hat{n}_{\partial k}(q_i, t) \cdot u_i^+(q_i, v_i) = -\hat{n}_{\partial k}(q_i, t) \cdot u_i^-(q_i, v'_i),$$

for each $i \in k$ and $i \in l$. This result states that the component of the individual particle flux normal to the wall is conserved at a deterministic wall. Using (13.4) in (13.1) gives the result

$$(13.5) \qquad W_{kl}^{(\star)} = W_{kl}^{(\mathrm{det})} = \delta(P'_k - P_k^\star(P_k, P_l))\delta(P'_l - P_l^\star(P_k, P_l))$$

which means, in accord with definition (6.47), that $W_{kl}^{(\star)}$ is a deterministic wall.

Equations (13.4) are valid, for example, for both the specular reflection law and the reverse reflection law. Both of these laws have been used in wall calculations in the literature.

13.2 Stochastic Walls

For stochastic walls, the objective is to determine the form of states that support reversible processes. To illustrate the physical and historical issues, an elementary proof, based originally on Boltzmann's work, will be presented to show that these states take a small set of special forms.

The results of this section have been known for some time. Boltzmann's original work was done in 1876 and Maxwell mentioned similar results in 1879. A slightly different approach was used by Fowler [**1936**], pp. 671–672, who investigated these issues in the context of Boltzmann's equation. He focused on distributions that support detailed balancing in Boltzmann's equation, imposed the equilibrium condition as $f_r f_s - f_r^* f_s^* = 0$, and equated coefficients, to obtain a particular form for the equilibrium particle probability distribution density. Fowler showed that only rigid body motions, which is a uniform rotation superposed on a translation, are allowed for the equilibrium state.

The results of this chapter are obtained for the general TIS theory. In this case, the conditions that L_k must meet if it is to be a time-independent solution of (5.16) are computed. These conditions determine the structure of the equilibrium system distribution. It is easy to show that similar results apply to solutions of Boltzmann's evolution equation as well. From a mathematical perspective, these conditions are obtained by first showing that L_k must be a function of $\mathcal{H}_k$ alone and requiring that $d\mathcal{H}_k/dt = 0$. The results of this chapter are then a trivial mathematical consequence of the symmetries of the Hamiltonian. However, the abstract mathematical methods used to demonstrate this offer little physical insight into what is meant by these results, so Boltzmann's approach is followed here.

Turning now to condition (9.22b), the fact that $R_k(Q_k, t)$ is arbitrary in (9.33) can be used without loss of generality to set $M_k(Q_k, t) = 1$. Then, for (9.22b) to hold, the relation $F_k = L_k$ must hold. This means in essence that

L_k must be a possible state of the system and must satisfy (5.14), (5.15), (5.16), and (5.29).

It is clear that L_k satisfies (5.14). Since the constraints in (9.22) are satisfied by L_k, the boundary condition (6.39b) will be met and (5.29) will be valid. Let us turn then to the question of the conditions under which L_k satisfies (5.16).

Taking the total time derivative of L_k given in (9.33) shows that to satisfy (5.16) it is necessary and sufficient that

$$(13.6) \qquad \frac{d}{dt}\left[\sum_{i\in k} b_k(q_i,t)K_i(p_i - p_k^*(q_i,t)) + R_k(Q_k,t) + \ln Z_k^L(t)\right] = 0.$$

Following a procedure of Boltzmann [1876], the arbitrariness of v_i^μ in the formula that results from this calculation is used to require that the coefficients of different powers of v_i^μ be zero separately. The arbitrariness of the particle coordinates q_i^μ will also be used to indicate which other combinations of terms must be zero.

Let us begin with the equations for the third power of v_i^μ, which is the highest power. For each $i \in k$, the third degree equation in v_i^μ is

$$(13.7) \qquad v_i \cdot \frac{b_k(q_i,t)}{\partial q_i} K_i(p_i) = 0.$$

This implies that

$$(13.8) \qquad \frac{\partial b_k(q_i,t)}{\partial q_i^\mu} = 0$$

and, hence,

$$(13.9) \qquad b_k(q_i,t) = b_k(t).$$

This relation can be used to replace $b_k(q_i,t)$ by $b_k(t)$ in (13.6) and then the representation $P_k^* = P_k^*(Q_k,t) = \times_{i\in k} p_k^*(q_i,t) = \times_{i\in k} m_i v_i^*(q_i,t)$ can be used for P_k^*. The condition (13.6) then takes the form

$$(13.10) \qquad \frac{d}{dt}[b_k(t)\mathcal{K}_k(P_k - P_k^*(Q_k,t)) + R_k(Q_k,t) + \ln Z_k(t)] = 0.$$

Carrying out the differentiation in (13.10), the second degree equation in v_i^μ is

$$(13.11) \qquad \frac{1}{2}\frac{db_k(t)}{dt} v_i \cdot p_i - b_k(t)v_i^\mu \frac{\partial v_i^{*,\nu}(q_i,t)}{\partial q_i^\mu} p_{i,\nu} = 0,$$

where $v_k^*(q_i,t) = p_k^*(q_i,t)/m_i$. Using the definition

$$(13.12) \qquad A_k(t) = 2\int_0^t du\, C_k(u),$$

this equation has the solution

(13.13a) $$b_k(t) = b_k(0)e^{A_k(t)},$$

(13.13b)
$$v_k^{*\mu}(q_i, t) = C_k(t)q_i^{\mu} + \epsilon^{\mu\nu\xi}s_{k,\nu}(t)q_{i,\xi} + c_k^{\mu}(t),$$

where $C_k(t)$, $s_{k,\nu}(t)$, and $c_k^{\mu}(t)$ do not depend on i because of the boundary condition (6.39b). The function $C_k(t)$ is integrable and once continuously differentiable; $s_{k,\nu}(t)$ and $c_k^{\mu}(t)$ are once continuously differentiable vector functions.

Using the definition $V_k^* = V^*(Q_k, t) = \times_{i \in k} v_k^*(q_i, t)$, the first degree equation for $v_{i,\mu}$ can then be written

(13.14a)
$$-\frac{\partial}{\partial t}[b_k(t)P_k^{*\mu}(Q_k, t)] - b_k(t)\frac{\partial \Phi_k(Q_k, t)}{\partial Q_{k,\mu}} + b_k(t)V_{k,\nu}^*\frac{\partial P_k^{*\nu}}{\partial Q_{k,\mu}} + \frac{\partial R_k(Q_k, t)}{\partial Q_{k,\mu}} = 0$$

and the zero degree equation is

(13.14b)
$$\frac{\partial}{\partial t}\left[\tfrac{1}{2}b_k(t)P_k^* \cdot V_k^* + R_k(Q_k, t)\right] + b_k(t)V_k^* \cdot \frac{\partial \Phi_k(Q_k, t)}{\partial Q_k} + \frac{d\ln Z_k(t)}{dt} = 0.$$

Since the q_j, for $j \neq i$, can be varied independently, equations (13.14) must not depend on q_j. Part of this dependence can be removed, simplifying the equations further, by making use of the arbitrary function $R_k(Q_k, t)$ and defining it as

(13.15a) $$R_k(Q_k, t) = b_k(t)[\Phi_k(Q_k, t) - \tfrac{1}{2}P_k^* \cdot V_k^* + R_k^0(Q_k, t)],$$

where $R_k^0(Q_k, t)$ is defined for later convenience as the particle sum function

(13.15b) $$R_k^0(Q_k, t) = \sum_{i \in k} r_i(q_i, t)$$

with $r_i(q_i, t)$ arbitrary.

Using (13.15) in (13.14) gives the result

(13.16a) $$-\frac{\partial}{\partial t}[b_k(t)P_k^{*\mu}(Q_k, t)] + b_k(t)\frac{\partial R_k^0(Q_k, t)}{\partial Q_{k,\mu}} = 0$$

for the i-component of the first degree equation and

(13.16b)
$$\frac{\partial}{\partial t}\left[b_k(t)\{\Phi_k(Q_k, t) + R_k^0(Q_k, t)\}\right] + b_k(t)V_k^* \cdot \frac{\partial \Phi_k(Q_k, t)}{\partial Q_k} + \frac{d\ln Z_k(t)}{dt} = 0.$$

for the zero degree equation.

By (13.13) and (4.20), it follows that

(13.17) $$V_k^* \cdot \frac{\partial \Phi_k(Q_k, t)}{\partial Q_k} = C_k(t)Q_k \cdot \frac{\partial \Phi_{k,n}(Q_k)}{\partial Q_k} + V_k^* \cdot \frac{\partial \Phi_{k,x}(Q_k, t)}{\partial Q_k}.$$

The relation $V_k^* \cdot (\partial \Phi_{k,x}(Q_k, t)/\partial Q_k) = \sum_{i \in k} v_i^* \cdot (\partial \Phi_{i,x}(q_i, t)/\partial q_i)$ means that the i-particle term in the sum depends only on the q_i coordinate and not any others. The remaining q_j dependence in the i-particle component of (13.16b) is in the $\Phi_{k,n}(Q_k)$ term. This q_j dependence is removed next by using (13.17) and requiring

$$(13.18) \qquad b_k(t) C_k(t) Q_k \cdot \frac{\partial \Phi_{k,n}(Q_k)}{\partial Q_k} + \frac{db_k(t)}{dt} \Phi_{k,n}(Q_k) = 0.$$

Using $db_k(t)/dt = 2b_k(t) C_k(t)$ and the relation

$$(13.19) \qquad Q_k \cdot \frac{\partial \Phi_{k,n}(Q_k)}{\partial Q_k} = \frac{1}{2} \sum_{i \in k} \sum_{j \in k-i} (q_i - q_j) \cdot \frac{\partial \phi_{ij}(q_i - q_j)}{\partial q_i},$$

it follows that the result

$$(13.20) \qquad b_k(t) C_k(t) \left[\frac{1}{2}(q_i - q_j) \cdot \frac{\partial \phi_{ij}(q_i - q_j)}{\partial q_i} + \phi_{ij}(q_i - q_j) \right] = 0$$

is obtained for each i, j. Since $b_k(t) > 0$, it follows that either

$$(13.21a) \qquad\qquad\qquad C_k(t) = 0$$

or

$$(13.21b) \qquad\qquad \phi_{ij}(q_i - q_j) = a_{ij} |q_i - q_j|^{-2},$$

for some constant a_{ij}. Potentials of the form (13.21b) are excluded by (4.2). Therefore, condition (13.21a) must be true. This leads immediately to[1]

$$(13.22a) \qquad\qquad b_k(t) = b_k = \text{const.},$$
$$(13.22b) \qquad\qquad v_k^{*\mu}(q_i, t) = \epsilon^{\mu\nu\xi} s_{k,\nu}(t) q_{i,\xi} + c_k^\mu(t).$$

The q_i's are independent, so the first and zero degree equations can be separated into single components. Using (13.21a) with (13.16a) yields for the i-component of the first degree equation

$$(13.23a) \qquad\qquad -\frac{\partial p_i^{*\mu}(q_i, t)}{\partial t} + \frac{\partial r_i(q_i, t)}{q_{i,\mu}} = 0.$$

For the zero degree equation, note that the last term of (13.16b) does not contain q_i. This fact and the fact that $\partial r_i(q_i, t)/\partial t$ is arbitrary as a function

---

[1] Equation (13.22b) is the equation of a "rigid motion" (rotation plus translation) of the system. Dilation has been excluded by condition (13.21a). See equation (84.2) in Truesdell and Toupin [**1960**], p. 350.

of the time are used to write (13.16b) as two equations:

$$(13.23\text{b}) \quad b_k \sum_{i \in k} \left\{ \frac{\partial r_i(q_i, t)}{\partial t} + \frac{\partial \Phi_{i,x}(q_i, t)}{\partial t} + v_i^* \cdot \frac{\partial \Phi_{i,x}(q_i, t)}{\partial q_i} \right\} = 0,$$

$$(13.23\text{c}) \qquad\qquad \frac{d \ln Z_k(t)}{dt} = 0.$$

Due to the special form of $v_k^{*\mu}$ in (13.22b), it is clear that equation (13.23a) will be integrable only under certain conditions. These conditions are satisfied in two cases. The first case is

$$(13.24\text{a}) \qquad\qquad \text{Case (i):} \qquad \frac{\partial p_i^{*\mu}(q_i, t)}{\partial t} = 0,$$

$$(13.24\text{b}) \qquad\qquad \frac{\partial r_i(q_i, t)}{\partial q_i} = 0.$$

The second case uses a version of a theorem by Ferdinand Frobenius[2] to show that there exists a function $h_i(q_i, t)$ such that

$$(13.25\text{a}) \qquad \text{Case (ii):} \qquad p^{*\mu}(q_i, t) = \frac{\partial h_i(q_i, t)}{\partial q_{i,\mu}},$$

$$(13.25\text{b}) \qquad\qquad r_i(q_i, t) = \frac{\partial h_i(q_i, t)}{\partial t}.$$

Beginning with Case (i), it follows immediately that

$$(13.26\text{a}) \qquad v_i^{*\mu}(q_i, t) = v_i^{*\mu}(q_i) = \epsilon^{\mu\nu\xi} s_{k,\nu} q_{i,\xi} + c_k^\mu,$$
$$(13.26\text{b}) \qquad\qquad r_i(q_i, t) = 0.$$

The result (13.26b) follows from the fact that $r_i(q_i, t) = r_i(t)$ and $r_i(t)$ can be absorbed into the $\ln Z_k(t)$ term. Next, for $i \in k$ in (13.23b), it follows that

$$(13.27\text{a}) \qquad \frac{\partial \Phi_{i,x}(q_i, t)}{\partial t} + v_i^*(q_i) \cdot \frac{\partial \Phi_{i,x}(q_i, t)}{\partial q_i} = 0,$$

$$(13.27\text{b}) \qquad\qquad Z_k(t) = Z_k = \text{const.}$$

Equation (13.27a) means that the external potentials are constant on particle trajectories moving with the stream velocity v_i^*. A consequence of this is that if $\partial \Phi_{i,x}(q_i, t)/\partial t = 0$, then $\partial \Phi_{i,x}(q_i, t)/\partial q_i$ must be orthogonal to v_i^*. A vector orthogonal to $v_i^*(q_i)$ must lie in the (s_k, q_i) plane and be orthogonal to the fixed vector c_k^μ. Since q_i is arbitrary, this requirement means in effect that (a) $c_k^\mu = 0$ or (b) $s_{k,\mu} = 0$.

---

[2]Frobenius' theorem states that Leonard Euler's criterion is both necessary and sufficient for the existence of a function such as h_i defined in the text. See, for example, the references given in James Partington [**1949**], p. 38, footnote 2, or Jean Dieudonné [**1960**], Section X.9.

Equation (13.27a) can be solved by separating the variables with the result

(13.28a)
$$\frac{\partial \ln \Phi^t_{i,x}(t)}{\partial t} = -a,$$

(13.28b)
$$v^*(q_i) \cdot \frac{\partial \ln \Phi^q_{i,x}(q_i)}{\partial q_i} = a,$$

where the factorization $\Phi_{i,x}(q_i, t) = \Phi^t_{i,x}(t)\Phi^q_{i,x}(q_i)$ was also used. Let $g_i(\cdot)$ be a once continuously differentiable function of its argument. Then (13.22b) allows us to write (13.28b) in the form

(13.29)
$$\frac{\partial \ln \Phi^q_{i,x}(q_i)}{\partial q_{i,\mu}} = a\frac{v^{*\mu}(q_i)}{[v^*(q_i)]^2} + \frac{\partial g_i([v^*(q_i)]^2)}{\partial q_{i,\mu}}.$$

If $a \neq 0$ and $s_{k,\nu} \neq 0$, equation (13.29) is not integrable. If $a = 0$, the solution, apart from a constant, is

(13.30)
$$\Phi^{(1)}_{i,x}(q_i) = e^{g_i([v^*_i(q_i)]^2)}.$$

If $s_{k,\nu} = 0$ so that $v^{*,\mu}(q_i) = c^\mu$ is constant, it is easy to show that the solution of (13.29) is

(13.31)
$$\Phi^{(2)}_{i,x}(q_i, t) = Ae^{-a[t-q_i\cdot(c/c^2)]},$$

where A is some constant and a is the constant defined in (13.28).

With the solutions (13.30) or (13.31) for the external potential, the first set of results is[3]

Result (i)

(13.32a)
$$b_k(q_i, t) = b_k = \text{const.},$$

(13.32b)
$$v^{*\mu}(q_i, t) = \epsilon^{\mu\nu\xi}s_{k,\nu}q_{i,\xi} + c^\mu_k,$$

(13.32c)
$$R_k(Q_k, t) = b_k[\Phi_{k,n}(Q_k) - \tfrac{1}{2}P^*_k(Q_k) \cdot V^*_k(Q_k) + \Phi^{(s)}_{k,x}(Q_k, t)],$$

(13.32d)
$$Z_k(t) = Z_k = \text{const.}$$

The symbol $\Phi^{(s)}_{k,x}(Q_k, t)$ in (13.32c) is valid for $s = 1$ or $s = 2$ and represents either $\Phi^{(1)}_{k,x}$ or $\Phi^{(2)}_{k,x}$. Result (i) corresponds to the system as a whole rotating with a constant angular velocity about a fixed axis that is moving with a constant uniform stream velocity. If both $s_{k,\nu} = 0$ and $c^\mu_k = 0$, then $a = 0$ and $v^{*\mu}_i(q_i, t) = 0$. It then follows from (13.27a) that $\Phi_{i,x}(q_i, t) = \Phi_{i,x}(q_i)$, where the external potential $\Phi_{i,x}(q_i)$ is otherwise arbitrary. This corresponds to a system at rest in a time-independent, spatially arbitrary field of force.

---

[3]A system of described in this way was called a "free system" by Maxwell [**1879b**], (Maxwell [**1890**], p. 715).

Using Result (i) with the definition of L_k in (13.6) gives the result

$$(13.33a) \qquad L_k(Q_k, P_k, t) = e^{-b_k \mathcal{H}_k^L(Q_k, P_k, t)},$$

where
(13.33b)
$$H_k^L(Q_k, P_k, t) = \mathcal{K}_k(P_k - P_k^*(Q_k)) + \Phi_{k,n}(Q_k) + \Phi_{k,x}^{(s)}(Q_k, t) - \tfrac{1}{2} P_k^*(Q_k) \cdot V_k^*(Q_k).$$

Consider next Case (ii). Unless $s_{k,\nu}(t) = 0$, equation (13.23a) cannot be integrated; to make progress, set $s_{k,\nu}(t) = 0$ and satisfy (13.25) with the function $h_i(q_i, t) = m_i c_k(t) \cdot q_i$. Using these substitutions gives the results

$$(13.34a) \qquad v_i^{*\mu}(q_i, t) = v_i^{*\mu}(t) = c_k^\mu(t)$$

and

$$(13.34b) \qquad r_i(q_i, t) = m_i \frac{dc_k(t)}{dt} \cdot q_i.$$

The results (13.34) are used in (13.23b) to obtain

$$(13.35) \qquad \frac{\partial \Phi_{i,x}(q_i, t)}{\partial t} + m_i \frac{d^2 c_k^\mu(t)}{dt^2} q_{i,\mu} + c_k(t) \cdot \frac{\partial \Phi_{i,x}(q_i, t)}{\partial q_i} = 0.$$

As the first solution to this equation, define

$$(13.36) \qquad \Phi_{i,x}^{(3)}(q_i) = \tfrac{1}{2} m_i \omega^2 q_i^2$$

and use this to obtain the following solution of (13.23)

$$(13.37a) \qquad c^\mu(t) = a^\mu \cos \omega t + b^\mu \sin \omega t,$$
$$(13.37b) \qquad Z_k(t) = Z_k = \text{const.},$$

where a^μ and b^μ are constant. If $a \cdot b = 0$ and $a^2 = b^2$, there is a more specialized solution

$$(13.38) \qquad \Phi_{i,x}^{(4)}(q_i, t) = \tfrac{1}{2} A m_i \omega^2 q_i^2 - B m_i q_i \cdot (a \sin \omega t - b \cos \omega t),$$

where $A + B = 1$. Summing either (13.36) or (13.38) over $i \in k$ leads to $\Phi_{k,x}^{(3)}(Q_k, t)$ or $\Phi_{k,x}^{(4)}(Q_k, t)$, respectively.

Collecting the formulas (13.22), (13.34), (13.36) or (13.38), and (13.37), leads to the second set of results:

Result (ii)

$$(13.39a) \qquad b_k(q_i, t) = b_k = \text{const.},$$
$$(13.39b) \qquad v^{*\mu}(q_i, t) = a^\mu \cos \omega t - b^\mu \sin \omega t,$$
$$(13.39c) \quad R_k(Q_k, t) = b_k[\Phi_{k,n}(Q_k) - \tfrac{1}{2} P_k^*(Q_k) \cdot V_k^*(Q_k) + \Phi_{k,x}^{(s)}(Q_k)],$$
$$(13.39d) \qquad Z_k(t) = Z_k = \text{const.}$$

where $\Phi_{k,x}^{(s)}$ stands for $\Phi_{k,x}^{(3)}$ or $\Phi_{k,x}^{(4)}$. This result corresponds to the system as a whole acting as a harmonic oscillator.[4] The Hamiltonian energy and pressure of a harmonic oscillator will be discussed as part of an approximation procedure in Chapter 14.

Mathematically, these results follow from an analysis of the representations of the invariant group of the Galilean transformation extended by a dilation transformation.

13.3 Uniqueness of the Factorization Property

The derivation of the reference distribution inequality, (9.25), presented in Chapter 9 depended on the existence of a pair of bounded reference distributions, L_k and L_l, which are invariant under the interface transition probability chosen. The special form (9.33) of the reference distributions is close to that of an equilibrium state. It was shown in Chapter 9 that this form for these distributions is preserved by the unbiased interface transition probability, $W_{kl}^{(\text{unb})}$. For this reason, $W_{kl}^{(\text{unb})}$ was adopted to describe system interfaces.

The unbiased interface transition probability assigns an equal probability to each outgoing kl state with the same total momentum, angular momentum and energy as the incoming kl state. In this section, the main concern is with questions of uniqueness. That is, given $W_{kl}^{(\text{unb})}$ and the momentum, angular momentum, and energy preserving transitions from the mechanical state $(Q_k, P_k') \times (Q_l, P_l')$ to the state $(Q_k, P_k) \times (Q_l, P_l)$, the set of all system states for which the relation

$$(13.40) \qquad L_k(Q_k, P_k', t)L_l(Q_l, P_l', t) = L_k(Q_k, P_k, t)L_l(Q_l, P_l, t)$$

holds are sought. An elementary proof that is based in part on Boltzmann [1876] and in part on Truesdell and Muncaster [1980] will be presented. As before, more powerful but less transparent mathematical methods are available.

Certain conclusions were drawn earlier in the chapter concerning the conditions on the parameters of L_k that follow from requiring that it meet the condition $dL_k/dt = 0$ of being a state. These will be set aside for the moment and the question of uniqueness approached without the requirement that L_k be a possible state of the k system. Many of these same conditions will follow from imposing the requirement (13.40) on the functions $L_k(Q_k, P_k, t)$.

Let us begin with a quasilocal domain that intersects the volume sets of the systems k and l. Recall that each quasilocal domain behaves as an

---

[4] A solution of this type excited some interest in the 19th century since it does not lead to a time-independent equipartition of the energy. See Culverwell [1890], Thomson [1891], Bryan [1891], p. 118–119. See also Uhlenbeck and Ford [1963], pp. 80–81.

independent system. At any time t, the particles are partitioned into non-overlapping quasilocal domains in accord with the quasilocal forces approximation. At the same time t, the reference distribution is required to factor for these domains so that $L_k = \prod_{a \in \kappa} L_k^a$.

The question under consideration is what requirements need to be placed on the quasilocal functions b_k and g_k^μ used in the representations of L_k and L_l so that the relation (13.40) between systems is valid. These considerations indicate that there is no loss of generality if a single quasilocal domain a that intersects the kl boundary $w_{kl}(t)$ is used for the analysis at time t. The variables Q_k and P_k will be used in computations with the understanding that the functions stated below depend only on that subset of the particle variables that is appropriate to the quasilocal domain under consideration.

Let us use (9.1) to define $S_k^{aL}(Q_k, P_k, t)$ as the phase entropy of the L_k^a distribution. The relation between the system phase entropy and the quasilocal domain entropies is $S_k^L(Q_k, P_k, t) = \sum_{a \in \pi(t)} S_k^{aL}(Q_k, P_k, t)$. The factorization of L_k and L_l in (13.40) is then used to obtain for each $a \in \pi(t)$ the relation[5]

$$(13.41) \quad S_k^{aL}(Q_k, P_k', t) + S_l^{aL}(Q_l, P_l', t) = S_k^{aL}(Q_k, P_k, t) + S_l^{aL}(Q_l, P_l, t).$$

For convenience, the label a on these formulas will be dropped. It will be reintroduced when needed later and a sum over $a \in \pi(t)$ will be taken to combine the results for each domain. As before, $G_k(Q_k, P_k, t)$ is abbreviated as G_k and $G_k(Q_k, P_k', t)$ as G_k' for any phase function, G_k.

Boltzmann proved a theorem concerning the representation of the entropy for two particle collisions.[6] A proof was also given in Truesdell and Muncaster [1980], Chapter VI. The strategy used here will extend a commonly used method to cover the multiparticle interactions envisioned in TIS.

Let us assume that the only conserved quantities in the interaction of two systems are the total energy, total momentum, and total angular momentum. The most general conserved scalar quantity that can be constructed from these conserved quantities for a single system takes the form[7]

$$(13.42) \quad R_k^\star(Q_k, P_k, t) = A_k \mathcal{H}_k(Q_k, P_k, t) + B_k \cdot \mathcal{P}_k + C_k \cdot \mathcal{J}_k(Q_k, P_k) + D_k,$$

where A_k and D_k are scalar constants and B_k^μ and C_k^μ are constant three vectors. The function $R_k^\star(Q_k, P_k, t)$ will be abbreviated as $R_k^\star$ and $R_k^\star(Q_k, P_k', t)$ as $R_k^{\star\prime}$. Because of the conservation of these quantities in a multiparticle

---

[5] Condition (9.18a) guarantees that this relation is always valid.

[6] See Boltzmann [1876], pp. 76–79, for a version of the theorem and its proof. See also Maxwell [1866] (Maxwell [1890], pp. 45–48), and Boltzmann [1877b], pp. 167–172.

[7] For a suitably defined phase velocity $V_k^*(q, t)$ and its associated phase momentum $P_k^*(q, t)$, it is not hard to show that the function $R_k^\star(Q_k, P_k, t)$ defined in (13.42) can be written in the form $R_k^\star(Q_k, P_k, t) = \mathcal{H}_k(Q_k, P_k - P_k^*(q, t), t) + D_k$, where D_k is a constant.

interaction between systems k and l, it follows that

$$(13.43) \qquad R_k^\star + R_l^\star = R_k^{\star'} + R_l^{\star'}.$$

Next, using the fact that $S_k^L = -k_B \ln L_k$, it follows from (13.41) and (13.43) that

$$(13.44) \qquad \ln L_k(R_k^\star) + \ln L_l(R_l^\star) = \ln L_k(R^{\star'k}) + \ln L_l(R_l^{\star'}).$$

Using (13.43) to replace $R_l^{\star'}$ in (13.44) and taking the derivative of the resulting equation with respect to $R_k^\star$ gives the result[8]

$$(13.45) \qquad \frac{d\ln L_k(R_k^\star)}{dR_k^\star} = \frac{d\ln L_l(R_k^\star + R_l^\star - R_k^{\star'})}{dR_k^\star}.$$

Because the left side of (13.45) does not depend on $R_l^\star$ or $R_k^{\star'}$ and the right side does, the derivative on each side is a constant independent of its argument. This is expressed as

$$(13.46) \qquad \frac{d\ln L_k(R_k^\star)}{dR_k^\star} = X_k.$$

The solution of this is

$$(13.47) \qquad L_k(R_k^\star(Q_k, P_k, t)) = e^{X_k R_k^\star(Q_k, P_k, t)}.$$

Comparing this result with (9.33) and allowing the possible time dependence of the coefficients, indicates first that a solution requires $b_i^a(q_i, t) = b_k^a(t)$. This implies that $X_k = -1$, $A_k = b_k(t)$, $B_k = g_k^a(t)$, and $C_k = s_k^a(t)$, where g_k^a and s_k^a are constant 3-dimensional vectors. The constant D_k, in the form $\ln K_k$, can be expressed as a multiplicative constant in (13.47) and, if desired, associated with the normalization of L_k. Because normalization is handled separately, the choice $D_k = 0$ will be used here.

Using these results in the definition of S_k^{aL}, and reintroducing the quasi-local domain designation, yields the result

$$(13.48) \quad S_k^{aL}(Q_k, P_k, t) = -k_B[b_k^a(t)\mathcal{H}_k^a(Q_k, P_k, t) + g_k^a \cdot \mathcal{P}_k^a + s_k^a \cdot \mathcal{J}_k^a(Q_k, P_k)],$$

which is valid at a given time t. These results will now be extended from the shared quasilocal domain a to the whole k and l systems. Within each quasilocal domain, the relations (13.43) are valid if p_i is defined by $p_i = p_i'$ for each particle i not in the domain under consideration. In the quasilocal case, b_k^a, g_k^a and s_k^a are quasilocal step functions required to be constant within a given quasilocal domain, but they may take different values in different quasilocal domains. This means that they are domain dependent quasilocal step functions written $b_k^a(q, t)$, $g_k^a(q, t)$ and $s_k^a(q, t)$.

---

[8]This proof was introduced by Maxwell [**1860**]. It has been called *Maxwell's theorem* by Bailyn [**1994**], pp. 428–431. It was also used by Pascual Jordan [**1933**], p. 8, among others.

If the whole system k is taken as the quasilocal domain, these functions must be constant over the whole k volume set and written $b_k(t)$, $g_k(t)$, and $s_k(t)$. If the combined kl system is isolated, the quantities $\mathcal{P}_k + \mathcal{P}_l$, $\mathcal{J}_k(Q_k, P_k) + \mathcal{J}_l(Q_l, P_l)$, and $\mathcal{H}_k(Q_k, P_k) + \mathcal{H}_l(Q_l, P_l)$ are constants. If it is also the case that the analysis is performed in the center of momentum frame of the combined kl system, then $\mathcal{P}_k + \mathcal{P}_l = 0$ and $\mathcal{J}_k(Q_k, P_k) + \mathcal{J}_l(Q_l, P_l) = 0$.

This is as far as the investigation of uniqueness can carry us. Requiring, as in Section 13.2, that the function L_k be a state of the system implies that the functions b_k, g_k, and s_k are constants.

13.4 The Equilibrium State

For a system to be at equilibrium, it is necessary that (i) there is a Galilean reference frame, called the *system rest frame* above, in which all of the walls or boundaries of the system are at rest with respect to each other; (ii) the frame in which the system is represented, observed, and analyzed, is the system rest frame; and (iii) the *particle rest frame* coincides with the system rest frame. A frame that meets criteria (i)–(iii) is called the *equilibrium rest frame* of the system.

The process of bringing a system to equilibrium can be understood informally in terms of these criteria. Suppose the system and observation frame meet conditions (i) and (ii). Suppose also that the walls are rigidly attached to massive supports or to the earth, so that the walls can absorb momentum and angular momentum without moving appreciably. Suppose finally that the particle rest frame does not coincide with the system rest frame. In this situation, the particles reaching the walls will carry part of the total momentum and total angular momentum of the particle system, which are non-zero with respect to the system rest frame. Particles carrying some of the excess total momentum and total angular momentum reaching the walls will dissipate these excess quantities into these walls and their supports. Similarly, particles that are deficient in their ultimate share of the total momentum and total angular momentum will tend to pick up these quantities from other particles and the walls and their supports. This process will continue and the system will evolve until the particle rest frame coincides with the system rest frame and criterion (iii) is satisfied. The system is then at equilibrium.

In accord with the above results, the k *asymptotic equilibrium state*, F_k^ϵ, is defined in the equilibrium rest frame of system k by[9]

$$(13.49a) \qquad F_k^\epsilon(Q_k, P_k) = e^{-\beta_k \mathcal{H}_k^\epsilon(Q_k, P_k)}.$$

---

[9]For more information on the historical development of this form for the equilibrium state, see Chapter 18 and EIS. Requiring that the equilibrium state is defined only in the equilibrium rest frame of the system is important in avoiding paradoxes that have come up in the literature when the equilibrium state has been transformed into other frames and calculations are made using this transformed state.

Equilibrium phase averages are normalized using the k *equilibrium partition function* defined by

$$(13.49b) \qquad Z_k^\epsilon = \int dQ_k \int dP_k \, e^{-\beta_k \mathcal{H}_k^\epsilon(Q_k, P_k)} \chi_{\Omega_k(t)}(Q_k, P_k).$$

The equilibrium Hamiltonian energy $\mathcal{H}_k^\epsilon$ is defined below.

Let $\mathcal{C}_k^\epsilon$ be the *macroscopic equilibrium state,* which is the set of macroscopic parameters required to fully characterize the equilibrium state. It is defined by the ordered set of macroscopic system quantities

$$(13.49c) \qquad \mathcal{C}_k^\epsilon = (\mathcal{N}_k, \beta_k, d_k(t), \mathcal{H}_k^u, \mathcal{H}_k^l).$$

An examination the definition (4.27) of Ω_k, the definitions in Chapter 4 of the Hamiltonian, and their use in the integral stated in (13.49b) shows that the list $\mathcal{C}_k^\epsilon$ of parameters describing the system and its components is sufficient to determine the value of the integral defining the partition function for the equilibrium state. The abbreviated form of the equilibrium state, expressed as $\mathcal{C}_k^\epsilon = (\beta_k, \delta_k)$ and valid for a closed system with a fixed boundary, no surface effects, and a fixed thermature, will be used in most circumstances. Unless it is stated otherwise, macroscopic equilibrium thermodynamic quantities will be assumed from now on to be functions of the state (β_k, δ_k).

It will be shown in Part II that all other equilibrium quantities can be derived from the partition function using the TIS formalism in parallel with the results of standard equilibrium theory.

The equilibrium version of the Hamiltonian function, given by

$$(13.49d) \qquad \mathcal{H}_k^\epsilon(Q_k, P_k) = \mathcal{K}_k(P_k) + \Phi_{k,n}(Q_k) + \Phi_{k,x}^\epsilon(Q_k),$$

is used in (13.49a) and (13.49b). The function $\Phi_{k,x}^\epsilon(Q_k)$ is the potential defined in (13.32) that takes the form (13.30) or (13.31). It follows that

$$(13.49e) \qquad \frac{d\mathcal{H}_k^\epsilon(Q_k, P_k)}{dt} = 0.$$

To establish equilibrium between subsystems, they must be at rest in a common equilibrium rest frame.

The normalized version of the equilibrium state is written

$$(13.49f) \qquad F_k^{(n)\epsilon}(Q_k, P_k) = \frac{1}{Z_k^\epsilon} F_k^\epsilon(Q_k, P_k).$$

When a function $G_k(Q_k, P_k)$ is phase averaged over the F_k^ϵ state, this is written in the bracket notation as

$$(13.50)$$
$$G_k^\epsilon = G_k(\mathcal{C}_k^\epsilon) = \langle G_k(Q_k, P_k) \rangle_{k\epsilon}$$
$$= \frac{1}{Z_k^\epsilon} \int dQ_k \int dP_k \, F_k^\epsilon(Q_k, P_k) \chi_{\Omega_k(t)}(Q_k, P_k) G_k(Q_k, P_k).$$

The form of the equilibrium functions labeled Result (i) and presented in (13.32), with the external potential stated in (13.30), will be used for F_k^ϵ exclusively from now on. The harmonic oscillator case stated in Result (ii) will be of concern only as an intermediate distribution in the discussion of the absolute zero state in the next chapter.

The state F_k^ϵ also satisfies

$$(13.51) \qquad \frac{\partial F_k^\epsilon(Q_k, P_k)}{\partial t} = 0,$$

so it is a stationary state. Further, using F_k^ϵ in (7.57) and (7.58), it follows by symmetry that the i particle working and heating do not depend on the time and

$$(13.52a) \qquad W_{\partial i}^\epsilon(q) = 0,$$
$$(13.52b) \qquad Q_{\partial i}^\epsilon(q) = 0.$$

The state F_k^ϵ is asymptotic in the sense that, for F_k^ϵ to exist, the temperature at the boundary and the location of the boundary must be rigidly fixed by outside devices. This is equivalent to a heat and pressure "bath" that is usually defined in terms of the thermodynamic limit. It will be shown in Chapter 14 that the entropy S_k^ϵ of the equilibrium state is maximal so that $S_k(t) \leq S_k^\epsilon$.

13.5 The Khinchin Equilibrium State

Khinchin [1943] proposed a somewhat different state as an equilibrium state distribution. Extending his work slightly, the *Khinchin equilibrium state*, F_k^κ, can be defined by decomposing the t system microcanonical state into subsystem states. He defined $F_k^\kappa(Q_k, M_k, J_k, H_k)$ as a function of the $3\mathcal{N}_k$-dimensional coordinate Q_k and the 3-dimensional total system momentum vector M_k and total system angular momentum vector J_k, and the total Hamiltonian energy H_k. It is a solution of a convolution equation that partitions the total t momentum, angular momentum, and energy, between the systems $t - k$ and k. Because the microcanonical state is the state that assigns equal probability to any mechanical state on the shell in phase space defined by I_t, this convolution equation is written

$$(13.53) \quad \int dM_k \int dJ_k \int dH_k\, F_{t-k}^\kappa(Q_{t-k}, \mathcal{C}_{\mathcal{P},t} - M_k, \mathcal{C}_{\mathcal{J},t} - J_k, \mathcal{C}_{\mathcal{H},t} - H_k)$$
$$\times F_k^\kappa(Q_k, M_k, J_k, H_k) = F_t^\gamma(Q_t, P_t).$$

The Khinchin equilibrium state keeps track of the momentum, angular momentum and energy assigned to a subsystem. It is considerably more complicated than the asymptotic equilibrium state F_k^ϵ. In addition, because of the connection between F_k^κ and F_{t-k}^κ in (13.53), equation (13.40) does not hold for any two arbitrary Khinchin states F_k^κ and F_l^κ that share a common interface. This is another way of saying that they are not independent in the sense of probability, which makes these states less useful than the asymptotic equilibrium state for calculations and they cannot be a part of TIS.

In EIS, Khinchin's treatment of this state is reviewed and it is shown that the asymptotic result $F_k^\kappa \to F_k^\epsilon$ follows if it is true and remains true that $\mathcal{N}_k \ll \mathcal{N}_{t-k}$ as the total number of particles $\mathcal{N}_t$ increases. See Khinchin [1943] or Khinchin [1951], pp. 180–230, for further details. The F_k^κ state will not be pursued further here.

13.6 Justifying the Use of the Equilibrium Distribution

The statistical aspects of the question of when the equilibrium distribution can be used to describe a system have not received adequate attention in the literature. The usual thermodynamic approach is to note that systems with the proper conditions at their boundaries (fixed boundaries, uniform temperature and pressure at the boundary) do go to equilibrium. They are said to have reached it when the values of a few macroscopic functions are spatially uniform in the proper way. More careful treatments observe that some macroscopic functions may take much longer than others to reach their uniform values and may not reach full uniformity for some time. In either case, once equilibrium is deemed to have been reached, the equilibrium distribution (13.49) is then applied freely to the underlying particle system and the thermodynamic properties of the system are computed. Deviations from the equilibrium state are treated as fluctuations and stability theorems are investigated in an attempt to determine how likely a particular deviation of the system from equilibrium is before it returns to equilibrium. The implication of stability theorems is that there is some kind of 'thermodynamic force', in the sense of Le Chatelier's restoring forces that are analogous to Hooke's restoring forces in springs, that returns a system to equilibrium. [10]

From the TIS perspective, it is necessary to begin with the exact state F_k^δ, which represents the system specific path of the system in phase space as a consequence of its mechanics, and understand the relation between it and any other state that may be assigned to the system. One method of doing this is statistical This method partitions the phase space and uses observations to determine the system coarse-grained distribution for this collection of cells. The approximate exact state is then inferred from measurements of the number of particles in each of these cells.

Conceptually, the problem is that of determining the unknown exact population point distribution, F_k^δ, of the underlying particle system at a given time. To do this we have at our disposal the results of a few macroscopic measurements. The observed data is clearly not ever sufficient to determine

---

[10] For a brief discussion of Le Chatelier's principle and its applications in thermodynamics, see Dickerson [1969], pp. 225, 231, 263, 319, 344. Le Chatelier's principle is not used in TIS.

the appropriate exact state. However, in the special case when the k system contains a small sample of particles taken from a large number of particles comprising the t system, it can be shown, following Khinchin [1943], that the equilibrium state F_k^ϵ is the appropriate description for the k system. The central limit theorem of probability, an extension of the law of large numbers, is used to justify this choice as an asymptotic consequence as the number of particles in the t system grows without limit.[11]

When the number of particles in the t system is finite, the equilibrium distribution is only one of a vast number of distributions that could be attributed to the underlying system based on the same sample data. While in practice this multiplicity of possibilities does not usually lead to problems, because the macroscopic predictions are nearly the same for most of these states, this approach will fail for systems with few particles.

The movement of a system to the equilibrium state as an asymptotic consequence of the evolution of the system in time from an arbitrary initial state, when its boundary conditions are fixed appropriately and the wall operator is stochastic, is addressed in Chapter 26. In this situation, specific information on the system, which allows us to describe the system using a more detailed distribution, is gradually erased by being dumped into the environment.

---

[11]Khinchin's work on this asymptotic result and the use of the central limit theorem of probability in asymptotic thermodynamics is discussed in EIS.

CHAPTER 14

The Absolute Zero State

This chapter is focused on systems near absolute zero. Nernst's theorem, often called the third law of thermodynamics, and Planck's extension of it were conjectures concerning the behavior of the entropy in systems near absolute zero. These conjectures are examined in detail in relation to classical particle mechanics. As part of this, the TIS classical absolute zero state is defined and used to calculate some thermodynamic quantities at absolute zero.

287

14.1 The Nernst and Planck Conjectures

A major issue for those working in thermodynamics at the end of the nineteenth century was the fact that a definite value for the thermodynamic entropy of substances could not be computed from first principles. A number of workers addressed this question along with the problem that the values for the heat capacity calculated from the formulas of classical mechanics differed considerably from the experimental values at low temperatures. Theoretical investigations by Max Planck, Walter Nernst, J. H. van't Hoff, and Heike Kammerlingh Onnes were published on these problems in the early twentieth century. Nernst, [**1906a**], [**1906b**], in particular, concerned himself with the heat capacity at absolute zero in the hope that the value of the entropy at absolute zero could be fixed so that the quantities of importance to thermodynamics would all have definite values.

Nernst used the free energy as the basis for his calculations. The free energy is the energy available to do work after the change in heat is subtracted from the change in the total energy. The *equilibrium free energy* is defined in TIS notation by $A_k^{\text{fe}} = \mathcal{H}_k^\epsilon - T_k S_k^{(\text{n})\epsilon}$, where $\mathcal{H}_k^\epsilon$ and $S_k^{(\text{n})\epsilon}$ are the equilibrium Hamiltonian and entropy functions. Some of the formalism for the free energy will be presented in this chapter so that the ideas expressed by Nernst and Planck can be given formal expression. The equilibrium thermature and volume partial derivatives of the free energy are computed in Chapter 18. The free energy and other thermodynamic potentials will be discussed systematically from the TIS viewpoint in Chapter 19.

On the basis of experimental work, Nernst [**1906a**], p. 2, rejected the conjecture of Marcellin Berthelot that the free energy is the same as the total energy of a substance.[1] In terms of the TIS formalism, Berthelot's conjecture states that $A_k^{\text{fe}} = \mathcal{H}_k$ at all temperatures. Nernst rejected this because the thermodynamic relation (14.3) below then implies that A_k^{fe} is independent of the temperature for all values of T_k. However, Nernst did feel that A_k^{fe} is independent of the temperature in the neighborhood of absolute zero. To study this question, he computed the differences in the free energy and Hamiltonian energy of two thermodynamic phases of a substance near absolute zero and examined the consequences for thermodynamics.

In order to review Nernst's analysis, let us divide the k system into two domains, $k1$ and $k2$. Let one phase of a substance occupy domain $k1$ and another phase occupy $k2$ and define differences in the free energy and Hamiltonian energy between these phases by

(14.1a) $$A_{k,12}^{\text{fe}} = A_{k1}^{\text{fe}} - A_{k2}^{\text{fe}},$$

(14.1b) $$\mathcal{H}_{k,12} = \mathcal{H}_{k1} - \mathcal{H}_{k2}.$$

---

[1]See also Nernst [**1926**].

Let $\mathcal{Q}_{k,12}$ designate the heat transferred from $k2$ to $k1$ during a process. Nernst used these elements in a version of the Clapeyron equation expressing the heat transfer in terms of the change in the free energy with respect to the change in the temperature. It is expressed symbolically in the form (TIS notation)[2]

$$(14.2) \qquad dA^{\text{fe}}_{k,12} = \mathcal{Q}_{k,12}\frac{d\beta_k}{\beta_k}.$$

This equation was written as $dA_{12} = -\mathcal{Q}_{12}dT/T$ in the standard notation that Nernst used. It is based on the change in the system energy due to the work $-P\Delta V$ done on a system with pressure P when there is a change in volume ΔV. This connection is expressed formally as $dA/dT = -d(P\Delta V)/dT = -\Delta V\,dP/dT$. Clapeyron's equation $\Delta V\,dP/dT = \Delta \mathcal{Q}/T$ is then used to obtain the final result.

Nernst employed the relation $\mathcal{Q}_{k,12} = [k_B\beta_k]^{-1}S^{(n)\epsilon}_{k,12} = -[A^{\text{fe}}_{k,12} - \mathcal{H}_{k,12}]$ to obtain from (14.2) the fundamental equilibrium equation[3]

$$(14.3) \qquad A^{\text{fe}}_{k,12} - \mathcal{H}_{k,12} = -\beta_k\frac{\partial A^{\text{fe}}_{k,12}}{\partial\beta_k}.$$

Alternatively, the expression of $\mathcal{Q}_{k,12}$ in terms of the entropy can be used to obtain from (14.2) the relation[4]

$$(14.4) \qquad S^{(n)\epsilon}_{k,12} = k_B\beta_k^2\frac{dA^{\text{fe}}_{k,12}}{d\beta_k}.$$

---

[2]The usual form for the Clapeyron equation in connection with a phase change, expressed in the modern form used in standard thermodynamics is (Dickerson [**1969**], p. 229)

$$\frac{dP}{dT} = \frac{\Delta H^{(en)}_{\text{vap}}}{T\Delta V_{\text{vap}}},$$

where $\Delta H^{(en)}_{\text{vap}}$ is the enthalpy change due to vaporization, representing the latent heat required to release the vapor, and ΔV_{vap} is the change in volume associated with the vaporization. See also the discussion of Clapeyron's equation in Chapter 17.

[3]Equation (14.3) is expressed as

$$\frac{\partial[\beta_k A^{\text{fe}}_k]}{\partial\beta_k} = \mathcal{H}^{\epsilon}_k$$

in (18.45a).

[4]See equation (18.43a).

Nernst went on to say that equation (14.3) "may be regarded as a summary of the older thermodynamics."[5]

Using the temperature derivative of equation (14.3) as the basis for his investigation, Nernst conjectured, in TIS notation and using β_k instead of T_k, that

$$
(14.5) \quad \lim_{\beta_k \to \infty} k_B \beta_k^2 \frac{\partial^2 [\beta_k A_{k,12}^{\mathrm{fe}}]}{\partial \beta_k^2} = \lim_{\beta_k \to \infty} k_B \beta_k^2 \left[2 \frac{\partial A_{k,12}^{\mathrm{fe}}}{\partial \beta_k} + \beta_k \frac{\partial^2 A_{k,12}^{\mathrm{fe}}}{\partial \beta_k^2} \right]
$$

$$
= \lim_{\beta_k \to \infty} k_B \beta_k^2 \left[\frac{\partial \mathcal{H}_{k1}}{\partial \beta_k} - \frac{\partial \mathcal{H}_{k2}}{\partial \beta_k} \right] = - \lim_{\beta_k \to \infty} [C_{k1,V} - C_{k2,V}] = 0,
$$

where 1 and 2 refer to any two phases or "modifications" of the same substance. The relation (12.53) and then the definition of the conventional heat capacity given in (12.54) were used to obtain the third equality in (14.5). The first consequence of this equation is that $\partial A_{k,12}^{\mathrm{fe}}/\partial \beta_k \sim O(1/\beta_k^{2+\delta})$ with $\delta > 0$. This in turn implies by (14.3) that $\lim_{\beta_k \to \infty} A_{k,12}^{\mathrm{fe}} - \mathcal{H}_{k,12} = 0$.[6]

The next step was taken by Planck in 1910.[7] He extended Nernst's conjecture by stating that in the $T_k \to 0$ limit the entropy for each phase of a substance approaches a definite finite value. This removes the possibility, allowed by Nernst's conjecture, that while the differences between the entropies of all phases of a substance approach 0 as $T_k \to 0$, the entropies themselves may all go to $-\infty$ in this limit. Planck used the fact that the entropy is defined up to an arbitrary additive constant to choose 0 as the constant at absolute zero; this means that his conjecture for each phase, with $i \in \{1, 2\}$, takes the form

$$
(14.6) \qquad\qquad \lim_{\beta_k \to \infty} S_{ki}^{(\mathrm{n})\epsilon} = 0.
$$

This has the further implication that the specific heats of all substances vanish at absolute zero. Nernst quickly adopted Planck's extension of his work and

---

[5]Nernst [**1926**], p. 3, is echoing the statement of Massieu [**1869**], which is quoted in Section 19.1. In Nernst's original notation equations (14.3) and (14.4) are written

$$
A - U = T \frac{dA}{dT}, \qquad S = -\frac{dA}{dT}.
$$

Equation (14.3) is often called the Gibbs-Helmholtz equation, but for reasons of historical accuracy, the name "Thomson-Massieu (Gibbs-Helmholtz) equation" will be used here. It was first deduced by Thomson in 1855, expressed in terms of a potential by Massieu [**1869**], and used later by Gibbs [**1875**] and Helmholtz [**1882**]. See Partington [**1949**], pp. 182–183, for Thomson's derivation of the equation.

[6]For an excellent discussion of what is usually called the "Nernst Heat theorem" and many historical references, see Partington [**1949**], pp. 212–230.

[7]See Planck [**1922**], Preface to the Third Edition (1910), p. xi, and Chapter VI.

replaced (14.5) as the statement of his "theorem" by (TIS notation)[8]

(14.7)
$$\lim_{\beta_k \to \infty} k_B \beta_k^2 \frac{\partial^2 [\beta_k A_k^{\text{fe}}]}{\partial \beta_k^2} = \lim_{\beta_k \to \infty} k_B \beta_k^2 \left[\frac{\partial A_k^{\text{fe}}}{\partial \beta_k} + \beta_k \frac{\partial^2 A_k^{\text{fe}}}{\partial \beta_k^2} \right]$$
$$= - \lim_{\beta_k \to \infty} C_{k,V} = 0.$$

If there is more than one phase present, as in the case above, this equation is valid for each phase. Nernst then took on the task of showing that the experimental data was consistent with this point of view.

Nernst was aware that the molar heat capacity of an ideal gas is $3R/2$, where R is the molar gas constant defined in terms of Boltzmann's constant and Avogadro's number by $R = k_B \mathcal{A}_k$. He was also aware of the experimental results of Dulong and Petit that the molar heat capacity of solids is about $3R$.[9] These facts were problems for his original 1906 conception of the heat theorem. However, the quantum theory of the molar heat capacities of solids developed in 1907 by Einstein and extended in 1912 by Debye showed that the energy of a solid is proportional to T_k^4 and that the molar heat capacity therefore follows a T_k^3 law—which does vanish as $T_k \to 0$. On the other hand, the experimental results on monatomic gases showed that their molar heat capacity is more like that of the classical ideal gas. To deal with this problem, Nernst [**1926**] proposed a theory of the "degeneracy of gases" at absolute zero that claimed that the molar heat capacity suddenly drops to zero as the temperature approaches absolute zero.

14.2 Formulations of The Third Law of Thermodynamics

For completeness, Nernst's statement of what he called the "third law of thermodynamics" will be mentioned. He [**1926**], Chapter VII, enunciated the *"Principle of the Unattainability of the Absolute Zero: There cannot be any process taking place in finite dimensions by means of which a body can be cooled to the absolute zero."* He, pp. 92–93, stated his version of the three laws of thermodynamics laws in terms of what cannot be done:

(1) It is impossible to construct a machine that continuously produces heat or external work out of nothing.

(2) It is impossible to construct a machine that continuously converts the heat of its surroundings into external work.

(3) It is impossible to devise an arrangement by which a body may be completely deprived of its heat, i.e. cooled to the absolute zero."

---

[8]See Nernst [**1926**]. Planck was aware that his conjecture was a significant step beyond the original conjecture of Nernst. In Planck [**1922**], p. 276, footnote, he stated that if the conjectures (14.6) or (14.7) were wrong, Nernst's original conception of the theorem might still be correct.

[9]See Dulong and Petit [**1820**], pp. 222-233.

Nernst's justifications for the third law were based on Carnot cycles with one reservoir at absolute zero along with his conjecture concerning the heat capacities of substances at absolute zero.[10] While statement (3) may be true, the reasoning Nernst presented does not seem compelling. Discussions of Nernst's theorem in current textbooks do not fare much better. In addition, the result obtained in Section 14.5 below indicates that the classical entropy diverges as the temperature approaches absolute zero.

In modern terms, the third law is often taken as the statement that Berthelot's conjecture holds at absolute zero. By (14.3), this will be true if the relation $\lim_{\beta_k \to \infty} \beta_k (\partial A_k^{fe}/\partial \beta_k) = 0$ is true. By the single phase version of the relation (14.4), this is equivalent to the truth of the relation $\lim_{\beta_k \to \infty} [\beta_k]^{-1} S_k^{(n)\epsilon} = 0$.[11] This last relation will be examined in Section 14.7. There is another serious problem concerning the use of the function $\partial A_k^{fe}/\partial \beta_k$ in Nernst's reasoning. The discussion of the free energy and free enthalpy in Chapter 19 shows that they, and their derivatives with respect to temperature, cannot be generalized to the nonequilibrium case. This means that the equations (14.1)–(14.7) are only valid for equilibrium theory and cannot be a part of the general TIS theory.

14.3 Critical Points of the Hamiltonian

The term *critical point* is most often used in thermodynamics to refer to a point at which the densities of the liquid and gaseous states of a substance are the same. It can also be used in other thermodynamic cases, such as a solid-solid transition from one particle arrangement to another in substances under increasing pressure, when some parameter becomes the same for two distinct phases. In mathematics, the same term is used to describe a point in a manifold defined by a function at which the defining function is degenerate to some degree. In the simplest case, the first derivative of the function defining the manifold vanishes at that point, so the implicit function theorem cannot be used to define its inverse function there. When there is the possibility of confusion, the term *thermodynamic critical point* will be used to refer to the critical point of thermodynamics and *mechanical critical point* will be used to refer to the mathematical critical points of the system Hamiltonian function.

The values of the Hamiltonian of an isolated system define a family of manifolds in phase space. The differential of the Hamiltonian defines a flow on each manifold. In terms of the differential geometry of these manifolds, a mathematical critical point (also called a *singular point*) is a point at which the differential of the Hamiltonian vanishes. For simplicity, the concerns in this chapter are restricted to the case in which the Hessian (or Gram

---

[10] As Charles Brunold [**1930**], p. 116, pointed out, cycles with the lower limit at absolute zero would be 100% efficient in converting heat to work.

[11] For more information on the modern approach to the third law in terms of free energy difference equations, see Dickerson [**1969**], pp. 196–208.

determinant) of the Hamiltonian does not vanish at the critical points, so that the critical points are *nondegenerate*.[12] In TIS notation, $(Q_t^\circ, 0)$ is a critical point for system t if and only if $\nabla_{2t} \mathcal{H}_t(Q_t^\circ, 0) = 0$. As Lagrange observed, if $\mathcal{H}_t(Q_t^\circ, 0)$ is also a local minimum of $\mathcal{H}_t(Q_t, P_t)$, then the critical point $(Q_t^\circ, 0)$ is a *stable equilibrium point*.[13] In terms of the TIS notation in Chapter 4, the stable equilibrium points of the t system are the members of $\Omega_{t,\min}$. For an isolated system, the facts that the system trajectory must fall within $\Omega_t(t)$, that the t system conserved quantities are bounded, and that the critical points on this manifold are isolated, mean that a system trajectory cannot move into a stable equilibrium point unless the system mechanical state is located at one to begin with. If the mechanical state of an isolated system is already at a stable equilibrium point, it cannot leave that point.

For subsystems of an isolated system, there are corresponding local stable equilibrium points in the subsystem phase space that are the local minima of the subsystem Hamiltonian. Because the k subsystem Hamiltonian is not the same as the k marginalization of the t system Hamiltonian due to the use of averaged external potentials in the subsystems, the sets of critical points may differ. At a subsystem critical point, the net total force, including both internal and external forces, on each particle in the system is zero. A system located at a local critical point can move away from the point only if the external conditions change.

14.4 The Absolute Zero State

The local critical points of the subsystems of the total system are of interest next. It will be shown that the TIS formalism for computing the temperature computed at these points gives absolute zero. The analysis is performed in the equilibrium rest frame of the k system. The use of this frame is required, as illustrated by the example of calculation of the temperature of a monochromatic beam given in Section 12.2 above and similar examples for the pressure, so that the thermature and pressure computed from their analogs will be independent of the observer's Galilean reference frame.

Since, in general, the particles in a system are interacting with each other, a system can remain at a single point in phase space if and only if the system state is concentrated at a phase point of the type $(Q_k^\circ, 0)$, where Q_k° is a local minimum of the Hamiltonian. For a system at a local minimum,

---

[12]See Ralph Abraham and Jerrold Marsden [**1978**], pp. 72–73, 136–140, 189–207, 231–235, or Vladimir Arnold [**1974**], pp. 371–380, for a discussion of some of the issues in mechanics concerning critical points. See also Shlomo Sternberg [**1964**], p. 47, for the definition of a critical point in differential geometry. For more general types of critical points, see the mathematical work on Morse Theory and René Thom's Catastrophe Theory. The thermodynamic implications of the more general theory of critical points are discussed in EIS.

[13]See Lefschetz [**1963**], pp. 76–77. For a modern approach, see Abraham and Marsden [**1978**], pp. 73, 207.

the mechanical state will remain fixed at the local stable equilibrium point $(Q_k^\circ, 0)$.[14] Anticipating the results below, this is called the k-*absolute zero mechanical state* and it is designated by F_k^0. Since F_k^0 is normalized, it must take the singular form (TIS notation)

$$(14.8) \qquad F_k^0(Q_k, P_k) = \delta(Q_k - Q_k^\circ)\delta(P_k).$$

The partial time derivative of F_k^0 is zero. The facts that $\delta(P_k)V_k = 0$ and $\delta(Q_k - Q_k^\circ)(\partial\Phi_{k,n}(Q_k)/\partial Q_k) = 0$ can be used to show $V_{2k} \cdot \nabla_{2k} F_k^0 = 0$. These facts imply that F_k^0 is a state because conditions (5.14), (5.15), (5.16), and (5.29), are met, so it is a stationary state of the system.

The next goal is to demonstrate, as energy is extracted from the k system, that the $\beta_k \to \infty$ limit of the asymptotic equilibrium state F_k^ϵ is the F_k^0 state. It is not hard to show (see Chapter 18) that the average kinetic energy of a system at equilibrium is asymptotically proportional to β_k^{-1} as the range of available energy $C_{k,H}$ increases. Conversely, β_k will increase as the average kinetic energy of the system is lowered and the range of available energy is reduced. Since energy can be transferred from potential to kinetic and back, it is necessary to lower the total system energy.[15]

An elementary proof of these assertions concerning isolated critical points and the $\beta_k \to \infty$ limit of F_k^ϵ will be given with an emphasis on the physical point of view. Assume for simplicity that the net external force acting on the k system is constant in time at each point $q \in d_k$. As kinetic energy is extracted from the system, the phase point (Q_k, P_k) will become "trapped" in the neighborhood of a relative minimum, $(Q_k^\circ, 0) \in \Omega_{k,\min}$, of the potential energy.[16] Let us first note that the minimum energy for the system in the neighborhood of the point $(Q_k^\circ, 0)$ is $\Phi_{k,n}(Q_k^\circ)$, which is the potential energy at the critical point $(Q_k^\circ, 0)$. Then, by the definition (4.27) of $\Omega_k(t)$ and the fact that trajectories around local minima are stable, an upper limit $\mathcal{H}_k^u$ for any $(Q_k^\circ, 0) \in \Omega_{k,\min}$ can be found such that $C_{\mathcal{H},k} = \mathcal{H}_k^u - \mathcal{H}_k^l$ is small enough so that all trajectories initially in a set $\Theta^* \subset \Omega_k(t)$ around the local minimum $(Q_k^\circ, 0)$ remain within a bounded set $\Upsilon_k \subset \Omega_k(t)$, with $\Theta^* \subset \Upsilon_k$. Assume from now on that a suitable $\mathcal{H}_k^u$ has been chosen and that the phase point representing the system is initially in the set Θ^*.

For any $\eta > 0$, define the closed neighborhood of $(Q_k^\circ, 0)$ by

$$(14.9) \qquad \Xi_k^\eta = \{\, (Q_k, P_k) \mid |(Q_k, P_k) - (Q_k^\circ, 0)| \le \eta \,\},$$

---

[14]Unstable equilibrium points at maxima of the Hamiltonian are excluded from consideration because the stable equilibrium points are the physically more significant states.

[15]Gibbs [**1902**], Chapter X, used a similar method, expressed in terms of the energies, to obtain the microcanonical distribution from the canonical distribution.

[16]In general, these local minima are determined by the system boundary conditions, (β_k, δ_k) and perhaps long range electromagnetic or gravitation forces. They can change if the boundary conditions change. Critical points that are initially isolated may disappear or coalesce into a degenerate point. The more powerful mathematical methods discussed in EIS are needed to deal with these cases.

where $|\cdot|$ is the Euclidean norm in $\mathbf{R}^{6\mathcal{N}_k}$. Let us choose $\mathcal{H}_k^u - \mathcal{H}_k^l$ smaller, if necessary, so that $\partial^2 \Phi_{k,n}(Q_k)/\partial Q_k^2 \geq 0$ for $(Q_k, P_k) \in \Upsilon_k$, with equality only at $Q_k = Q_k^\circ$. Because $\Phi_{k,n}(Q_k)$ is continuous, these assumptions mean, for each $\eta > 0$ such that $\Xi_k^\eta \subset \Upsilon_k$, that there exists two small positive numbers, $\epsilon_2(\eta) > \epsilon_1(\eta) > 0$, for which

(14.10a) $\mathcal{H}_k(Q_k, P_k) - \Phi_{k,n}(Q_k^\circ) \leq \epsilon_1(\eta),$ for $(Q_k, P_k) \in \Xi_k^{\eta/2},$

(14.10b) $\mathcal{H}_k(Q_k, P_k) - \Phi_{k,n}(Q_k^\circ) \geq \epsilon_2(\eta),$ for $(Q_k, P_k) \in \Upsilon_k - \Xi_k^\eta.$

Using the relations expressed in (14.10), the next step is to show that the equilibrium state F_k^ϵ is concentrated at $(Q_k^\circ, 0)$ as $\beta_k \to \infty$. Assume η is chosen so that Ξ_k^η is a proper subset of Υ_k. Then, because Ξ_k^η is a convex $6\mathcal{N}_k$-dimensional ball in phase space, the properties of the Lebesgue measure defined in (2.10) are used to show $\mu_k(\Upsilon_k) > \mu_k(\Xi_k^{\eta/2} \cap \Upsilon_k) = \mu_k(\Xi_k^{\eta/2}) > 0$. Using (14.10), A_1 and A_2 are the phase volumes defined by

(14.11)
$$
A_1 = \int dQ_k \int dP_k \, \chi_{\Xi_k^\eta}(Q_k, P_k) \chi_{\Upsilon_k}(Q_k, P_k) e^{-\beta_k[\mathcal{H}_k(Q_k, P_k) - \Phi_{k,n}(Q_k^\circ)]}
$$
$$
> \int dQ_k \int dP_k \, \chi_{\Xi_k^{\eta/2}}(Q_k, P_k) e^{-\beta_k[\mathcal{H}_k(Q_k, P_k) - \Phi_{k,n}(Q_k^\circ)]}
$$
$$
> \mu_k(\Xi_k^{\eta/2}) e^{-\beta_k \epsilon_1(\eta)} > 0
$$

and

(14.12)
$$
A_2 = \int dQ_k \int dP_k \, (1 - \chi_{\Xi_k^\eta}(Q_k, P_k)) \chi_{\Upsilon_k}(Q_k, P_k) e^{-\beta_k[\mathcal{H}_k(Q_k, P_k) - \Phi_{k,n}(Q_k^\circ)]}
$$
$$
\leq \mu_k(\Upsilon_k) e^{-\beta_k \epsilon_2(\eta)}.
$$

From (14.11) and (14.12), it follows that

(14.13)
$$
0 \leq A_2 \leq \frac{\mu_k(\Upsilon_k)}{\mu_k(\Xi_k^{\eta/2})} e^{-\beta_k \epsilon(\eta)} A_1,
$$

where

(14.14)
$$
\epsilon(\eta) = \epsilon_2(\eta) - \epsilon_1(\eta) > 0.
$$

Since the results of this theory are independent of the choice of the origin of the energy, the origin is shifted to $-\Phi_{k,n}(Q_k^\circ)$ for convenience and the normalized equilibrium state is written as

(14.15a) $\quad F_k^{(n)\epsilon}(Q_k, P_k) = [Z_k^\epsilon]^{-1} \chi_{\Upsilon_k}(Q_k, P_k) e^{-\beta_k[\mathcal{H}_k(Q_k, P_k) - \Phi_{k,n}(Q_k^\circ)]}$

with

(14.15b) $\quad Z_k^\epsilon = \int dQ_k \int dP_k \, \chi_{\Upsilon_k}(Q_k, P_k) e^{-\beta_k[\mathcal{H}_k(Q_k, P_k) - \Phi_{k,n}(Q_k^\circ)]}.$

From (14.15a) is obtained the result

(14.16) $$F_k^{(n)\epsilon}(Q_k^\circ, 0) = [Z_k^\epsilon]^{-1}.$$

By (14.11), (14.12), the localization of the support of $F_k^{(n)\epsilon}$ in Υ_k, and its normalization, it follows

(14.17) $$1 = \langle 1 \rangle_{k\epsilon} = F_k^{(n)\epsilon}(Q_k^\circ, 0)[A_1 + A_2].$$

This in turn leads to the result

(14.18) $$F_k^{(n)\epsilon}(Q_k^\circ, 0)A_1 \leq 1 \leq F_k^{(n)\epsilon}(Q_k^\circ, 0)A_1 \left[1 + \frac{\mu_k(\Upsilon_k)}{\mu_k(\Xi_k^{\eta/2})} e^{-\beta_k\epsilon(\eta)} \right].$$

From this it follows for any $\eta > 0$ that

(14.19a) $$\lim_{\beta_k \to \infty} F_k^{(n)\epsilon}(Q_k^\circ, 0)A_1 = 1$$

and, hence,

(14.19b) $$\lim_{\beta_k \to \infty} F_k^{(n)\epsilon}(Q_k^\circ, 0)A_2 = 0.$$

It then follows from (14.19b) that, for any $\delta > 0$ and fixed $\eta > 0$, there is a $\beta_k(\delta, \eta)$ such that for $\beta_k > \beta_k(\delta, \eta)$, the relation

(14.20) $$\left\langle 1 - \chi_{\Xi_k^\eta}(Q_k, P_k) \right\rangle_{k\epsilon} < \delta$$

is valid.

Suppose now that $G_k(Q_k, P_k)$ is any phase function continuous on the domain Υ_k. Then, $G_k(Q_k, P_k)$ is uniformly continuous and bounded on Υ_k. This means that for each $\delta > 0$, there is an $\eta(\delta)$ such that $(Q_k, P_k) \in \Xi_k^{\eta(\delta)}$ implies

(14.21) $$|G_k(Q_k, P_k) - G_k(Q_k^\circ, 0)| < \delta.$$

For convenience below, let us define

(14.22) $$G_k(\beta_k) = \langle G_k(Q_k, P_k) \rangle_{k\epsilon}.$$

The next step is to choose $\delta > 0$ followed by $\eta(\delta)$ and then $\beta_k(\delta, \eta)$ such that (14.21) and (14.20) hold. In this case, it follows that

(14.23) $$|G_k(\beta_k) - G_k(Q_k^\circ, 0)| \leq \langle |G_k(Q_k, P_k) - G_k(Q_k^\circ, 0)| \rangle_{k\epsilon}$$

$$\leq \left\langle \chi_{\Xi_k^\eta}(Q_k, P_k) \right\rangle_{k\epsilon} \sup_{(Q_k, P_k) \in \Xi_k^\eta} |G_k(Q_k, P_k) - G_k(Q_k^\circ, 0)|$$

$$+ 2 \left\langle \left(1 - \chi_{\Xi_k^\eta}(Q_k, P_k) \right) \right\rangle_{k\epsilon} \sup_{(Q_k, P_k) \in \Upsilon_k} |G_k(Q_k, P_k)|.$$

Since $\left\langle \chi_{\Xi_k^\eta}(Q_k, P_k) \right\rangle_{k\epsilon} \leq 1$, (14.23) leads to

$$(14.24) \qquad |G_k(\beta_k) - G_k(Q_k^\circ, 0)| \leq \delta \left[1 + 2 \sup_{(Q_k, P_k) \in \Upsilon_k} |G_k(Q_k, P_k)| \right].$$

Because δ is arbitrary, the result

$$(14.25) \qquad \lim_{\beta_k \to \infty} G_k(\beta_k) = G_k(Q_k^\circ, 0) \qquad \text{a.e.}$$

follows for any phase function continuous on Υ_k. Since $G_k(Q_k, P_k)$ is an arbitrary continuous function, a further conclusion is that[17]

$$(14.26) \qquad \lim_{\beta_k \to \infty} F_k^\epsilon(Q_k, P_k) = F_k^0(Q_k, P_k) \qquad \text{a.e.}$$

This confirms the statements above concerning the critical points of the system and the absolute zero state.

The energy analog of the absolute zero state is $\Phi_{k,n}(Q_k^\circ)$ and the system number density analog is $\nu_k^0(Q_k) = \sum_{i \in k} \delta(q_i - q_i^\circ)$. The kinetic energy is clearly zero. The exact state pressure is given by

$$(14.27) \qquad \mathrm{P}_k^\delta(t) = \frac{1}{3\delta_k(t)} \left[2\mathcal{K}_k(P_k^{(s)}(t)) - Q_k^{(s)}(t) \cdot \frac{\partial \Phi_k(Q_k^{(s)}(t), t)}{\partial Q_k} \right].$$

Using the facts that $\partial \Phi_k(Q_k, t)/\partial Q_k|_{Q_k = Q_k^\circ} = 0$ and $\mathcal{K}_k(0) = 0$, it is easy to see that $\mathrm{P}_k^{\delta,0}(t) = 0$. This implies, by the definition (12.28) of $\beta_k^\delta(t)$, that $\beta_k^\delta = \infty$. The average of $-d\mathcal{H}_k(Q_k, P_k, t)/d\Omega_k$ over F_k^0 shows that the momentum current density tensor, and the virial tensors, are also zero at absolute zero as expected. The heat flow, working, and net forces, are all zero which means that $dp_k^{0\mu}(q, t)/dt = 0$ and $dE_k^0(q, t)/dt = 0$.

14.5 The Lower Bound on the Entropy

Using methods similar to those of the last section, it will be shown that the entropy attains the lower bound $-\infty$ for any F_k^δ state. Let us approximate the F_k^δ state, concentrated at the point $(Q_k^{(s)}(t), P_k^{(s)}(t))$, for $n = 1, 2, \ldots$, by the normalized function[18]

$$(14.28) \qquad F_k^{\delta,n}(Q_k, P_k, t) = \prod_{i \in k} \left[\frac{n}{\pi q_i^d p_i^d} \right]^3 e^{-n \sum_{i \in k} \left[\frac{(q_i - q_i^{(s)}(t))^2}{[q_i^d]^2} + \frac{(p_i - p_i^{(s)}(t))^2}{[p_i^d]^2} \right]}.$$

---

[17]This follows from a uniqueness theorem for Radon measures with support in Υ_k. See Treves [**1967**], pp. 223–224.

[18]See Treves [**1967**], Chapter 15, for approximations of the delta measure in this form. To avoid complications, the Q_k and P_k integrations here are each over all of $\mathbf{R}^{3\mathcal{N}_k}$.

The quantities $q_i^d > 0$ and $p_i^d > 0$ are included for dimensional reasons; in essence, they set the coordinate and momentum scales. It is easy to show that $< 1 >_{k\delta,n} = 1$ for each $n > 0$. The fact that $\chi_{\Omega_k(t)}(Q_k, P_k) = 1$ in the neighborhood of the point $(Q_k^{(s)}(t), P_k^{(s)}(t))$ is then used to show that as $n \to \infty$, $F_k^{\delta,n}$ converges to F_k^δ.

The $F_k^{\delta,n}(Q_k, P_k, t)$ function has been specially constructed so that it yields the $F_k^\delta(Q_k, P_k, t)$ state in the limit as $n \to \infty$, but it is not a solution of the system evolution equation and not a state of the system. This implies that the entropy function obtained from it next is not the entropy of the k system either. In spite of being introduced for mathematical and not physical reasons, the distribution $F_k^{\delta,n}(Q_k, P_k, t)$ becomes a state asymptotically as $n \to \infty$ under the assumptions operating here. This is due to the fact that with the proper choice of q_i^d and p_i^d the argument of the exponential in (14.28) is asymptotically the Hamiltonian of a harmonic oscillator. Setting $q_i^d = \sqrt{2/k_i}$ and $p_i^d = \sqrt{2m_i}$ yields the harmonic oscillator Hamiltonian

$$(14.29) \qquad \mathcal{H}_k(Q_k, P_k) = \sum_{i \in k} \left[\frac{1}{2m_i} [p_i - p_i^{(s)}(t)]^2 + \frac{k_i}{2} [q_i - q_i^{(s)}(t)]^2 \right].$$

The point $(Q_k^{(s)}(t), P_k^{(s)}(t)) = (Q_k^\circ, 0)$ is the equilibrium rest point for this oscillator. The asymptotic validity of this representation of the Hamiltonian in the neighborhood of a stable equilibrium point $(Q_k^\circ, 0)$ is a consequence of the fact that $\partial \Phi_k(Q_k)/\partial Q_k > 0$, for $Q_k \neq Q_k^\circ$ in a small enough neighborhood of Q_k°, and $\partial \Phi_k(Q_k^\circ)/\partial Q_k = 0$, so that the lowest order nonzero term that depends on Q_k in the expansion of the potential in a Taylor series based at $(Q_k, P_k) = (Q_k^\circ, 0)$ is the second order term.

The entropy function $S_k^{\delta,n}(Q_k, P_k, t)$ associated with the $F_k^{\delta,n}(Q_k, P_k, t)$ distribution density is computed using the definition (9.10) of the phase entropy. This yields the i-particle phase entropy function and the k phase entropy function in the form

(14.30a)

$$S_i^{\delta,n}(q_i, p_i, t) = -k_B \left\{ 3 \ln \left[\frac{n}{\pi q_i^d p_i^d} \right] - n \left[\frac{(q_i - q_i^{(s)}(t))^2}{(q_i^d)^2} + \frac{(p_i - p_i^{(s)}(t))^2}{(p_i^d)^2} \right] \right\},$$

$$(14.30b) \qquad S_k^{\delta,n}(Q_k, P_k, t) = \sum_{i \in k} S_i^{\delta,n}(q_i, p_i, t).$$

Computing the expectation function of $S_k^{\delta,n}(Q_k, P_k, t)$ over $F_k^{\delta,n}(Q_k, P_k, t)$ gives the result

$$(14.31) \qquad S_k^{\delta,n} = \left\langle S_k^{\delta,n}(Q_k, P_k, t) \right\rangle_{k\delta,n} = -3k_B \left\{ \sum_{i \in k} \ln \left[\frac{n}{\pi q_i^d p_i^d} \right] - \mathcal{N}_k \right\}.$$

Taking the $n \to \infty$ limit yields[19]

(14.32)
$$S_k^\delta = \lim_{n \to \infty} S_k^{\delta,n} = -\infty.$$

Since the absolute zero state is an F_k^δ state and is the unique representation for a substance at this temperature, it follows that the entropy for any substance takes the value $-\infty$ at absolute zero. This result confirms the lower bound for the entropy stated in Axiom 6.1.

Using the quantities $q_i^d = \sqrt{2/k_i}$ and $p_i^d = \sqrt{2m_i}$ introduced above, the volume derivative of this entropy analog is

(14.33)
$$\frac{dS_k^{\delta,n}(Q_k, P_k, t)}{d\Omega_k} = \frac{nk_B}{3\delta_k(t)} \sum_{i \in k} \left[k_i[q_i - q_i^{(s)}(t)] \cdot q_i - \frac{1}{m_i}[p_i - p_i^{(s)}(t)] \cdot p_i \right]$$
$$= \frac{2nk_B}{3\delta_k(t)} \sum_{i \in k} \left\{ \left[\frac{k_i(q_i - q_i^{(s)}(t))^2}{2} - \frac{(p_i - p_i^{(s)}(t))^2}{2m_i} \right] \right.$$
$$\left. + \left[\frac{k_i(q_i - q_i^{(s)}(t)) \cdot q_i^{(s)}(t)}{2} - \frac{(p_i - p_i^{(s)}(t)) \cdot p_i^{(s)}(t)}{2m_i} \right] \right\}.$$

It is not hard to show that

(14.34) $\left\langle q_i - q_i^{(s)}(t) \right\rangle_{k\delta,n} = 0$,   and $\left\langle p_i - p_i^{(s)}(t) \right\rangle_{k\delta,n} = 0$.

It then follows by an easy calculation that

(14.35)
$$\left\langle \frac{dS_k^{\delta,n}(Q_k, P_k, t)}{d\Omega_k} \right\rangle_{k\delta,n} = 0.$$

Similarly, it is not hard to show using the fact that the first term after the second equal sign in (14.33) is the volume derivative of the phase Hamiltonian for the state $F_k^{\delta,n}$ when it is viewed as a harmonic oscillator equilibrium state of the form presented in (13.39). It should be noted, however, that this state is an artifact of the approximation and does not have dynamical significance because of the presence of the trajectory $(q_i(t), p_i(t))$ in the "Hamiltonian".

---

[19]This lower limit for the entropy is not new. It is the same, for example, as that stated by Joseph Mayer and Maria Mayer [**1940**], p. 103, using a different approach. See also Dickerson [**1969**], equations (4-25), p. 160, with (2-75), p. 50, for the change in the entropy of an ideal gas between two points (T_1, V_1) and (T_2, V_2). The $T_1 \to 0$ limit in Dickerson's formula (4-25) for the change in the entropy of any process that begins at absolute zero yields $\Delta S = \infty$, which represents the change in entropy for any process that begins at absolute zero. The $T \to 0$ limit of the Sakur-Tetrode equation for the entropy of a quantum ideal gas, in Dickerson, p. 55, equation (2-92), also yields $S^{(0)} = -\infty$. The Sakur-Tetrode equation is also discussed in QTS.

Computing the pressure using this Hamiltonian by the same methods as were used for (14.33) and (14.35) has the consequence

$$(14.36) \qquad\qquad P_k^{\delta,n}(t) = 0.$$

These results show that the pressure associated with a harmonic oscillator vanishes. This general result for the average pressure of a harmonic oscillator is expected because a harmonic oscillator does not require an external force or pressure to maintain it. These results are consistent with the view of pressure as a measure of the "escaping tendency" of a substance.[20] As mentioned after the definition of the thermature analog (12.12b), it cannot be used in cases in which the system pressure is zero. It follows that the analog (12.12b) cannot be used to compute the thermature for the $F_k^{\delta,n}$ distribution. Gibbs' analog (12.7) will be used instead for the calculation of the thermature of this distribution in the next section.

The calculation of the entropy or other physical properties associated with the exact state can be done by starting with an equilibrium state at a finite temperature and then taking the $\beta_k \to \infty$ limit. As will be demonstrated in the next section in the calculation of the conventional heat capacity at absolute zero, the $n \to \infty$ limit in (14.31) corresponds to the $\beta_k \to \infty$ limit.

14.6 Heat Capacity at Absolute Zero

In order to investigate the classical theory of thermodynamic quantities near absolute zero, the distribution defined in (14.28) will be used to compute the temperature and the heat capacity in the neighborhood of the stable equilibrium point $(Q_k^\circ, 0)$ and then the $n \to \infty$ limit will be taken. The results will be obtained by elementary methods that emphasize the physical aspects of the situation.

The k system is described by the distribution $F_k^{\delta,n}(Q_k, P_k, t)$ with $p_i^d = \sqrt{2m_i}$ and $q_i^d = \sqrt{2/k_i}$. Gibbs' analog (12.7) will be used to compute the thermature $\beta_k^{\delta,n}(t)$ at time t relative to the instantaneous system rest frame for a system with the system specific trajectory $(Q_k^{(s)}(t), P_k^{(s)}(t))$. Using an unbounded phase space integration as before, this calculation takes the form

$$(14.37)$$

$$
\beta_k^{\delta,n} = \frac{3\mathcal{N}_k - 2}{2} \left\langle [K_k(P_k - \mathcal{P}_k^{(s)}(t))]^{-1} \right\rangle_{k\delta,n}
$$

$$
= \frac{3\mathcal{N}_k - 2}{2} \int dQ_k \int dP_k \, \frac{F_k^{\delta,n}(Q_k, P_k, t)}{\mathcal{K}_k(P_k - \mathcal{P}_k^{(s)}(t))}
$$

$$
= \frac{3\mathcal{N}_k - 2}{2} \left[\prod_{i \in k} \left(\frac{n}{2\pi m_i} \right)^{\frac{3}{2}} \right] \int dP_k \, \frac{e^{-n \sum_{i \in k} \frac{(p_i - p_i^{(s)}(t))^2}{2m_i}}}{\mathcal{K}_k(P_k - P_k^{(s)}(t))},
$$

---

[20] See, for example, Dickerson [**1969**], p. 232, 253, for this characterization of pressure.

where $\mathcal{K}_k(P_k - P_k^{(s)}(t))$ is the kinetic energy in this instantaneous rest frame.

To do the integration in (14.37), let us use the principal value for the integration across the momentum phase point $P_k = P_k^{(s)}(t)$. This gives the result

(14.38)
$$\int dP_k \frac{e^{-n\sum_{i\in k}[2m_i]^{-1}(p_i - p_i^{(s)}(t))^2}}{\mathcal{K}_k(P_k - P_k^{(s)}(t))} = \int dP_k \int_n^\infty dx\, e^{-x\mathcal{K}_k(P_k - P_k^{(s)}(t))}$$
$$= \int_n^\infty dx \int dP_k\, e^{-x\mathcal{K}_k(P_k - P_k^{(s)}(t))}$$
$$= \int_n^\infty dx \prod_{i\in k}\left[\frac{2\pi m_i}{x}\right]^{\frac{3}{2}} = \frac{2n}{3\mathcal{N}_k - 2}\prod_{i\in k}\left[\frac{2\pi m_i}{n}\right]^{\frac{3}{2}}$$

is obtained. It is used with (14.38) to show

(14.39)
$$\beta_k^{\delta,n} = n \qquad \text{for } \mathcal{N}_k \geq 1.$$

This shows that $\beta_k^{\delta,n} = O(n)$ and

(14.40)
$$\lim_{n\to\infty} \beta_k^{\delta,n} = \infty$$

as claimed. The result (14.39) shows that the $n \to \infty$ limit is the same as the $\beta_k^{\delta,n} \to \infty$ limit.

Let us consider first an ideal gas of harmonic oscillators with the Hamiltonian given by (14.29). The averages of $K_k(P_k)$ and $[K_k(P_k)]^2$ are computed first for the distribution $F_k^{\delta,n}$ along with some results needed for the calculation of the heat capacity. This will give us results for an ideal gas of harmonic oscillators described by the distribution $F_k^{\delta,n}$. In effect, the rest frame for this calculation is the instantaneous frame in which $(Q_k, P_k) = (Q_k^{(s)}(t), P_k^{(s)}(t))$.

The symmetry of the phase averaging integral leads to

(14.41a)
$$\left\langle \mathcal{K}_k(P_k - P_k^{(s)}(t)) \right\rangle_{k\delta,n} = \prod_{i\in k}\left(\frac{n}{2\pi m_i}\right)^{\frac{3}{2}} \int dP_k\, e^{-n\mathcal{K}_k(P_k - P_k^{(s)}(t))}$$
$$\times \mathcal{K}_k(P_k - P_k^{(s)}(t))$$
$$= \frac{3\mathcal{N}_k}{2n}$$

and, similarly,

(14.41b)
$$\left\langle [\mathcal{K}_k(P_k - P_k^{(s)}(t))]^2 \right\rangle_{k\delta,n} = \frac{3\mathcal{N}_k}{2n^2}\left(\frac{3\mathcal{N}_k + 2}{2}\right).$$

With these results, (12.53) is used to obtain the k-system heat capacity for this collection of non-interacting harmonic oscillators in the form

$$(14.42) \qquad\qquad \mathcal{K}_{k,V}^{\delta,n} = -\frac{3\mathcal{N}_k}{2n^2}.$$

The result (14.40) is used with (14.42) to compute of the heat capacity at absolute zero with the result

$$(14.43) \qquad\qquad \mathcal{K}_{k,V}^0 = \lim_{n\to\infty} \mathcal{K}_{k,V}^{\delta,n} = 0.$$

This result is consistent with (12.53) when an exact state is used to calculate the phase averages.

To obtain the conventional heat capacity in molar terms, set $\mathcal{N}_k = \mathcal{A}$, where $\mathcal{A}$ is Avogadro's number, and use the gas constant $R = k_B\mathcal{A}$ with the relation between the conventional heat capacity and the TIS heat capacity given in (12.54) to obtain the conventional specific heat

$$(14.44) \qquad\qquad C_{k,V}^{\delta,n} = -k_B n^2 \mathcal{K}_{k,V}^{\delta,n} = \frac{3R}{2}$$

per mole. It follows from (14.42) that the $n \to \infty$ limit of the conventional specific heat is

$$(14.45) \qquad\qquad C_{k,V}^0 = \lim_{n\to\infty} C_{k,V}^{\delta,n} = \frac{3R}{2}.$$

per mole at absolute zero. This temperature independent result for an ideal gas of harmonic oscillators is in accord with the usual thermodynamic result for ideal gases.

The absolute zero limit for the heat capacity in the general interacting particles case will be examined next. This involves the more difficult calculation of the interparticle potential averages. The strategy will be to expand $\Phi_{k,n}(Q_k)$ in a Taylor's series around the stable equilibrium coordinate value Q_k°. A lattice model will be used in which the locations of particles are expressed in terms of displacements from their equilibrium points. This change requires that the $F_k^{\delta,n}$ distribution defined in (14.28) is modified with q_i° replacing $q_i^{(s)}(t)$ and $p_i^\circ = 0$ replacing $p_i^{(s)}(t)$, where q_i° is the location of lattice node i and p_i° is the momentum of the particle at this lattice node. In addition, the interparticle potential function $\phi_{ij}(z)$ is assumed to be an analytic function of the complex variable z in the neighborhood of the point $z^\circ = |q_i^\circ - q_j^\circ|$ for each $i, j \in k$. The resulting normalized $F_k^{L,n}$ lattice distribution is written in terms of the momentum and *displacement* vectors $\xi_i^\mu = q_i^\mu - q_i^{\circ\mu}$ as

$$(14.46) \qquad F_k^{L,n}(Q_k, P_k) = \prod_{i\in k} \left[\frac{n^2 k_i}{4\pi^2 m_i}\right]^{\frac{3}{2}} e^{-n\sum_{i\in k}\left[\frac{p_i^2}{2m_i}+\frac{k_i\xi_i^2}{2}\right]}.$$

The function $\Phi_{k,n}(Q_k)$ will be averaged over the modified $F_k^{L,n}$ distribution term by term. This procedure will then be repeated with $[\Phi_{k,n}(Q_k)]^2$. Since $\Phi_{k,n}(Q_k)$ has a local minimum at Q_k°, it follows immediately that

(14.47)
$$\frac{\partial \Phi_{k,n}(Q_k^\circ)}{\partial q_i} = 0$$

for each $i \in k$. It is necessary to expand $\Phi_{k,n}(Q_k)$ and $[\Phi_{k,n}(Q_k)]^2$ only to the fourth order because each order, as will be seen below, contributes an additional factor of $1/\sqrt{n}$ to the expansion. Since $[\beta_k^{L,n}]^2 \sim O(n^2)$, it is clear that terms of order higher than the fourth will not contribute in the $n \to \infty$ limit.

In the Taylor expansions of $\Phi_{k,n}(Q_k)$ and $[\Phi_{k,n}(Q_k)]^2$, the summation convention is used for both Greek vector and Latin particle indices for notational simplicity. Then the Taylor expansion of $\Phi_{k,n}(Q_k)$ to the fourth order is

(14.48)
$$\Phi_{k,n}(Q_k) \approx \Phi_{k,n}(Q_k^\circ) + \frac{1}{2}\frac{\partial^2 \Phi_{k,n}(Q_k^\circ)}{\partial q_r^\mu \partial q_s^\nu}\xi_r^\mu \xi_s^\nu + \frac{1}{6}\frac{\partial^3 \Phi_{k,n}(Q_k^\circ)}{\partial q_r^\mu \partial q_s^\nu \partial q_u^\sigma}\xi_r^\mu \xi_s^\nu \xi_u^\sigma$$
$$+ \frac{1}{24}\frac{\partial^4 \Phi_{k,n}(Q_k^\circ)}{\partial q_r^\mu \partial q_s^\nu \partial q_u^\sigma \partial q_v^\lambda}\xi_r^\mu \xi_s^\nu \xi_u^\sigma \xi_v^\lambda.$$

Squaring this gives the Taylor expansion of $[\Phi_{k,n}(Q_k)]^2$ to fourth order as

(14.49) $\quad [\Phi_{k,n}(Q_k)]^2 \approx [\Phi_{k,n}(Q_k^\circ)]^2 + \Phi_{k,n}(Q_k^\circ)\dfrac{\partial^2 \Phi_{k,n}(Q_k^\circ)}{\partial q_r^\mu \partial q_s^\nu}\xi_r^\mu \xi_s^\nu$

$$+ \frac{2}{6}\Phi_{k,n}(Q_k^\circ)\frac{\partial^3 \Phi_{k,n}(Q_k^\circ)}{\partial q_r^\mu \partial q_s^\nu \partial q_u^\sigma}\xi_r^\mu \xi_s^\nu \xi_u^\sigma + \frac{2}{24}\Phi_{k,n}(Q_k^\circ)\frac{\partial^4 \Phi_{k,n}(Q_k^\circ)}{\partial q_r^\mu \partial q_s^\nu \partial q_u^\sigma \partial q_v^\lambda}\xi_r^\mu \xi_s^\nu \xi_u^\sigma \xi_v^\lambda$$
$$+ \frac{1}{4}\frac{\partial^2 \Phi_{k,n}(Q_k^\circ)}{\partial q_r^\mu \partial q_s^\nu}\xi_r^\mu \xi_s^\nu \frac{\partial^2 \Phi_{k,n}(Q_k^\circ)}{\partial q_u^\sigma \partial q_v^\lambda}\xi_u^\sigma \xi_v^\lambda.$$

With the distribution (14.46), it can be shown that every coordinate integration in $\langle \Phi_{k,n}(Q_k) \rangle_{kL,n}$ and $\langle [\Phi_{k,n}(Q_k)]^2 \rangle_{kL,n}$ is a product of integrals, each referring to single particle and a single coordinate direction, of the form

(14.50)
$$\left(\frac{nk_i}{2\pi}\right)^{\frac{1}{2}} \int_{-\infty}^{\infty} d\xi_i^\mu \, e^{-\frac{nk_i}{2}[\xi_i^\mu]^2} (\xi_i^\mu)^w = \begin{cases} \dfrac{\Gamma\left(\frac{w+1}{2}\right)}{\sqrt{\pi}}\left(\dfrac{2}{nk_i}\right)^{\frac{w}{2}}, & \text{for } w \text{ even,} \\ 0, & \text{for } w \text{ odd,} \end{cases}$$

for $i \in k$ and $0 \le \mu \le 3$. Integrals over products of displacements for different particles and for different directions vanish. This is expressed as $\langle \xi_r^\mu \xi_s^\nu \rangle_{kL,n} = \delta^{\mu\nu}\delta_{rs} \langle [\xi_r^\mu]^2 \rangle_{kL,n}$. The non-zero averages of the displacements are therefore

(14.51a)
$$\langle \xi_r^\mu \xi_s^\nu \rangle_{kL,n} = \delta^{\mu\nu}\delta_{rs}\frac{1}{nk_r}$$

and

(14.51b) $\left\langle \xi_r^\mu \xi_s^\nu \xi_u^\sigma \xi_v^\lambda \right\rangle_{kL,n} = \dfrac{1}{n^2} \left[\delta^{\mu\nu} \delta_{rs} \delta^{\sigma\lambda} \delta_{uv} \dfrac{1}{k_r k_u} + \delta^{\mu\sigma} \delta_{ru} \delta^{\nu\lambda} \delta_{sv} \dfrac{1}{k_r k_s} \right.$

$$\left. + \delta^{\mu\lambda} \delta_{rv} \delta^{\nu\sigma} \delta_{su} \dfrac{1}{k_r k_s} \right].$$

This result is valid for the case in which $\mu = \nu = \sigma = \lambda$ and $r = s = u = v$ as well.

By (14.47), the $F_k^{L,n}$ average of $\Phi_{k,n}(Q_k)$ is

(14.52)

$$\left\langle \Phi_{k,n}(Q_k) \right\rangle_{kL,n} \approx \Phi_{k,n}(Q_k^\circ) + \frac{1}{2} \frac{\partial^2 \Phi_{k,n}(Q_k^\circ)}{\partial q_r^\mu \partial q_s^\nu} \left\langle \xi_r^\mu \xi_s^\nu \right\rangle_{kL,n}$$

$$+ \frac{1}{24} \frac{\partial^4 \Phi_{k,n}(Q^\circ)}{\partial q_r^\mu \partial q_s^\nu \partial q_u^\sigma \partial q_v^\lambda} \left\langle \xi_r^\mu \xi_s^\nu \xi_u^\sigma \xi_v^\lambda \right\rangle_{kL,n} + O\left(\frac{1}{n^3} \right).$$

For $[\Phi_{k,n}(Q_k)]^2$, it follows similarly that

(14.53)

$$\left\langle [\Phi_{k,n}(Q_k)]^2 \right\rangle_{kL,n} \approx [\Phi_{k,n}(Q_k^\circ)]^2 + \Phi_{k,n}(Q_k^\circ) \frac{\partial^2 \Phi_{k,n}(Q_k^\circ)}{\partial q_r^\mu \partial q_s^\nu} \left\langle \xi_r^\mu \xi_s^\nu \right\rangle_{kL,n}$$

$$+ \frac{2}{24} \Phi_{k,n}(Q_k^\circ) \frac{\partial^4 \Phi_{k,n}(Q^\circ)}{\partial q_r^\mu \partial q_s^\nu \partial q_u^\sigma \partial q_v^\lambda} \left\langle \xi_r^\mu \xi_s^\nu \xi_u^\sigma \xi_v^\lambda \right\rangle_{kL,n}$$

$$+ \frac{1}{4} \frac{\partial^2 \Phi_{k,n}(Q_k^\circ)}{\partial q_r^\mu \partial q_s^\nu} \frac{\partial^2 \Phi_{k,n}(Q_k^\circ)}{\partial q_u^\sigma \partial q_v^\lambda} \left\langle \xi_r^\mu \xi_s^\nu \xi_u^\sigma \xi_v^\lambda \right\rangle_{kL,n} + O\left(\frac{1}{n^3} \right).$$

The square of (14.52) is used with (14.52) and then (14.51) is applied to show that

(14.54) $\quad [\langle \Phi_{k,n}(Q_k) \rangle_{kL,n}]^2 - \langle [\Phi_{k,n}(Q_k)]^2 \rangle_{kL,n} \approx$

$$\frac{1}{4} \frac{\partial^2 \Phi_{k,n}(Q_k^\circ)}{\partial q_r^\mu \partial q_s^\nu} \frac{\partial^2 \Phi_{k,n}(Q_k^\circ)}{\partial q_u^\sigma \partial q_v^\lambda} \left\{ \langle \xi_r^\mu \xi_s^\nu \rangle_{kL,n} \langle \xi_u^\sigma \xi_v^\lambda \rangle_{kL,n} - \langle \xi_r^\mu \xi_s^\nu \xi_u^\sigma \xi_v^\lambda \rangle_{kL,n} \right\}$$

$$= -\frac{1}{4 k_r k_s n^2} \frac{\partial^2 \Phi_{k,n}(Q_k^\circ)}{\partial q_r^\mu \partial q_s^\nu} \frac{\partial^2 \Phi_{k,n}(Q_k^\circ)}{\partial q_u^\sigma \partial q_v^\lambda} \left[\delta^{\mu\sigma} \delta_{ru} \delta^{\nu\lambda} \delta_{sv} + \delta^{\mu\lambda} \delta_{rv} \delta^{\nu\sigma} \delta_{su} \right].$$

Making use of the definition of the heat capacity (12.53) again, it is easy to show

(14.55)

$$\mathcal{K}_{k,V}^{L,n} = [\langle \mathcal{H}_k(Q_k, P_k) \rangle_{kL,n}]^2 - \langle [\mathcal{H}_k(Q_k, P_k)]^2 \rangle_{kL,n}$$

$$= [\langle \mathcal{K}_k(P_k) \rangle_{kL,n}]^2 - \langle [\mathcal{K}_k(P_k)]^2 \rangle_{kL,n}$$

$$+ [\langle \Phi_{k,n}(Q_k) \rangle_{kL,n}]^2 - \langle [\Phi_{k,n}(Q_k)]^2 \rangle_{kL,n}.$$

Using (14.42) and (14.54) in this formula gives

$$(14.56) \quad \mathcal{K}_{k,V}^{L,n} \approx -\frac{1}{n^2}\left\{\frac{3\mathcal{N}_k}{2} + \frac{1}{4k_r k_s}\frac{\partial^2\Phi_{k,n}(Q_k^\circ)}{\partial q_r^\mu \partial q_s^\nu}\frac{\partial^2\Phi_{k,n}(Q_k^\circ)}{\partial q_u^\sigma \partial q_v^\lambda}\right.$$

$$\left. \times\left[\delta^{\mu\sigma}\delta_{ru}\delta^{\nu\lambda}\delta_{sv} + \delta^{\mu\lambda}\delta_{rv}\delta^{\nu\sigma}\delta_{su}\right]\right\}.$$

It follows from this that $\mathcal{K}_{k,V}^{L,n} \leq 0$, as required. For simplicity, let us assume that the particles are identical and that $k_r = k_s = k_0$ for all $r, s \in k$. To obtain the conventional heat capacity, the quantity $\beta_k \sim n$ is used for the thermature and then (14.56) with (12.53) which yields

$$(14.57) \quad \mathcal{C}_{k,V}^{L,n} \approx k_B\left\{\frac{3\mathcal{N}_k}{2} + \frac{1}{4k_0^2}\frac{\partial^2\Phi_{k,n}(Q_k^\circ)}{\partial q_r^\mu \partial q_s^\nu}\frac{\partial^2\Phi_{k,n}(Q_k^\circ)}{\partial q_u^\sigma \partial q_v^\lambda}\right.$$

$$\left. \times\left[\delta^{\mu\sigma}\delta_{ru}\delta^{\nu\lambda}\delta_{sv} + \delta^{\mu\lambda}\delta_{rv}\delta^{\nu\sigma}\delta_{su}\right]\right\}.$$

An examination of (14.57) shows that the partitioning of the energy between the kinetic and potential forms depends on the ratio of the first and second terms. These quantities are both required to be finite and greater than zero but are otherwise arbitrary. To fix their ratio, let us require that each function in the sequence $F_k^{L,n}(Q_k, P_k)$ defined in (14.49) is a system state. Taking the total time derivative of $F_k^{L,n}$, the facts that v_i^μ is arbitrary and external forces vanish are used to show that

$$(14.58a) \qquad \frac{dF_k^{L,n}(Q_k, P_k)}{dt} = -n\sum_{i\in k} v_i \cdot \left[k_i\xi_i - \frac{\partial\Phi_{k,n}(Q_k)}{\partial q_i}\right] = 0$$

if and only if

$$(14.58b) \qquad \frac{\partial\Phi_{k,n}(Q_k)}{\partial q_{i,\mu}} = k_i\xi_i^\mu,$$

that is, if Hooke's law is obeyed. This is expected to be the case in some solids at moderate to low temperatures. Since the potential depends on the square of the displacement, the law of equipartition of energy is also expected to be valid.[21] The force law (14.58b) is used to compute the second derivative

---

[21] The law of the equipartition of energy applies to the rotational energy as well. The rotational energy is expressed as $p_i^2/2I_i$, where I_i is the moment of inertia of particle i. For simplicity in the development of the TIS theory, it was assumed at the outset that the particles do not have any internal degrees of freedom, so vibrational and rotational energy have been ignored throughout this work. For further information on the equipartition of energy, see the discussion in Jeans [1925], pp. 80–110. For historical notes on the equipartition of energy and a discussion of its application to rotating molecules, see Partington [1949], pp. 250–251; 330–332, 842–844.

of $\Phi_{k,n}(Q_k)$ with respect to q_s as

(14.59)
$$\frac{\partial^2 \Phi_{k,n}(Q_k^\circ)}{\partial q_r^\mu \partial q_s^\nu} = k_r \delta_{\mu\nu} \delta^{rs}.$$

It is not hard to show using the summation convention for both vector indices and particle indices that

(14.60)
$$\delta_{\mu\nu}\delta^{rs}\delta_{\sigma\lambda}\delta^{uv}\left[\delta^{\mu\sigma}\delta_{ru}\delta^{\nu\lambda}\delta_{sv} + \delta^{\mu\lambda}\delta_{rv}\delta^{\nu\sigma}\delta_{su}\right] = 6\mathcal{N}_k.$$

Using this in (14.57) yields the result

(14.61a)
$$C_{k,V}^{L,n} = 3k_B\mathcal{N}_k + O\left(\frac{1}{n}\right).$$

In molar terms, set $\mathcal{N}_k = \mathcal{A}$ to obtain the specific heat in the form

(14.61b)
$$C_{k,V}^{L,n} = 3R + O\left(\frac{1}{n}\right),$$

which is the law of Dulong and Petit for the molar specific heat of a solid.[22]

Next, observe that if the interparticle potential depends on a coupling parameter, such as the ϵ, for example, in the interaction potential of Lennard-Jones discussed in EIS, it is easy to show that the contribution of the potential energy disappears and the conventional heat capacity in (14.61) becomes that of an ideal gas as $\epsilon \to 0$.[23] Furthermore, if (14.58) is valid, $F_k^{L,n}$ can be converted into the form of an equilibrium state by using $\beta_k^L = n$.

Mathematically, the results of this section are a straightforward consequence of Morse's theorem for real-valued functions satisfying (4.2a) in the neighborhood of a nondegenerate critical point.[24] The validity of Hooke's law in this case is a consequence of the assumption that the critical point is nondegenerate.

An issue that is left unresolved here is the question of which minimum the system settles into as heat is extracted. In the absence of outside perturbations, the initial conditions determine which local minimum will trap the

---

[22]This can be obtained directly for a harmonic oscillator. The equipartition of energy is valid for a harmonic oscillator and each coordinate of the oscillator contributes an energy $k_B T_k/2$. For the $6\mathcal{N}_k$ coordinates of a system, this is $< E >= 3k_B\mathcal{N}_k T_k$, which leads to the heat capacity (14.61) observed by Dulong and Petit. The same result was obtained by Boltzmann [**1871b**], p. 731, who used a solid with particles bound to their equilibrium position by a linear force law (Hooke's law). See Bailyn [**1994**], pp. 448–449, for the harmonic oscillator calculation leading to this result. Although Bailyn stated that this result is valid only for high temperatures, this is a low temperature result for solids that neglects quantum effects.

[23]See also J. O. Hirschfelder, C. F. Curtiss and R. B. Bird [**1954**] on the Lennard-Jones interaction potential.

[24]See Morse and Cairns [**1969**], pp. 14, 16, 20–28.

system. However, if there are external perturbations that provide sufficient "noise", the system can move from one local minimum to another and eventually move into a global minimum. This question is important to the theory of phase boundaries and will be discussed further in EIS.

14.7 Assessing the Nernst and Planck Conjectures

The results of Section 14.4 show that Planck's conjecture cannot be maintained from a classical viewpoint and that the entropy in any thermodynamics based on classical mechanics can only be defined in terms of differences. However, it is clear from (14.31) and (14.39) that

$$(14.62) \qquad \lim_{n \to \infty} [\beta_k^{\delta,n}]^{-1} S_k^{\delta,n} = 0.$$

It follows that Berthelot's conjecture holds at absolute zero under the assumptions made in Section 14.6.[25]

In Section 14.6, it was shown that the heat capacities do not vanish at absolute zero. The only question remaining is Nernst's original conjecture stating that the difference in the heat capacities of two phases of a substance vanish at absolute zero. If there are two phases of a substance at absolute zero, they must occupy different critical points because each absolute zero state, if defined by a nondegenerate critical point, is unique up to an interchange of particles with the same set of interaction potentials.[26] If the critical point of phase 1 is designated by (Q_{k1}°) and phase 2 by (Q_{k2}°), then the difference in the conventional heat capacities is obtained using (14.57) as

$$(14.63) \quad C_{k2,V}^{\delta,n} - C_{k1,V}^{\delta,n}$$
$$\approx \frac{k_B}{4k_0^2} \left[\frac{\partial^2 \Phi_{k,n}(Q_{k,2}^\circ)}{\partial q_r^\mu \partial q_s^\nu} \frac{\partial^2 \Phi_{k,n}(Q_{k,2}^\circ)}{\partial q_u^\sigma \partial q_v^\lambda} - \frac{\partial^2 \Phi_{k,n}(Q_{k,1}^\circ)}{\partial q_r^\mu \partial q_s^\nu} \frac{\partial^2 \Phi_{k,n}(Q_{k,1}^\circ)}{\partial q_u^\sigma \partial q_v^\lambda} \right]$$
$$\times \left[\delta^{\mu\sigma} \delta_{ru} \delta^{\nu\lambda} \delta_{sv} + \delta^{\mu\lambda} \delta_{rv} \delta^{\nu\sigma} \delta_{su} \right] \bigg\}.$$

If (14.58b) is assumed to be a valid approximation and k_0 is independent of location in the lattice, it can be used in (14.63) to give immediately

$$(14.64) \qquad \lim_{\beta_k \to \infty} [C_{k2,V}^{\delta,n} - C_{k1,V}^{\delta,n}] = 0.$$

This follows from the fact that the second derivative of the version of $\Phi_{k,n}(Q_k)$ given in (14.58b) with respect to q in the Taylor expansion does not depend on the point Q_k° at which it is evaluated. The result (14.64) confirms the validity of Nernst's original conjecture for the case of classical solids obeying Hooke's law near absolute zero.

---

[25]See Kurth [**1960**] for a derivation of this relation based on similar assumptions.

[26]This is a consequence of Morse's theorem on the canonical form of functions at a nondegenerate critical point.

CHAPTER 15

The Galilean Transformation
of Thermodynamic Quantities

The close connection between TIS thermodynamics and mechanics is pursued in this chapter. The focus is on how the transformation properties of the components of thermodynamics follow from the transformation properties of their analogs.

15.1 Transformations in Thermodynamics

While the Galilean transformations of thermodynamic quantities are the central concern of this chapter, these are not the only kind of transformations that will eventually be needed. Legendre transformations of the Lagrangian, for example, will play an important role in some of the work in EIS and QTS. For historical information on mechanics, see Ernst Mach [1893]. For general background information on various aspects of classical mechanics, see Goldstein [1950], Abraham and Marsden [1978], and other texts. A good conceptual background is provided by David Oliver [1994].

Issues that have been raised in previous chapters concerning transformations will be addressed systematically in this chapter.

Reference Frames.

The significance of the frame of reference was first addressed in Chapters 2 and 4 where it was pointed out that the usual formalism for the Hamiltonian energy in the mechanics of a particle system is not invariant under translations of the phase momentum frame in the same way it is for translations of the phase coordinate frame. Two special frames, the system rest frame and the particle rest frame, were chosen to give the thermodynamic pressure and thermature, which depend on the system kinetic energy, values that are independent of the observer's Galilean frame. The use of these rest frames for computing the pressure and thermature was supported by an analysis of the behavior of the concepts themselves under a change of frame.[1]

The fact that Galilean frames of reference are chosen by macroscopic observers implies that a number of quantities that play an auxiliary role in the TIS theory need to be considered when making Galilean transformations between frames. For a single system, these are the macroscopic quantities describing the arena in which the system operates. The primary quantities in this category, listed in Chapter 2, are the system volume set, $d_k(t)$, the volume boundary set, $\partial d_k(t)$, the volume, $\delta_k(t)$, the phase volume set, $\Omega_k(t)$, the system projection operator, $\chi_{\Omega_k(t)}(Q_k, P_k)$, and the derivatives of the system projection operator with respect to time, volume, and thermature. For systems interacting with each other, this set is extended to include the fluxes through the boundary, which are described using the boundary vectors $\hat{n}_{\partial k}(q, t)$ and $v_{\partial k}(q, t)$ and the wall transition probability.

In Chapter 4, the origins of the particle number, momentum, and angular momentum scales, were fixed by the choice of the t or k particle rest frame as the TIS reference frame. If this reference frame is also inertial, Newton's second law, expressed as the relation $F = ma$ connecting force, mass and acceleration, is valid for all forces in that frame. In addition, all transformations

---

[1] For a discussion of some of the issues concerning thermodynamics in different reference frames in special relativity, see David van Dantzig [1939] and the references cited there. This, and more recent work on thermodynamics in relativistic spaces, will be discussed in RIS.

on the microscopic level are required to be compatible with the Hamiltonian mechanics of the particles, so they must also be canonical. This means in particular that Galilean transformations between reference frames and transformations generated by the time, volume, and thermature partial derivative operators must be canonical.

Accelerations a_i^μ and coordinate differences $q_i^\mu - q_j^\mu$ are invariant under Galilean transformations. This implies that interparticle forces of the form $m_i a_i^\mu = F_i^\mu = -\sum_{j \in k} \partial \phi_{ij}(q_i - q_j)/\partial q_i$, and therefore Newton's second law, are also invariant under Galilean transformations. In an arbitrary reference frame, forces that can be expressed in the form $F_i^\mu = m_i a_i^\mu$, where F_i^μ is the sum of forces between particle i and other particles, are called *inertial forces;* other forces are called *non-inertial forces.* Those external forces that are not tied to particles in TIS are frame-dependent. It follows that every system at rest in a frame-dependent external force field, whether gravitational, electromagnetic, or something else, is in fact at rest in a non-inertial frame. On the other hand, a charged system that is freely accelerated in an electromagnetic field or a system freely falling in an external gravitational field is in an inertial frame in the absence of other forces impeding this motion.

The frame of the fixed stars is usually presumed to be inertial in classical mechanics and special relativity. Stepping beyond the need for inertial frames, a holdover from Newton's absolute space and time, required the generalization of the notions of force, kinematics, and geometry. Einstein's key insight concerning the relation of gravitation and mechanics was to see the equivalence of local gravitational forces and the local acceleration of the reference frame. This allowed him to express energy as a shaper of geometry and geometry as a determining factor in motion.

Both a gravitational field and the acceleration of the reference frame give rise to deviations in the observed motion of particles from Newtonian paths that indicate that the reference frame is non-inertial. While frame-dependent forces in rotating reference frames are often referred to as "fictitious" or "apparent" forces, they can be used to determine the rotation of the reference frame with respect to the fixed stars. In addition, the total force acting at various points on a system can always be divided unambiguously in any reference frame into frame independent inertial forces, which are associated with particle interactions, and the frame-dependent non-inertial forces that comprise the rest of the total.

Taken together, these facts imply that measurements and definitions made in one frame may not be valid in another frame. In particular, it was found that the pressure and thermature do not have the same values in all reference frames, whether Galilean or not. Their frame dependence stems from the fact that they both depend in part on the frame-dependent kinetic energy. An analysis of this situation above led to the requirement that they must be defined in the system rest frame. For several interacting systems, it was also required that a single rest frame must be chosen for these calculations so that the thermatures and pressures of the individual systems have the proper relation to each other. Otherwise, the mechanical aspects of the pressure, for example, cannot be compared for two interacting systems.

Canonical Transformations.

If Hamilton's equations and calculations based on them are valid in one frame, they are valid in all frames related to this one by a canonical transformation. A canonical transformation of a phase quantity such as P_k preserves the value of the Poisson bracket for that phase quantity. This preserves the total time derivative of a phase quantity and therefore the phase space aspects of the time evolution of the system. It will be shown in passing below that Galilean transformations and the volume derivative operator transformation are canonical.[2]

The study of Hamilton's equations from the standpoint of canonical transformations leads to a more abstract viewpoint than that provided by Galilean transformations. One consequence of this more abstract viewpoint, as discussed by Goldstein [**1950**], p. 245, is that the distinction between the particle coordinates and their momenta is broken down. He illustrated this with a generating function that interchanges the roles of the coordinate and momentum variables and went on to say

> "This simple example should emphasize the independent status of generalized coordinates and momenta. They are both needed to describe the motion of a system in the Hamiltonian formulation, and the distinction between them is practically one of nomenclature. One can shift the names around with at most no more than a change in sign. There is no longer present in the theory any lingering remnant of the concept of q_i as a spatial coordinate and p_i as a mass times a velocity."

Galilean frame transformations are the primary concern here, so the more abstract theory will not be considered again until EIS, where it will be developed in more detail.

The Coordination of Microscopic and Macroscopic Frames.

Because TIS operates with parallel descriptions on the macroscopic and microscopic levels, the transformation of any macroscopic thermodynamic quantity, such as a change in the frame of reference by a Galilean transformation, is required to be consistent with the corresponding transformation of the analog function that represents this quantity microscopically. In addition, attention must be paid to the differences in the transformation of those quantities that can be represented as particle sum functions and localized at

---

[2]See Goldstein [**1950**], Chapter 8, for example, for a discussion of canonical transformations and their expression in terms of a generating function. Goldstein also discussed the Galilean transformation of Newton's laws.

each particle, such as energy, momentum, and angular momentum, and those purely thermodynamic quantities, such as entropy and temperature, which require the system distribution for their calculation and cannot be represented as particle sum functions.

The parallel macroscopic/microscopic transformations are represented in TIS by pairs of associated operators. The requirement that these pairs of operators, and others that may be introduced, operate in a parallel fashion on the thermodynamic functions and their corresponding analog phase functions is expressed as the requirement that the following operator diagram is commutative:

$$
\begin{array}{ccc}
O^1_{k,\mathrm{mac}} & \xrightarrow{\;Tr^{(\mathrm{ind})}_{\mathrm{mac}}\;} & O^2_{k,\mathrm{mac}} \\[4pt]
{\scriptstyle <\cdot>^1_{kt}}\Big\uparrow & & \Big\uparrow{\scriptstyle <\cdot>^2_{kt}} \\[4pt]
O^1_{k,\mathrm{mic}} & \xrightarrow[\;Tr_{\mathrm{mic}}\;]{} & O^2_{k,\mathrm{mic}}
\end{array}
$$

FIGURE 15.1.   Diagram of Operator Relations

In this diagram, the macroscopic transformation between thermodynamic functions is $Tr^{(\mathrm{ind})}_{\mathrm{mac}}$ and the microscopic transformation between analog functions is Tr_{mic}. The microscopic and macroscopic levels of description are connected by the expectation value or phase averaging measure both before and after the transformations. Three such pairs of transformation operators have already been introduced: the thermodynamic partial time derivative operator and the phase total time derivative operator, the thermodynamic partial thermature derivative operator and the phase total thermature derivative operator, and the thermodynamic partial volume derivative operator and the phase total volume derivative operator.

To define precisely what is meant by parallel transformations, consider a macroscopic operator that acts on thermodynamic observable quantities and a microscopic operator that acts on their analogs. Most such transformations, such as a Galilean frame transformation or a TIS derivative operator, will also affect the system distribution density as well as the thermodynamic analog function. In Figure 15.1, $Tr^{(\mathrm{ind})}_{\mathrm{mac}}$ is a macroscopic operator that transforms the local macroscopic observable $O^1_{k,\mathrm{mac}}$ into another local macroscopic observable $O^2_{k,\mathrm{mac}}$. Similarly, Tr_{mic} is a microscopic operator that transforms the microscopic analog $O^1_{k,\mathrm{mic}}$, corresponding to $O^1_{k,\mathrm{mac}}$, into the analog $O^2_{k,\mathrm{mic}}$, corresponding to $O^2_{k,\mathrm{mac}}$. It may also transform the state $F^1_k(Q_k, P_k, t)$ to $F^2_k(Q_k, P_k, t)$. The operator relationships expressed in Figure 15.1 are required to be valid for all transformations or mappings allowed in the theory.

In practice, the commutativity of the diagram is guaranteed by defining the macroscopic operator $Tr_{\text{mac}}^{(\text{ind})}$ in terms of the microscopic operator Tr_{mic}. That is, the result of the action of the transformation $Tr_{\text{mac}}^{(\text{ind})}$ on $O_{k,\text{mac}}^1$, where $O_{k,\text{mac}}^1 = \left\langle O_{k,\text{mic}}^1 \right\rangle_{kt}^{(1)}$, is defined to be $O_{k,\text{mac}}^2$, where $O_{k,\text{mac}}^2 = \left\langle O_{k,\text{mic}}^2 \right\rangle_{kt}^{(2)} = \left\langle Tr_{\text{mic}} O_{k,\text{mic}}^1 \right\rangle_{kt}^{(2)}$. In this way, the macroscopic transformation $Tr_{\text{mac}}^{(\text{ind})}$ is *induced* by the microscopic transformation. The precise form of the combination of the analog $O_{k,\text{mic}}^2$ and the state $F_k^2(Q_k, P_k, t)$ used in the expectation calculation depends on the type of operator that Tr_{mic} is, e.g., whether it is a frame transformation, a differential operator, or something else. This was the way that the macroscopic transformations associated with the macroscopic time, volume, and thermature partial derivative operators were defined in terms of the microscopic operators in previous chapters.

15.2 The Transformation of Densities and Volumes

Let us turn now from global considerations to the transformation of specific quantities within the theory and begin with the macroscopic quantities describing the arena in which a system operates. The primary quantities in this category are the system volume set $d_k(t)$, the volume boundary set $\partial d_k(t)$, the volume $\delta_k(t)$, the phase volume set $\Omega_k(t)$, and the system projection operator $\chi_{\Omega_k(t)}(Q_k, P_k)$. The derivatives of the system projection operator with respect to time, volume, and thermature, will be considered afterward. For systems interacting with each other, this set will also be extended in the discussion below to include the fluxes through the boundary and the wall transition probability.

The symbol T_a will designate the 6-dimensional Galilean transformations of coordinate and momentum phase space and T_A will designate the $6\mathcal{N}_k$-dimensional Galilean transformations of phase space defined in Section 2.5. It is easy to show that the inverse transformations, $[T_a]^{-1}$ and $[T_A]^{-1}$, also exist. In the analysis of these transformations, the frame in which a phase average map is computed will be designated by a superscript such as $\sharp$ or $\flat$.

In terms of this notation, the main focus of this chapter is to show that the following diagram of frame transformation mappings is commutative:

$$
\begin{array}{ccc}
G_k^{(\flat)}(q^{(\flat)}, t) & \xrightarrow{\;T_a^{(\text{ind})}\;} & G_k^{(\sharp)}(q^{(\sharp)}, t) \\[4pt]
{\scriptstyle <\cdot>_{kt}^{(\flat)}}\big\uparrow & & \big\uparrow{\scriptstyle <\cdot>_{kt}^{(\sharp)}} \\[4pt]
G_k^{(\flat)}(Q_k^{(\flat)}, p_i^{(\flat)}, t) & \xrightarrow[\;T_A\;]{} & G_k^{(\sharp)}(Q_k^{(\sharp)}, P_k^{(\sharp)}, t)
\end{array}
$$

FIGURE 15.2.   Galilean Frame Transformations

The operator $T_a^{(\text{ind})}$ is the transformation of thermodynamic quantities induced by the microscopic Galilean transformation of the analogs.

Since the Jacobian of the a_r frame rotation defined in Section 2.5 is the determinant $|a_r|$, and $|a_r| = 1$, the proper transformation of the 6-dimensional phase volume differential is

(15.1a) $$d^3 q_i^{(\sharp)} = |a_r| d^3 q_i^{(\flat)} = d^3 q_i^{(\flat)},$$

(15.1b) $$d^3 p_i^{(\sharp)} = |a_r| d^3 p_i^{(\flat)} = d^3 p_i^{(\flat)}.$$

Next, let $d_k^{(\flat)}(t)$ represent a volume set defined in the $\flat$ frame. A corresponding set $d_k^{(\sharp)}(t)$ is defined in the $\sharp$ frame in terms of $d_k^{(\flat)}(t)$ by

(15.2) $$d_k^{(\sharp)}(t) = \{\, q^{(\sharp)} \mid [T_{aq}]^{-1} q^{(\sharp)} \in d_k^{(\flat)}(t) \,\}.$$

Using this result with (15.1) shows that the volume is invariant under proper Galilean transformations:

(15.3) $$\delta_k^{(\sharp)}(t) = \int d^3 q^{(\sharp)} \chi_{d_k^{(\sharp)}(t)}(q^{\sharp}) = |a_r| \int d^3 q^{(\flat)} \chi_{d_k^{(\flat)}(t)}(q^{\flat}) = \delta_k^{(\flat)}(t).$$

Finally, because $r_{ij} = |q_i - q_j|$ is invariant under Galilean transformations, the central forces between particles, which depend only on r_{ij}, are also invariant. The transformations of $\Omega_k(t)$ and $\chi_{\Omega_k(t)}(Q_k, P_k)$ will be taken up in Section 15.5 below.

The results of this section indicate that the macroscopic volume set relations (2.26a) and (2.26b) are consistent with the microscopic description. The transformation properties of $\hat{n}_{\partial k}(q, t)$ and $v_{\partial k}(q, t)$ shown in (2.26c) and (2.26d) follow directly from the definition of the Galilean transformations.

15.3 The Transformation of Phase Space

Let us next turn to the $6\mathcal{N}_k$-dimensional case and use the constant 3-vectors $q_a^{(\flat)}$ and $v_a^{(\flat)}$ to define the following $3\mathcal{N}_k$-dimensional constant vectors

(15.4a) $$Q_A^{(\flat)} = \times_{i \in k} q_a^{(\flat)},$$

(15.4b) $$V_A^{(\flat)} = \times_{i \in k} v_a^{(\flat)},$$

and

(15.5a) $$p_{a,i}^{(\flat)} = m_i v_a^{(\flat)},$$

(15.5b) $$P_A^{(\flat)} = \times_{i \in k} p_{a,i}^{(\flat)}.$$

These are used to display the $6\mathcal{N}_k$-dimensional transformation, T_A, which can be written in terms of its $3\mathcal{N}_k$-dimensional coordinate and momentum transformations (T_{AQ}, T_{AP}), as

(15.6a) $$Q_k^{(\sharp)} = T_{AQ} Q_k^{(b)} = \times_{i \in k} T_{aq} q_i^{(b)},$$

(15.6b) $$P_k^{(\sharp)} = T_{AP} P_k^{(b)} = \times_{i \in k} T_{ap} p_i^{(b)},$$

The transformations T_{AQ} and T_{AP} are clearly invertible. The transformation of the phase volume differential is

(15.7a) $$dQ_k^{(\sharp)} = |a_r|^{N_k} dQ_k^{(b)} = dQ_k^{(b)},$$

(15.7b) $$dP_k^{(\sharp)} = |a_r|^{N_k} dP_k^{(b)} = dP_k^{(b)}.$$

15.4 The Transformation of Phase Analog Functions

The Galilean frame transformations of phase functions are considered next. Suppose $(Q_k^{(\sharp)}, P_k^{(\sharp)})$ and $(Q_k^{(b)}, P_k^{(b)})$ refer to the same physical point in phase space and are related by the frame transformation (15.6): $(Q_k^{(\sharp)}, P_k^{(\sharp)}) = T_A(Q_k^{(b)}, P_k^{(b)})$. As a physical quantity, the value of a phase function should transform in accord with its units and tensor functions should be rotated appropriately to compensate for the change of frame.

Intuitively, the value of the scalar function representing the particle number density, say, should not depend on the frame in which it is computed as long as the argument of the function $\nu_k(q, t)$ refers to the same point q as it is expressed in each frame. Because the number density is the number of particles in a volume divided by the volume, and both the number of particles and the volume are invariant under a Galilean transformation, the invariance of $\nu_k(q, t)$ under a Galilean transformation is expected. Using $q^{(\sharp)} = T_{aq} q^{(b)}$, this invariance is expressed as the relation $\nu_k^{(\sharp)}(q^{(\sharp)}, t) = \nu_k^{(b)}(T_{aq}^{-1} q^{(\sharp)}, t) = \nu_k^{(b)}(q^{(b)}, t)$. Other quantities, such as the momentum, angular momentum, and energy, are frame-dependent.

Rather than attach a frame designation to each physical function, the standard approach is to define physical functions in some chosen frame and express their values in other frames in terms of a different function of the coordinates defined in the original frame. This means that the frame transformation $b \to \sharp$ will induce a transformation of a function $G_k^{(b)}(Q_k^{(b)}, P_k^{(b)})$ defined at $(Q_k^{(b)}, P_k^{(b)})$ in the original b frame to a function $G_k^{(\sharp)}(Q_k^{(\sharp)}, p_i^{(\sharp)}, t)$ that expresses the transformed value of $G_k^{(b)}$ at the corresponding point $(Q_k^{(\sharp)}, P_k^{(\sharp)})$ in the new $\sharp$ frame. The transformation $G_k^{(b)} \to G_k^{(\sharp)}$ was defined above as the transformation induced by the Galilean frame transformation. It is represented as

(15.8)
$$G_k^{\mu \cdots}(Q_k^{(\sharp)}, P_k^{(\sharp)}, t) = a_{r, \nu}^{\mu} \ldots G_k^{\nu \cdots}(T_{AQ} Q_k^{(b)}, T_{AP} P_k^{(b)}, t)$$
$$= [T_A^{(ind)} G_k]^{\mu \cdots}(Q_k^{(b)}, P_k^{(b)}, t).$$

The *induced transformation* $[T_A^{(\text{ind})}G_k]^{\mu\cdots}(Q_k^{(b)}, P_k^{(b)}, t)$ of $G_k^{\mu\cdots}(Q_k^{(b)}, P_k^{(b)}, t)$ is determined in this way by the transformation properties of the components of $G_k^{\mu\cdots}$ and ultimately by the frame transformation. In spite of the unfamiliar terminology, this is a standard perspective on transformations.[3]

The induced transformations of various physical quantities of importance to the theory will be worked out next. In the transformations of phase functions below, the coordinates will retain their frame designations and the induced transformations of the functions will be shown explicitly. The transformations used are those of Section 2.5. The objective is to verify that the induced transformations presented there are consequences of the parallel transformations of microscopic analog functions in conjunction with the phase averaging formalism.

The representation (2.8) of the Dirac delta measure and the fact that the determinant of the transformation satisfies $|a_r| = 1$ are used with (15.1) and the fact that, as a tensor transformation, $T_{aq}(q^{(b)} - q_i^{(b)}) = 0$ if and only if $q^{(b)} - q_i^{(b)} = 0$, to show for the microscopic density analog that

$$(15.9) \quad \nu_i^{(\sharp)} = \delta(q^{(\sharp)} - q_i^{(\sharp)}) = \delta(T_{aq}q^{(b)} - T_{aq}q_i^{(b)}) = \delta(q^{(b)} - q_i^{(b)}) = \nu_i^{(b)}.$$

The Galilean transformations of some mechanical analog functions are presented next. These formulas will be used in conjunction with the transformation properties of the system distribution density below to calculate the Galilean transformations of thermodynamic quantities and confirm the results summarized in Chapter 2.

Particle Number and Particle Current.

$$(15.10a) \qquad \nu_i^{(\sharp)} = \nu_i^{(b)},$$

$$(15.10b) \qquad \nu_i^{(\sharp)\mu} = a_{r,\nu}^{\mu}[\nu_i^{(b)\nu} - v_a^{(b)\nu}\nu_i].$$

Momentum and Momentum Current.

$$(15.11a)$$
$$p_i^{(\sharp)\mu} = a_{r,\nu}^{\mu}[p_i^{(b)\nu} - m_i v_a^{(b)\nu}],$$

$$(15.11b)$$
$$p_i^{(\sharp)\mu\nu} = a_{r,\rho}^{\mu}a_{r,\sigma}^{\nu}[p_i^{(b)\rho\sigma} - 2p_i^{(b)(\rho}v_a^{|(b)|\sigma)} + m_i v_a^{(b)\rho}v_a^{(b)\sigma}].$$

---

[3] For an early discussion of this kind of transformation formalism, see Pauli [**1933**], p. 177. For more modern work on induced representations, see George W. Mackey [**1968**]. The transformation of a phase function such as $G_k(Q_k, P_k, t)$ induced by a Galilean or other frame transformation of its arguments is discussed by Mackey in Chapter 10.

Angular Momentum and Angular Momentum Current.

$$j_i^{(\sharp)\mu}(q_i^{(\sharp)}, p_i^{(\sharp)}) = |a_r| a_{r,\nu}^\mu [j_i^{(b)\nu}(q_i^{(b)}, p_i^{(b)}) - \epsilon^{\nu\eta\xi}([v_{a,\eta}^{(b)}t + q_{a,\eta}^{(b)}]p_{i,\xi}^{(b)}$$

(15.12a)
$$+ m_i(q_{i,\eta}^{(b)} - v_{a,\eta}^{(b)}t - q_{a,\eta}^{(b)})v_{a,\xi}^{(b)})],$$

$$j_i^{(\sharp)\mu\nu}(q_i^{(\sharp)}, p_i^{(\sharp)}) = |a_r| a_{r,\xi}^\mu a_{r,\eta}^\nu [j_i^{(b)\xi\eta}(q_i^{(b)}, p_i^{(b)}) - j_i^{(b)\xi}(q_i^{(b)}, p_i^{(b)})v_a^{(b)\eta}$$

$$- \epsilon^{\xi\sigma\lambda}[(v_{a,\sigma}^{(b)}t + q_{a,\sigma}^{(b)})(p_{i,\lambda}^{(b)\ \eta} - v_{a,\lambda}^{(b)}p_i^{(b)\eta}$$

$$- p_{i,\lambda}^{(b)}v_a^{(b)\eta} + m_i v_{a,\lambda}^{(b)}v_a^{(b)\eta})$$

(15.12b)
$$+ q_{i,\sigma}^{(b)}v_{a,\lambda}^{(b)}(p_i^{(b)\eta} - m_i v_a^{(b)\eta})\}].$$

Kinetic Energy and Kinetic Energy Current.

(15.13a)
$$K_i^{(\sharp)}(p_i^{(\sharp)}) = K_i^{(b)}(p_i^{(b)}) - v_a^{(b)} \cdot p_i^{(b)} + \tfrac{1}{2} m_i v_a^{(b)} \cdot v_a^{(b)},$$

$$K_i^{(\sharp)\mu}(p_i^{(\sharp)}) = a_{r,\nu}^\mu [K_i^{(b)\nu}(p_i^{(b)}) - K_i^{(b)}(p_i^{(b)})v_a^{(b)\nu} - p_i^{(b)\rho\nu}v_{a,\rho} + (p_i^{(b)} \cdot v_a^{(b)})v_a^{(b)\nu}$$

(15.13b)
$$+ \tfrac{1}{2} m_i v_a^{(b)} \cdot v_a^{(b)}(v_i^{(b)\nu} - v_a^{(b)\nu})].$$

Potential Energy.

(15.14b)
$$\Phi_{i,n}^{(\sharp)}(Q_k^{(\sharp)}) = \Phi_{i,n}^{(b)}(Q_k^{(b)}),$$

(15.14c)
$$\Phi_{i,x}^{(\sharp)}(q_i^{(\sharp)}, t) = \Phi_{i,x}^{(b)}(q_i^{(b)}, t).$$

The transformations of the i-particle Hamiltonian energy, energy, and their currents can be constructed from the transformation properties of these components. Except for the Hamiltonian energy, the system quantities are obtained by summing over $i \in k$. The transformations of other quantities are obtained similarly.

15.5 The Transformation of System Distributions

To determine the transformation of thermodynamic quantities, it is necessary to determine the effect of a Galilean transformation on a system distribution density. The presentation of this transformation is based on the results for the transformations of two sets of phase points central to the theory: the phase coordinate volume set $D_k(t)$, and the system phase volume set $\Omega_k(t)$. The first step is to define the transformation of the system projection operator.

It has been shown already in (15.2) how the volume sets $d^{(\sharp)}(t)$ and $d_k^{(b)}(t)$ are related under a Galilean transformation. That definition can be extended to the set $D_k(t)$ by

(15.15) $$D_k^{(\sharp)}(t) = \{\, Q_k^{(\sharp)} \mid T_{AQ}^{-1} Q_k^{(\sharp)} \in D_k^{(\flat)}(t)\,\}.$$

Similarly, $\Omega_k^{(\sharp)}(t)$ is defined in terms of $\Omega_k^{(\flat)}(t)$ by

(15.16) $$\Omega_k^{(\sharp)}(t) = \{\, (Q_k^{(\sharp)}, P_k^{(\sharp)}) \mid [T_A]^{-1} (Q_k^{(\sharp)}, P_k^{(\sharp)}) \in \Omega_k^{(\flat)}(t)\,\}.$$

This connection is used to show that the system projection operator transforms as

(15.17) $$\chi_{\Omega_k^{(\sharp)}(t)} (Q_k^{(\sharp)}, P_k^{(\sharp)}) = \chi_{\Omega_k^{(\flat)}(t)} (Q_k^{(\flat)}, P_k^{(\flat)}).$$

This relation means that the function on the left is 1 when the function on the right is 1 and is 0 when the function on the right is 0 when the phase coordinates are connected by a Galilean frame transformation.

For the transformation of the system probability distribution density, let us use the fact that the coordinates $(Q_k^{(\sharp)}, P_k^{(\sharp)})$ and $(Q_k^{(\flat)}, P_k^{(\flat)})$ refer to the same phase point represented in different frames. Consider a neighborhood of this point in the respective $\sharp$ and $\flat$ frames. The volume of a neighborhood is preserved by a Galilean transformation and the transformed value $(Q_k^{(\sharp)1}, P_k^{(\sharp)1})$ of any event $(Q_k^{(\flat)1}, P_k^{(\flat)1})$ within the $\flat$ neighborhood around $(Q_k^{(\flat)}, P_k^{(\flat)})$ will also appear within the corresponding transformed $\sharp$ neighborhood around $(Q_k^{(\sharp)}, P_k^{(\sharp)})$. This implies that the probability or relative probability that an event will fall within the neighborhood will be independent of the frame in which it is computed.

Using $dQ_k^{(\flat)} dP_k^{(\flat)}$ as the neighborhood, the relative probability that the state $(Q_k^{(\flat)}, P_k^{(\flat)})$ will fall in this neighborhood is $F_k^{(\flat)}(Q_k^{(\flat)}, p_i^{(\flat)}, t) dQ_k^{(\flat)} dP_k^{(\flat)}$. It follows that the relationship between the relative probability expressed in the $\flat$ frame and in the $\sharp$ frame is

(15.18) $$F_k^{(\sharp)}(Q_k^{(\sharp)}, P_k^{(\sharp)}, t) dQ_k^{(\sharp)} dP_k^{(\sharp)} = F_k^{(\flat)}(Q_k^{(\flat)}, P_k^{(\flat)}, t) dQ_k^{(\flat)} dP_k^{(\flat)}.$$

The invariance of the phase volume elements, as stated in (15.7), implies that the system distribution densities are frame independent in the sense that the value of $F_k^{(\sharp)}$ at $(Q_k^{(\sharp)}, P_k^{(\sharp)}, t)$ is related to the value of $F^{(\flat)}$ at $(Q_k^{(\flat)}, P_k^{(\flat)}, t)$ by the symbolic relation

(15.19) $$F_k^{(\sharp)}(Q_k^{(\sharp)}, P_k^{(\sharp)}, t) = F_k^{(\flat)}(Q_k^{(\flat)}, P_k^{(\flat)}, t),$$

where it is understood that $(Q_k^{(\sharp)}, P_k^{(\sharp)}) = T_A(Q_k^{(\flat)}, P_k^{(\flat)})$. The results (15.17) and (15.18) imply immediately that the partition functions for F_k in these two frames are related by

$$(15.20) \quad Z_k^{(\sharp)}(t) = \int dQ_k^{(\sharp)} \int dP_k^{(\sharp)} \, \chi_{\Omega_k^{(\sharp)}(t)}(Q_k^{(\sharp)}, P_k^{(\sharp)}) F_k^{(\sharp)}(Q_k^{(\sharp)}, P_k^{(\sharp)}, t)$$

$$= \int dQ_k^{(b)} \int dP_k^{(b)} \, \chi_{\Omega_k^{(b)}(t)}(Q_k^{(b)}, P_k^{(b)}) F_k^{(b)}(Q_k^{(b)}, P_k^{(b)}, t) = Z_k^{(b)}(t) = Z_k(t).$$

This shows that the k local norm of F_k in $\Omega_k(t)$, defined as the partition function in Chapter 5, is preserved under a Galilean transformation. It follows from (15.18) and (15.20) that the actual probability densities, and not just the relative probability densities determined by the system distribution densities, are the same in these two frames. In other words, $[Z_k^{(\sharp)}(t)]^{-1}F_k^{(\sharp)}$ is normalized in $\Omega_k^{(\sharp)}(t)$ if and only if $[Z_k^{(b)}(t)]^{-1}F_k^{(b)}$ is normalized in $\Omega_k^{(b)}(t)$.

It follows easily that $F_k^{(\sharp)}$ is non-negative if and only if $F_k^{(b)}$ is non-negative. This and the previous statement about the partition function mean that $F_k^{(\sharp)}$ meets the two conditions (5.14) for the $\sharp$ frame if and only if $F_k^{(b)}$ meets them for the b frame. Another consequence is that $[Z_k(t)]^{-1}F_k$ remains a probability distribution density under Galilean transformations. Finally, it is not hard to show that the support of $F_k^{(\sharp)}$ falls within $\Omega_k^{(\sharp)}(t)$ if the support of $F_k^{(b)}$ falls within $\Omega_k^{(b)}(t)$. However, this is not a general TIS requirement because the support of $F_k(Q_k, P_k, t)$ for an open system is not required to be limited to $\Omega_k(t)$.

Most of the thermodynamic formulas that will be discussed refer to one system only, so most of the work below will be done in the system rest frame for that system. When it is necessary to consider more than one system, the convention adopted is that the rest frame of the collection of systems taken together is the reference frame and thermodynamic functions for all systems involved are computed in this frame. This will be indicated explicitly when it occurs.

15.6 The Transformation of Phase Averages

To work out the formal details of the relation between the computation of phase averages in different frames, let $G_k^{\mu\cdots}(Q_k, P_k, t)$ represent a phase analog function where $\mu\cdots$ represents none, one, or more, possible tensor indices. The transformation of these indices is written in the form $a_{r,\nu}^{\mu} \ldots G_k^{\nu\cdots}(Q_k, P_k, t)$, where the ellipsis in $a_{r,\nu}^{\mu} \ldots$ implies one corresponding factor of a_r for every tensor index. This notation is used, along with the above results on Galilean transformations, (15.8), (15.9), (15.18), (15.20), and the formula (5.9) for the local phase average of a phase function, to express a local thermodynamic function in the $\sharp$ frame in terms of its definition in the

b frame as

$$(15.21) \quad G_i^{(\sharp)\mu\cdots}(q^{(\sharp)}, t) = \left\langle \delta(q^{(\sharp)} - q_i^{(\sharp)}) G_i^{(\sharp)\mu\cdots}(Q_k^{(\sharp)}, p_i^{(\sharp)}, t) \right\rangle_{kt}^{(\sharp)}$$

$$= \frac{1}{Z_k^{(\sharp)}(t)} \int dQ_k^{(\sharp)} \int dP_k^{(\sharp)} \, F_k^{(\sharp)}(Q_k^{(\sharp)}, P_k^{(\sharp)}, t) \chi_{\Omega_k^{(\sharp)}(t)}(Q_k^{(\sharp)}, P_k^{(\sharp)}) \delta(q^{(\sharp)} - q_i^{(\sharp)})$$

$$\times \, G_i^{(\sharp)\mu\cdots}(Q_k^{(\sharp)}, p_i^{(\sharp)}, t)$$

$$= \frac{|a_r|^{2\mathcal{N}_k}}{Z_k^{(b)}(t)} \int dQ_k^{(b)} \int dP_k^{(b)} \, F_k^{(b)}(Q_k^{(b)}, P_k^{(b)}, t) \chi_{\Omega_k^{(b)}(t)}(Q_k^{(b)}, P_k^{(b)}) \delta(q^{(b)} - q_i^{(b)})$$

$$\times \, [T_A^{(\mathrm{ind})} G_i^{(b)}]^{\mu\cdots}(Q_k^{(b)}, p_i^{(b)}, t)$$

$$= \left\langle \delta(q^{(b)} - q_i^{(b)}) [T_A^{(\mathrm{ind})} G_i^{(b)}]^{\mu\cdots}(Q_k^{(b)}, p_i^{(b)}, t) \right\rangle_{kt}^{(b)} = [T_a^{(\mathrm{ind})} G_i^{(b)}]^{\mu\cdots}(q^{(b)}, t).$$

The induced macroscopic transformation from the function $G_i^{(\sharp)\mu\cdots}(q^{(\sharp)}, t)$ to $[T_a^{(\mathrm{ind})} G_i^{(b)}]^{\mu\cdots}(q^{(b)}, t)$ is defined here by the induced microscopic transformation $G_i^{(\sharp)\mu\cdots}(Q_k^{(\sharp)}, p_i^{(\sharp)}, t) \to [T_A^{(\mathrm{ind})} G_i^{(b)}]^{\mu\cdots}(Q_k^{(b)}, p_i^{(b)}, t)$. This result shows that the transformation of the phase averages is given by

$$(15.22a) \qquad\qquad G_i^{(\sharp)\mu\cdots}(q^{(\sharp)}, t) = [T_a^{(\mathrm{ind})} G_i^{(b)}]^{\mu\cdots}(q^{(b)}, t),$$

$$(15.22b) \qquad\qquad G_i^{(\sharp)\mu\cdots}(t) = [T_a^{(\mathrm{ind})} G_i^{(b)}]^{\mu\cdots}(t).$$

The induced transformations of various physical quantities are stated in Section 15.4 for the phase analogs and in Section 2.5 for their corresponding thermodynamic functions. This result guarantees that the induced transformations of macroscopic quantities are in accord with the induced transformations of the microscopic quantities that are determined by the scalar and tensor properties of the original transformation of the coordinate and momentum frame and that the diagram in Figure 15.2 will be commutative.

Let us next apply (15.22), the transformations of Section 15.4, and the results of Section 15.5, to verify the thermodynamic transformation laws exhibited in Chapter 2. Using (15.9) and (15.10a) with $\nu_i(t) = 1$ and $\nu_k(t) = \mathcal{N}_k$, it is easy to show that the number of particles and the local density for corresponding points are invariant in accord with equations (2.26e). The relation $\mu_i(q, t) = m_i \nu_i(q, t)$ is used next to show that (2.26f) is valid also. It is also easy to show by (15.10b) that (2.26g) is valid as well.

For the momentum, the phase function transformation (15.11a) is substituted in (15.21) to show that the relation (2.26h) is correct. For the momentum current density, (15.11b) and (15.21) are employed again to obtain

$$(15.23)$$

$$\mathcal{P}_k^{(\sharp)\mu\nu}(q^{(\sharp)}, t) = a_{r,\xi}^\mu a_{r,\eta}^\nu [\mathcal{P}_k^{(b),\xi\eta}(q^{(b)}, t) - 2v_a^{(b)(\xi} \mathcal{P}_k^{(b)\eta)}(q^{(b)}, t)$$

$$+ v_a^{(b)\xi} v_a^{(b)\eta} \mu_k^{(b)}(q^{(b)}, t)].$$

This implies that (2.26i) is correct as well. Similarly, for the angular momentum, (15.12a) is used in (15.21) to demonstrate (2.26j) and again a simple extension gives (2.26k) as well.

Since the scalar $r_{ij} = |q_i - q_j|$ is invariant under a Galilean transformation, (7.14) allows us to conclude for the internal forces that

$$(15.24) \qquad \mathcal{F}_{k,n}^{(\sharp)\mu}(q^{(\sharp)}, t) = a_{r,\nu}^{\mu} \mathcal{F}_{k,n}^{(b)\nu}(q^{(b)}, t).$$

Finally, using (7.17) with (15.11), it follows from (5.29) that

$$(15.25) \qquad \mathcal{F}_{\partial k}^{(\sharp)\mu}(q^{(\sharp)}, t) = a_{r,\nu}^{\mu} \mathcal{F}_{\partial k}^{(b)\nu}(q^{(b)}, t).$$

These results confirm (2.26l) and (2.26n). The analysis of external forces requires showing that the forces between systems k and l, expressed in the external potentials $\Phi_{k,x}(Q_k, t)$ and $\Phi_{l,x}(Q_l, t)$, are unchanged in a Galilean frame transformation in which all systems are transformed simultaneously. This is done by performing the Galilean transformation in the t system and then dividing it into the same subsystems again. Because the forces in the t system are all internal, the external force experienced by a particle in system k or l will be the same after the transformation. This yields the invariance of the external force expressed in equation (2.26m).

Using the phase function transformations (15.13) and (15.14) for the energy in (15.21) yields the equations (2.26o), (2.26p), (2.26q), (2.26r), and (2.26s) immediately. Since $K_i(p_i - \bar{p}_i(q, t))$ is invariant under the velocity boost transformation (2.24b), (2.26t) follows.

Turning next to the working, applying a Galilean transformation to the surface working functions defined in (7.56) or (7.57) yields (2.26v) immediately when the closure condition (5.29) and the transformation (15.9) are used. To obtain the transformation of the body working, (7.70) is used first to obtain the transformed total phase time derivative of the energy in the form

$$(15.26) \qquad \frac{dE_k^{(\sharp)}(Q_k^{(\sharp)}, P_k^{(\sharp)}, t)}{dt} = \frac{dE_k^{(b)}(Q_k^{(b)}, P_k^{(b)}, t)}{dt} + V_A^{(b)} \cdot \frac{\partial \Phi_{k,x}^{(b)}(Q_k^{(b)}, t)}{\partial Q_k^{(b)}},$$

where $V_A^{(b)} = \times_{i \in k} v_a^{(b)}$ and (4.20a) was used to show $V_A^{(b)} \cdot (\partial \Phi_{k,n}^{(b)}(Q_k)/\partial Q_k) = 0$. The local phase averaging formula (15.21), the definition (7.15) for the external force density, and the definition (7.73) of the body force are then applied to obtain

$$(15.27) \qquad W_{k,b}^{(\sharp)}(q^{(\sharp)}, t) = W_{k,b}^{(b)}(q^{(b)}, t) - v_a^{(b)} \cdot \mathcal{F}_{k,x}^{(b)}(q^{(b)}, t).$$

This demonstrates (2.26u).

It follows that the Galilean transformations of the phase averaged quantities exhibited in Section 2.4 are the same as the phase averages of the Galilean transformations of the phase quantities. This means that the diagram in Figure 15.2. is commutative.

15.7 The Invariance of the Entropy

Transformations of the purely thermodynamic quantities entropy and temperature differ from the transformations of the mechanical quantities. Because mechanical quantities are particle sum functions, they can be "seated" in the individual particles themselves as properties. The purely thermodynamic quantities, on the other hand, depend explicitly on the distribution and therefore require separate consideration.

The frame-independence of the k system probability distribution density $F_k(Q_k, P_k, t)$, established in (15.19), stated that $F_k^{(\sharp)}(Q_k^{(\sharp)}, P_k^{(\sharp)}, t) = F_k^{(b)}(Q_k^{(b)}, P_k^{(b)}, t)$ whenever $(Q_k^{(\sharp)}, P_k^{(\sharp)}) = T_A(Q_k^{(b)}, P_k^{(b)})$. It follows immediately that

$$(15.28) \quad S_k^{(\sharp)}(Q_k^{(\sharp)}, P_k^{(\sharp)}, t) = [T_A^{(\mathrm{ind})} S_k^{(b)}](Q_k^{(b)}, P_k^{(b)}, t) = S_k^{(b)}(Q_k^{(b)}, P_k^{(b)}, t).$$

It also follows from this and the previous results that

$$(15.29)$$

$$
\begin{aligned}
S_k^{(\sharp)}(t) &= \left\langle S_k^{(\sharp)}(Q_k^{(\sharp)}, P_k^{(\sharp)}, t) \right\rangle_{kt}^{(\sharp)} \\
&= \frac{1}{Z_k^{(\sharp)}(t)} \int dQ_k^{(\sharp)} \int dP_k^{(\sharp)} \, F_k^{(\sharp)}(Q_k^{(\sharp)}, P_k^{(\sharp)}, t) \chi_{\Omega_k^{(\sharp)}(t)}(Q_k^{(\sharp)}, P_k^{(\sharp)}) \\
&\quad \times S_k^{(\sharp)}(Q_k^{(\sharp)}, P_k^{(\sharp)}, t) \\
&= \frac{|a_r|^{6N_k}}{Z_k^{(b)}(t)} \int dQ_k^{(b)} \int dP_k^{(b)} \, F_k^{(b)}(Q_k^{(b)}, P_k^{(b)}, t) \chi_{\Omega_k^{(b)}(t)}(Q_k^{(b)}, P_k^{(b)}) \\
&\quad \times S_k^{(b)}(Q_k^{(b)}, P_k^{(b)}, t) \\
&= \left\langle S_k^{(b)}(Q_k^{(b)}, P_k^{(b)}, t) \right\rangle_{kt}^{(b)} = S_k^{(b)}(t) = S_k(t).
\end{aligned}
$$

This shows that the thermodynamic entropy is independent of the frame and verifies (2.26α).

15.8 The Pressure and Thermature

From a macroscopic perspective, the pressure is connected with the mechanical balances in a system in that a system boundary is fixed at a location if, in the absence of other forces acting in the boundary itself, such as surface tension, the pressures on both sides of the boundary are equal. However, in a Galilean reference frame moving at a constant velocity orthogonal to the boundary, this relation will be viewed differently. To see this, consider two systems sharing a plane boundary. The pressures of each system can be computed in a reference frame at rest with respect to the boundary. Suppose these pressures are balanced and there are no other forces that would keep the particles on both sides of the boundary from moving. In a frame moving at a constant velocity that is orthogonal to the boundary, the boundary will be moving but not accelerating. This means that the pressures, as force per unit area, should still be balanced at the boundary from the point of view of the moving reference frame.

Suppose next that the internal virial is transformed to a new velocity frame that is translated and rotated with respect to the original frame. The internal coordinate virial component of the pressure, $\Xi_{k,n}(Q_k) = Q_k \cdot (\partial \Phi_{k,n}(Q_k)/\partial Q_k)$, is invariant under this transformation. This follows from the fact that $\Phi_{k,n}(Q_k)$ depends on $|q_i - q_j|$, $|T_a q_i - T_a q_j| = |q_i - q_j|$, and the easily demonstrated result that

$$(15.30) \qquad Q_k \cdot \frac{\partial \Phi_{k,n}(Q_k)}{\partial Q_k} = \frac{1}{2} \sum_{i \in k} \sum_{j \in k-i} (q_i - q_j) \cdot \frac{\partial \phi_{ij}(q_i - q_j)}{\partial q_i}.$$

This implies in turn that
(15.31)

$$\Xi_{k,n}^{(\sharp)}(Q_k^{(\sharp)}, t) = Q_k^{(\sharp)} \cdot \frac{\partial \Phi_{k,n}^{(\sharp)}(Q_k^{(\sharp)})}{\partial Q_k^{(\sharp)}} = Q_k^{(b)} \cdot \frac{\partial \Phi_{k,n}^{(b)}(Q_k^{(b)})}{\partial Q_k^{(b)}} = \Xi_{k,n}^{(b)}(Q_k^{(b)}, t).$$

Let us next extend these considerations to $\Xi_{i,n}^{\mu\nu}(Q_k, t)$. The transformation from the b to the $\sharp$ frame is expressed by

$$(15.32) \qquad \Xi_{i,n}^{(\sharp)\mu\nu}(Q_k^{(\sharp)}) = a_{r,\xi}^\mu a_{r,\eta}^\nu \left[(q_i^{(b),\xi} - v_a^{(b)\xi} t - q_a^{(b)\xi}) \frac{\partial \Phi_{i,n}^{(b)}(Q_k^{(b)})}{\partial q_{i,\eta}^{(b)}} \right].$$

Using (4.20a), it is easy to show that $\sum_{i \in k}(v_a^{(b)\xi} t + q_a^{(b)\xi})(\partial \Phi_{i,n}^{(b)}(Q_k^{(b)})/\partial q_{i,\eta}^{(b)}) = 0$. Summing (15.32) over $i \in k$ therefore yields immediately

$$(15.33) \qquad \Xi_{k,n}^{(\sharp)\mu\nu}(Q_k^{(\sharp)}) = a_{r,\xi}^\mu a_{r,\eta}^\nu \sum_{i \in k} \left[q_i^{(b),\xi} \frac{\partial \Phi_{i,n}^{(b)}(Q_k^{(b)})}{\partial q_{i,\eta}^{(b)}} \right].$$

The local phase average of the internal virial is obtained from this result in the form

$$(15.34) \quad \Xi_{k,n}^{(\sharp)\mu\nu}(q^{(\sharp)}, t) = -a_{r,\xi}^\mu a_{r,\eta}^\nu q^{(b)\xi} \mathcal{F}_{k,n}^{(b)\eta}(q^{(b)}, t) = a_{r,\xi}^\mu a_{r,\eta}^\nu \Xi_{k,n}^{(b)\xi\eta}(q^{(b)}, t),$$

which verifies (2.26x).

It was observed above that a Galilean transformation, viewed as a frame transformation, must affect all of phase space, so the external forces after the transformation must be the same as before the transformation because the physical situation is the same. Moreover, because the particle theory is invariant under coordinate translations, the thermodynamic version must be invariant under these translations as well. The macroscopic choice of where to put boundaries between systems, however, establishes a macroscopic reference frame that does not depend on relative particle coordinates. This will spoil the translation invariance of the theory unless special precautions are taken when dealing with the external virial and external forces.

To illustrate this, consider the transformation of the i-particle external virial tensor $\Xi_{i,x}^{\mu\nu}(q_i, t)$ from the $\flat$ to the $\sharp$ frame. A naive application of the transformation results in the relation

(15.35)

$$
\Xi_{i,x}^{(\sharp)}(q_i^{(\sharp)}, t) = (q_i^{(\flat)} - v_a^{(\flat)}t - q_a^{(\flat)}) \cdot \frac{\partial \Phi_{i,x}^{(\flat)}(q_i^{(\flat)}, t)}{\partial q_i}
$$

$$
= \Xi_{i,x}^{(\flat)}(q_i^{(\flat)}, t) - (v_a^{(\flat)}t + q_a^{(\flat)}) \cdot \frac{\partial \Phi_{i,x}^{(\flat)}(q_i^{(\flat)}, t)}{\partial q_i}.
$$

The last term on the right of (15.35) is the apparent work done on particle i due to the external forces exerted by the particles in other systems because of the coordinate translation $v_a^{(\flat)}t + q_a^{(\flat)}$ in the transformation. Its local phase average is $(v_a^{(\flat)\xi}t + q_a^{(\flat)\xi}) \cdot F_{i,x}^{(\flat)}(q, t)$. However, as mentioned above, this result does not take account of the fact that the locations of the particles exerting the forces in the l system have also been transformed into the same new frame and exert an equal and opposite force at the boundary. This gives rise to a term $-(v_a^{(\flat)\xi}t + q_a^{(\flat)\xi}) \cdot F_{i,x}^{(\flat)}(q, t)$ that represents work done on the system that cancels the apparent work done by the system.

To illustrate this cancellation, assume for simplicity that there are only two systems k and l. For these systems taken together, the analogs for the external forces become internal forces and the previous calculation applies. In terms of the external virial formalism itself, it is easy to show

(15.36)

$$
\Xi_{k,x}^{(\sharp)}(Q_k^{(\sharp)}, t) + \Xi_{l,x}^{(\sharp)}(Q_l^{(\sharp)}, t) = Q_k^{(\sharp)} \cdot \frac{\partial \Phi_{k,x}^{(\sharp)}(Q_k^{(\sharp)}, t)}{\partial Q_k} + Q_l^{(\sharp)} \cdot \frac{\partial \Phi_{l,x}^{(\sharp)}(Q_l^{(\sharp)}, t)}{\partial Q_l}
$$

$$
= \sum_{i \in k} \sum_{j \in l} \left[q_i^{(\sharp)} \cdot \frac{\partial \phi_{ij}(q_i^{(\sharp)} - q_j^{(\sharp)})}{\partial q_i} + q_j^{(\sharp)} \cdot \frac{\phi_{ji}(q_i^{(\sharp)} - q_j^{(\sharp)})}{\partial q_j} \right]
$$

$$
= \sum_{i \in k} \sum_{j \in l} (q_i^{(\sharp)} - q_j^{(\sharp)}) \cdot \frac{\partial \phi_{ij}(q_i^{(\sharp)} - q_j^{(\sharp)})}{\partial q_i}
$$

$$
= \sum_{i \in k} \sum_{j \in l} (q_i^{(\flat)} - q_j^{(\flat)}) \cdot \frac{\partial \phi_{ij}(q_i^{(\flat)} - q_j^{(\flat)})}{\partial q_i} = \Xi_{k,x}^{(\flat)}(Q_k^{(\flat)}, t) + \Xi_{l,x}^{(\flat)}(Q_l^{(\flat)}, t).
$$

It follows from this that the sum $\Xi_{k,x}(t) + \Xi_{l,x}(t)$ is invariant under a Galilean transformation. This calculation also demonstrates that the k external virial analog $\Xi_{k,x}(Q_k, t)$ preserves its form $Q_k \cdot (\partial \Phi_{k,x}(Q_k, t)/\partial Q_k)$ under the transformation. In addition, $\Xi_{k,x}^{(\sharp)}(t) = \Xi_{k,x}^{(b)}(t)$ for each $k \in [t]$ is valid as well. A similar calculation shows that $\Xi_{k,x}^{(\sharp)\mu\nu}(q^{(\sharp)}, t) = a_{r,\xi}^{\mu} a_{r,\eta}^{\nu} \Xi_{k,x}^{(b)\xi\eta}(q^{(b)}, t)$ for the transformation of the external virial tensor, which demonstrates (2.26y). A similar argument was given above to show that the transformation of external forces satisfies equation (2.26m).

One of the major reasons for selecting a special reference frame in which to define the pressure and thermature was that the volume derivative operator, used in their definitions, does not give the same result in two Galilean frames that differ by a velocity translation (velocity boost) transformation. To show this, let us use the definition of the volume derivative in terms of the Λ transformation and consider a Galilean velocity boost transformation from the b frame to the $\sharp$ frame. Computing the momentum virial derivative of the kinetic energy in each frame gives the result

$$(15.37) \quad \lim_{\lambda \to 1} \frac{\partial K_i(\lambda^{-1} p_i^{(\sharp)})}{\partial \lambda} = -2K_i(p_i^{(\sharp)}) \neq -2K_i(p_i^{(b)}) = \lim_{\lambda \to 1} \frac{\partial K_i(\lambda^{-1} p_i^{(b)})}{\partial \lambda},$$

where $p_i^{(\sharp)\mu} = a_{r,\nu}^{\mu}[p_i^{(b)\nu} - m_i v_a^{(b)\nu}]$, $|v_a^{(b)}| > 0$, and $[T_a^{(\mathrm{ind})} K_i] = K_i(p_i^{(b)}) - p_i^{(b)} \cdot v_a^{(b)} + K_i(p_a^{(b)})$ with $K_i(p_a^{(b)}) = (m_i[v_a^{(b)}]^2/2)$.

This problem is avoided in TIS by always calculating the kinetic virial derivative in the system rest frame. The local thermodynamic pressure $P_k(q, t)$ is fixed at a definite value by the calculation performed in the system rest frame and is therefore invariant under Galilean transformations. This result confirms the pressure transformation (2.26z). A similar argument establishes the validity of the pressure tensor transformation (2.26w).

For similar reasons the local thermature, which is based on the pressure and the volume derivative of the entropy, is unaffected by a Galilean transformation. This confirms (2.26β) and (2.26γ). The definition of the thermature for F_k^{δ} states, stated in the formula (12.28), is also required to be defined in the system rest frame and is therefore not subject to change by a Galilean transformation either.

15.9 Canonical Transformations

Under a Galilean frame transformation, the coordinates and velocities in every system are transformed simultaneously. By the definitions (4.7), (4.8), and the invariance of $r_{ij} = |q_i - q_j|$ for each $i, j \in k$, the invariance of $\Phi_i(Q_k, t)$ follows. It is easy, using the transformations for the components defined above, to show that the transformation of the Hamiltonian $\mathcal{H}_k$ is canonical, i.e., that $\mathcal{H}_k(Q_k^{(\sharp)}, P_k^{(\sharp)}, t)$ satisfies the Hamiltonian equations (4.14) or (4.15) in the $\sharp$ frame if $\mathcal{H}_k(Q_k^{(b)}, P_k^{(b)}, t)$ satisfies the Hamiltonian equations in the b frame. A formal demonstration can also be given using a generating function for the transformation similar to the one discussed next for the dilation transformation.

Let us reconsider the dilation transformation introduced in Chapter 10 and the claim that it is a canonical transformation. The transformation $Q_k \to \Lambda Q_k$, used originally to obtain the volume phase derivative operator by parametric differentiation, was extended to the combined dilation transformation $(Q_k, P_k) \to (\Lambda Q_k, \Lambda^{-1} P_k)$ because the former transformation is not canonical.

Let us pursue the difference between a canonical and a non-canonical transformation by comparing the canonical transformation of (Q_k, P_k) to the corresponding transformation of $(Q_k, \dot{Q}_k)$. The velocity of particle i is defined as the tangent vector to the coordinate trajectory $q_i(t)$ at some time t. Under the Λ transformation, it is transformed from $v_i = dq_i(t)/dt$ to $\lambda_i v_i = \lambda_i dq_i(t)/dt$. This change of scale for the velocity from v_i to $\lambda_i v_i$ means that the particle will traverse the distance $\lambda_i |q_{i,2} - q_{i,1}|$ in the same time interval t_{12} that the untransformed velocity v_i traverses the untransformed distance $|q_{i,2} - q_{i,1}|$. The difference between the canonical Λ transformation $(Q_k, P_k) \to (\Lambda_k Q_k, \Lambda_k^{-1} P_k)$ and the Λ transformation $(Q_k, \dot{Q}_k) \to (\Lambda_k Q_k, \Lambda_k \dot{Q}_k)$ points up an important distinction between a canonical and a non-canonical variable in Hamiltonian mechanics as illustrated by the difference between the transformation of the momentum p_i and the transformation of the mass times the velocity $m_i v_i$.

The momentum p_i is the cotangent vector to the trajectory at time t, which means that it is a function of the tangent vector. In the Hamiltonian formulation of mechanics, the momentum is independent of the coordinates and can be varied separately, whereas in the Lagrangian formulation the variation of the velocity $\dot{q}_i$ is given by the time derivative of the variation of the coordinate q_i. The respective Λ transformations of P_k and $\dot{Q}_k$ reflect this difference.

Preserving the Hamiltonian basis of the theory under a transformation $(Q_k, P_k) \to (Q_k^\star, P_k^\star)$ means that there is a Legendre transformation from the Hamiltonian function $\mathcal{H}_k(Q_k, P_k, t)$ to a new Hamiltonian $\mathcal{H}_k^\star(Q_k^\star, P_k^\star, t)$ expressed in the new coordinates, such that Hamilton's equations $\dot{Q}_k^\star = \partial \mathcal{H}_k^\star/\partial P_k^\star$ and $\dot{P}_k^\star = -\partial \mathcal{H}_k^\star/\partial Q_k^\star$ are valid. It is shown in many standard works that the canonical transformation from (Q_k, P_k) to $(Q_k^\star, P_k^\star)$ is given by a Legendre transformation expressed in terms of a suitable generating function that connects the two sets of coordinates.[4]

A generating function for the canonical dilation transformation from coordinates (Q_k, P_k) to $(Q_k^\star, P_k^\star)$ is given by

(15.38) $$G_2(Q_k, P_k^\star) = \lambda Q_k \cdot P_k^\star.$$

---

[4]Legendre transformations are discussed further in EIS.

Following the notation and formalism of Goldstein [1950], pp. 237–243, it follows for each $i \in k$ that

(15.39a)
$$p_i = \frac{\partial G_2(Q_k, P_k^\star)}{\partial q_i} = \lambda p_i^\star,$$

(15.39b)
$$q_i^\star = \frac{\partial G_2(Q_k, P_k^\star)}{\partial p_i^\star} = \lambda q_i,$$

Solving this set of equations for the new coordinates $(Q_k^\star, P_k^\star)$ in terms of the old coordinates yields $(Q_k^\star, P_k^\star) = (\Lambda Q_k, \Lambda^{-1} P_k)$, which is the form of the transformation used in Chapter 10.

15.10 Thermodynamics in a Non-Inertial Frame

The well-known fact that the rest frame in a terrestrial laboratory is not an inertial frame was mentioned above. From the standpoint of the frame defined by the sun, the earth is rotating about its own axis and is also revolving around the sun. The sun itself is rotating about its axis and is revolving around the center of our Galaxy in the frame of the "fixed stars".

Since the eighteenth century, experiments on earth have been used to demonstrate that a rest frame on the surface of the earth is not an inertial frame. The nineteenth century gyroscope experiments and the pendulum experiments of Léon Foucault are the most famous of these. The rotation the earth is mirrored in the rotation of the plane of oscillation of a Foucault pendulum about a circle during a twenty-four hour period of the earth's rotation. This can be observed directly without any need to refer to the sun or the fixed stars. Another demonstration of the earth's rotation is provided by the fact that a plumb line does not point towards the center of the earth. The centrifugal force due to the earth's rotation acts on the bob at the end of the line to pull it away from the direction of the center of the earth. The centripetal acceleration ranges up to about 0.3% of the acceleration of gravity depending on the latitude of the experiment.[5]

A question connected with thermodynamics in a non-inertial frame came up in the middle of the nineteenth century—although it was not framed in these terms at the time. An objection was raised by Joseph Loschmidt [1876] to the conclusion that the temperature of a vertical column of gas in a gravitational field is independent of the altitude at which it is measured. The barometer formula stated that the pressure at an altitude $z \geq 0$ above sea level is given in terms of the pressure $P(0)$ at sea level by $P(z) = P(0)e^{-mgz/k_B T}$. Loschmidt felt that the temperature should vary with height and that this barometer formula could not be correct.

---

[5]The ratio of the centrifugal force due to the earth's revolution about the sun to the centrifugal force due to its rotation is about 0.2. These numbers and some of the formalism presented in the discussion below are based on Goldstein [1950], pp. 132–140.

This claim of Loschmidt conflicted with the results of calculations that showed intensive variables in equilibrium thermodynamics have a constant value at different points in the system volume. Investigating Loschmidt's claim, Bailyn [**1994**], pp. 231–232, reviewed Gibbs variational proof that the intensive variables in a system at equilibrium are uniform. To show the uniformity of the intensive variables representing the pressure, temperature, and chemical potential of an equilibrium system, Gibbs had considered an isolated macroscopic thermodynamic system, divided it into two subsystems, and used a variational method. He computed the total variation of the system energy by varying the energy in each subsystem separately with respect to volume, entropy, and particle number—subject to the constraints that the total variation of the volume, entropy, and particle number vanish. Gibbs applied his stability principle for the system energy, which states for the total energy variation $\delta^T U$ that $\delta^T U \geq 0$, to the resulting variation equations. Using the fact that the variations of the individual variables are arbitrary, Gibbs then demonstrated that the pressure, temperature, and chemical potential, must be the same in both subsystems.

After discussing Gibbs' [**1875**] proof of the uniformity of intensive thermodynamic variables, Bailyn, pp. 254–256, later echoed Loschmidt's objections and raised the question of whether the variation of the equilibrium pressure with height in a gravitational field violates Gibbs' result that the pressure is uniform in an equilibrium system. He observed that Gibbs had extended his thermodynamic variational method to include gravitational and electromagnetic fields to accommodate situations in which there are external fields. For the analysis of Loschmidt's objection in terms of Gibbs' extended formalism, Bailyn defined the z-axis as orthogonal to the surface of the earth and parallel to a radius vector from the earth's center. Gibbs' extension in this case requires using the local interaction energy in the form $\phi^{(G)}(q,t)\mu_k(q,t)$, where $\phi^{(G)}(q,t) = gq_z$ is the gravitational potential energy along the z-axis and $\mu_k(q,t)$ is the local mass density for the k system. Integrating this interaction energy density over the volume of the system gives the total interaction energy. Gibbs then recomputed the total variation of the system energy with this added component, but this time worked locally with infinitesimal volume components. He considered a mixture of component particle types indexed by a and having a mass m_a. The mass density of particle type a is $\mu_a(q,t) = m_a \nu_a(q,t)$ and the total local mass density is $\mu_k(q,t) = \sum_a \mu_a(q,t)$.

Gibbs expressed the infinitesimal energy change in an infinitesimal volume of one of these elements by $du_k = T_k ds_k + \Sigma_a \mu_a^{(\text{fe})} dn_a$, where u_k is the local energy density, s_k is the local entropy density, $\mu_a^{(\text{fe})}$ is the chemical potential for particles of type a, and n_a is the number of particles of type a. Applying his stability principle to the total variation, and keeping the gravitational potential fixed because it is not part of the system,

Gibbs was able to show that $T_k = $ const., so the temperature is uniform as before. He also showed that the chemical potential is modified so that $\mu_k^{(\text{fe})(\text{tot})} = \mu_k^{(\text{fe})} + \phi_k^{(G)}(q,t)\partial\mu_k(q,t)/\partial n_a = $ const.

Using the Thomson-Massieu (Gibbs-Duhem) relation $d\mathrm{P}_k = s_k dT_k + \sum_a n_a d\mu_a^{(\text{fe})}$, Bailyn answered the question he had raised above concerning the pressure. Assuming $dT_k = 0$ in this case and that $d\mu_k^{(\text{fe})}(q,t) = -m_a d\phi_k^{(G)}(q,t)$, it follows that $d\mathrm{P}_k = -\mu_k(q,t)\phi_k^{(G)}(q,t)$ and, when specialized to the coordinate system above, this leads to the barometer formula by integration. The same method can be used with an electric field by defining the local interaction energy density as $e\nu_k(q,t)\phi_k^{(em)}(q,t)$ and proceeding as above.

The implications for TIS of the fact that we are never in an inertial frame on earth has been skirted up to now. The question arises as to whether the theory presented in this book, and the versions of thermodynamics and statistical mechanics presented by other authors, are devoid of application on earth, in the neighborhood of the sun, or inside the Milky Way galaxy. A simple inquiry will suffice here to show how the formulas valid in an inertial frame can be extended in some cases to non-inertial frames. A more complete answer requires the abandonment of Newtonian mechanics and the use of relativity theory.[6]

Let us consider a body k that is rotating about a fixed axis with a constant angular velocity $\omega_k^{(\natural)}$ in a fixed coordinate frame $\natural$ that is presumed to be inertial. The $\flat$ coordinate system is a system rest frame that is at rest in the rotating body. A reference frame located in a laboratory on the surface of the earth will serve as our example. The effect of the earth's rotation on particles moving on the surface of the earth is to cause their paths to curve away from the expected Newtonian path due to the frame-dependent Coriolis and centrifugal forces. This change in the particle paths is interpreted as a force, but one which does not originate in interactions with other particles and can therefore not be represented by Newton's second law. These non-inertial forces cannot be transformed away by a Galilean transformation because a Galilean transformation cannot be used to transform a system into an accelerated frame. In short, a Galilean transformation cannot transform a non-inertial frame into an inertial frame.

Determining the interaction energy between the particles and the fields and making this interaction energy part of the total Hamiltonian energy solved the problem of including gravitation and electromagnetic fields in the thermodynamic formalism. This same procedure can be used for the terrestrial laboratory in the rotating frame of the earth. To show this, let us consider the description of the motion of a particle in both the $\natural$ inertial frame of the fixed stars and the $\flat$ frame fixed in the rotating earth. The particle

---

[6]That generalization will be taken up in RIS.

velocity $v_i^{(\natural)\mu}$ in the $\natural$ frame is expressed in terms of the rotating $\flat$ frame using the angular velocity $\omega_{k,\nu}^{(\natural)}$ by

$$(15.40) \qquad v_i^{(\natural)\mu} = v_i^{(\flat)\mu} + \epsilon^{\mu\nu\xi}\omega_{k,\nu}^{(\natural)}q_{i,\xi}^{(\flat)}.$$

The relation between the total time derivatives of the vector quantity $G_k^\mu(t)$ in the two coordinate systems is given by

$$(15.41) \qquad \frac{d_\natural G_k^\mu(t)}{dt} = \frac{d_\flat G_k^\mu(t)}{dt} + \epsilon^{\mu\nu\xi}\omega_{k,\nu}^{(\natural)}G_{k,\xi}(t),$$

where, in an obvious notation, $d_\natural/dt$ is the total time derivative in the $\natural$ inertial frame and $d_\flat/dt$ is the total time derivative in the $\flat$ rotating frame.

   This formalism is used to compute the acceleration in the inertial frame of a particle located on the surface of the earth in terms of its acceleration in the body frame of the earth. The body frame is centered at the center of the earth and all coordinates are measured in this frame. Using (15.40), the acceleration of a particle i in the $\natural$ frame is expressed in terms of quantities in the body frame $\flat$ by

$$(15.42)$$

$$a_i^{(\natural)\mu} = \frac{d_\natural v_i^{(\natural)\mu}}{dt} = \frac{d_\flat v_i^{(\flat)\mu}}{dt} + \epsilon^{\mu\nu\xi}\omega_{k,\nu}^{(\natural)}v_{i,\xi}^{(\flat)}$$

$$= a_i^{(\flat)\mu} + 2\epsilon^{\mu\nu\xi}\omega_{k,\nu}^{(\natural)}v_{i,\xi}^{(\flat)} + \epsilon^{\mu\nu\xi}\omega_{k,\nu}^{(\natural)}\epsilon_{\xi\rho\sigma}\omega_k^{(\natural)\rho}q_i^{(\flat)\sigma}.$$

In accord with Newton's second law, the 3-dimensional i-particle force and $3\mathcal{N}_k$-dimensional k force in the $\natural$ frame associated with the acceleration $a_{i,\natural}^\mu$ are

$$(15.43a) \qquad F_i^{(\natural)\mu} = m_i a_i^{(\natural)\mu},$$

$$(15.43b) \qquad F_k^{(\natural)\mu} = \times_{i\in k} F_i^{(\natural)\mu}.$$

The 3-dimensional i-particle effective force, experienced by the particle in the rotating system, the $3\mathcal{N}_k$-dimensional k effective force, and the 3-dimensional total effective force are expressed as

$$(15.44a)$$

$$F_i^{(\flat)\mu} = F_i^{(\natural)\mu} - 2\epsilon^{\mu\nu\xi}\omega_{k,\nu}^{(\natural)}p_{i,\xi}^{(\flat)} - m_i\epsilon^{\mu\nu\xi}\omega_{k,\nu}^{(\natural)}\epsilon_{\xi\rho\sigma}\omega_k^{(\natural)\rho}q_i^{(\flat)\sigma},$$

$$(15.44b)$$

$$F_k^{(\flat)\mu} = \times_{i\in k} F_i^{(\flat)\mu},$$

$$(15.44c)$$

$$\mathcal{F}_k^{(\flat)\mu} = \sum_{i\in k} F_i^{(\flat)\mu}.$$

The last term on the right in (15.44a) is the *i-particle centrifugal force*. It, and the corresponding *k centrifugal force*, are written in the $\flat$ frame as

(15.45a)
$$F_{i,f}^{(\text{cf})\mu}(q_i^{(\flat)}) = -m_i \epsilon^{\mu\nu\xi} \omega_{k,\nu}^{(\natural)} \epsilon_{\xi\rho\sigma} \omega_k^{(\natural)\rho} q_i^{(\flat)\sigma}$$
$$= -m_i [\omega_k^{(\natural)\mu}(\omega_k^{(\natural)} \cdot q_i^{(\flat)}) - q_i^{(\flat)\mu}(\omega_k^{(\natural)})^2],$$

(15.45b)
$$F_{k,f}^{(\text{cf})\mu}(Q_k^{(\flat)}) = \times_{i \in k} F_{i,f}^{(\text{cf})\mu}(q_i^{(\flat)}).$$

The label f refers to the fact that this is a frame-dependent quantity. The result $q_i \cdot F_{i,f}^{(\text{cf})}(q_i) \geq 0$ shows that $F_{i,f}^{(\text{cf})}(q_i)$ is directed outward along the radius vector from the center of the earth. In the $\flat$ coordinate system, $F_{i,f}^{(\text{cf})}(q_i)$ has the magnitude $m_i |q_i| [\omega_k^{(\natural)}]^2 \sin\theta$, where θ is the angle between the radius vector q_i and the north pole of the earth.

Corresponding to this force, an *i-particle centrifugal potential* and a *k centrifugal potential* can be defined in the $\flat$ frame in the form

(15.46a)
$$\Phi_{i,f}^{(\text{cf})}(q_i^{(\flat)}) = \tfrac{1}{2} m_i \left\{ (\omega_k^{(\natural)} \cdot q_i^{(\flat)})^2 - [q^{(\flat)}]_i^2 (\omega_k^{(\natural)})^2 \right\},$$

(15.46b)
$$\Phi_{k,f}^{(\text{cf})}(Q_k^{(\flat)}) = \sum_{i \in k} \Phi_{i,f}^{(\text{cf})}(q_i^{(\flat)}).$$

When $v_i^{(\flat)} = 0$, the centrifugal force is the only non-inertial force. When $v_i^{(\flat)} \neq 0$, there is an additional force called the Coriolis force. The frame-dependent *i-particle Coriolis force* and the *k Coriolis force* are defined in terms of the frame angular velocity $\omega_{k,\nu}^{(\natural)}$ in the $\flat$ frame by

(15.47a)
$$F_{i,f}^{(\text{Co})\mu}(p_i^{(\flat)}) = -2\epsilon^{\mu\nu\xi} \omega_{k,\nu}^{(\natural)} p_{i,\xi}^{(\flat)},$$

(15.47b)
$$F_{k,f}^{(\text{Co})\mu}(P_k^{(\flat)}) = \sum_{i \in k} F_i^{(\text{Co})\mu}(q_i^{(\flat)}, p_i^{(\flat)}).$$

The Coriolis force is orthogonal to both $\omega_k^{(\natural)\mu}$ and $p_i^{(\flat)\mu}$ and points in opposite directions in the northern and southern hemispheres of the earth.

The Coriolis force can also be obtained in the $\flat$ frame from a potential called the *i-particle Coriolis potential* and the *k Coriolis potential* that are written

(15.48a)
$$\Phi_{i,f}^{(\text{Co})}(q_i^{(\flat)}, p_i^{(\flat)}) = 2\epsilon^{\mu\nu\xi} q_{i,\mu}^{(\flat)} \omega_{k,\nu}^{(\natural)} p_{i,\xi}^{(\flat)},$$

(15.48b)
$$\Phi_{k,f}^{(\text{Co})}(Q_k^{(\flat)}, P_k^{(\flat)}) = \sum_{i \in k} \Phi_{i,f}^{(\text{Co})}(q_i^{(\flat)}, p_i^{(\flat)}).$$

The upshot of this analysis is that the equation of motion for a particle in the non-inertial rotating $\flat$ frame considered here can be expressed in terms of a potential that depends on both the particle velocity and its location. In this case, the *i-particle frame potential* and the *k frame potential* can be represented as the sum of the Coriolis potential and the centrifugal potential by[7]

(15.49a)
$$\Phi_{i,f}(q_i^{(b)}, p_i^{(b)}) = \Phi_{i,f}^{(Co)}(q_i^{(b)}, p_i^{(b)}) + \Phi_{i,f}^{(cf)}(q_i^{(b)}),$$

(15.49b)
$$\Phi_{k,f}(Q_k^{(b)}, P_k^{(b)}) = \sum_{i \in k} \Phi_{i,f}(q_i^{(b)}, p_i^{(b)}).$$

The *i-particle total potential* and *k total potential* in the rotating non-inertial frame (nif) are

(15.50a)
$$\Phi_{i,f}^{(nif)}(Q_k^{(b)}, p_i^{(b)}, t) = \Phi_{i,n}(Q_k^{(b)}) + \Phi_{i,x}(q_i^{(b)}, t) + \Phi_{i,f}(q_i^{(b)}, p_i^{(b)}).$$

(15.50b)
$$\Phi_{k,f}^{(nif)}(Q_k^{(b)}, P_k^{(b)}, t) = \Phi_{k,n}(Q_k^{(b)}) + \Phi_{k,x}(Q_k^{(b)}, t) + \Phi_{k,f}(Q_k^{(b)}, P_k^{(b)}).$$

The $3\mathcal{N}_k$-dimensional *k force* vector and 3-dimensional *k total force* vector associated with this potential are

(15.51a)
$$F_{k,f}^{(nif)\mu}(Q_k^{(b)}, P_k^{(b)}, t) = -\frac{\partial \Phi_{k,f}^{(nif)}(Q_k^{(b)}, P_k^{(b)}, t)}{\partial Q_{k,\mu}},$$

(15.51b)
$$\mathcal{F}_{k,f}^{(nif)\mu}(Q_k^{(b)}, P_k^{(b)}, t) = -\sum_{i \in k} \frac{\partial \Phi_{k,f}^{(nif)}(Q_k^{(b)}, P_k^{(b)}, t)}{\partial q_{i,\mu}}.$$

Computing the derivative in (15.51b) yields the effective force in the ♭ frame displayed in (15.44).

In this simple example, the frame potential $\Phi_{k,f}(Q_k^{(b)}, P_k^{(b)})$ defined in (15.49) appears in (15.50) as a potential field that can be viewed as "distorting" the coordinate frame in such a way that the noninertial force $F_{k,f}^{(nif)\mu}$ experienced by particle i is the negative gradient of the total potential that includes this potential. In that sense, the potential $\Phi_{k,f}(Q_k^{(b)}, P_k^{(b)})$ gives rise to a force field that allows the use of Newtonian mechanics in a non-inertial frame.

---

[7] Velocity dependent potentials also play a role in electromagnetic theory. These are discussed in EIS.

The *i-particle Hamiltonian* and *k Hamiltonian* functions in this non-inertial frame are written

(15.52a)
$$\mathcal{H}_i^{(\mathrm{nif})}(Q_k^{(\flat)}, p_i^{(\flat)}, t) = \mathcal{H}_i(Q_k^{(\flat)}, p_i^{(\flat)}, t) + \Phi_{i,f}(q_i^{(\flat)}, p_i^{(\flat)}),$$

(15.52b)
$$\mathcal{H}_k^{(\mathrm{nif})}(Q_k^{(\flat)}, P_k^{(\flat)}, t) = \mathcal{H}_k(Q_k^{(\flat)}, P_k^{(\flat)}, t) + \Phi_{k,f}(Q_k^{(\flat)}, P_k^{(\flat)}).$$

The derivative of $\mathcal{H}_k^{(\mathrm{nif})}(Q_k^{(\flat)}, P_k^{(\flat)}, t)$ with respect to $Q_k^{(\flat)}$ is

(15.53)
$$\frac{\partial \mathcal{H}_k^{(\mathrm{nif})}(Q_k^{(\flat)}, P_k^{(\flat)}, t)}{\partial Q_{k,\mu}^{(\flat)}} = \times_{i \in k} \left[\frac{\partial \Phi_k(Q_k^{(\flat)}, t)}{\partial q_{i,\mu}^{(\flat)}} + \frac{\partial \Phi_{k,f}(Q_k^{(\flat)}, P_k^{(\flat)})}{\partial q_{i,\mu}^{(\flat)}} \right]$$

$$= \times_{i \in k} \left[\frac{\partial \Phi_k(Q_k^{(\flat)}, t)}{\partial q_{i,\mu}^{(\flat)}} + 2\epsilon^{\mu\nu\xi} \omega_{k,\nu}^{(\natural)} p_{i,\xi}^{(\flat)} \right.$$

$$\left. + m_i \left\{ \omega_k^{(\natural)\mu}(\omega_k^{(\natural)} \cdot q_i^{(\flat)}) - q_i^{(\flat)\mu}(\omega_k^{(\natural)})^2 \right\} \right].$$

Similarly, the derivative of $\mathcal{H}_k^{(\mathrm{nif})}(Q_k^{(\flat)}, P_k^{(\flat)}, t)$ with respect to $P_k^{(\flat)}$ is

(15.54)
$$\frac{\partial \mathcal{H}_k^{(\mathrm{nif})}(Q_k^{(\flat)}, P_k^{(\flat)}, t)}{\partial P_{k,\xi}^{(\flat)}} = V_k^{(\flat)\xi} + 2\epsilon^{\mu\nu\xi} Q_{k,\mu}^{(\flat)} \omega_{k,\nu}^{(\natural)}.$$

Using these results, the pressure analog is expressed in terms of the volume derivative of $H_k^{(\mathrm{nif})}(Q_k^{(\flat)}, P_k^{(\flat)}, t)$ in the laboratory system rest frame $\flat$ by

(15.55)
$$\mathrm{P}_k^{(\mathrm{nif})}(Q_k^{(\flat)}, P_k^{(\flat)}, t) = -\frac{d\mathcal{H}_k^{(\mathrm{nif})}(Q_k^{(\flat)}, P_k^{(\flat)}, t)}{d\Omega_k}$$

$$= \mathrm{P}_k(Q_k^{(\flat)}, P_k^{(\flat)}, t) + \frac{1}{3\delta_k(t)} \sum_{i \in k} m_i \left[[q_i^{(\flat)}]^2 (\omega_k^{(\natural)})^2 - (q_i^{(\flat)} \cdot \omega_k^{(\natural)})^2 \right].$$

This version of the pressure differs from that in an inertial frame by the terms that depend on $\omega_k^{(\natural)}$. These terms reflect the mechanical effects of the rotation in modifying the pressure.

The total time derivative operator in the non-inertial frame is

(15.56)

$$
\begin{aligned}
\frac{d_{(\mathrm{nif})}}{dt} &= \frac{\partial}{\partial t} + \frac{\partial \mathcal{H}_k^{(\mathrm{nif})}}{\partial P_k} \cdot \frac{\partial}{\partial Q_k} - \frac{\partial \mathcal{H}_k^{(\mathrm{nif})}}{\partial Q_k} \cdot \frac{\partial}{\partial P_k} \\
&= \frac{d}{dt} + 2\epsilon^{\mu\nu\xi} \sum_{i\in k} \left[q_{i,\mu}^{(\flat)} \omega_{k,\nu}^{(\natural)} \frac{\partial}{\partial q_i^\xi} - \omega_{k,\mu}^{(\natural)} p_{i,\nu}^{(\flat)} \frac{\partial}{\partial p_i^\xi} \right] \\
&\quad - \sum_{i\in k} m_i \left[\omega_k^{(\natural)\xi}(\omega_k^{(\natural)} \cdot q_i^{(\flat)}) - q_i^{(\flat)\xi}(\omega_k^{(\natural)})^2 \right] \frac{\partial}{\partial p_i^\xi}.
\end{aligned}
$$

This formula can be used with the results (15.53) and (15.54) to define a Poisson Bracket operator $V_{2k}^{(\mathrm{nif})} \cdot \nabla_{2k}$ if desired. Let us assume next for simplicity that $\partial \mathcal{H}_k(Q_k, P_k, t)/\partial t = 0$. Using the definition (15.56), it follows immediately that the total time derivative of the non-inertial Hamiltonian vanishes:

(15.57)
$$
\frac{d\mathcal{H}_k^{(\mathrm{nif})}(Q_k^{(\flat)}, P_k^{(\flat)}, t)}{dt} = 0.
$$

The derivatives of $J_k^\mu(Q_k^{(\flat)}, P_k^{(\flat)})$ with respect to $Q_k^{(\flat)}$ and $P_k^{(\flat)}$ are

(15.58a)
$$
\frac{\partial J_k^\mu(Q_k^{(\flat)}, P_k^{(\flat)})}{\partial Q_{k,\nu}^{(\flat)}} = \times_{i\in k} \epsilon^{\mu\nu\xi} p_{i,\xi}^{(\flat)},
$$

(15.58b)
$$
\frac{\partial J_k^\mu(Q_k^{(\flat)}, P_k^{(\flat)})}{\partial P_{k,\xi}^{(\flat)}} = \times_{i\in k} \epsilon^{\mu\nu\xi} q_{i,\nu}^{(\flat)},
$$

These results are used with (15.54) and (15.55) to show that the total time derivatives of the k total momentum and k total angular momentum are the total force and total torque

(15.59a)
$$
\frac{d_{(\mathrm{nif})} \mathcal{P}_k^{(\flat)\mu}}{dt} = -\sum_{i\in k} \frac{\partial \Phi_{k,f}^{(\mathrm{nif})}(Q_k^{(\flat)}, P_k^{(\flat)}, t)}{\partial q_{i,\mu}^{(\flat)}} = \mathcal{F}_{k,f}^{(\flat)(\mathrm{nif})\mu}(Q_k^{(\flat)}, P_k^{(\flat)}, t),
$$

(15.59b)
$$
\frac{d_{(\mathrm{nif})} J_k^\mu(Q_k^{(\flat)}, P_k^{(\flat)})}{dt} = -\sum_{i\in k} \epsilon^{\mu\nu\xi} q_{i,\nu}^{(\flat)} \frac{\partial \Phi_{k,f}^{(\mathrm{nif})}(Q_k^{(\flat)}, P_k^{(\flat)}, t)}{\partial q_i^{(\flat)\xi}}.
$$

The relation (15.53) implies that a stationary equilibrium state, written $\mathcal{H}_k^{(\mathrm{nif})\epsilon}(Q_k^{(\flat)}, P_k^{(\flat)})$, can be defined in a non-inertial frame. A statement of the requirements on the definition of an equilibrium state is given in (13.32) with the form (13.30) or (13.31) for the external potential. In Chapter 13, velocity dependent potentials were not considered, so it is necessary to evaluate whether the potential $\Phi_{k,f}^{(\mathrm{nif})}(Q_k^{(b)}, P_k^{(b)}, t)$, defined in (15.50), is in a suitable form to be an equilibrium potential.

An examination of the derivation of the equilibrium potential in Chapter 13 shows that the rotating coordinate frame examined here can be expressed in the form required for equilibrium. This is demonstrated by setting $v_i^*(q_i^{(b)}) = 2\epsilon^{\mu\nu\xi}q_{i,\mu}^{(b)}\omega_{k,\nu}^{(\natural)}$, using the form of $R_k(Q_k, t)$ given in (13.32), and choosing $g([v^*(q_i^{(b)})]^2) = \ln\{\frac{1}{2}[(\omega_k^{(\natural)} \cdot q_i^{(b)}) - [q_i^{(b)}]^2(\omega_k^{(\natural)})^2]\}$ so that the equilibrium potential $\Phi_k^{(1)}(Q_k^{(b)})$ defined in (13.30) is given by $\Phi_k^{(1)}(Q_k^{(b)}) = \Phi_k^{(\mathrm{cf})}(Q_k^{(b)})$ and $-p_i^{(b)} \cdot v_i^*(q_i^{(b)}) = \Phi_{i,f}^{(\mathrm{Co})}(q_i^{(b)}, p_i^{(b)})$ is the Coriolis potential. Using these results with (13.33) yields the non-inertial Hamiltonian function in the form (15.52).

The *k non-inertial equilibrium state* and the *k non-inertial equilibrium partition function* take the form

$$(15.60a) \qquad F_k^{(\mathrm{nif})\epsilon}(Q_k^{(b)}, P_k^{(b)}) = e^{-\beta_k \mathcal{H}_k^{(\mathrm{nif})\epsilon}(Q_k^{(b)}, P_k^{(b)})},$$

$$(15.60b)\, Z_k^{(\mathrm{nif})\epsilon} = \int dQ_k^{(b)} \int dP_k^{(b)} \, \chi_{\Omega_k(t)}(Q_k, P_k) e^{-\beta_k \mathcal{H}_k^{(\mathrm{nif})\epsilon}(Q_k^{(b)}, P_k^{(b)})},$$

where the equilibrium noninertial frame Hamiltonian is

$$(15.61) \qquad \mathcal{H}_k^{(\mathrm{nif})\epsilon}(Q_k^{(b)}, P_k^{(b)}) = \mathcal{H}_k^\epsilon(Q_k^{(b)}, P_k^{(b)}) + \Phi_{k,f}(Q_k^{(b)}, P_k^{(b)}).$$

In other non-inertial frames, such as a rotating frame in which the speed of rotation varies, there will clearly not be an equilibrium state.

This brief sketch indicates that for some non-inertial frames the formalism for the Theory of Interacting Systems can be applied in the form that has been presented here. This is a consequence of the fact that the forces in the non-inertial frame of this example could be represented by a potential function, so it is easy to extend the formalism to include this frame.

CHAPTER 16

Entropy and Information

The basic formalism of the classical version of the Theory of Interacting Systems is complete. In this chapter, some of the remaining issues will be sorted out concerning the interpretation of the formalism and its fundamental principles. The main concern in this chapter is the TIS perspective on the entropy and the other thermodynamic axioms of the theory.

16.1 A Recapitulation

Past approaches to interpreting the entropy in terms of concrete particle models were reviewed in Chapter 8. The most popular of these particle models interpret the entropy in terms of the mixing of particles of different types or in terms of the ordering of these particles relative to each other. Other interpretations depend on the number of particles, their degrees of freedom, the use of the thermodynamic limit, or the multiplicity and complexity of particle interactions. It was concluded that some of these models led to paradoxes, others were simply inappropriate to the task, and none was acceptable as an explanation of the entropy. In support of this assessment, Maxwell's misgivings concerning the possibility of a physical interpretation of the entropy were mentioned.

The TIS version of the entropy was defined in Chapter 9. It was demonstrated there that the entropy in the general TIS theory is not a particle sum function, so there is no natural interpretation of the entropy as a property of individual particles in the same way there is for the momentum, say. Contrary to current thermodynamic beliefs, this implies immediately that the entropy is not an extensive quantity. One important innovation is the separation of the partition function, defined as the local norm of the system distribution, from the system distribution it normalizes and from the entropy. This step distinguishes TIS thermodynamics from standard thermodynamics in which the logarithm of the partition function is included as part of the definition of the entropy.

The consistency of TIS thermodynamics with respect to the particle dynamics will be examined as a first step. The main issues involved are connected with time reversal properties of the underlying particle dynamics and the recurrence properties of a system of particles in a fixed volume. Attempts over the years by authors to modify the particle dynamics to obtain irreversibility and a steady increase in the system entropy are then examined. This inquiry is continued with a consideration of a few alternative nonequilibrium versions of thermodynamics. Some of the mathematical and conceptual issues concerning the role of thermodynamic quantities as statistical measures and the representation of particle systems are discussed. Finally, the definition of information in this context is addressed and used in the discussion of the relation claimed between entropy, intelligence, and information.

16.2 Time Reversal and Recurrence

While most of those working on the relation of statistical mechanics to thermodynamics attempted to find a way to obtain irreversible behavior from the reversible evolution equation of Hamiltonian mechanics, not everyone felt this was possible. As Bailyn [1994], p. 426, mentioned, Hermann von Helmholtz [1886] stated that it is very likely that Hamiltonian mechanics governs the evolution of nature and that irreversibility in thermodynamics is not an essential aspect of things but is a consequence of our inability, as macroscopic beings, to reverse all velocities and put all particles back where they were.

One of the foundations of the statistical mechanics of isolated systems is the Liouville theorem. The Liouville theorem uses the system evolution equation to show that the Lebesgue measure of any measurable set in phase space is preserved under the transformation of phase space that occurs in time via Hamilton's equations.[1] Equation (5.16) is the measure preserving evolution equation for the classical TIS theory. For an isolated system, the evolution equation $dF_t/dt = 0$ is valid in $\Omega_k(t)$ and at the t system boundary, so it is clearly invariant under time reversal. Moreover, the appropriate version of Axiom 6 states that $dS_t(t)/dt = 0$, which is time reversal invariant as well. Taking these facts together, it follows that the thermodynamics of an isolated system is reversible. Thus there is no conflict between the TIS particle mechanics and thermodynamics for an isolated system on the issue of irreversibility because both are reversible.

The interacting system state, F_k, by contrast, does not satisfy equation (5.16) at the system boundary and in general $dS_k/dt \neq 0$ except at equilibrium. It is not hard to show, using the stochastic property of the wall operator W_k, that the evolution of any system $k \in [t]$ is not generally time reversal invariant. As a consequence, there is again no conflict between statistical mechanics and thermodynamics because both are irreversible.

Another important consequence of Liouville's theorem is the Poincaré recurrence theorem. The Poincaré recurrence theorem states (in its modern form) that an isolated mechanical system with a fixed, finite total energy in a fixed, bounded volume will, in a finite time, return to the neighborhood of almost any initial mechanical state.[2] In terms of the TIS formalism, the requirements of a fixed total energy and a fixed, bounded volume in the theorem imply that the system remains within a fixed and bounded phase set, Ω_t^α, such that $\Omega_t(t) \subset \Omega_t^\alpha$ for all times $t \in [0, \alpha)$. The requirements for applying Poincaré's recurrence theorem are that $\Omega_t^\infty = \limsup_{\alpha \to \infty} \Omega_t^\alpha$ is bounded and that $\Omega_t(t) = \Omega_t^0$, where Ω_t^0 is a constant set in phase space, and $\Omega_t^0 \subset \Omega_t^\infty$. Let us use $|\cdot|$ for the Euclidean norm and define a phase space

---

[1]Liouville's theorem was published in 1838 in Liouville's journal. It was used in Jacobi's *Vorlesungen über Dynamik* in 1843. Boltzmann used it after that. For a more recent presentation of Liouville's theorem, see, for example, Khinchin [**1943**], pp. 15–19. As a terminological point, in mathematics "Liouville's theorem" refers to Cauchy's result that a uniformly bounded analytic function is constant. The Liouville theorem in physics is mathematically equivalent to the statement that Hamiltonian flows are measure preserving automorphisms. The volume of phase space is also one of the Poincaré integral invariants for an isolated system. See Goldstein [**1950**], pp. 247–250, or Epstein [**1931**], pp. 483–485.

[2]On Poincaré's recurrence theorem, see pp. 67–72 of Poincaré [**1890**]. See also Carathéodory's [**1919**] improvement of his proof using the theory of Lebesgue measures. For a physical discussion, see Kurth [**1960**], pp. 57–60.

neighborhood of the point (Q_t^0, P_t^0) in system t by $\{ (Q_t, P_t) \mid |(Q_t, P_t) - (Q_t^0, P_t^0)| < \eta \}$ for some $\eta > 0$. Then the recurrence theorem states that there exists a time $0 < T_\eta < \infty$ such that $|(Q_t(T_\eta), P_t(T_\eta)) - (Q_t(0), P_t(0))| < \eta$.

Poincaré's theorem is true for any isolated system in TIS described by the exact state $F_t^\delta(Q_t, P_t, t)$ with support in a fixed, bounded volume set. The value of η that defines the original neighborhood at $t = 0$ may be chosen small enough so that, for any $\epsilon > 0$ at $t = T_\eta$, each member s of any finite set of exact state thermodynamic observables will satisfy the condition $|G_t^{\delta,s}(\Delta(Q, T_\eta)) - G_t^{\delta,s}(\Delta(Q, 0))| < \epsilon$, where $\mu(\Delta(Q, t)) = \mu(\Delta(Q, 0))$ for $0 \le t \le T_\eta$. It follows that the thermodynamics of an isolated system described by the exact state is recurrent.

The stationary microcanonical state F_t^γ, the stationary canonical equilibrium state F_k^ϵ, and the stationary absolute zero state F_k^0 are asymptotic idealizations without dynamical significance and recurrence does not apply to them.[3] For an isolated system, the exact F_t^δ state will generally recur and the general state F_t will not. For a subsystem, aside from the trivial case of the asymptotic stationary F_k^ϵ state with permanently fixed boundary conditions, the F_k state will also not recur. The non-recurrence of F_k is due to the fact that mechanical states in a neighborhood of a given state at a given time will usually diverge exponentially from each other at later times.[4] For this reason, the recurrence times of two nearby exact states will usually differ by a large finite amount. The question of whether the F_k^δ state will recur is indeterminate. Because F_k^δ is based on the system specific trajectories $(Q_k^{(s)}(t), P_k^{(s)}(t))$, which depend on averaged k system external potentials reflecting the effects of particles in other systems, their behavior is different from the recurring trajectories $(Q_k(t), P_k(t))$, which are the k component of the F_t^δ trajectories $(Q_t(t), P_t(t))$.

In summary, the thermodynamics of any isolated system described by an exact state F_t^δ is both reversible and recurrent. An isolated system described by a non-stationary F_t state is reversible but not recurrent. An interacting system described by a nonequilibrium state F_k is neither reversible nor recurrent. Finally, it was shown above that the contrast between the recurrence properties of F_t^δ and F_t does not lead to any conflicts between the statistical mechanics and the thermodynamics of the system. This means that there are no paradoxes in TIS associated with recurrence and time reversal.

---

[3]This is also true of states defined in the thermodynamic limit.

[4]For further details on the mathematical aspects of some of these issues, see Arnold and Avez [**1968**], p. 53, for a discussion of dynamical systems called "C-systems." The C-systems form an open set in the space of classical dynamical systems. Discussions of the related topics of ergodicity and mixing, the KAM (Kolmogorov-Arnold-Moser) theorem, and infinite systems are in Oliver Penrose [**1979**].

16.3 Modifying the Dynamics

The fact that systems seem to go to equilibrium and remain there if the external conditions are constant has always been hard to reconcile with the reversible laws of mechanics. Several attempts have been made to account for this by proposing a change in the dynamics of particles that would lead to behavior consistent with macroscopic irreversibility.

Abstract versus Physical Dynamics.

One of the earliest proposals was made by Maxwell [1873b], who proposed that a non-mechanical element must be added to particle dynamics in order to derive thermodynamics, and the second law in particular, from statistical mechanics. This idea was expressed more formally by Thomson [1873], p. 325, who made a distinction between reversible "abstract dynamics" and the irreversible "physical dynamics" of bodies with the attendant friction and inequalities of temperature, etc.[5] Maxwell [1873b] spoke of this in terms of "irregular" molecular motions in discussions of irreversibility. His use of the term irregular in this context was similar to Krönig's [1856] assumption of irregular molecular motions as the basis for his calculations in kinetic theory. In essence, this is the assumption that particle motions are always "random," with respect to the direction and magnitude of their velocities and their locations when used in computations of physical quantities in kinetic theory.[6]

The notion of irregularity is one member of a family of assumptions that have been used in this setting. The Stosszahlansatz of Maxwell and Boltzmann and Boltzmann's "molecular chaos" assumptions are similar in origin and employment.[7] When the formal expressions of these assumptions are examined, it is clear that they are randomizing assumptions that give a particular special form to the particle distribution at each point in time. This has the effect of disconnecting the particle distribution at one moment from previous distributions and future ones. As Maxwell and Thomson were aware, these assumptions are inconsistent with the dynamics of the particles and the evolution of a particle distribution out of previous particle distributions.

In the late nineteenth century, a number of authors tried different physical hypotheses that would have the effect of introducing a stochastic element into the equations of mechanics. Max Planck, for example, introduced the concept of (random) "natural radiation" as an aspect of all radiation to obtain thermodynamic irreversibility from radiation theory.[8] Boltzmann soon showed that natural radiation is not consistent with Maxwell's equations and

---

[5]See Brush [1976], Chapter 14, for further details.

[6]See the discussion of various early assumptions concerning particle distributions in P. and T. Ehrenfest [1912], pp. 4–5.

[7]See Chapter 22 for more on Boltzmann's molecular chaos idea.

[8]Planck's concept of natural radiation and his use of it in a proof of an electromagnetic H theorem is discussed in Oliver Darrigol [1992], Chapter 2.

Planck abandoned the idea. In general, these modifications of classical mechanics lacked plausibility and were not pursued further. With the advent of quantum mechanics in the early twentieth century, still other authors tried to invoke random phases in quantum mechanics to prove the point. However, von Neumann [1932c] showed that unobserved systems are reversible in quantum mechanics, so the concept of random phases also failed to solve the reversibility problem.

The Brussels School.

A more modern attempt to modify the dynamics in order to provide a nonequilibrium thermodynamics appears in the work of I. Prigogine, C. George, F. Henin, and L. Rosenfeld [1973], Prigogine [1973], and subsequent books. Because Prigogine and his associates were located in Brussels, this approach was called the Brussels School.

The entropy was discussed from this perspective by Prigogine [1980], p. 5, who argued that the flow of heat in an isolated body (always?) leads to a uniform state. He went on to maintain that the irreversibility due to this flow of heat is a consequence of the second law of thermodynamics that in turn implies the existence of a function S such that $dS/dt \geq 0$. Prigogine subdivided the entropy function into an internal and an external component and expressed it in the form $\frac{dS_k}{dt} = \frac{d_i S_k}{dt} + \frac{d_e S_k}{dt}$, where d_i represents the differential with respect to "internal" changes and d_e the differential with respect to "external" changes.[9] This division of the change in the entropy into an externally and an internally generated change, has been used by a number of other authors. These treatments usually include an entropy flux vector and an entropy balance equation that is valid at equilibrium. The perspective associated with these equations views entropy "as a fluid that can be destroyed, or created, or produced."[10] In the TIS approach, the entropy changes only at the boundary and the division of the change in the entropy into internal and external components is rejected because there is no particle mechanism internal to the system to change it.

Prigogine, p. 44, continued his argument in support of modifying the particle dynamics by observing that systems with three or more particles are in general not integrable. He maintained, p. 166, further that classical trajectories are unobservable idealizations. Based on his analysis of Boltzmann's ideas, he concluded, p. 176, that his concept of irreversibility is "quite similar

---

[9]A division of this sort was expressed by Poincaré [1906], for example, who stated it in relation to heat as $dQ = d_i Q + d_e Q$. On the other hand, Clausius [1866], p. 459, rejected internally generated heat as a source of entropy. However, as mentioned in Chapter 8, Clausius' rejection of internal entropy production is not consistent with his maxim that the entropy of the universe always increases toward a maximum—whether a system of particles is isolated or not.

[10]See Wolfgang Yourgrau, Alwyn van der Merwe, and Gough Raw [1966], Chapter 1. See also the treatment of internally and externally generated entropy, along with entropy balance equations, fluxes, and currents, in Bailyn [1994], pp. 168–171.

to that put forward by Boltzmann." In addition, "Irreversibility is the man-ifestation *on a macroscopic scale* of 'randomness' *on a microscopic scale."* (Prigogine's italics.) Prigogine stated that the difference between his view and Boltzmann's is that Boltzmann's irreversibility is due to superimposing "molecular chaos" on the dynamical equations, whereas from his point of view,[11] both "randomness *and* irreversibility are consequences of the struc-ture of the equations of motion."

Prigogine and his associates modified the microscopic dynamics by adding an irreversible component to an operator they called the "Liouville operator" in their system evolution equation. This operator employs a modified Hamil-tonian, with an added "friction" or "dissipation" term, in the Poisson bracket operator $\{\mathcal{H}_k, F_k\}_{pb}$, or commutator, that appears in the (quantum) system evolution equation. In TIS terminology, this is a modification of the Hamil-tonian bracket operator defined in Chapter 4. Prigogine and others defined various projection operators to pick out the states of physical interest.

Whenever a stochastic operator or a projection operator is part of the evolution equation of a system, the system distribution will usually evolve to equilibrium. For this reason, Prigogine's evolution equation will in fact evolve into a stationary state. This evolution, however, is an artifact of the modifications of the evolution equation and his use of projection operators. This is the same mathematical mechanism that leads the time derivative of the "course grained entropy" to predict the evolution of any system into a stationary state. In TIS, the introduction of the stochastic wall operator as part of the independent systems approximation had this effect. However, Prigogine's modifications of the evolution equation do not operate in the same way and do not have the same physical justification as the use of the wall operator in TIS. In addition, the physical interpretation of Prigogine's theory is obscure.

To take one example, Prigogine's new Liouville operator acting on the distribution of an isolated system connects system states with different total energies. Although energy is conserved overall, it is difficult to see how this theory can be understood or justified within the framework of an underlying particle model. Another issue is the interpretation, again in terms of the par-ticle model, of the division of the entropy into two parts: one concerned with internal energy flows and one concerned with flows to the outside. Prigogine has not presented acceptable particle analog functions for the internal entropy generation and the entropy flux vector. These facts undermine Prigogine's effort to replace current quantum or classical dynamics with something else. Prigogine's conclusion that the Hamiltonian approach is of limited usefulness in describing microscopic systems, in either the classical or quantum version of dynamics, is therefore rejected as not well founded.

---

[11]See the dynamical theory presented in Prigogine [**1973**], Chapter 8, and elsewhere.

Maximal Entropy Formalism.

Another example of modifying the laws of mechanics is the "Maximal Entropy Formalism" of Jaynes [1978], pp. 91–102. The system state at any time is defined by the requirement that the entropy be maximal with respect to the constraints imposed at that time by a small set of macroscopic parameters characterizing the system. This is a prescription for assigning a new probability distribution density to the system at each point in time. For an isolated system, the total energy is fixed and Jaynes formalism is equivalent to assigning the microcanonical state to the system. For a subsystem of a larger system, on the other hand, it amounts to assigning at each instant a canonical or equilibrium state to the subsystem. The parameters of the state at a given time are computed from the properties of the system measured at that time.

Although a new state may be assigned to a system when desired using thermodynamic information, this state is not required by any physical principle to be an equilibrium state (or a maximal entropy state) as argued by Jaynes [1957], p. 623. Once an initial state is assigned to the system, there is no mechanical justification for using Jaynes"s procedure to determine subsequent states. Jaynes' procedure also falters when we have information on the local values of system thermodynamic quantities and not just global values. It is not clear how to incorporate this information and what to maximize. In short, Jaynes' theory of how the sequence of states of the system is to be determined is inconsistent with the time evolution of the underlying particle model from the assigned initial state. It is also clear that Jaynes's formalism is inconsistent with the evolution of the particle system in accord with Hamiltonian mechanics.

16.4 Macroscopic Nonequilibrium Thermodynamics

There are quite a few macroscopic nonequilibrium theories that are currently being pursued. A good survey of current work on nonequilibrium thermodynamics, and criticism of some of this work, has been given by Stanislaw Sieniutycz and Peter Salamon [1990a]. Discussions of the theories themselves appear in a recent series of volumes on this topic edited by Sieniutycz and Salamon.

Taken together, these theories have postulated a large variety of thermodynamic forces, reversible and irreversible fluxes, entropy production processes, diffusion kinetic energy, etc., in addition to the mechanical kinetic energy, thermal momentum, and metrics on thermodynamic surfaces. There are few constraints on this proliferation of thermodynamic components and equations. The complaints of Truesdell [1969] concerning the lack of a physical basis for the introduction of various assumptions concerning the relaxation rates of components of physical systems to equilibrium apply to all of these new components as well.

From the TIS point of view, almost all of these quantities and their relations, which are introduced as part of phenomenological equations, do not have an obvious particle interpretation and cannot be given one. This means that they are far from satisfying the TIS requirements for an adequate thermodynamics. To demonstrate these points, a few of the more prominent versions of these theories will be discussed briefly.

Irreversible Thermodynamics.

Irreversible Thermodynamics is an approach to modifying the dynamics that has been gaining adherents in recent years. In a review article, Jou, Casas-Vázquez, and Lebon [1988], discussed the ideas behind the theory and proposed an extension of it. The classical version of the theory is related to the Navier-Stokes and Euler-Cauchy equations. It is based on the introduction of dissipative fluxes into the formalism of thermodynamics. These fluxes include a heat flux, a viscous pressure tensor, diffusion flux, and an electric current. Evolution equations for these fluxes are obtained using standard procedures of irreversible thermodynamics. This approach has sometimes been called *mesoscopic* because it involves both microscopic variables and macroscopic variables to bridge the gap between the $6\mathcal{N}_k$-dimensional set of variables in microscopic statistical mechanics and the 5 variables (3 local velocity components and two independent thermodynamic variables, such as pressure and temperature) required to describe an ordinary fluid in standard thermodynamics.

The mesoscopic approach makes a distinction between conserved quantities, which are "slow" variables that retain their values, and other variables, considered "fast" that are assumed to decay rapidly. If the fluxes decay rapidly, they are considered dependent variables and are expressed in terms of gradients of the basic macroscopic thermodynamic quantities. If they happen to be slow in a given setting, they are elevated to the status of independent variables. The authors maintained that this approach is a useful check on some of the standard assumptions of current theories in nonequilibrium thermodynamics, such as the postulate of local equilibrium. A very brief précis and critique of this theory will be given next.

The balance equations of mass, momentum, and energy depend on the kinematic velocity field v^μ, the mass density ρ, the thermodynamic specific volume $\nu = 1/\rho$, and the specific internal energy per unit mass u. These balance equations are

(16.1)
$$\rho\dot{v}^\mu = -\nabla_\nu P^{\mu\nu} + \rho F^\mu, \qquad \rho\dot{\nu} = \nabla \cdot v, \qquad \rho\dot{u} = -\nabla \cdot \mathcal{Q} - P^{\mu\nu}\nabla_\mu v_\nu + \rho E.$$

The dot over a quantity represents the material derivative discussed in Chapter 11. These equations employ the pressure tensor $P^{\mu\nu}$, the heat flow vector $\mathcal{Q}^\mu$, the specific body force per unit mass F^μ, and the specific energy supply E in addition to the variables v^μ, ν, and u, describing the fluid. The pressure tensor is usually decomposed into

(16.2) $$P^{\mu\nu} = P\delta^{\mu\nu} + P^{(v)\mu\nu},$$

where P is the pressure and $P^{(v)\mu\nu}$ is the "viscous part" of the pressure tensor. The viscous pressure tensor $P^{(v)\mu\nu}$ in turn is usually decomposed into a scalar bulk viscous pressure $P^{(v)}$ and a traceless tensor $P^{(d)\mu\nu}$ called the "deviator" in the form

(16.3) $$P^{(v)\mu\nu} = P^{(v)}\delta^{\mu\nu} + P^{(d)\mu\nu}.$$

It follows immediately that $P^{(v)} = \frac{1}{3}\delta_{\mu\nu}P^{(v)\mu\nu}$.

If the external force and energy supply terms are ignored, the equations (16.1) involve P, Q^μ and $P^{(v)\mu\nu}$ in addition to the variables describing the fluid, so they are not self-contained. This deficiency is remedied by the introduction of an equation of state and constitutive equations. These represent significant further assumptions on the behavior of the matter involved. The entropy differential, under the name "Gibbs equation", was used as an equality in accord with the hypothesis of local thermodynamic equilibrium (LTE) in the form

(16.4) $$dS = T^{-1}[du + Pd\nu].$$

It is employed to give the equations of state for T^{-1} and P as

(16.5a) $$T^{-1}(u,\nu) = \frac{\partial s}{\partial u},$$

(16.5b) $$T^{-1}(u,\nu)P(u,\nu) = \frac{\partial s}{\partial \nu}.$$

Let us use (16.1) with (16.5) to put the entropy balance equation into the form $\rho\dot{s} + \nabla \cdot J_S = \sigma$. This leads to the results

(16.6a) $$J_S^\mu = T^{-1}Q^\mu,$$

(16.6b) $$\sigma = -T^{-1}Q \cdot \nabla T - T^{-1}P^{(v)}\nabla \cdot v - T^{-1}P^{(d)\mu\nu}V_{\mu\nu}^\circ,$$

where the traceless velocity gradient tensor is $V_{\mu\nu}^\circ = \nabla_\mu v_\nu - \frac{1}{3}\delta_{\mu\nu}\nabla \cdot v$.

The quantities Q^μ, $P^{(v)}$, and $P^{(v)\mu\nu}$ are usually called the *dissipative fluxes* and the quantities $T^{-1}\nabla T$, $T^{-1}\nabla \cdot v$, and $T^{-1}V_{\mu\nu}^\circ$ are usually called the *thermodynamic forces*. It is next assumed that the second law is locally valid, which means that $\sigma \geq 0$. This is guaranteed by the constitutive relations

(16.7a) $$Q^\mu = -\mu_1 T^{-2}\nabla T,$$

(16.7b) $$P^{(v)} = -\mu_0 T^{-1}\nabla \cdot v,$$

(16.7c) $$P^{(d)\mu\nu} = -\mu_2 T^{-1}V^{\circ,\mu\nu}.$$

with positive semi-definite transport coefficients: $\mu_1 \geq 0$, $\mu_0 \geq 0$ and $\mu_2 \geq 0$. In terms of the Fourier heat flow equation and the Navier-Stokes equations, these transport coefficients are related to the thermal conductivity λ by $\mu_1 = \lambda T^2$, the bulk viscosity ζ by $\mu_0 = \zeta T$ and the shear viscosity η by $\mu_2 = 2\eta T$. Taken together, the constitutive relations (16.7), the balance equations (16.1), and the equations of state (16.5), form 16 equations in 16 unknowns: the 5 scalars u, ν, T, P, and $P^{(v)}$; two 3-vectors v^μ and Q^μ; and the 5 independent components of the traceless tensor $P^{(d)\mu\nu}$. These equations are completely determined when appropriate boundary and initial conditions are employed.

A number of arbitrary assumptions are required to make Irreversible Thermodynamics work. First, there are the assumptions of local thermodynamic equilibrium (LTE) and the local validity of the second law. Second, the pressure, the pressure tensor deviator, and the heat flow vector are assumed to satisfy the arbitrary constitutive equations (16.7). There is no attempt to justify these equations based on the underlying particle mechanics. That would be difficult anyway because of the presence of the macroscopic temperature in the constitutive relations—which means that these equations mix macroscopic thermodynamic and microscopic mechanical concepts. Finally, as Jou, Casas-Vázquez, and Lebon [**1988**], p. 1112, point out, introducing the constitutive equations (16.7) into the balance equations (16.1) leads to a parabolic set of equations implying an infinite speed for the propagation of thermal and viscous signals.[12]

To deal with these problems, Jou, Casas-Vázquez, and Lebon summarized proposals for a further generalization of this formalism that make Q^μ, $P^{(v)}$ and $P^{(d)\mu\nu}$ independent variables and the entropy a function not only of the classical variables u and ν, but also of Q^μ, $P^{(v)}$ and $P^{(d)\mu\nu}$. They used these variables in an extended form of the entropy relation (16.4) to create an equation of state for each of the variables involved.

Except for a few points of criticism, the extended version irreversible thermodynamics theory presented by Jou, Casas-Vázquez, and Lebon will not be discussed in any detail. The main point is that the generalized version of the theory requires a much more elaborate set of assumptions to make it go and generates a plethora of coefficients. Some of these coefficients are not related to thermodynamic quantities or relations, so other criteria must be used to determine them. Jou, Casas-Vázquez, and Lebon, p. 1116, point out in one case that the principle of frame indifference, proposed by Truesdell and Noll, leads to the determination of two coefficients as $+1$ each—while the comparison of the generalized constitutive equations to the Chapman-Enskog equations leads to the values -1 for these coefficients. They noted that this problem had led to a controversy in the literature and proceeded, somewhat uneasily, to ignore these coefficients because they are associated with physical quantities that are too small to be readily observed.

---

[12]The criticisms presented in Chapter 11 concerning the Navier-Stokes equations and theories based on the material derivative apply here as well.

With respect to the entropy, the authors admitted that it is difficult to assess what portion of the generalized entropy production term σ is subject to the local second law, which is itself another assumption of the extended theory. The hypothesis of local thermodynamic equilibrium is retained for the linearized version of the theory, but the authors acknowledge that this assumption may not be well founded. Finally, the coefficients of the components of the extended theory are determined using various fluctuation schemes, relaxation assumptions, and linearized relations. When the basis for these assumptions is examined, however, they are ad hoc and implausible. For example, Einstein's formula for a fluctuation, given in the form $e^{\delta^2 S}$, is used in conjunction with the assumed local second law to support statements about the range of values that variables must take. However, this formula cannot be given a meaningful interpretation from the TIS perspective in the nonequilibrium case. With regard to relaxation assumptions, there is little evidence that the system relaxes to equilibrium at all in the various ways assumed and no way to judge whether this relaxation is adiabatic, isobaric, isothermal, or something else. Finally, the linearized constitutive relations are usually assumed on an ad hoc basis to solve a mathematical problem and are not physically motivated in terms of the underlying mechanics. In short, these manifold schemes, assumptions, and relations, piled one on another, are all very suspect.

It follows that the extended irreversible thermodynamics presented by Jou, Casas-Vázquez, and Lebon, which is based on the work of many authors in the developing literature in this area, is unconvincing. Assumptions are heaped on assumptions and the whole scheme rests on some very dubious assumptions concerning the local behavior of the system distribution and the local behavior of the entropy. The theory is far from being expressible in terms of the properties of the underlying particle system and it is unlikely that the dynamic predictions of this theory will be in accord with the underlying mechanics.

Other Theories.

Some chaos theorists have suggested that an application of chaos theory may give an account of the second law of thermodynamics. Aside from the perspective developed by Prigogine, these ideas do not seem to have been worked out into a full treatment of thermodynamics and statistical mechanics. The chaotic phenomena discussed in the literature are macroscopic phenomena, but the theorists invoking chaos have provided no theoretical relation that connects the macroscopic chaotic description with the underlying classical or quantum particle description.

The Theory of Interacting Systems.

The irreversible evolution of interacting systems in the Theory of Interacting Systems is due to the replacement of the interactions between particles in different systems with averaged, rather than exact, external potentials, and a stochastic boundary formalism. The boundary formalism is expressed in terms of a stochastic wall transition operator W_k that approximates the influence of one system on another. The stochastic aspect of the theory acts only at the system boundaries—a fact reflected in equations (9.15) and (9.16) for the change in the entropy in time. It is precisely this stochastic element in the theory that leads to Clausius inequality. The inclusion of an averaged external potential and a boundary formalism in the TIS theory is a significant modification of the Hamiltonian mechanics as far as subsystems are concerned.

It was argued at length in RSO that this division of the world is in accord with physical practice and the use of the independence and quasilocality approximations should be viewed as an epistemological step, required to separate a knowing subject from a studied object and the objects from each other, rather than a physical approximation procedure. From this perspective, the TIS theory is not subject to the criticisms of the other theories presented here.

16.5 Homogeneous Systems and Equilibrium States

Discussions of the thermodynamics of isolated systems usually refer to the "empirical fact" that such systems go to equilibrium if left alone for a long enough time. It is true that an isolated or reasonably well-isolated system does exhibit a "homogenizing" effect with respect to the spatial variation of local thermodynamic densities. This homogenization is also facilitated by stirring (Gibbs [1902], Chapter XII). The homogenization of a system due to diffusion or stirring is a result of the mixing of particles, momentum, angular momentum, and energy belonging originally to different regions of the system. This effect shows up clearly, for example, in the calculation of the time evolution of a system of non-interacting "petites planetes" in Poincaré [1906], Section 2, and in a system of non-interacting particles reflected specularly at the wall in the work of Lebowitz and Frisch [1957] and Grad [1961], Section 9.[13] It is with reference to such homogenous states that the arguments of Gibbs [1902], Chapter XII, and Boltzmann [1896a], pp. 310–312, that most states are "equilibrium states" make sense.[14] Thus, although an isolated system may be in a state with uniform densities most of the time, and this state may be macroscopically indistinguishable from the asymptotic equilibrium state, the above examples show that it cannot be concluded that the system will remain this way forever.[15]

---

[13] See also the discussion in Nikolai Krylov [1950], pp. 114–116.

[14] See Uhlenbeck and Ford [1963], pp. 5–16, for a discussion of the arguments of Gibbs and Boltzmann.

[15] See John Blatt's [1959], pp. 749–750, discussion of Otto Hahn's "spin-echo" experiment in this regard.

The conclusion drawn from these points is that although most situations we encounter involve an apparent irreversibility—in that macroscopic systems of mixed molecules seldom unmix themselves, etc.—this is quite different from saying that the particles in these situations *must* behave this way or that there is a physical function that guarantees that this will be so.[16]

16.6 Statistical Considerations

Throughout this book the tension has been noted pointed out the precise mechanical description of a system, as a point in $6\mathcal{N}_k$-dimensional phase space that is described by the exact state F_k^δ, and a statistical description, in terms of a distribution with an extension in phase described by the general state F_k. The ultimate expression of this tension lies in the fact that the entropy is defined for general states and undefined for the exact state. In addition, from the statistical point of view, the connection between the Maxwell equilibrium momentum distribution and the Gaussian normal distribution of statistics is no accident. To begin with, the Gaussian distribution was suggested to Maxwell by analogy to the law of random errors.[17] The modern justification for the Gaussian law considers particle momentum measurements to be small samples from a large pool, and the momenta of the particles are assumed to be equally distributed independent random variables. A Gaussian one-particle momentum distribution is then expected as a consequence of the central limit theorem of probability.

The TIS choice of the mean (phase average) as its major statistic rests on the fact that it is a good point estimator. In statistical terms, it is (1) *unbiased,* (2) *consistent,* (3) *efficient,* and (4) *sufficient.*[18] These properties imply that the mean of a sample is a good choice for estimating the mean of the underlying population. With respect to an isolated system, the fact that the total energy is fixed implies that attributing the canonical distribution F_t^ϵ to the system is the same as attributing the F_t^γ state to the system was discussed above. If all that is known about the isolated system is that it has a given particle number, total momentum, total angular momentum, and total energy, attributing the F_t^γ state to the system is an example of using the "principle of insufficient reason," as recommended by Jaynes [**1957**], with regard to choosing an initial state consistent with what we know. However, we are not compelled to use F_t^γ in this case. A different state F_t can be chosen for the system, if desired, as long as F_t has the proper total energy, momentum, and angular momentum. According to the results of Chapter 5,

---

[16]I will return to a discussion of irreversibility in thermodynamics in EIS, Part II.

[17]See, for example, Maxwell [**1866**], [**1870**](Maxwell [**1890**], p. 154) and Maxwell [**1875**] ([**1890**], p. 408).

[18]For further information on statistics in statistical mechanics, see Benoit Mandelbrot [**1962**], Laslo Tisza and Paul M. Quay [**1963**], and Gilbert N. Lewis [**1930**]. For an elementary discussion of these statistical notions concerning point estimators, see any good statistics text.

the *time average* of the state F_t will evolve, in the limit of infinite time, into a state proportional to F_t^γ on a subset of the manifold of states accessible to F_t^γ from a given initial mechanical state.[19]

Sometimes the system distribution F_k is reified and treated as if it is a picture of some sort of physical "fuzziness" of the system. This happens when the equilibrium distribution, for example, is taken to mean that there are literally n_1 particles in the system in the mechanical state (q_1, p_1), etc. In this situation, the probabilistic meaning of the distribution has been lost sight of and the distribution has taken on a statistical meaning that is not appropriate to its origin.[20] This viewpoint corresponds to a switch in the interpretation of the system state from a probability distribution to a statistical population distribution. The proper interpretation from the TIS standpoint is expressed by statements of the form: "the system is in the mechanical state (Q_k, P_k) at time t with probability density $F_k(Q_k, P_k, t)$."[21]

The important point, from a statistical point of view, is that we can choose any time to be the initial time for a system and the distribution we chose to represent the system at this time is a free choice from among all those that are in accord with the information we have about the system.[22] The point first made by Zermelo [**1896b**] in connection with Boltzmann's work, and presented above in the discussion of Jaynes' theory, was that this is allowed only for the initial state of the system. The states that evolve from this one cannot be chosen freely in the same sense. Some have tried to remove the arbitrariness associated with the initial choice of a state description for the system by proposing a rule. Jaynes' [**1957**] maximum entropy rule was discussed above. A similar rule was proposed by Katz [**1967**], p. 40, who considered the case in which we have limited, self-consistent and independent information about a system: *"We assign to our system that statistical element f that reproduces all available information and has the maximal missing information."* The maximal missing information referred to by Katz is supposed to be taken account of by using Jaynes' [**1957**] procedure for determining the state of maximal entropy. In practice, however, rules such as those of Jaynes or Katz are not used. The state attributed to an actual system depends on the use to which we want to put the system and the calculations we want to make.

---

[19]The set of phase points that are passed through by a system trajectory that starts at a given point and moves forever is called a *metrically transitive* set in ergodic theory. There may be many mutually disjoint metrically transitive sets for a given system.

[20]This is the problem, for example, in the discussion of the "objective" properties of entropy and the system distribution in Kenneth J. Denbigh and Jonathan Stafford Denbigh [**1985**].

[21]See Chapter 22 for a further discussion of the interpretation of the system state in the context of Boltzmann's equation.

[22]Krylov [**1950**], Jaynes [**1957**], and Katz [**1967**], have emphasized the significance of this choice although from somewhat different points of view.

These points show that any assignment of a state to the system, when there is not sufficient information to assign an exact F_k^δ state to it, is a conventional choice that can be made determinate by an arbitrary rule. It is not forced on us by an intrinsic physical property of the system. This arbitrariness will accordingly be reflected in the thermodynamic quantities, and the entropy in particular, computed for the system. It follows that the entropy is not an intrinsic property of the system itself, but of the choices we have made concerning how to describe it. The usefulness of describing systems in probabilistic terms and deriving the thermodynamic functions as the mean values of phase analog functions depends, as Jaynes [1957], p. 626, emphasized, on the variance of the quantity sought. If the variance is large, the actual value of the thermodynamic quantity under discussion for a given exact state may deviate considerably from the mean value obtained by phase averaging. In addition, Jaynes [1957], p. 171, has emphasized that while information can be lost in a variety of ways, it can only be increased by measurement.[23] These issues are the same for every other theory based on probability, including quantum mechanics.

16.7 Alternative Descriptions

The theory presented in this volume has been limited to an important, but idealized, case: simple classical point particles with mass, without internal energy, which interact only by central forces, do not interact with radiation, do not undergo nuclear or chemical reactions, and do not change their identity.[24] Any conclusions to be drawn from the results of this theory apply only to this simplified model and cannot be extrapolated to other situations without reanalysis. For this reason, conclusions cannot be drawn to the effect that the universe will recur or, on the other hand, that "The entropy of the universe strives to attain a maximum value."[25]

The concept of independence is central to the TIS formulation of both thermodynamics and statistical mechanics. Describing one part of the universe as separate from, and independent of, the rest of the universe is basic to posing and solving physical problems. However, as mentioned above, the choice of an independent systems description itself has consequences as far as entropy is concerned. To show this, theorem VII of Gibbs [1902], p. 133, is used with the initial condition (6.17) to demonstrate that a result of the

---

[23]Jaynes [1957], p. 186, makes the important point that false information can be injected into the situation by an approximation procedure. This point is illustrated below when it is shown that the entropy of the [t] system in the initial state is greater than the entropy of the t system in its initial state. This result reflects the fact that the partition of the t system to obtain the [t] system is an example of "coarse graining" the t system entropy. See also Katz [1967], pp. 40–50.

[24]In EIS and QIS, electrodynamics will be introduced and some of these restrictions will be lifted.

[25]See Clausius [1865] or Jaynes [1978], p. 193.

"coarse graining" $F_t \to F_{[t]}$ is $S_{[t]}(t) \geq S_t(t)$, with equality if and only if $F_{[t]}(t) = F_t(t)$.[26] This shows again that the entropy attributed to a system clearly depends on the description chosen for the system and is not an independent physical aspect of the system in the same way the energy is.

As discussed in Section 16.1, Liouville's theorem does not apply to the case in which the system is not isolated. In addition, the distributions attributed to systems interacting with others are not required to exhibit Poincaré recurrences. Further, reversing the velocities of particles colliding with the wall leads again to the same inequality of Clausius since the same stochastic wall operator is used for the reversed velocities as for the original case. Thus, the system evolution, *in terms of the F_k description*, is irreversible.[27] The conclusion also follows that the increase in entropy associated with interacting systems that are not in equilibrium states is genuine. If, on the other hand, the microscopic description of the interaction between the particles of two systems sharing a common boundary is expressed in terms of the reversible mechanical laws, both of the systems must be considered parts of one system that is described microscopically. The concept of the wall does not play a role in this case. One point of view or the other must consistently be chosen if contradictions and inconsistencies are to be avoided.

The open systems version of this theory differs from the closed version presented here in several significant ways. Open systems are also analyzed by assuming that they are subsystems of a larger isolated, closed system. For open systems, the fundamental relation (5.29) and the important consequences obtained from it are no longer valid. The thermodynamics of open systems will be presented in QIS.

16.8 The Measure of Information

Another perspective that has been proposed off and on for many years is the interpretation of the entropy as a measure of information in a system. To evaluate this viewpoint, let us first examine the mathematical theory of the measure of information, which was developed independently in response to the needs of communications systems.[28]

---

[26]See also Poincaré [**1906**], pp. 372–373.

[27]See Maxwell [**1878b**] (Maxwell [**1890**] pp. 669–671), for a discussion of this point that is similar to the one given here in several respects. As a terminological point, macroscopic irreversibility is designated by the term 'unreturnability' in EIS, Chapter 11, to distinguish it from the microscopic mechanical notion of irreversibility in reference to the underlying particle system. It was also mentioned in Chapter 8 that aspects of this concept were designated by the term 'unrecoverability' in the older literature.

[28]Many books and papers on entropy and information have appeared over the years. See, for example, E. P. Wigner and Mutsuo Yanase [**1963**], Katz [**1967**], Denbigh and Denbigh [**1985**], Leff and Rex [**1990**]. The TIS perspective differs in a number of significant ways from the views presented in these books and papers.

Work on the information content of messages originated with the work of R. Hartley on the transmission of telegraph messages. Hartley [**1928**], p. 538–541, was concerned with an objective quantitative measure of the information in a message. He observed that the number of possible distinct sequences of symbols in a message of length n using an alphabet of s symbols is s^n. With s fixed, Hartley noted that the amount of information in a message increased exponentially with the length of the message. Furthermore, the contribution of a symbol selected from the alphabet that appeared late in the message would make a far greater contribution to the value of the information than the selection of an earlier symbol. Since a late symbol in a message is as easy to transmit as is an earlier one, Hartley proposed that the amount of information be proportional to the to the length of the message (i.e., the number of selections involved in creating the message). The formula Hartley obtained for his measure of information, H, is then[29]

$$(16.8) \qquad\qquad H = \ln s^n = n \ln s.$$

Hartley's approach to information lay dormant for a number of years until it was taken up again by Claude E. Shannon, and then pursued by many authors. Shannon and Warren Weaver [**1949**] generalized Hartley's approach and considered a set of n possible events (i.e., an n symbol alphabet) for which the probability of occurrence for each event i is p_i. Shannon proposed that the measure H associated with choice of one of these symbols, which is analogous to the choice of one of Hartley's messages of length 1, is proportional to the information this choice conveys and should meet the following criteria:[30]

1. H should be continuous in the p_i's;
2. if all the p_i's are equal, so that $p_i = 1/n$, then H should be a monotonically increasing function of n;
3. if a choice is broken down into two successive choices, the original H should be the weighted sum of the individual values of H.

Shannon and Weaver [**1949**], p. 19, showed that the only measure of information content in a message of length k that met their criteria in the discrete case is given by the function

$$(16.9) \qquad\qquad H = -K \sum_{i \in k} p_i \ln p_i,$$

where K is a positive constant, and the quantity p_i represents the probability that symbol i will appear in the message. As a set of probabilities, the set

---

[29]See also Harry Nyquist [**1024**], pp. 332–333, 342–344.
[30]Shannon and Weaver [**1949**], p. 19. See also Amnon Katz [**1967**], Chapter 2.

$\{p_i\}$ is required to satisfy the following conditions:

(16.10a) $0 \leq p_i \leq 1,$

(16.10b) $\sum_{i \in k} p_i = 1.$

Shannon and Weaver [**1949**], pp. 54–58, then extended this formalism to the continuous case with the formula

(16.11) $H = -K \int dx\, p(x) \ln p(x),$

where the dimension of the random variable x corresponds to the length of the message. Shannon and Weaver [**1949**], p. 20, noted the similarity between this formula and Boltzmann's thermodynamic entropy, H.

Taking up the notion of information at this point, Brillouin [**1962**] recommended that the information be identified with the negative of the entropy, or "negentropy" as he called it. In terms of TIS notation, the information associated with system k at time t, $I_k(t)$, can be written

(16.12) $I_k(t) = -k_B^{-1} S_k(t).$

Eugene P. Wigner and Mutsuo M. Yanase [**1963**] analyzed the information of distributions. They began by stating that entropy is "a measure of our ignorance", or, if multiplied by -1, "a measure of our knowledge of the state of a system."[31] Wigner and Yanase approached the problem of discovering the proper form for the formal expression of the information content by presenting a set of requirements that an expression for the information content of a distribution must meet. These requirements were

(a) "If two different ensembles are united, the information content of the resulting ensemble should be smaller than the average information content of the component ensembles."

(b) "The information content of the union of two systems should be the sum of the information contents of the components."

(c) "The information content of an isolated system, or of an ensemble of isolated systems, should be independent of the time."

(d) "In the process which is the opposite of that considered under (b), when a joint system is separated into two parts, the information content should, in general, drop because any knowledge of statistical correlations between the properties of the two systems will be lost by considering them separate."

---

[31] For a condensed history of the recognition of the relation between distributions and information, Wigner and Yanase mentioned in a footnote an article by Weaver in 1949 and the last few pages of an article by Marian von Smoluchowski [**1913**].

The union of ensembles considered in (a) is a single system obtained by combining the two original systems that still retain their separate identities. The union considered in (b) is a formal union of the two ensembles that remain independent. Wigner and Yanase considered both the classical and quantum cases. Only the classical case will be considered here.[32]

The information content of a classical distribution was given by Wigner and Yanase, p. 911, as (TIS notation)

$$(16.13) \qquad I_c(F_k) = \int dQ_k \int dP_k \, F_k^{(n)}(Q_k, P_k, t) \ln F_k^{(n)}(Q_k, P_k, t),$$

where $F_k^{(n)}(Q_k, P_k, t)$ is a normalized probability distribution density in this discussion. They remarked that it is well known that this expression in invariant under a canonical transformation and referred to Gibbs [**1902**] and Richard C. Tolman [**1938**], p. 52, for proofs. To verify requirement (a), consider two ensembles of systems that have weights a and $1-a$, for $0 \le a \le 1$ in the combination of systems envisioned in the union (a). Because $F_k^{(n)} \ln F_k^{(n)}$ is a convex function for $F_k^{(n)} \ge 0$, the result

$$(16.14) \qquad I_c(aF_k^{(1)} + (1-a)F_k^{(2)}) \le aI_c(F_k^{(1)}) + (1-a)I_c(F_k^{(2)})$$

follows immediately. For case (b), the distribution of the combined system was given by Wigner and Yanase, p. 914, as $F_k(Q_k, P_k, Q_k', P_k', t) = F_k^{(1)}(Q_k, P_k, t)F_k^{(2)}(Q_k', P_k', t)$. This factorization in combination with the logarithmic form for the information content of a distribution in (16.13) implies that the information content of the combined distribution is the sum of the information contents of the distributions (1) and (2). The requirement (c), that the information content of an isolated distribution is independent of the time, is easy to verify by a direct computation of the time derivative of the macroscopic functional I_c using (5.16).

To prove the classical version of requirement (d), Wigner and Yanase defined the marginalizations of the combined distribution by

(16.15a)

$$F_k^{(1)}(Q_k, P_k, t) = \int dQ_k' \int dP_k' \, F_k(Q_k, P_k, Q_k', P_k', t),$$

(16.15b)

$$F_k^{(2)}(Q_k', P_k', t) = \int dQ_k \int dP_k \, F_k(Q_k, P_k, Q_k', P_k', t),$$

Using these definitions, the joint distribution was then expressed in the form
(16.16)
$$F_k(Q_k, P_k, Q_k', P_k', t) = F_k^{(1)}(Q_k, P_k, t)F_k^{(2)}(Q_k', P_k', t) + G_k(Q_k, P_k, Q_k', P_k', t),$$

---

[32]The quantum case is dealt with in QTS, Part III.

where
(16.17)
$$\int dQ_k \int dP_k\, G_k(Q_k, P_k, Q'_k, P'_k, t) = \int dQ'_k \int dP'_k\, G_k(Q_k, P_k, Q'_k, P'_k, t) = 0.$$

In terms of this notation, requirement (d) can be written

$$(16.18) \quad \int dQ_k \int dP_k \int dQ'_k \int dP'_k\, F_k \ln F_k$$
$$\geq \int dQ_k \int dP_k\, F_k^{(1)} \ln F_k^{(1)} + \int dQ'_k \int dP'_k\, F_k^{(2)} \ln F_k^{(2)}.$$

To demonstrate (16.18), the decomposition (16.16) of the joint distribution F_k is used in the left side of (16.18) and the logarithm is expanded into a series. The fact that $F_k \ln F_k$ is greater than the sum of the first terms in the expansion up to terms linear in G_k implies that $F_k \ln F_k > F_k^{(1)} F_k^{(2)} \ln F_k^{(1)} F_k^{(2)} + G_k \left(1 + \ln F_k^{(1)} F_k^{(2)}\right)$. It follows from this that

$$(16.19) \quad \int dQ_k \int dP_k \int dQ'_k \int dP'_k\, F_k \ln F_k$$
$$\geq \int dQ_k \int dP_k \int dQ'_k \int dP'_k\, [F_k^{(1)} F_k^{(2)} \ln F_k^{(1)} F_k^{(2)} + G_k(1 + \ln F_k^{(1)} F_k^{(2)})].$$

The last term on the right in (16.19) is zero. This is a consequence of the fact that $\ln F_k^{(1)} F_k^{(2)} = \ln F_k^{(1)} + \ln F_k^{(2)}$ followed by an application of (16.17). This result implies that the inequality (16.18) is valid.

The demonstration by Wigner and Yanase that the requirements (a)–(d) on the representation of the information content of a distribution are met in the classical case by the functional $I_c(F_k)$ indicates that it is an acceptable measure of the information content of a distribution. They stated, but did not consider, that the change in information due to a measurement is also an important aspect of the problem.

16.9 Entropy and Intelligence

A number of approaches to understanding the entropy in relation to life and to intelligence have been proposed. Because these approaches have a long-standing history and the TIS perspective is so different, some aspects of this issue will be reviewed to highlight these differences.

The relation between entropy, intelligence, and information was investigated first by Maxwell who considered a thought experiment that employed the services of a small "finite being" or demon that was "very observant and neat-fingered". The purpose of this demon was to show that an intelligence could violate the second law of thermodynamics by selecting which molecules to allow through a door between two chambers so that one chamber could become warmer and the other cooler without the expenditure of work.[33]

Various authors between 1880 and 1913 also noticed that fluctuations in the values of the system energy or momentum might be used to construct a perpetual motion machine. This possibility was examined in terms of physical requirements on the demon at a conference at Göttingen sponsored by the Wolfskehl Commission. In a discussion of automatic machines to capture energy from fluctuations, Marian Smoluchowski [1913], p. 119, observed that the mechanical fluctuations of the system were not coordinated with the thermal fluctuations and stated that this is the reason these automatic machines could not work. Smoluchowski, pp. 119–120, 'naturalized' the demon, to use the terminology of John Earman and John D. Norton [1998], in his inquiry by assuming that the demon is subject to the laws of thermodynamics just as all other matter. However, "our lack of knowledge of the processes of life" prevented Smoluchowski from concluding that the fluctuations in the demon would then guarantee on the average that its activities would obey the second law. In this were the case, in spite of its intelligence, the demon could not extract work at the expense of heat from the system.[34]

Another approach to the question of whether living organisms are subject to the second law of thermodynamics, discussed in Chapter 8, was taken by Leo Szilard. He investigated the idea that an intelligence might be able to violate the second law of thermodynamics. In a pioneering work that was written in 1925 as his Habilitation paper at Göttingen and published in 1929, Szilard [1925b], p. 121, noted that Smoluchowski had left open the possibility that the second law could be evaded by an intelligent being. To avoid irrelevant issues connected with entropy and life, Szilard disembodied the intelligence. He created a thought experiment that illustrated in a semi-realistic way the effects that a measuring and decision making system (an "intelligence") could have on the entropy of a system. Szilard showed that the second law could be preserved only if the measurements made by the demon generate an "average increase in entropy" that is greater than or equal to the decrease in the entropy of the system caused by the demon using this information. This idea offered a way around the conceptual difficulties and gaps in the theory relating statistical mechanics and thermodynamics that Maxwell had drawn attention to.

Szilard's [1925b] thought experiment consists of a horizontal hollow cylinder fitted inside with a thin piston that can move back and forth along the cylinder. The piston has a hole in its center and a cover for the hole. The cylinder is sealed at each end and there is one molecule trapped in the

---

[33]See the discussion of Maxwell's demon in Chapter 6 and the references cited there.

[34]The work of these authors and the subsequent work of Szilard has been analyzed in Earman and Norton [1008]. They have reviewed both the historical highlights and recent discussions of problems in Szilard's work. I will draw on their work and present the conclusions of their review below.

system. This molecule can be made to do work by lifting a weight, which is accomplished by attaching a rope to one side of the piston or the other, depending on which side of the piston the molecule is on when the door is closed. The molecule does work by making the piston pull the weight up by the rope as the one-molecule gas expands to fill the container. This experiment is reversible, in the standard thermodynamic sense of that term, because it is set up so that the weight is lifted very slowly (quasistatically). Szilard pointed out that to fully account for the entropy in this reversible experiment, an observer needs to know which side of the cylinder the molecule is on so that the entropy of expansion can be calculated properly.[35] Szilard showed, using semipermiable membranes and the methods of phenomenological thermodynamics, that there must be entropy created by a measurement, and that this entropy must be greater than the minimum he identified, if the second law is to be preserved.

Some of Szilard's ideas were taken up and pursued further by several workers, notably Leon Brillouin [1962].[36] Brillouin painstakingly analyzed a number of different measurement methods that the demon could use and showed that in each case the entropy increase associated with the measurement exceeded the entropy decrease of the system the demon is working with.

Szilard's view of the connection between entropy and information was challenged by Josef M. Jauch and J. G. Báron [1972] on the grounds that formal similarities and a reliance on Szilard's [1925b] analysis had fooled workers into identifying the two concepts. They noted that ter Haar [1955], for example, had rejected this association between the two concepts on the grounds that the entropy in information theory is not the thermodynamic entropy and that the use of the same name was due to a rather loose use of the term "information."

Jauch and Báron objected to Szilard's reasoning on two grounds. First, they claimed that closing the door in the piston results in a compression of the one-molecule gas to one-half its volume without an expenditure of energy and thereby violates Gay-Lussac's law. This violation makes Szilard's idealized procedure illegitimate and unacceptable in a thought experiment. Second, they claimed that the observer does not need to know which side of the piston in the cylinder the molecule is on because an automatic device can sense the motion and attach a weight to a rope being pulled up one side or the other.

The paper of Jauch and Báron was answered by O. Costa de Beauregard and Myron Tribus [1972]. These authors observed that opening the door in

---

[35] A fuller account of some of the other aspects of Szilard's work is given in QTS in connection with von Neumann's employment of Szilard's ideas in the development of his approach to quantum thermodynamics and the definition of quantum entropy.

[36] See, in addition, Lewis [1930], pp. 572–573 for a discussion of entropy as information in the context of Szilard's work. For other work on the physical aspects of this issue, see the references in Brillouin [1962]. See also Rothstein [1951] and Katz [1967].

the piston when it is at the center of the cylinder results in the pressure dropping to one-half the value it would have when the door was closed because the molecule, even though its kinetic energy is unchanged, now spends only half its time in each half of the cylinder. Closing the door in the piston is not forbidden from a microscopic point of view, which means that Gay-Lussac's law and other macroscopic laws are not relevant. De Beauregard and Tribus concluded that Szilard's idealization of such a frictionless door is acceptable. They emphasized that part of the conceptual issue at the root of the disagreement is the distinction between macroscopic irreversibility and microscopic reversibility. They finally observed that accounting for the entropy change in the reversible process of lifting a weight very slowly by means of heat absorbed by the molecule from the reservoir requires understanding which half of the cylinder the molecule was in when the door was closed as Szilard claimed. They concluded that it is irrelevant to this calculation whether this measurement was made by an observer or by an automatic device.

John Earman and John D. Norton [**1998**] have reviewed this work. They drew on the recent literature on Maxwell's demon in a reexamination of how Maxwell introduced the demon and the steps leading to Szilard's work. They, p. 455–459, showed the importance of fluctuations to Szilard's conclusion that there is an inevitable hidden cost in the acquisition of information. Because of their analysis, Earman and Norton, p. 460–464, maintained that Szilard's mathematical relations depended on fluctuations, and therefore referred to entropy changes averaged over many cycles of operation of his apparatus and not to just one cycle. In discussing these results, they observed that most, but not quite all, of the subsequent commentators discussed these results as if they referred to the actual entropy obtained in each cycle of the instrument. Earman and Norton showed that there are problems in the modern literature on entropy because it was built on these "uncertain foundations." They observed that Maxwell's original idea of using the demon to circumscribe the validity of the second law had been replaced by the felt need to save the second law from violation by the demon. Most of these new papers missed Szilard's original point and misunderstood the role of fluctuations in the operation of Szilard's engine.

There were two approaches to exorcizing the demon in Earman and Norton's account. The first, proposed initially by Smoluchowski, was that as a thermal system, fluctuations in the demon would on average defeat its attempt to violate the second law. The second, adopted by Szilard, is that there are hidden costs associated with the operation of the intelligence because there are costs associated with a measurement used for acquiring the required information. Earman and Norton observed also that Szilard felt that the costs associated with the acquisition of information was a consequence of the second law and not a separate postulate. Modern work, however, has often treated the costs associated with acquiring information as a separate postulate.

Earman and Norton, pp. 461–464, examined the papers by Jauch and Báron and by de Beauregard and Tribus to look for why there was a discrepancy between using the average entropy computed over many cycles by Szilard and the individual entropy of one cycle used in these papers. They concluded that the quantity $k_B \ln 2$ that Szilard had assigned to his lower bound $e^{-\bar{S}_1/k_B} + e^{-\bar{S}_2/k_B}$ for a measurement was a convenient minimum but that the actual lower bound was arbitrary in any given measurement. Earman and Norton noted that other authors had treated the entropy increase as if $k_B \ln 2$ was the quantity of interest and the minimum entropy increase in each measurement. Von Neumann [**1932c**], p. 400, for example, viewed Szilard's $k_B \ln 2$ as the minimum entropy increase in a cycle of operation of the machine and called it the 'thermodynamic value of knowledge'—the consequence of an either/or choice. Several other authors, notably Léon Brillouin and D. Gabor, were cited by Earman and Norton, p. 460, as making this mistake.

Earman and Norton observed that both Jauch and Báron [**1972**] and the reply to it by Costa de Beauregard and Tribus [**1972**] had missed the significance of the fluctuations in Szilard's work. They went on to object to the treatment of this issue by Żurek [**1984**] who stated that quantum mechanics was required for its solution and claimed that fluctuations can be evaded.

Earman and Norton concluded that Szilard's view of the second law was statistical and that it could only be demonstrated in the long run by a sequence of many trials. Moreover, they returned to their observation that Maxwell had originally introduced the demon to show the limitations on the second law and its statistical nature, but that it had evolved in the literature into a threat against which the second law must be protected. To get at this question, Earman and Norton, p. 462, recommended that the microphysics of the demon be examined as a way of limiting the possible demons to those with physical significance. They also returned to the question of which version of the second law the discussions of Maxwell's demon are trying to protect—the absolute or the statistical version. Initially, Smoluchowski and Szilard were concerned with the second law in the face of fluctuation phenomena and how an intelligent being might turn fluctuations into long-term violations of the second law. However Earman and Norton noted that the discussions in the literature did not refer to fluctuations at all. They dismissed the objections of Jauch and Báron to Szilard's work as having missed the point in this regard. They did not resolve, however, the question of whether the demon is exorcized by observing that he is a thermodynamic system and that his fluctuations defeat his purposes, as Smoluchowski maintained, or he is an intelligent system that has to pay hidden costs for acquiring and using knowledge, as Szilard claimed.

16.10 Entropy and Information

Maxwell's understanding that the entropy may involve an element of our knowledge was mentioned above. His demon provides a graphical representation of how knowledge and measurement could have an effect on the entropy attributed to a system. He also observed that a being with powers of following each molecule in its flight would have no need of the concept of entropy. Maxwell concluded that the second law could have only a statistical validity for systems based on microscopic particles.

These insights of Maxwell were the first step in detaching entropy from the viewpoint that it is a physical aspect of matter that can be possessed as a property. This represents a change from the view of the system state as a physical entity in its own right towards the viewpoint of a state similar to that advocated by Bohr for quantum mechanics. The core of the reconciliation between the reversibility and recurrence aspects of the underlying particle theory and the thermodynamics constructed out of the description of these particles and their interactions lies in the understanding that it is the choice of the state we use to describe a system that determines its entropy. This epistemological point does not diminish the usefulness of the concept of entropy in thermodynamics. It simply changes our understanding of what it is and how it works.

The association of an increase in entropy with the loss of information in an irreversible process was made quite clearly by Lewis [1930], p. 573: "The increase in entropy comes when a *known* distribution goes over into an *unknown* distribution. The loss which is characteristic of an irreversible process, is *loss of information*. ... Gain in entropy always means loss of information, and nothing more." This statement is strikingly close to the point of view presented in TIS. From a physical point of view, the system distribution evolves in time in accord with the mechanical laws governing the motion of the underlying particle system in conjunction with the boundary conditions. It was shown in the TIS formalism that the entropy, and therefore the information we have of the system, will change only when an individual particle encounters a stochastically described boundary. When this is the case, the rate of change of the information is given by equation (9.15) or (9.16). In a pictorial sense, this means that when a particle with a sharply defined trajectory encounters a fuzzy (stochastic) wall, the reflected trajectory will be fuzzy. Under constant and uniform boundary conditions, this process will continue until the whole volume is filled with a uniform probability fog distinguished only by differences in the probability of finding the system at different energy levels (the equilibrium distribution). This illustrates the situation in which our information about the system has been reduced to a minimum.

Another statement of a similar viewpoint was made by Edwin T. Jaynes [1965] in a comparison of the entropies defined by Gibbs and Boltzmann. He presented the viewpoint that entropy is an anthropomorphic concept. He, pp. 397–398, began by observing that there is no unique thermodynamics for a macroscopic system because the same system can be described in a number of ways for different experiments. He argued that these version of the entropy are not simply different aspects of a "true entropy" that underlies them all because in macroscopic systems we can always introduce new degrees of freedom when we please in the description of the system. This can continue until an exact description is given of the locations and velocities of each particle in the system. At this point, he argued, the description is no longer appropriate for thermodynamics, but belongs to mechanics. He concluded that *"Even at the purely phenomenological level, entropy is an anthropomorphic concept. For it is a property, not of the physical system, but of the experiments you or I choose to perform on it."* He acknowledged that he had first heard this view from E. P. Wigner.

The presentation and discussion in the last section of Szilard's attempt to fill the gap left by Smoluchowski's analysis concerning the possibility that a living entity can violate the second law illustrates the reasoning that has been used to assess the significance of the second law in the face of living organisms that have knowledge and can make choices. Comparing the conceptual basis and methodology employed by Szilard, Brillouin, and others, in the analysis of this matter highlights the significant differences between the physical viewpoint of those approaching issues of thermodynamics from this perspective and the TIS approach.

From the TIS perspective, recall that it has been emphasized on a number of occasions in this book that the entropy of a system can only be defined in terms of the system distribution. From the TIS point of view, therefore, the entropy is not required to increase and does not always increase in the way that standard thermodynamics predicts. It can decrease, for example, when a different choice is made of which distribution to use to describe a system without requiring or causing a change in the energy or entropy of any other system. This choice, therefore, is an act by an intelligent being that changes the entropy in an arbitrary way that does not preserve standard versions of the second law. This means that the work of Maxwell, Szilard, Brillouin, and others who tried to uncover the implications of the second law by using a demon or to demonstrate how to preserve it from intelligent demons is not relevant to the TIS perspective.

In terms of the TIS entropy analog $-k_B \ln F_k(Q_k, P_k, t)$, the association of entropy with information is equivalent to saying the system distribution represents our information about the system. If this information is complete and precise, a particular F_k^δ state can be attributed to the system with an entropy of $-\infty$ or an information content of $+\infty$. Otherwise, a more diffuse F_k state, or a maximally diffuse F_k^ϵ state with information content $-k_B \beta_k \mathcal{H}_k^\epsilon$, should be used. This information about the system state is not limited to the standard mathematical distributions and can be represented in a variety of ways and in a variety of contexts. From the perspective of "Algorithmic Information Theory", for example, the system distribution and its boundary conditions can be represented for some m, n as a computer program of length less than or equal to m that will output this information in less than n steps.[37] In this context, a computer program that requires more than about $6\mathcal{N}_k$ binary numbers to represent it will offer no advantage over the exact mechanical description of the underlying particle system, (Q_k, P_k).

Upon reflection, the appearance of an aspect of information in an essential role in physical theory is not surprising. We, as biological organisms, are above all information processing entities. It is the information content of our theories that helps us to understand and manipulate nature and contributes to our survival. It follows then, in the light of the case presented above concerning the definition and properties of the entropy, that equation (16.12) is an appropriate way of interpreting its connection with information. Viewing the entropy as a measure of information also helps to resolve many of the conflicts that have plagued our attempts to understand it as a physical property of a system. To be sure, this point of view means that there is no direct physical model for entropy in the same way there is a model for the energy that is seated in the particles. However, entropy mediates the relationship between our world of information and the description of the particle world that underlies it. It therefore plays a very different role than energy does and a different epistemological status is warranted for it.

In keeping with the overall TIS perspective, the system state is not accorded an independent reality or existence as something possessed by the system. This viewpoint contrasts with many discussions of systems at equilibrium in which it is implied that the number of particles with particular energies is determined by the system equilibrium distribution. The statistical view of the system state in these texts is distinct from the probabilistic view in TIS in which only probabilities can be stated that a particle or the system as a whole are in some particular mechanical state. The statistical and probabilistic viewpoints are not equivalent and it has been pointed out previously that significant confusion has arisen from conflating the two.

Briefly, TIS denies that the entropy is a physical property of any system or the particles in it. It is an aspect of the description chosen for a system in the context of the other systems with which it is interacting. There are a number of ways to represent it as a thermodynamic analog using convex functions and there is no criterion for choosing one of these analog functions as the "right" one. A non-logarithmic convex function, for example, will be

---

[37]For a review of Algorithmic Information Theory, see G. Chaitlin [1977] and the references cited there. For applications to physics, see C. Bennett [1982] and R. Landauer [1961].

used in Chapter 26 to demonstrate that a system in certain circumstances will evolve from an arbitrary state to the equilibrium state.

The introduction of quantum mechanics with its requirement that probability be used as a fundamental aspect of the theory, and the changes that its formalism and theory of measurement entailed for the physical description of particle systems, has significantly complicated the issues. Nevertheless, the discussion of entropy and information in the quantum context in QTS supports the conclusions drawn here concerning the relation of entropy and information.

Part II

Classical and Equilibrium
Thermodynamics

It will not escape the reader's notice, that while from one point of view the operations which are here described are quite beyond our powers of actual performance, on account of the impossibility of handling the immense number of systems which are involved, yet from another point of view the operations described are the most simple and accurate means of representing what actually takes place in our simplest experiments in thermodynamics. The states of the bodies which we handle are not known to us exactly. What we know about a body can generally be described most accurately and most simply by saying that it is taken at random from a great number (ensemble) of bodies which are completely described.

Josiah Willard Gibbs, *Statistical Mechanics*, 1902

CHAPTER 17

Classical Thermodynamics

In Part II, classical and equilibrium thermodynamics are studied in the context of the general TIS theory. Thermodynamic potentials and operators are discussed as well.

17.1 Classical Thermodynamics

One of the abiding goals of thermodynamics has been the development of equations of state that connect a set of macroscopic physical quantities describing a system with the value of another macroscopic physical quantity associated with the system. An equation of state connecting the pressure and volume of a substance was first given a mathematical form in the seventeenth century as the Boyle-Towneley law: $PV = $ const. It was extended by Gay-Lussac and Mariotte to include the effect of the temperature and written in the form $PV = RT$, where R is the gas constant. These laws are exact for ideal gases and approximately true for very dilute gases. The next steps were based on the growing recognition in the late seventeenth century that the concepts of heat and temperature are distinct. Studying this distinction led to the discovery of the latent heats of fusion and evaporation and the specific heats of substances by Joseph Black in the middle of the eighteenth century.[1]

Thermodynamics from the late seventeenth century up to the advent of equilibrium thermodynamics and kinetic theory in the middle of the nineteenth century is called *classical thermodynamics*.[2] Classical thermodynamics began with the formal study of how substances respond to heating or cooling. It was based on the observation that the heat entering a substance will either raise its temperature, cause a change in the volume, or increase its pressure. Black observed the reproducible capacity of a substance in a fixed volume or at a given pressure to absorb or give off a particular amount of heat with a specific change in its temperature. The change in the heat of a mole of a given substance with a particular change in temperature was called the *specific heat* of the substance. The change of the volume of a substance in response to the absorption or emission of heat was associated with what was called the latent heat with respect to volume or pressure. Since heat flow, volume, pressure, and temperature of a substance are all measurable, it was possible to establish an experimental and theoretical study of macroscopic processes involving heat transfer between thermodynamic bodies and the effect of this heat change on the volumes, pressures, and temperatures of the bodies. The formalism associated with this study was called the *theory of calorimetry*.[3]

The widespread introduction of steam engines in the early nineteenth century fostered an extension of these ideas into a consideration of the relation between heat and work. The quantitative application of the theory

---

[1] For a discussion of Black's discovery of latent and specific heats, see Douglas McKie and Niels H. de V. Heathcote [**1935**], Chapters I and II. For more information on this and other topics covered in this chapter, see also the excellent survey of thermodynamics from the standard perspective in Bailyn [**1994**].

[2] For a conceptual and historical analysis of classical theory, see Truesdell and Bharatha [**1977**] and Truesdell [**1980**].

[3] See Truesdell [**1980**], Chapter 2, for a conceptual discussion of the early equations of state and the theory of calorimetry.

was further stimulated by the development of accurate thermometers and calorimeters in the course of the nineteenth century.

The most popular theory of heat in the early nineteenth century was the caloric theory that explained heat phenomena in terms of the behavior of small particles called caloric particles. The analysis of the relation of heat and work by Carnot [1824] was within the framework of the caloric theory of heat. While the caloric theory was soon superseded by the emerging kinetic theory, Carnot's theory survived and served as the foundation for the further development of thermodynamics. Clapeyron [1834] took up the mathematical analysis of Carnot's ideas, followed by Thomson [1849], [1851], and Clausius [1856], who removed it from the context of caloric theory.

17.2 The Theory of Calorimetry

Formally, classical thermodynamics studied physical quantities associated with macroscopic objects and the responses of these objects to changes in the values of physical quantities that described them. The macroscopic objects were called *thermodynamic bodies* and the sequence of changes of a body was called a *thermodynamic process*. Some experiments associated heat flow with temperature differences and these were described theoretically by Newton's law of cooling or Fourier's law of heat conduction. The theory of calorimetry went beyond this by associating changes in both the temperature and volume of a body with a flow of heat into or out of it.

The theory of calorimetry is based on what Truesdell [1980], pp. 15–27, has called the "doctrine of latent and specific heats". This doctrine states that the flow of heat into or out of a body can be described entirely in terms of two functions that are called the 'latent heat with respect to changes in the system volume' and the 'specific heat with respect to changes in the system temperature'. When the rates of change of the heat or work produced or received by a body are needed, the terms heating and working will be used. The thermature rather than the temperature will continue to be the primary representative in the formalism for issues concerning the temperature. The gain in the uniformity of approach to thermodynamic issues and their representation is worth the loss of historical accuracy in the notation.

In this formalism, the volume falls within the range $0 < \delta_k(t) < \infty$, the pressure in the range $0 \leq P_k < \infty$ and the thermature falls within the range $0 < \beta_k(t) \leq \infty$. Each of these quantities is assumed to be a continuous function of the time or some convenient parameter. The state of a body at equilibrium is a function of these macroscopic parameters and the number of particles. An example is the equation of state of Gay-Lussac and Mariotte for a classical ideal gas at equilibrium. In TIS notation, the ideal gas equation of state is $\delta_k P_k = k_B \mathcal{N}_k T_k$. This relation shows that three independent variables are needed to specify the state of an ideal gas. In general, for systems at equilibrium, any three macroscopic parameters from the set $\{ \mathcal{N}_k, \beta_k, \delta_k, \mathrm{P}_k \}$ are sufficient to determine the fourth for a body. The assumption that systems are closed implies that $\mathcal{N}_k$ does not vary for a given system, so $\mathcal{N}_k$ is assumed from now on to be a fixed system parameter and will not be stated as an independent variable in representations of the equilibrium state. As a consequence, two parameters from the set $\{ \beta_k, \delta_k, \mathrm{P}_k \}$, called the set of *state parameters,* are sufficient to fix the third for a body.[4]

The *state* of a body is defined as a set of parameters sufficient for a full characterization of the body within the context of a theory. Two parameters selected from the set of three state parameters in a given setting are therefore called the *equilibrium state* of the body for that setting. The choice of which two independent variables to use as the system state is usually determined by which two of the three above quantities are controlled at the boundary of the system. Once a state has been defined, all other quantities of the equilibrium thermodynamics of a body, including the energy and entropy, can be defined as functions of the state of the body.

An equation of state establishes relations between the members of the set of state parameters introduced above. This allows moving between thermodynamic representations. If the pressure, for example, is given as a function of the state (β_k, δ_k) by the function $\mathrm{P}_k(\beta_k, \delta_k)$, and the requirements of the implicit function theorem are met for this function, the volume may be expressed as a function of the parameters (β_k, P_k) by $\delta_k(\beta_k, \mathrm{P}_k)$, where $\delta_k(\beta_k, \mathrm{P}_k)$ is computed from $\mathrm{P}_k(\beta_k, \delta_k)$ by means of the implicit function theorem. In other settings, any two other quantities defined in equilibrium thermodynamics, such as the energy and entropy, can be used as independent variables to represent the equilibrium state.

In analyzing thermodynamic states and processes, the macroscopic quantities $\{ \beta_k, \delta_k, \mathrm{P}_k \}$ can be viewed as coordinate axes in a 3-dimensional space called *equilibrium space.* An equilibrium state is represented as a point in this space. It will be assumed for now that thermodynamic functions, such as $E_k(\delta_k, \beta_k)$, are at least once continuously differentiable in their arguments. As in the general case, a sequence of changes of the equilibrium state of a body is called a thermodynamic *process.* Because of the continuity assumptions on the functions representing the thermature, volume, and pressure, a process is a path in equilibrium space. For a system described by the independent variables (β_k, δ_k), this path can be represented by the trajectory $(\mathrm{P}_k(\beta_k(t), \delta_k(t)), \beta_k(t), \delta_k(t))$ in equilibrium space, where t is the time or some other convenient path parameter.

The assumptions on the physical parameters describing a thermodynamic system mean the two-dimensional space described by (β_k, δ_k) is a manifold

---

[4]This assumption is not valid in circumstances in which the surface area of the system and the forces of external systems play an important role. This is the case, for example, in the study of capillary action.

in the equilibrium space.[5] Because of the bounds on the thermature and the volume stated above, thermodynamic processes are restricted to a subset of the positive quadrant of the equilibrium manifold chosen. The process $(\beta_k(s), \delta_k(s))$ can also be seen as the projection on the (β_k, δ_k) equilibrium manifold of the path $(P_k(s), \beta_k(s), \delta_k(s))$ in equilibrium space.[6]

The theory of calorimetry is developed using these ideas by defining the specific heat and latent heat as functions of either the state (β_k, δ_k) or the state (β_k, P_k). The specific and latent heat functions depend on the time or path parameter only through the dependence of the system state parameters on the time or path parameter. This formalism is connected to the actual constitution of a given body by functions representing the constitutive relations and properties of the matter it contains. These constitutive relations determine how much the thermodynamic functions of a body change during a given process. The constitutive elements that determine the actual heat capacity of a particular body, associated with an increase in the average kinetic energy of the particles, and the actual latent heat of the body, associated with changes in the internal potential energy, are the particle interaction potentials and the particle masses.

The basic concepts of the theory of calorimetry are listed in Table 17.1:[7]

Table 17.1: Notation

Quantity	Truedell's Notation	TIS Notation
Volume	$V(t)$	$\delta_k(t)$
Temperature	$\theta(t)$	$T_k(t)$
Thermature	$1/k_B\theta(t)$	$\beta_k(t)$
Pressure	$P(t)$	$P_k(t)$
Heating	$Q(t)$	$Q_{\partial k}(t)$

For the independent variables (β_k, δ_k), the most important constitutive functions (TIS notation) are:

1. the *pressure,* $P_k(t)$, expressed in the form

$$(17.1) \qquad P_k(t) = \varpi_k(\beta_k(t), \delta_k(t));$$

---

[5] The mathematical issues concerning manifolds defined in equilibrium space by various choices of the independent variables are discussed in EIS, Part IV.

[6] This approach has its roots in Gibbs [**1873**] and is explicitly used in Duhem [**1891**], [**1903**]. A generalized approach is taken in EIS, Part IV, and the definitions of equilibrium manifolds in equilibrium space by imposing conditions on generalized versions of the thermodynamic potentials are investigated. Equilibrium space also has an associated geometry that is not necessarily Euclidean, so the differential geometry of equilibrium manifolds will be examined there as well.

[7] For this list, see Truesdell and Bharatha [**1977**], pp. vii, 148.

2. the *heat capacity at constant volume*, $K_{k,V}(\beta_k(t), \delta_k(t))$; and
3. the *latent heat with respect to volume*, $\Lambda_{k,V}(\beta_k(t), \delta_k(t))$.

The rate of the flow of heat into the body as a function of the time is expressed by the *calorimetry equation* as a function of the process $(\beta_k(t), \delta_k(t))$ as[8]

$$(17.2) \qquad Q_{\partial k}(t) = K_{k,V}(\beta_k(t), \delta_k(t))\dot{\beta}_k(t) + \Lambda_{k,V}(\beta_k(t), \delta_k(t))\dot{\delta}_k(t).$$

As *constitutive relations*, the following inequalities are usually assumed:[9]

$$(17.3\text{a}) \qquad\qquad \frac{\partial P_k(t)}{\partial \delta_k} < 0,$$

$$(17.3\text{b}) \qquad\qquad \frac{\partial P_k(t)}{\partial \beta_k} < 0,$$

$$(17.3\text{c}) \qquad\qquad \Lambda_{k,V}(\beta_k(t), \delta_k(t)) > 0,$$

$$(17.3\text{d}) \qquad\qquad K_{k,V}(\beta_k(t), \delta_k(t)) < 0.$$

Note that the directions of the inequalities in (17.3b) and (17.3d) are reversed from those of the usual formulation due to the use here of $\beta_k(t)$ rather than $T_k(t)$ as the basic variable. This notational difference will also appear in some of the relations defined below.

By (17.1) and the usual assumption that the pressure $\varpi(\beta_k, \delta_k)$ satisfies the requirements of the implicit function theorem, equation (17.2) can be rewritten to express the heating of a body in terms of the thermature and pressure:

$$(17.4) \qquad Q_{\partial k}(t) = K_{k,\text{P}}(\beta_k(t), \delta_k(t))\dot{\beta}_k(t) + \Lambda_{k,\text{P}}(\beta_k(t), \delta_k(t))\dot{P}_k(t).$$

The function $K_{k,\text{P}}(\beta_k, \delta_k)$ is called the *heat capacity at constant pressure* and $\Lambda_{k,\text{P}}(\beta_k, \delta_k)$ is called the *latent heat with respect to pressure*. When the thermature and pressure are used as the independent variables, the relations $K_{k,\text{P}}(\beta_k, P_k) = K_{k,\text{P}}(\beta_k, \delta_k(\beta_k, P_k))$ and $\Lambda_{k,\text{P}}(\beta_k, P_k) = \Lambda_{k,\text{P}}(\beta_k, \delta_k(\beta_k, P_k))$ are used to express the (β_k, P_k) forms of the functions in terms of the (β_k, δ_k) forms used here. Comparing (17.4) with (17.2) yields

$$(17.5) \qquad\qquad \Lambda_{k,\text{P}}(\beta_k, \delta_k) = \left[\frac{\partial P_k}{\partial \delta_k}\right]^{-1} \Lambda_{k,V}(\beta_k, \delta_k)$$

and

$$(17.6) \qquad K_{k,\text{P}}(\beta_k, \delta_k) - K_{k,V}(\beta_k, \delta_k) = -\left[\frac{\partial P_k}{\partial \delta_k}\right]^{-1} \frac{\partial P_k}{\partial \beta_k} \Lambda_{k,V}(\beta_k, \delta_k).$$

---

[8] For a conceptual and historical discussion of $\Lambda_{k,V}(\beta_k, \delta_k)$, see Truesdell [**1980**], pp. 198–203.

[9] Truesdell and Bharatha [**1977**], pp. 9, 60, have emphasized that the relations (17.3b) and (17.3c) are not universally true. These relations are violated in the anomalous behavior of water near freezing at atmospheric pressure, for example. The work in subsequent chapters will not depend on these assumptions.

By (17.3) the constitutive relations that result from this are

(17.7a) $\Lambda_{k,P}(\beta_k, \delta_k) < 0,$

(17.7b) $K_{k,P}(\beta_k, \delta_k) < K_{k,V}(\beta_k, \delta_k).$

The standard definition of the working, the rate at which work is being done on the body, in the context of processes defined by the changing values of (β_k, δ_k) is[10]

(17.8) $W_{\partial k}^{(c)}(t) = -P_k(t)\dot{\delta}_k(t).$

Let us express the pressure in terms of the time t by $P_k(t) = P_k(\beta_k(t), \delta_k(t))$. According to the theory of calorimetry, the net heat absorbed and the net work done during the process $(\beta_k(t), \delta_k(t))$ operating between times t_1 and t_2 on the path $\mathcal{P} = (P_k(t), \beta_k(t), \delta_k(t))$ are:

(17.9a)

$$Q(\mathcal{P}) = \int_{t_1}^{t_2} dt\, Q_{\partial k}(t) = \int_{\mathcal{P}} [d\delta_k\, \Lambda_{k,V}(\beta_k, \delta_k) + d\beta_k\, K_{k,V}(\beta_k, \delta_k)],$$

(17.9b)

$$W(\mathcal{P}) = \int_{t_1}^{t_2} dt\, W_{\partial k}^{(c)}(t) = \int_{\mathcal{P}} d\delta_k\, \varpi_k(\beta_k, \delta_k).$$

Note that for the reversed process path, that is, a process following the path $-\mathcal{P}$, the definitions of the integrals imply

(17.10a) $Q(-\mathcal{P}) = -Q(\mathcal{P}),$

(17.10b) $W(-\mathcal{P}) = -W(\mathcal{P}).$

Ideal gases play an important role in testing theoretical notions in thermodynamics because of the mathematical tractability of the formulas representing them. The pressure function of an ideal gas is expressed as

(17.11) $P_k^{(ig)} = \dfrac{\mathcal{N}_k}{\beta_k \delta_k}.$

In standard notation, this is expressed as nRT/V for n moles of gas. For ideal gases, it is easy to show that Karl Holtzmann's assumption (1845) that $\Lambda_{k,V} = P_k^{(ig)}$ and Julius Robert Mayer's assumption (1845) that $K_{k,P} - K_{k,V} = -N_k\beta_k^{-2}$ are equivalent.[11]

---

[10]Recall that the discussion in Chapter 11 concluded that this definition of $W_{\partial k}^{(c)}(t)$ is not compatible with the TIS surface working $W_{\partial k}(t)$ defined in (7.49) because (17.8) is only an approximation in a particle based approach.

[11]See Truesdell and Bharatha [**1977**], pp. 154–158, on this.

17.3 Cyclic Processes

Carnot's theory of heat engines was mentioned in Chapter 8.[12] Using steam engines as a model, he considered heat engines that convert heat to work in a cyclic operation. The basis for his analysis was a heat engine that uses a working substance to take in heat from a reservoir at a high temperature, convert part of that to work, and discharge heat to a reservoir at a lower temperature. This engine operates cyclically, so that the moving parts and the working substance in the engine are returned at the end of a cycle to their initial macroscopic states.

Carnot limited his discussion to cases in which the difference between the temperature of the heat reservoirs and the working substance in the engine is vanishingly small. He recognized that finite temperature differences, for which the flow of heat is not accompanied by a corresponding change in the volume of the working substance, would result in the less efficient operation of the machine. Kuhn [1960], p. 252–253, has drawn a parallel between this efficiency theorem of Sadi Carnot and the engineering theorem of his father Lazare Carnot that states "In order that [machines] produce their maximum possible effect, there must be no percussion, i.e., the motion must always change by insensible steps." The importance of both the heat source and the heat sink, fundamental to Sadi Carnot's work, was also recognized in these engineering texts.

Let us consider a cyclic process $\mathcal{C}$ as a closed path in the positive quadrant of the (β_k, δ_k) equilibrium manifold.[13] Assume that $\mathcal{C}$ is a closed oriented curve that does not intersect itself. Assume also that the relation $\Lambda_{k,V} > 0$ is valid on and within this curve.[14] This is illustrated in Figure 17.1, where the arrowheads indicate the direction in which the curve is being traversed.

Along segment (1,2), the body absorbs heat at the fixed thermature β_k^{12}. By (17.2), the rate of heat absorption is

$$(17.14) \qquad\qquad Q_{\partial k}(t) = \Lambda_{k,V}(\beta_k^{12}, \delta_k)\dot{\delta}_k.$$

The total heat absorbed by the body along path (1,2) is

$$(17.15) \qquad\qquad \mathcal{Q}(1,2) = \int_1^2 d\delta_k\, \Lambda_{k,V}(\beta_k^{12}, \delta_k).$$

---

[12]Carnot's theory has been analyzed carefully in Truesdell and Bharatha [1977].

[13]Rankine [1854], p. 115, asserted that James Watt was the first to make use of pressure-volume diagrams in his Steam-Engine Indicator diagram. He, p. 116, also stated that Carnot and Clapeyron were probably the first to use them in conjunction with energy considerations, but complains that their work was marred by adherence to the caloric theory.

[14]See Truesdell and Bharatha [1977], pp. 6, 50, for a discussion of these aspects of Carnot cycles. See also the discussion of anomalous Carnot cycles there. Water near freezing is anomalous because $\Lambda_{k,V} < 0$ in the neighborhood of 4° C.

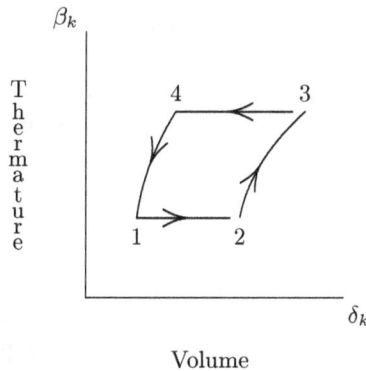

Volume

FIGURE 17.1. A Regular Carnot Cycle

Similarly, along (3,4), the heat emitted is

$$(17.16) \qquad \mathcal{Q}(3,4) = \int_3^4 d\delta_k \, \Lambda_k(\beta_k^{34}, \delta_k).$$

Along (2,3) the thermature is raised (the temperature T_k is lowered) from β_k^{12} to β_k^{34} by expanding the working substance adiabatically. Similarly, along (4,1) the thermature is lowered (the temperature is raised) from β_k^{34} to β_k^{12} by compressing the working substance adiabatically. This means

$$(17.17) \qquad \mathcal{Q}(2,3) = \mathcal{Q}(4,1) = 0.$$

Two important assumptions made by Carnot have already been mentioned: that the temperatures (thermatures) β_k^{12} and β_k^{34} are held fixed by reservoirs and that heat exchange is allowed only when the body thermature has been brought to the thermature of the reservoir. In Carnot's view, this avoids wasting motive power, which depends on temperature differences. Thus, only those changes in temperature that are associated with changes in volume are allowed.[15] Using calorimeter experiments as the basis, Carnot [**1824**], also stated in a footnote on p. 67, somewhat defensively, his further assumption that the heat evolved in going from $(\beta_k^{12}, \delta_k^1)$ to $(\beta_k^{12}, \delta_k^2)$ is independent of the path.

Carnot's requirement that no heat be allowed to flow between bodies that have a non-zero temperature difference between them is an idealization, but it is consistent with his search for theoretical limits.[16] A formula for the maximum efficiency for the conversion of the heat absorbed into work, which depends only on the ratio of the temperatures, was developed in part

---

[15]See Carnot [**1824**], pp. 51, 56–57, 58, 60.
[16]See Carnot [**1824**], pp. 58, 60.

by Thomson in 1851 followed by Rankine in 1851 and Clausius in 1856.[17] The maximum work $W(\mathcal{C})$ that can be obtained from the heat absorbed, $Q(1,2)$, in this cycle is the product of the heat absorbed and the efficiency. It does not depend on the working substance and takes the form[18]

$$(17.18) \qquad W(\mathcal{C}) = \left(1 - \frac{\beta^{12}}{\beta^{34}}\right) Q(1,2).$$

Carnot's theory of cyclical processes was moved from its dependence on caloric theory by Clausius [**1850**] and Thomson [**1851**]. Clausius showed that Carnot's assumption that the quantity of heat transferred across a boundary from one system to another is independent of the path is rooted in caloric theory and is incorrect. Clausius replaced this with the assumption that the entropy is path independent for reversible paths.

Clausius defined the rate of entropy change for reversible paths to be a quantity equivalent to (TIS notation)

$$(17.19) \qquad \frac{dS_k(t)}{dt} = k_B \beta_k(t) Q_{\partial k}(t),$$

where $Q_{\partial k}(t)$ is the rate at which heat is flowing across the boundary. Systems not at equilibrium are not on reversible paths and Clausius [**1862**] maintained for these systems that (TIS notation)

$$(17.20) \qquad \frac{dS_k(t)}{dt} > k_B \beta_k(t) Q_{\partial k}(t).$$

Clausius also included in his formalism the newly discovered conservation of energy, which is expressed in the form

$$(17.21) \qquad \frac{dE_k(t)}{dt} = Q_{\partial k}(t) + W_k(t),$$

where $W_k(t) = W_{k,b}(t) + W_{\partial k}(t)$ is the total working. The relations (17.19), (17.20) and (17.21), were presented in Clausius [**1854**] and provided an axiom set for thermodynamics.[19]

---

[17]See Truesdell and Bharatha [**1977**], pp. 90–91, and Chapter 13, for historical details and a further discussion of efficiency. See also Truesdell [**1980**], pp. 113–117, for a proof that Carnot cycles alone are the most efficient for given extremes of temperature.

[18]The standard form of this relation is

$$W(\mathcal{C}) = \left(1 - \frac{T^{34}}{T^{12}}\right).$$

See Bailyn [**1994**], pp. 83–90, for more on the efficiency of Carnot cycles and how Clausius removed the assumption that heat (as caloric) is conserved and replaced that with the conservation of work (energy). Bailyn has also discussed how the statements of the second law by Thomson and Clausius are related to Carnot's conceptions.

[19]See Truesdell and Bharatha [**1977**], pp. 137–139.

Using the state $(\beta_k(t), \delta_k(t))$ of a body as the fundamental variables describing it, Truesdell and Bharatha [1977], p. 149, have proposed a similar axiom set for classical thermodynamics:

1. the "caloric equation of state,"

$$(17.22a) \qquad\qquad E = \bar{E}(\beta_k, \delta_k),$$
$$(17.22b) \qquad\qquad S = \bar{S}(\beta_k, \delta_k);$$

2. the "balance of energy," equation (17.21) above;
3. the "balance of entropy," equation (17.19) above;
4. the "linear working," equation (17.8) with (17.1) above; and
5. an axiom that $\bar{E}$, $\bar{S}$ and ϖ satisfy equations (17.21), (17.19) and (17.8) when these are mathematically well defined.

Thomson [1852] introduced the notion that all thermodynamic processes are either *reversible* or *irreversible*. He defined reversible processes as those in which the mechanical energy can be restored to its earlier condition and irreversible processes as those for which there is a "*dissipation* of mechanical energy". He characterized reversible processes as those for which a quantity heat is transferred between a cold body and a hot body in "a definite proportion depending on the temperatures of the two bodies". This means that the net entropy of the heat transferred is zero in a reversible process. As a consequence of this viewpoint, Thomson stated that (1) There is a universal tendency at present in the material world to the dissipation of mechanical energy; (2) any restoration of mechanical energy is impossible without the dissipation of a greater amount of energy; and (3) that "Within a finite period of time past, the earth must have been, and within a finite period of time to come the earth must again be, unfit for the habitation of man as at present constituted, unless operations have been, or are to be performed, which are impossible under the laws to which the known operations going on at present in the material world are subject."[20] The prediction in (3) soon came to be called the "heat death of the universe" and excited a great deal of popular interest at the time.

From the standpoint of the Theory of Interacting Systems, it was established in Chapter 13 that the relation (3.15) for a reversible process can be used to describe the system on path segments (1, 2) and (3, 4) because the system and the reservoir are at equilibrium with equal temperatures during the transfer of heat into or out of the body on these paths. For the reversible adiabatic processes on path segments (2, 3) and (4, 1), there is no heat flow. The process is controlled macroscopically and the boundaries are at rest, so the quantity designated by $\lambda^*(\mathcal{C}_H)W_{\partial k}(q, t)$ in Chapter 9 can be neglected. In this situation, (3.15) implies that (3.16b) will hold and the process will

---

[20]See Thomson [1852] ([1882], Vol. 1, pp. 512, 514). See also the discussion in Bailyn [1994], pp. 96–100, and the remarks of Maxwell [1870] quoted in Section 8.4.

be reversible in time as well. A sequence of processes such as these, which occur entirely on the equilibrium manifold in Carnot's analysis, can only be approached asymptotically by real, time-dependent processes. However, the upper limit predicted for the conversion of heat into work and the other consequences of Carnot's approach to defining thermodynamic limits, established by making use of these idealized processes, are still valid within the general thermodynamics of the Theory of Interacting Systems.

17.4 The Equilibrium Manifold

Let $\beta_k(s)$ and $\delta_k(s)$ be once continuously differentiable real valued functions of the real variable s that define a path in the equilibrium manifold. Let $\mathcal{C}_k(s)$ be the path of the system parameterized by s and write it in the form

$$(17.23) \qquad \mathcal{C}_k(s) = (\beta_k(s), \delta_k(s)).$$

An implicit assumption in the discussion above of thermodynamic processes is that the equilibrium functions representing the physical quantities of the theory remain valid during the process. It was recognized early that this will not be the case for real processes. Clausius' result that the entropy of a body will increase during an irreversible processes, coupled with the fact that any real process was recognized to be irreversible in this sense, provided a concrete reminder of this.

In response, the theory of *quasistatic processes,* also called *quasistatic paths,* was invented. In accord with the usual terminology, this assumption maintains that if movement along the path $\mathcal{C}_k(s)$ in the equilibrium manifold is sufficiently slow, the assumptions underlying equilibrium theory will be valid and the functions of equilibrium thermodynamics can be used to describe the body as it evolves. Every process that follows a quasistatic path is reversible and *vice versa.* In modern form, the assumption that a process follows a quasistatic path is stated as the assumption of local thermodynamic equilibrium (LTE).

The requirement that all processes are quasistatic is a severe limitation on classical and equilibrium theory. It means, for example, that a piston compressing a gas is required to move infinitely slowly. Few real physical processes of interest come close to meeting this criterion. In addition, the transient effects of changing some system parameter, such as allowing the boundary to move, are often the effects in which we are interested.

To formalize this investigation, suppose the set $d_k(s)$ is a volume set function that depends on the parameter s and that $\partial d_k(s)$ is its boundary. These set functions are assumed to meet the requirements stated in Chapter 2. The volume, $\delta_k(s)$ is the measure of this volume set and is defined by

(2.10). Differentials of the state parameters are expressed using the notation

(17.24a)
$$d\delta_k = d\delta_k(s) = \frac{d\delta_k(s)}{ds}ds,$$

(17.24b)
$$d\beta_k = d\beta_k(s) = \frac{d\beta_k(s)}{ds}ds.$$

Let us define the *heating* done by a system moving on a quasistatic path $\mathcal{C}_k(s)$ from s to $s + ds$ using (17.2) by

(17.25)
$$Q_{\partial k}(\mathcal{C}_k(s))ds = K_{k,V}(\mathcal{C}_k(s))d\beta_k + \Lambda_{k,V}(\mathcal{C}_k(s))d\delta_k.$$

The Hamiltonian energy is used with the heating to define the working on the path $\mathcal{C}_k(s)$ by the form

(17.26)
$$W_k(\mathcal{C}_k(s))ds = d\mathcal{H}_k(\mathcal{C}_k(s)) - Q_{\partial k}(\mathcal{C}_k(s))ds.$$

Expanding the differential $d\mathcal{H}_k(\mathcal{C}_k(s))$ and using (17.25) gives

(17.27)
$$W_k(\mathcal{C}_k(s))ds = \left(\frac{\partial \mathcal{H}_k(\mathcal{C}_k(s))}{\partial \delta_k} - \Lambda_{k,V}(\mathcal{C}_k(s)) \right) d\delta_k$$
$$+ \left(\frac{\partial \mathcal{H}_k(\mathcal{C}_k(s))}{\partial \beta_k} - K_{k,V}(\mathcal{C}_k(s)) \right) d\beta_k.$$

Since the system working is a mechanical quantity that depends on the motion of the walls of the system, the expression (17.8) implies that the working depends only on the change in the volume and not on the change in the thermature. This means that the coefficient of $d\beta_k$ in (17.27) should vanish, which implies

(17.28a)
$$K_{k,V}(\mathcal{C}_k(s)) = \frac{\partial \mathcal{H}_k(\mathcal{C}_k(s))}{\partial \beta_k}$$

is used.[21] For (17.8) to be valid as well, so that $W_k(\mathcal{C}_k(s)) = W_{\partial k}^{(c)}(s)$, let us define

(17.28b)
$$\Lambda_{k,V}(\mathcal{C}_k(s)) = \frac{\partial \mathcal{H}_k(\mathcal{C}_k(s))}{\partial \delta_k} + P_k(\mathcal{C}_k(s)).$$

These relations show the consequences of the definition (17.8) of the working along a quasistatic path when the heating is defined in accord with the theory of calorimetry.

---

[21] $K_{k,V}$ is expressed in terms of the standard heat capacity $C_{k,V}$ by the conversion formula (12.54).

The derivatives of the entropy can be defined in terms of these quantities if, in accord with (17.19) and Axiom 6 of Chapter 3, the relation $dS_k(\mathcal{C}_k(s)) = k_B\beta_k Q_{\partial k}(\mathcal{C}_k(s))ds$ is used with (17.2). This gives the result

$$(17.29a) \qquad \frac{\partial S_k(\mathcal{C}_k(s))}{\partial \beta_k} = k_B\beta_k K_{k,V}(\mathcal{C}_k(s)),$$

$$(17.29b) \qquad \frac{\partial S_k(\mathcal{C}_k(s))}{\partial \delta_k} = k_B\beta_k \Lambda_{k,V}(\mathcal{C}_k(s)).$$

These results for the calorimetry equation differ from the β_k and δ_k derivatives of the entropy in equilibrium thermodynamics obtained in the next chapter.

Let us turn now to the case in which the independent variables are the thermature and pressure, (β_k, P_k). In the previous case, when the volume is held fixed, the heat flow into or out of the system is the same as the change in the energy of the system. In the current case, in which the pressure rather than the volume is held fixed, the work being done on the moving walls must be subtracted. It is convenient to use the enthalpy for this case. The *enthalpy* of a system on a path is defined using the global equilibrium version of the definition (7.81c) by

$$(17.30) \qquad \mathcal{H}_k^{(en)}(\mathcal{C}_k(s)) = \mathcal{H}_k(\mathcal{C}_k(s)) + \delta_k(s)P_k(\mathcal{C}_k(s)).$$

With reference to (17.26) and (17.8), the heat flow can then be expressed as[22]

$$(17.31) \qquad Q_{\partial k}(s)ds = d\mathcal{H}_k^{(en)}(\mathcal{C}_k(s)) - \delta_k(s)dP_k(s).$$

The constant pressure case satisfies the condition $dP_k = 0$ and the heat transfer is equal to the change in the enthalpy.

Let us next expand $d\mathcal{H}_k^{(en)}$ in (17.31) into its thermature and pressure derivatives and use (17.4) for $Q_{\partial k}(\mathcal{C}_k(s))ds$. The thermature and pressure are the independent variables. This leads to the identifications

$$(17.32a) \qquad \Lambda_{k,P}(\mathcal{C}_k(s)) = \frac{\partial \mathcal{H}_k^{(en)}(\mathcal{C}_k(s))}{\partial P_k} - \delta_k(s),$$

$$(17.32b) \qquad K_{k,P}(\mathcal{C}_k(s)) = \frac{\partial \mathcal{H}_k^{(en)}(\mathcal{C}_k(s))}{\partial \beta_k}.$$

As will be shown in Chapter 18, the entropy is independent of the path for reversible paths. This means that the entropy is a state function. It follows that the rate of change of the entropy along the reversible path, stated in (17.19) in terms of the thermature and heat flow, should satisfy Euler's

---

[22]See Dickerson [**1969**], formula (3-31), p. 94.

criterion when expressed as a differential of the thermature and pressure. Referring to (17.4), this means that

$$(17.33) \qquad \Lambda_{k,\mathrm{P}} + \beta_k \frac{\partial \Lambda_{k,\mathrm{P}}}{\partial \beta_k} = \beta_k \frac{\partial K_{k,\mathrm{P}}}{\partial P_k}.$$

Using (17.32) in (17.33) gives the standard result[23]

$$(17.34) \qquad \frac{\partial \mathcal{H}_k^{(en)}(\mathcal{C}_k(s))}{\partial \mathrm{P}_k} = \delta_k(s) + \beta_k \frac{\partial \delta_k(\beta_k(s), \mathrm{P}_k(s))}{\partial \beta_k}.$$

This formula will be useful below.

17.5 Heat and Work

Clapeyron [**1834**] used caloric theory and a Carnot cycle to deduce an important equation (TIS notation)[24]

$$(17.35) \qquad \beta_k \frac{\partial \mathrm{P}_k(t)}{\partial \beta_k} = -\frac{\Delta \mathcal{H}_k^{(en)}(t)}{\Delta \delta_k(t)}$$

for a substance undergoing a thermodynamic phase transition at constant pressure. Because the pressure is constant, the latent heat in this phase change is the change in enthalpy $\Delta \mathcal{H}_k^{(en)}(t)$ and $\Delta \delta_k(t)$ represents the associated change of volume. This equation was put on a sounder theoretical footing by Clausius [**1856**], p. 96. For a thermodynamic phase change in which the final phase is a gas, such as a vaporization or sublimation process, Clausius used the ideal gas law to express the volume of a mole of gas in terms of the pressure and temperature by $V_2 = V^{(ig)} = RT/\mathrm{P}^{(ig)}$, where R is the gas constant. Using $\Delta V = V^{(ig)} - V_1 \approx V^{(ig)}$ and expressing the change in latent heat by $\Delta \mathcal{H}^{(en)}$, Clausius rewrote Clapeyron's equation as the Clausius-Clapeyron equation in the form (TIS notation)

$$(17.36) \qquad \beta_k \frac{\partial \mathrm{P}_k(t)}{\partial \beta_k} \approx -\frac{\Delta \mathcal{H}_k^{(en)}(t)}{\Delta \delta_k(t)} \approx -\frac{\beta_k \mathrm{P}_k}{\mathcal{N}_k} \Delta \mathcal{H}_k^{(en)}(t).$$

---

[23]See Planck [**1922**], p. 124, or Dickerson [**1969**], formula (4-64), p. 177.

[24]In standard notation, Clapeyron's equation is equivalent to

$$T\frac{d\mathrm{P}}{dT} = \frac{L}{V_2 - V_1},$$

where L is the latent heat of a phase transition at pressure P and temperature T that is associated with a change in volume $V_2 - V_1$.

The usual derivation of Clapeyron's equation cannot be used in the TIS setting. A generalized version of it will be presented in the context of a particle based theory in EIS, Part IV.[25]

Anticipating the result (18.30), the relation (17.28b) can be used to express the latent heat with respect to volume in terms of the β_k derivative of the pressure. Similarly, the definition (18.34) gives the equilibrium heat capacity at constant volume. The equilibrium relations that result are

$$(17.37a) \qquad\qquad \Lambda_{k,V}^\epsilon = -\beta_k \frac{\partial \mathrm{P}_k^\epsilon}{\partial \beta_k},$$

$$(17.37b) \qquad\qquad K_{k,V}^\epsilon = \frac{\partial \mathcal{H}_k^\epsilon}{\partial \beta_k}.$$

Truesdell [1980], pp. 110–117, has called (17.37a) the Carnot-Clapeyron theorem.

For a process at constant pressure, the volume is defined as an implicit function of thermature and pressure (when it is valid to do so) and the differential

$$(17.38a) \qquad\qquad 0 = dP_k = \frac{\partial P_k}{\partial \beta_k} d\beta_k + \frac{\partial P_k}{\partial \delta_k} d\delta_k$$

is used to show

$$(17.38b) \qquad\qquad \frac{\partial \delta_k(\beta_k, \mathrm{P}_k)}{\partial \beta_k} = -\frac{\partial \mathrm{P}_k}{\partial \beta_k} \left[\frac{\partial \mathrm{P}_k}{\partial \delta_k} \right]^{-1}.$$

This result is used with (17.5) and (17.37a) to express the equilibrium version of $\Lambda_{k,\mathrm{P}}$ in the form

$$(17.39) \qquad\qquad \Lambda_{k,\mathrm{P}}^\epsilon = \beta_k \frac{\partial \delta_k(\beta_k, \mathrm{P}_k)}{\partial \beta_k}.$$

Using Euler's criterion again with $\beta_k Q_{\partial k}(\mathcal{C}_k(s))ds$ defined using (17.25) implies that the β_k derivative of (17.29b) must equal the δ_k derivative of

---

[25] See Dickerson [1969], p. 232, for the expression of the Clausius-Clapeyron equation in the standard form

$$\frac{dP}{dT} \approx \frac{\Delta \mathcal{H}^{(en)}}{TV^{(\mathrm{ig})}} \approx \frac{\mathrm{P}}{RT^2} \Delta \mathcal{H}^{(en)}$$

See also the discussion in Partington [1949], pp. 181–2. A related equation, which also appears in the literature, was called by Duhem [1891], p. 36, "l'équation de Clapeyron et de Clausius." It relates the heat of vaporization in a phase transition to the difference in the specific volumes of the two phases multiplied by $\Lambda_{k,V}^\epsilon$. Duhem expressed it (in his notation) as

$$Q_{\partial k}(t) = [\nu_1(\varpi(t), T) - \nu_2(\varpi(t), T)] T \frac{d\varpi(t)}{dT},$$

where $Q_{\partial k}(t)$ is the latent heat of vaporization, $\varpi(t)$ is the pressure as a function of thermature and $\nu_i(\varpi(t), T) = \delta_\xi^{-1}$ is the specific volume of phase i.

(17.29a). This yields the equilibrium relation

$$(17.40) \qquad \Lambda_{k,V}^\epsilon + \beta_k \frac{\partial \Lambda_{k,V}^\epsilon}{\partial \beta_k} = \beta_k \frac{\partial K_{k,V}^\epsilon}{\partial \delta_k}.$$

From (17.37) and (17.40), the important result

$$(17.41) \qquad \frac{\partial P_k^\epsilon}{\partial \beta_k} = \frac{\partial \Lambda_{k,V}^\epsilon}{\partial \beta_k} - \frac{\partial K_{k,V}^\epsilon}{\partial \delta_k}$$

is obtained.

Let us now consider the work done moving around a closed cycle $\mathcal{C}$ surrounding the surface $\mathcal{S}$ on the (β_k, δ_k) equilibrium manifold. Applying Stokes' theorem to (17.41) and (17.8) gives the result

$$
\begin{aligned}
(17.42) \qquad \mathcal{W}(\mathcal{C}) &= -\int_{\mathcal{C}} d\delta_k \, P_k^\epsilon \\
&= -\int_{\mathcal{S}} d\delta_k \, d\beta_k \frac{\partial P_k^\epsilon}{\partial \beta_k} \\
&= -\int_{\mathcal{C}} [\Lambda_{k,V}^\epsilon d\delta_k + K_{k,V}^\epsilon d\beta_k] \\
&= -Q(\mathcal{C}).
\end{aligned}
$$

This formula allows us to equate the work performed in a reversible cyclic process with the net heat absorbed in the cycle.[26]

The identification of $\Lambda_{k,V}^\epsilon$ and $K_{k,V}^\epsilon$ with functions of equilibrium thermodynamics shows that the fundamental quantities of classical thermodynamics can be expressed as phase averages and given an interpretation in terms of an underlying particle mechanics. This means that a thermodynamics based on Carnot's approach to heat engines can be constructed on the basis of the constitutive properties of materials—which are themselves determined by the set of interparticle potentials, $\{\phi_{ij}\}$.

---

[26]See Poincaré [1892], p. 140, on this calculation, which he calls the "theorem of Clausius." These points are discussed in detail in Truesdell [1980], pp. 198–199, 224–227.

CHAPTER 18

Equilibrium Thermodynamics

An extended, although slightly modified, form of standard equilibrium thermodynamics is shown to emerge naturally from the TIS formalism when the underlying particle distribution is the equilibrium distribution. A brief treatment of the formalism and some fundamental results will be presented. A detailed treatment of equilibrium thermodynamics is given in EIS.

18.1 Importance of the Equilibrium State

Equilibrium thermodynamics has been used as a touchstone for assessing the validity of formulas derived in the general TIS theory. One justification for the form (10.48) chosen for the thermodynamic volume partial derivative operator, for example, is that it gives the standard formulas of equilibrium thermodynamics when used with phase averages over the equilibrium phase state, $F_k^\epsilon(Q_k, P_k)$. That claim, and others, will be verified in this Chapter.

The notion of an equilibrium state was introduced in Chapter 13. In the last chapter, it was found that two macroscopic variables, such as δ_k and β_k, are sufficient to determine the equilibrium state when the system is closed and surface effects can be neglected.

For a system described by the equilibrium phase state, $F_k^\epsilon(Q_k, P_k)$, the average flow of moments of momentum through the boundary is zero. In accord with the principles of TIS, any normalized system probability distribution density $F_k^{(n)}(Q_k, P_k, t)$, including $F_k^{(n)\epsilon}(Q_k, P_k)$, represents the probability density of finding the system in some particular mechanical state (Q_k, P_k). This is equivalent to the probability density of finding the system in the exact state $F_k^\delta(Q_k, P_k, t)$.

The question of whether a system is at equilibrium or not is a macroscopic concern based on whether the walls are at rest in the observation frame, so the system is viewed in the system rest frame, the boundary conditions for the system are held fixed, and the measured rates of change of certain system quantities, such as the total energy, momentum, and angular momentum, are close enough to zero over some experimentally determined period of time that is sufficient for transient effects to die out. When these conditions are all met, the particles in the system are deemed to be in the equilibrium rest frame and the equilibrium phase state $F_k^\epsilon(Q_k, P_k)$ can be attributed to it.

Although the phase averaged flow of each moment of momentum through the boundaries vanishes for a system described by the equilibrium state, the actual flow of these moments of momentum through the boundary does not necessarily vanish for a given exact state $F_k^\delta(Q_k, P_k, t)$ at a given time. The phrases "close enough to zero" and "period of time sufficient for transient effects to die out" refer to judgments that are purposefully vague here because they depend on the tolerances desired by the experimenter and the fact that different experimental arrangements can have different settling times. Moreover, there is no guarantee that a system that meets the requirements for attributing the equilibrium state to it over one period of time will meet these requirements during another period of time. Because of the large number of particles, combinatorial calculations show that it is unlikely that the system will deviate very far from equilibrium behavior very often, but it is not forbidden. Mechanical states that have macroscopic consequences that deviate from the expected equilibrium behavior are referred to as "fluctuations away from equilibrium".

A fluctuation is usually described in terms of the deviation of a set of macroscopic physical quantities, obtained using a particular mechanical state F_k^δ, from the equilibrium values of these quantities. These differences can also be expressed in terms of the difference between the description of the system by an equilibrium state and by a general state $F_k(Q_k, P_k, t)$. To compare a system described by a general state with the equilibrium state, Einstein [1902] defined the relative probability of a given fluctuation from equilibrium by the formula $\mathcal{P}^\Delta(Q_k, P_k, t) = e^{-k_B^{-1}|\Delta S_k(Q_k, P_k, t)|}$, where $\Delta S_k(Q_k, P_k, t) = S_k(Q_k, P_k, t) - S_k^\epsilon(Q_k, P_k)$. Using the definition of the entropy, this is written $\Delta S_k(Q_k, P_k, t) = -k_B[\ln F_k(Q_k, P_k, t) - \ln F_k^\epsilon(Q_k, P_k)]$ in TIS notation, so that the difference of $\mathcal{P}^\Delta(Q_k, P_k, t)$ from 1 for each mechanical state (Q_k, P_k) is a measure of the deviation of the general phase state from the equilibrium phase state. To examine the meaning of this claim, suppose that $F_k(Q_k, P_k, t) < F_k^\epsilon(Q_k, P_k)$ is true for the mechanical state (Q_k, P_k) at time t. Einstein's definition can then be written using TIS definitions as $\mathcal{P}^\Delta(Q_k, P_k, t) = F_k(Q_k, P_k, t)/F_k^\epsilon(Q_k, P_k)$ with the inverse ratio if $F_k(Q_k, P_k, t) \geq F_k^\epsilon(Q_k, P_k)$. The problem with Einstein's definition is that there is no rationale for claiming that the ratio of two distribution densities is the probability of anything. It will not be used here.

On both the microscopic and macroscopic levels, the equilibrium state has played an important role in the description of systems because it is the only state with an explicit microscopic representation. It also has the following useful features:

(i) it is describable by a few macroscopic parameters;

(ii) it is reproducible by adjusting the macroscopic boundary conditions;

(iii) it is a stationary solution of the interacting systems evolution equation, (5.16);

(iv) the entropy of the state is linear in the energy (Gibbs' criterion).

Let us begin with these criteria and review what they imply concerning the equilibrium distribution. If the state $F_k^\star$ is assumed to be a stationary state, it satisfies the equation $\partial F_k^\star/\partial t = 0$. From this it follows immediately that the Poisson bracket of $F_k^\star(Q_k, P_k)$ vanishes, i.e.,

$$(18.1a) \qquad \frac{\partial F_k^\star(Q_k, P_k, t)}{\partial t} = 0,$$

$$(18.1b) \qquad V_{2k} \cdot \nabla_{2k} F_k^\star(Q_k, P_k) = 0.$$

To investigate this condition further, assume that the momentum, angular momentum, and Hamiltonian energy exhaust the set of conserved quantities for the k system. Assume also that the net external force on each particle of the k system vanishes, which means $\partial\Phi_{i,x}(Q_k, t)/\partial q_i = 0$, and the time-derivative of the external potential vanishes, so that $\partial\Phi_{i,x}(Q_k, t)/\partial t = 0$. It then follows from (4.22c) that the Hamiltonian is independent of the time and the function $\mathcal{R}_k(Q_k, P_k)$ defined as in (13.42) is the most general function of

(Q_k, P_k) that satisfies the differential equation $V_{2k} \cdot \nabla_{2k} \mathcal{R}_k(Q_k, P_k) = 0$. All other solutions are functions of $R_k(Q_k, P_k)$.

It follows that the solution of (18.1) is a function representing the state that can be written in the form $F_k^\star(Q_k, P_k) = f(\mathcal{R}_k(Q_k, P_k))$, where f is non-negative, belongs to the set $\mathcal{C}^2(\Omega_k)$ of twice continuously differentiable functions on the compact set Ω_k, and is bounded when its argument is bounded. This means that f is bounded for $(Q_k, P_k) \in \Omega_k(t)$ because $\mathcal{R}_k(Q_k, P_k)$ is bounded for $(Q_k, P_k) \in \Omega_k(t)$. Because TIS requires that the equilibrium distribution of a system be defined in its equilibrium rest frame, the function $\mathcal{R}_k(Q_k, P_k)$ reduces, by (13.42) and the definition of the particle rest frame, to $\mathcal{R}_k(Q_k, P_k) = \mathcal{H}_k(Q_k, P_k) + D_k$.

An additional requirement imposed by Gibbs was that the entropy of the equilibrium phase state be linear in the energy. Meeting this condition means that $\ln f = \alpha_k \mathcal{H}_k(Q_k, P_k) + D_k$ for some α_k and D_k that do not depend on (Q_k, P_k) or t. Using these assumptions with $F_k^\star$ leads to the k equilibrium phase state in the (unnormalized) form of Gibbs canonical state function $F_k^\epsilon(Q_k, P_k) = e^{-\beta_k \mathcal{H}_k^\epsilon(Q_k, P_k)}$ for a suitable time-independent Hamiltonian $\mathcal{H}_k^\epsilon(Q_k, P_k)$. This is a refinement of the result obtained in Chapter 13 where the requirements on the interactions of systems described by stationary states were analyzed from the standpoint of the conditions imposed by the conservation laws.

An alternative, and preferable, approach views the equilibrium system as a small part of a large system that plays the role of a heat bath. Khinchin [1943] showed that Gibbs canonical distribution is the asymptotic result for the state of a small system k when the central limit theorem of probability theory is used with the division of an isolated total system t into a small subsystem k and a large heat bath system l.

The microscopic equilibrium phase state is $\exp[-\beta_k \mathcal{H}_k^\epsilon(Q_k, P_k)]$ and the boundary conditions are determined by the macroscopic state parameters (β_k, δ_k). Having a specific form for the equilibrium state implies that the set of relations connecting various quantities in equilibrium thermodynamics is much richer than in the general case.

For a full treatment of equilibrium thermodynamics, including equations of state, thermodynamic surfaces and phase transitions, asymptotic thermodynamics and the central limit theorem of probability, and approximate equations of state, see EIS.

18.2 The Development of the Concept of Equilibrium

Historically, the concept of equilibrium in thermodynamic settings took shape in the late eighteenth century. An important step toward the modern conception was taken by Pierre Prévost [1791]. Prévost introduced the principle of the *movable equilibrium of heat* or the *theory of exchanges*. Although framed originally in terms of caloric particles of heat and conserved quantities of heat, the conception was transferable to succeeding theories.[1] The idea is that if one region contains more heat than another does, as measured by differences in temperature (by analogy to Volta's conception of electrical tension), more heat will leave the region with higher heat density than arrives from the region with lower heat density. This process continues until the heat densities become equal and equilibrium is achieved. This idea was fruitfully applied to radiant heat by Balfour Stewart and to radiation in general by Gustav Kirchhoff.

The definition of an equilibrium phase state in modern theories requires that the system be in contact with a reservoir. This reservoir must be large enough so that its temperature remains fixed no matter what the system temperature is. The system volume also needs to be fixed so the reservoir must maintain fixed wall locations no matter what forces the system is exerting on it. This can clearly be true only in an asymptotic sense, but can usually be closely approximated in practice. It will be assumed in this chapter that $\beta_k \mathcal{C}_{k,\mathcal{H}} \gg 1$ and $\mathcal{N}_t \gg \mathcal{N}_k$ so that sufficient particles, momentum, and energy, are available to the reservoir system to use in fixing the wall temperature and wall locations.

18.3 The Formalism of Equilibrium Thermodynamics

For convenience, let us consider system k with a fixed volume set d_k with boundary ∂d_k in the system rest frame and assume that sufficient time has past so that the particle rest frame coincides with the system rest frame. The system rest frame condition implies that $v_{\partial k}(q,t) = 0$ for each $q \in \partial d_k$. For this equilibrium phase state, the representation (13.49) is used. The phase average of any time-independent phase function $G_k(Q_k, P_k)$ over the equilibrium distribution F_k^ϵ is defined in (13.50) as $G_k^\epsilon = \langle G_k(Q_k, P_k) \rangle_{k\epsilon}$.

Let us now specialize the thermodynamic functions of Part I to the equilibrium case. The k *local equilibrium density* is

$$(18.2) \qquad \nu_k^\epsilon(q) = \sum_{i \in k} \langle \delta(q - q_i) \rangle_{k\epsilon}.$$

As a consequence of the choice of the equilibrium rest frame as the reference frame, the k *local equilibrium matter current density* and k *local equilibrium total momentum density* both vanish:

$$(18.3a) \qquad \nu_k^{\epsilon,\mu}(q) = 0,$$
$$(18.3b) \qquad \mathcal{P}_k^{\epsilon,\mu}(q) = 0.$$

---

[1] This was pointed out by D. B. Brace in his introductory essay on the work of Prévost, Balfour Stewart, and Gustav Kirchhoff, in the book containing the translation of Prévost [**1791**].

The k *local equilibrium momentum current density* is

(18.4) $$\mathcal{P}_k^{\epsilon,\mu\nu}(q) = 2\delta^{\mu\nu}\mathcal{K}_k^\epsilon(q),$$

where

(18.5) $$\mathcal{K}_k^\epsilon(q) = \sum_{i\in k} \langle \delta(q - q_i)K_i(P_i)\rangle_{k\epsilon}$$

is the k *local equilibrium kinetic energy density*. The k *local equilibrium potential energy* is defined by

(18.6) $$\Phi_k^\epsilon(q) = \Phi_{k,n}^\epsilon(q) + \Phi_{k,x}^\epsilon(q),$$

where

(18.7a) $$\Phi_{k,n}^\epsilon(q) = \sum_{i\in k} \langle \delta(q - q_i)\Phi_{i,n}(Q_k)\rangle_{k\epsilon},$$

(18.7b) $$\Phi_{k,x}^\epsilon(q) = \sum_{i\in k} \langle \delta(q - q_i)\Phi_{i,x}(Q_k)\rangle_{k\epsilon},$$

are the k *local equilibrium internal potential energy* and the k *local equilibrium external potential energy*, respectively. Recall that by (13.32), these potentials are required to have a very restricted form. The k *local equilibrium total energy* and k *local equilibrium Hamiltonian* are expressed in terms of the previous definitions by

(18.8a) $$E_k^\epsilon(q) = \mathcal{K}_k^\epsilon(q) + \Phi_{k,n}^\epsilon(q) + \tfrac{1}{2}\Phi_{k,x}^\epsilon(q),$$

(18.8b) $$\mathcal{H}_k^\epsilon(q) = \mathcal{K}_k^\epsilon(q) + \Phi_k^\epsilon(q).$$

The k *local equilibrium total energy current density* vanishes

(18.9) $$E_k^{\epsilon,\mu}(q) = 0.$$

The pressure is expressed in terms of the kinetic and coordinate virial terms. The k *local internal virial tensor* and the k *local external virial tensor* are

(18.10a) $$\Xi_{k,n}^\epsilon(q) = \sum_{i\in k} \langle \delta(q - q_i)\Xi_{k,n}(Q_k)\rangle_{k\epsilon},$$

(18.10b) $$\Xi_{k,x}^\epsilon(q) = \sum_{i\in k} \langle \delta(q - q_i)\Xi_{k,i}(q_i)\rangle_{k\epsilon}.$$

The k *local equilibrium pressure tensor* and k *local equilibrium pressure* are expressed in terms of these and the momentum current density by

(18.11a) $$\mathrm{P}_k^{\epsilon,\mu\nu}(q) = \frac{1}{\delta_k(t)}\left[\mathcal{P}_k^{\epsilon,\mu\nu}(q) - \Xi_{k,n}^{\epsilon,\mu\nu}(q) - \Xi_{k,x}^{\epsilon,\mu\nu}(q)\right],$$

(18.11b) $$\mathrm{P}_k^\epsilon(q) = \tfrac{1}{3}\delta_{\mu\nu}\mathrm{P}_k^{\epsilon,\mu\nu}(q) = \frac{1}{3\delta_k(t)}\left[\mathcal{P}_k^\epsilon(q) - \Xi_{k,n}^\epsilon(q) - \Xi_{k,x}^\epsilon(q)\right].$$

Turning now to the entropy, the definition (9.10) and the equilibrium phase state, (13.49), yield a special form for the equilibrium entropy phase function

$$(18.12) \qquad S_k^\epsilon(Q_k, P_k) = k_B \beta_k \mathcal{H}_k^\epsilon(Q_k, P_k).$$

This definition differs from the standard definition, which is based on a normalized distribution and takes the form

$$(18.13) \qquad S_k^{(n)\epsilon}(Q_k, P_k) = S_k^\epsilon(Q_k, P_k) + k_B \ln Z_k^\epsilon$$

in the TIS notation. This difference in definitions has the effect of modifying a few results derived below compared to the standard versions, but the calculations of expectation values lead to the same results. The equilibrium Hamiltonian, $\mathcal{H}_k^\epsilon(Q_k, P_k)$, is defined in equation (13.49d). The equilibrium phase average of $S_k^\epsilon(Q_k, P_k)$ is

$$(18.14) \qquad S_k^\epsilon = k_B \beta_k \mathcal{H}_k^\epsilon,$$

where S_k^ϵ and $\mathcal{H}_k^\epsilon$, the equilibrium values of the k system thermodynamic entropy and Hamiltonian energy, are defined by

$$(18.15a) \qquad S_k^\epsilon = \langle S_k^\epsilon(Q_k, P_k) \rangle_{k\epsilon},$$
$$(18.15b) \qquad \mathcal{H}_k^\epsilon = \langle \mathcal{H}_k^\epsilon(Q_k, P_k) \rangle_{k\epsilon}.$$

It is well known, as shown by a variational calculation similar to the one done in Section 9.7 for the reference distribution, that the formula for the entropy stated in equation (18.14) gives the entropy upper bound for a given equilibrium state $C_k^\epsilon = (\beta_k, \delta_k)$. This confirms the statement preceding (3.16a). In the quasilocal forces case, it is also true that $\mathcal{H}_k^\epsilon = E_k^\epsilon$ except near the boundary.

The time independence of the equilibrium Hamiltonian is demonstrated by using (13.49e) and (13.50) to show

$$(18.16) \qquad \frac{\partial \mathcal{H}_k^\epsilon}{\partial t} = \left\langle \frac{d\mathcal{H}_k^\epsilon(Q_k, P_k)}{dt} \right\rangle_{k\epsilon} + \langle \dot{\chi}_{\Omega_k(t)} \mathcal{H}_k^\epsilon(Q_k, P_k) \rangle_{k\epsilon} = 0.$$

The time derivative of (18.14), computed using (18.16), confirms (3.16) for the equilibrium case.

With the help of (18.12), and in accord with (12.38), the thermature derivative of the equilibrium entropy is

$$(18.17) \qquad \frac{\partial S_k^\epsilon(Q_k, P_k)}{\partial \beta_k} = k_B \mathcal{H}_k^\epsilon(Q_k, P_k).$$

This implies that the thermature derivative of the system distribution density is

$$(18.18) \qquad \frac{\partial F_k^\epsilon(Q_k, P_k)}{\partial \beta_k} = -\mathcal{H}_k^\epsilon(Q_k, P_k) F_k^\epsilon(Q_k, P_k).$$

Next, using the thermature derivative with $G_k = 1$ gives for the equilibrium case the result

$$(18.19) \qquad 0 = \frac{\partial \langle 1 \rangle_{k\epsilon}}{\partial \beta_k} = - \langle \mathcal{H}_k^\epsilon(Q_k, P_k) \rangle_{k\epsilon} - \frac{\partial \ln Z_k^\epsilon}{\partial \beta_k}.$$

It follows from this that

$$(18.20) \qquad \frac{\partial \ln Z_k^\epsilon}{\partial \beta_k} = -\mathcal{H}_k^\epsilon$$

in parallel to the calculation (12.37) for the general case. This result can also be computed directly from the definition (13.49b) of the partition function using (12.35) and (12.36).

The result

$$(18.21) \qquad S_k^\epsilon = -k_B \beta_k \frac{\partial \ln Z_k^\epsilon}{\partial \beta_k}.$$

is obtained from (18.14) and (18.20). This is used to obtain the standard result for the thermature derivative of the entropy in terms of the free energy $A_k^{(\text{fe})\epsilon} = -\beta_k^{-1} \ln Z_k^\epsilon$, which is introduced formally in the next chapter, as

$$(18.22) \qquad S_k^{(\text{n})\epsilon} = -k_B \beta_k^2 \frac{\partial \{[\beta_k]^{-1} \ln Z_k^\epsilon\}}{\partial \beta_k} = k_B \beta_k^2 \frac{\partial A_k^{(\text{fe})\epsilon}}{\partial \beta_k}.$$

By (18.12), the volume phase derivative of $S_k^\epsilon(Q_k, P_k)$ is

$$(18.23) \qquad \frac{dS_k^\epsilon(Q_k, P_k)}{d\Omega_k} = k_B \beta_k \frac{d\mathcal{H}_k^\epsilon(Q_k, P_k)}{d\Omega_k} = -k_B \beta_k \mathrm{P}_k^\epsilon(Q_k, P_k).$$

Using (10.47) with Z_k^ϵ and (18.23) leads to the standard result

$$(18.24) \qquad \frac{\partial \ln Z_k^\epsilon}{\partial \delta_k} = -\beta_k \left\langle \frac{d\mathcal{H}_k^\epsilon(Q_k, P_k)}{d\Omega_k} \right\rangle_{k\epsilon} = \beta_k \mathrm{P}_k^\epsilon.$$

A direct calculation confirms this result.

The relation

$$(18.25)$$
$$\frac{\partial \mathcal{H}_k^\epsilon}{\partial \delta_k} = \left\langle \frac{d\mathcal{H}_k^\epsilon(Q_k, P_k)}{d\Omega_k} \right\rangle_{k\epsilon} - \frac{1}{k_B} \left\langle \mathcal{H}_k^\epsilon(Q_k, P_k) \frac{dS_k^\epsilon(Q_k, P_k)}{d\Omega_k} \right\rangle_{k\epsilon} - \frac{\partial \ln Z_k^\epsilon}{\partial \delta_k} \mathcal{H}_k^\epsilon$$

follows from (10.44). The k *equilibrium pressure analog* is expressed as the phase volume derivative of the equilibrium Hamiltonian energy in the k equilibrium rest frame by

$$(18.26)$$
$$\mathrm{P}_k^\epsilon(Q_k, P_k) = -\frac{d\mathcal{H}_k^\epsilon(Q_k, P_k)}{d\Omega_k}$$
$$= -\frac{1}{3\delta_k} \left[2\mathcal{K}_k(P_k) - Q_k \cdot \frac{\partial \Phi_{k,n}^\epsilon(Q_k)}{\partial Q_k} - Q_k \cdot \frac{\partial \Phi_{k,\pi}^\epsilon(Q_k)}{\partial Q_k} \right],$$

where $P_k^\epsilon(Q_k, P_k)$ represents the equilibrium pressure analog in this equation. The k *equilibrium latent heat with respect to volume* is

(18.27)

$$\Lambda_{k,V}^\epsilon = -\frac{1}{k_B}\left\langle \mathcal{H}_k^\epsilon(Q_k, P_k)\frac{dS_k^\epsilon(Q_k, P_k)}{d\Omega_k}\right\rangle_{k\epsilon} - \frac{\partial \ln Z_k^\epsilon}{\partial \delta_k}\mathcal{H}_k^\epsilon$$
$$= -\beta_k[H_k^\epsilon P_k^\epsilon - \langle H_k^\epsilon(Q_k, P_k)P_k^\epsilon(Q_k, P_k)\rangle_{k\epsilon}].$$

To complete the calculation begun in (18.25), the relations (18.26) and (18.27) are used with (18.25) to show

(18.28)
$$\frac{\partial H_k^\epsilon}{\partial \delta_k} = -P_k^\epsilon + \Lambda_{k,V}^\epsilon.$$

The thermature derivative of the pressure is computed directly using the analog (18.26) with the result

(18.29)

$$\frac{\partial P_k^\epsilon}{\partial \beta_k} = -\langle \mathcal{H}_k^\epsilon(Q_k, P_k)P_k^\epsilon(Q_k, P_k)\rangle_{k\epsilon} - \frac{\partial \ln Z_k^\epsilon}{\partial \beta_k}P_k$$
$$= \mathcal{H}_k^\epsilon P_k^\epsilon - \langle \mathcal{H}_k^\epsilon(Q_k, P_k)P_k^\epsilon(Q_k, P_k)\rangle_{k\epsilon}.$$

Comparing (18.27) with (18.29), yields the result

(18.30)
$$\Lambda_{k,V}^\epsilon = -\beta_k\frac{\partial P_k^\epsilon}{\partial \beta_k}.$$

The relation (18.28) can then be rewritten as

(18.31)
$$\frac{\partial \mathcal{H}_k^\epsilon}{\partial \delta_k} = -P_k^\epsilon - \beta_k\frac{\partial P_k^\epsilon}{\partial \beta_k}.$$

Taking the β_k derivative of (18.24) yields by (18.31) and (18.20) the result

(18.32)
$$\frac{\partial^2 \ln Z_k^\epsilon}{\partial \beta_k \partial \delta_k} = -\frac{\partial \mathcal{H}_k^\epsilon}{\partial \delta_k} = \frac{\partial^2 \ln Z_k^\epsilon}{\partial \delta_k \partial \beta_k}.$$

This means that, as far as equilibrium states $\mathcal{C}_k^\epsilon$ are concerned, the quantity Z_k^ϵ and, by (18.20), (18.21) and (18.24), the quantities $\mathcal{H}_k^\epsilon$, S_k^ϵ, and P_k^ϵ, are all independent of the path connecting different equilibrium states $\mathcal{C}_k^\epsilon$ and $\mathcal{C}_k^{\epsilon'}$. It follows that the equilibrium functions Z_k^ϵ, $\mathcal{H}_k^\epsilon$, S_k^ϵ, and P_k^ϵ, are functions of the k thermodynamic equilibrium state. Observe that although Z_k^ϵ is a thermodynamic object, it is not an observable with an observable analog in the sense of Chapter 5.

With the previous results, it is easy to show that

(18.33a)
$$\frac{\partial S_k^\epsilon}{\partial \beta_k} = k_B\left[\mathcal{H}_k^\epsilon + \beta_k\frac{\partial \mathcal{H}_k^\epsilon}{\partial \beta_k}\right],$$

(18.33b)
$$\frac{\partial S_k^\epsilon}{\partial \delta_k} = -k_B\beta_k\left[P_k^\epsilon + \beta_k\frac{\partial P_k^\epsilon}{\partial \beta_k}\right].$$

These results are different from the those obtained in (17.29) for the calorimetry equation of classical thermodynamics. This is because the change in entropy is defined in terms of the change in the heat, as $dS_k = k_B \beta_k Q_{\partial k} ds$, in the calorimetry equation versus the definition (18.14) of the entropy in terms of the Hamiltonian in TIS equilibrium thermodynamics.

18.4 Equilibrium Heat Capacities

As was shown in Chapter 12, the heat capacity of a system depends on the distribution chosen to describe it. Physically, this is because the heat capacity of a body depends on the ability of that body to convert heat into internal potential energy and store it. Because the potential energy depends on the distances between particles, the system distribution and the boundary conditions will have an influence on how well the body can perform this task. In accounting for the results of experiments, it is the equilibrium version of the system state that is usually the proper choice because this is a definite state for which definite predictions can be made. In other specific cases, an appropriate approximation of the system distribution may be used for the same purpose in a nonequilibrium steady-state situation.

Let us begin with the formalism for the heat capacity at constant volume, $K_{k,V}$. Because the volume is fixed, there is a temperature change but no volume change, so moving boundaries do no work and all the energy entering the system may be considered heat. The heat capacity at constant volume for a single particle was defined in (12.52) as the rate of change in the particle Hamiltonian energy with respect to the change in the temperature with the volume fixed. When this is specialized to the equilibrium case and summed over $i \in k$, the result

$$(18.34) \qquad K_{k,V}^\epsilon = \frac{\partial \mathcal{H}_k^\epsilon}{\partial \beta_k} = [\mathcal{H}_k^\epsilon]^2 - \left\langle [\mathcal{H}_k^\epsilon(Q_k, P_k)]^2 \right\rangle_{k\epsilon} \leq 0$$

is obtained in parallel with (12.53). This result is consistent with taking the β_k derivative of the expectation integral defining $\mathcal{H}_k^\epsilon$, which is stated in (18.15b), followed by the use of (18.20). The thermature derivatives of all quantities defined as phase averages in this chapter are consistent with the use of the definition (12.41) of the thermature derivative when the probability distribution density is set to the equilibrium phase state. When it is necessary to make a connection with the standard formulas in the literature, the equilibrium conventional heat capacity at constant volume $C_{k,V}^\epsilon$ is used in place of $K_{k,V}^\epsilon$.

In cases of constant pressure rather than constant volume, the heat capacity at constant pressure $K_{k,\mathrm{P}}$ is used. Since the volume is not fixed, the heat added is not equal to the energy added. It follows that the definition is expressed in terms of the heat added to the system and not in terms of the change in the total energy of the system. The k *heat capacity at constant pressure* is therefore defined as the derivative of the enthalpy with respect to the thermature with the pressure held constant. Formally, the definition of the equilibrium enthalpy in (18.38) below is used with (18.31) to obtain the result

(18.35)

$$K_{k,\mathrm{P}}^\epsilon = \left.\frac{\partial \mathcal{H}_k^{(\mathrm{en})\epsilon}}{\partial \beta_k}\right|_{\mathrm{P}} = \frac{\partial \mathcal{H}_k^\epsilon}{\partial \beta_k} + \frac{\partial \mathcal{H}_k^\epsilon}{\partial \delta_k}\frac{d\delta_k(\beta_k, P_k)}{d\beta_k} + \mathrm{P}_k^\epsilon \frac{d\delta_k(\beta_k, P_k)}{d\beta_k}$$

$$= K_{k,V}^\epsilon - \beta_k \frac{\partial \mathrm{P}_k^\epsilon}{\partial \beta_k}\frac{d\delta_k(\beta_k, P_k)}{d\beta_k}$$

$$= K_{k,V}^\epsilon + \Lambda_{k,V}^\epsilon \frac{d\delta_k(\beta_k, P_k)}{d\beta_k}.$$

This definition of the heat capacity at constant pressure is justified by Gibbs' [1875] demonstration that $\Delta\mathcal{H}_k^{(\mathrm{en})}$ is the heat of reaction in an isobaric process. The final result obtained in (18.35) is in agreement with equation (17.6) obtained from the theory of calorimetry.[2]

By (18.35) and (17.38), the difference in the TIS heat capacities is

(18.36)
$$K_{k,\mathrm{P}}^\epsilon - K_{k,V}^\epsilon = \beta_k \left[\frac{\partial \mathrm{P}_k^\epsilon}{\partial \delta_k}\right]^{-1}\left[\frac{\partial \mathrm{P}_k^\epsilon}{\partial \beta_k}\right]^2.$$

Alternatively, the relation (18.35) can be expressed immediately in the standard form

(18.37)
$$K_{k,\mathrm{P}}^\epsilon - K_{k,V}^\epsilon = \left[\frac{\partial \mathcal{H}_k^\epsilon}{\partial \delta_k} + \mathrm{P}_k^\epsilon\right]\frac{\partial \delta_k(\beta_k, P_k^\epsilon)}{\partial \beta_k}.$$

18.5 Equilibrium Enthalpy

The *k equilibrium enthalpy* is defined by

(18.38)
$$\mathcal{H}_k^{(\mathrm{en})\epsilon} = \mathcal{H}_k^\epsilon + \mathrm{P}_k^\epsilon \delta_k = \left[\frac{\delta_k}{\beta_k}\frac{\partial}{\partial \delta_k} - \frac{\partial}{\partial \beta_k}\right]\ln Z_k^\epsilon.$$

The previous formalism is used to compute its derivatives with respect to the volume and thermature with the result

(18.39a)
$$\frac{\partial \mathcal{H}_k^{(\mathrm{en})\epsilon}}{\partial \beta_k} = K_{k,V}^\epsilon + \delta_k \frac{\partial \mathrm{P}_k^\epsilon}{\partial \beta_k},$$

(18.39b)
$$\frac{\partial \mathcal{H}_k^{(\mathrm{en})\epsilon}}{\partial \delta_k} = -\beta_k \frac{\partial \mathrm{P}_k^\epsilon}{\partial \beta_k} + \delta_k \frac{\partial \mathrm{P}_k^\epsilon}{\partial \delta_k}.$$

Observe that only the derivatives of $\ln Z_k^\epsilon$ with respect to the thermature and volume appear in the definition (18.38) of the enthapy. These quantities

---

[2]Compare the result (18.35) also with the standard equations in Clausius [1879], p. 180, 181 or equation (4-69) in Dickerson [1969], p. 178.

are both observables, so the enthalpy is an observable of the theory. An application of the enthalpy formalism will be presented in the next chapter in a discussion of the Joule-Thomson experiment.

18.6 Equilibrium Thermodynamic Potentials

The free energy was introduced informally in Chapter 14 in connection with the conjectures of Nernst and Planck. To complete this brief exposition of the fundamentals of the TIS version of equilibrium thermodynamics, the equilibrium free energy and a quantity called in TIS the equilibrium free enthalpy, which is usually called the Gibbs free energy in standard thermodynamics, will be introduced next and their volume and thermature derivatives will be computed. They will be introduced formally in the next chapter and analyzed there.

The standard thermodynamic definitions of the equilibrium free energy and the equilibrium free enthalpy in TIS notation are

(18.40a)
$$A_k^{(\text{fe})} = \mathcal{H}_k^{\epsilon} - [k_B \beta_k]^{-1} S_k^{(\text{n})\epsilon},$$

(18.40b)
$$G_k^{(\text{fe})} = \mathcal{H}_k^{(en)\epsilon} - [k_B \beta_k]^{-1} S_k^{(\text{n})\epsilon} = A_k^{(\text{fe})} + \delta_k \text{P}_k^{\epsilon}.$$

It is necessary to use the standard equilibrium entropy $S_k^{(\text{n})\epsilon}$ in these definitions instead of the TIS equilibrium entropy S_k^{ϵ} because using the TIS equilibrium entropy in place of the standard entropy in the definition of the free energy gives $A_k^{(\text{fe})} = 0$ identically. The k *equilibrium free energy* will therefore be defined directly in TIS in terms of the equilibrium partition function by

(18.41)
$$A_k^{(\text{fe})} = -[\beta_k]^{-1} \ln Z_k^{\epsilon}.$$

The k *equilibrium free enthalpy* is defined in similarly by

(18.42)
$$G_k^{(\text{fe})} = \frac{1}{\beta_k} \left[\delta_k \frac{\partial}{\partial \delta_k} - 1 \right] \ln Z_k^{\epsilon}.$$

The presence of the $\beta_k^{-1} \ln Z_k^{\epsilon}$ term in both of these definitions means that the free energy and free enthalpy are not observables.

With (18.41) and (18.42), it is easy to compute the thermature and volume derivatives of the free energy and free enthalpy with respect to the volume and the temperature. For the free energy, (18.20) and (18.24) imply immediately

(18.43a)
$$\frac{\partial A_k^{(\text{fe})}}{\partial \beta_k} = \frac{1}{k_B \beta_k^2} S_k^{(\text{n})\epsilon},$$

(18.43b)
$$\frac{\partial A_k^{(\text{fe})}}{\partial \delta_k} = -\text{P}_k^{\epsilon}.$$

Similarly, the thermature and volume derivatives of the free enthalpy are

(18.44a)
$$\frac{\partial G_k^{(\text{fe})}}{\partial \beta_k} = \frac{\partial A_k^{(\text{fe})}}{\partial \beta_k} + \delta_k \frac{\partial P_k^\epsilon}{\partial \beta_k},$$

(18.44b)
$$\frac{\partial G_k^{(\text{fe})}}{\partial \delta_k} = \delta_k \frac{\partial P_k^\epsilon}{\partial \delta_k}.$$

With a little rearrangement, the Thomson-Massieu (Gibbs-Helmholtz) relations can be written for the free energy and free enthalpy in the form[3]

(18.45a)
$$\frac{\partial [\beta_k A_k^{(\text{fe})}]}{\partial \beta_k} = \mathcal{H}_k^\epsilon,$$

(18.45b)
$$\frac{\partial [\beta_k G_k^{(\text{fe})}]}{\partial \beta_k} = \mathcal{H}_k^\epsilon - \delta_k \frac{\partial \mathcal{H}_k^\epsilon}{\partial \delta_k}.$$

If the pressure is held fixed, which is often the case for chemical reactions, then (18.45b) is written

(18.46)
$$\left. \frac{\partial [\beta_k G_k^{(\text{fe})}]}{\partial \beta_k} \right|_P = \mathcal{H}_k^{(en)}.$$

The free energy and free enthalpy, along with the chemical potential, will be discussed further in the next chapter.

18.7 Constitutive Relations

The differences between different kinds of matter are expressed in constitutive relations that describe the interactions of their particles.[4] In the general theory, it is the attributes of the interparticle potential functions in the set $\{\phi_{ij}(q_i - q_j)\}$ and the set of particle masses $\{m_i\}$ that determine the distinct properties of a given system. The interparticle potential functions are required to meet the conditions expressed in (4.2).

For the equilibrium case, more can be said. The first set of constitutive relations consists of the statements that $\ln Z_k^\epsilon$, $\mathcal{H}_k^\epsilon$, S_k^ϵ, and P_k^ϵ, are functions only of the state $\mathcal{C}_k^\epsilon$ of the system. The second set of constitutive relations consists of statements about the first derivatives of these quantities with respect to the state variables β_k and δ_k. These derivatives are concerned with heat capacity, pressure, working, and heating.

There are a number of relations between the derivatives of state functions that allow some to be expressed in terms of others. The volume derivative of the equilibrium Hamiltonian energy, for example, is expressed in terms of

---

[3]See the discussion of the designation of these relations by Thomson-Massieu versus Gibbs-Helmholtz in Chapter 14.

[4]See Truesdell and Toupin [**1960**], pp. 700–711, and Truesdell [**1969**], p. 19, for more information on constitutive relations.

the equilibrium pressure and the equilibrium latent heat in (18.31). For the heat capacity at constant volume, the relation $\partial \mathcal{H}_k^\epsilon / \partial \beta_k = K_{k,V}^\epsilon \leq 0$ implies that $\partial S_k^\epsilon / \partial \beta_k \leq k_B \mathcal{H}_k^\epsilon$ by (18.33a). The volume derivative of the entropy is expressed as $-k_B \beta_k (\partial (\beta_k \mathrm{P}_k^\epsilon) / \partial \beta_k)$ in (18.33b). It follows that the properties of the second set of constitutive relations can be expressed in terms of $\mathcal{H}_k^\epsilon$, $\partial \mathcal{H}_k^\epsilon / \partial \beta_k$, P_k^ϵ, $\partial \mathrm{P}_k^\epsilon / \partial \beta_k$ and $\partial \mathrm{P}_k^\epsilon / \partial \delta_k$ alone.

The relation $\partial \mathrm{P}_k^\epsilon / \partial \delta_k \leq 0$ introduced in (17.3a) is called the *stability condition*. To establish its meaning and validity requires an analysis of phase transitions. This question is discussed in a more general context in EIS, Part V. For now, it is viewed as a constitutive relation for the theory. In reference to $\partial \mathrm{P}_k^\epsilon / \partial \beta_k$, the anomalous behavior of water at freezing indicates that $\partial \mathrm{P}_k^\epsilon / \partial \beta_k > 0$ can occur in limited regions of equilibrium space.

The constitutive relations of the TIS version of equilibrium theory are summarized in the following list

1. The thermodynamic state is $\mathcal{C}_k^\epsilon$;
2. Z_k^ϵ, and therefore $\mathcal{H}_k^\epsilon$, S_k^ϵ, and P_k^ϵ, are state functions;
3. $\partial \mathrm{P}_k / \partial \delta_k \leq 0$; $\partial \mathrm{P}_k^\epsilon / \partial \beta_k \leq 0$ (usually).

It is easy to show that there is equality in item 3 when the density is zero.

CHAPTER 19

Thermodynamic Potentials and Operators

While thermodynamic potentials have played an important role in standard equilibrium thermodynamics, it was pointed out in the last chapter that the equilibrium versions of the free energy and free enthalpy are not observables. The question of whether microscopic analog functions exist for them, so that they can play a role in TIS thermodynamics, is addressed in this chapter. A related inquiry examines the relation between the thermodynamic derivative operators by calculating the commutation relations between the $\partial/\partial t$, $\partial/\partial\delta_k$, and $\partial/\partial\beta_k$, operators in both the general and equilibrium cases.

398

19.1 Thermodynamic Potentials

Thermodynamic potentials have been introduced informally and used in calculations in Chapters 14 and 18. It is time now to introduce them formally and examine their properties and status in the Theory of Interacting Systems.

Thermodynamic potentials have a long history. In 1855, Thomson [**1882**], Vol. I., p. 297, introduced a quantity $w = U - Q$, which is the difference of the system energy and the system heat. He used the second law in the form $\sum(Q/T) = 0$ for a cycle of a reversible process to obtain the equation $U = w - T(dw/dT)$. The quantity w and others like it were soon called thermodynamic potentials. While they are not usually considered observables, the idea is that thermodynamic potentials play a role that is similar to that of a potential in the theory of electricity. The concept of a thermodynamic potential in general was discussed by Massieu [**1869**].[1] Massieu, pp. 858–859, adopted the temperature T and volume V as independent variables and chose as one of his potentials the energy function $U(T, V)$. He, pp. 858–859, made use of Clausius' relation $dS = dQ/T$ for the entropy change in a reversible process and the relation $dQ = dU + PdV$ to obtain $dU = TdS - PdV$. With these elements, he showed that there exists another function $\psi(T, V)$ from which one may obtain "all the properties of bodies that one considers in thermodynamics." He called this function the *indicator function* of a body and, p. 1058, expressed the entropy in terms of it by

$$(19.1) \qquad S(T, V) = \psi(T, V) + T\frac{d\psi(T, V)}{dT} = \frac{d(T\psi(T, V))}{dT}.$$

Massieu showed that $P(T, V) = T(d\psi/dV)$, $U = T^2(d\psi/dT)$ and $S = \psi + U/T$.

If P and T are chosen as the independent variables instead of V and T, Massieu then showed that there is another indicator function $\psi'(T, P) = S(T, P) - U'(T, P)/T$ with similar properties. In this case, $U'(T, P)$ is defined by $U'(T, P) = U(T, P) + PV(T, P)$ and it follows from the definition of $\psi'(T, P)$ that $S = \psi' + T(d\psi'/dT) = d(T\psi'/dT) = \psi' + U'/T$ in parallel with (19.1). It also follows that $U' = T^2(d\psi'/dT)$ and $V = -T(d\psi'/dP)$.

In accord with modern notation, $U = \mathcal{H}$ and the potential $-T\psi = A^{(\mathrm{fe})} = \mathcal{H} - TS$ is called, following Helmholtz [**1882**], the *free energy*. Other names that have been used are the "available energy", the "maximal work", and the "Helmholtz free energy". For the potential $-T\psi' = G^{(\mathrm{fe})} = A^{(\mathrm{fe})} + PV = $

---

[1]Truesdell [**1980**] credits F. Reech with discovering the thermodynamic potentials, but the work of Reech was little noticed at the time. Others have objected to using the word "potential" to describe what are in fact generating functions. This issue will be revisited in a more general context in EIS, Part III. Massieu's work has also been discussed in Bailyn [**1994**], pp. 205–209.

$\mathcal{H} - TS + PV$, the name *free enthalpy* will be used. It has also been called the "thermodynamic potential" and the "Gibbs free energy".[2]

Helmholtz [**1882**] made use of Massieu's ψ potential that he stated bears the "same relation to the external work that the energy does to the sum of the external heat and work".[3] In stating Massieu's equations in terms of Gibbs' [**1902**] equilibrium theory, Massieu's energy U is interpreted as the equilibrium Hamiltonian energy $\mathcal{H}_k^\epsilon$. To obtain the expression for the free energy stated above, the "heat", expressed in the form $T_k S_k^{(n)\epsilon}$, is subtracted from the Hamiltonian energy. The definition of the free energy is then written in the standard form at equilibrium in TIS notation as

$$(19.2) \qquad\qquad A_k^{(fe)} = \mathcal{H}_k^\epsilon - [k_B \beta_k]^{-1} S_k^{(n)\epsilon}.$$

This was the form used in Chapter 14 in the discussion of Nernst's work.

Gibbs [**1875**] also made significant use of Massieu's potentials, which he designated ψ and η, in his analysis of phase phenomena at equilibrium and to facilitate the computation of thermodynamic relationships. In addition, he used the symbol χ for the enthalpy, which is designated in TIS by the symbol $\mathcal{H}_k^{(en)}$. Gibbs also showed that η represents the internal work done on the body itself when bonds between particles are either lengthened or shortened. In this case both the heat and the work at the boundary are subtracted from the energy to give the free enthalpy. This is equivalent to subtracting the heat from the enthalpy and is the justification for the use of the name "free enthalpy" in parallel with the name "free energy". The free enthalpy is

---

[2]The names 'Helmholtz free energy' and 'Gibbs free energy', often used for $A^{(fe)}$ and $G^{(fe)}$, are historically incorrect. See Partington [**1949**], p. 187. For the other names mentioned in the text, and references to the literature, see Partington [**1949**], pp. 162, 169–170, 182, 185–7. Partington [**1949**], p. 185, footnote 1, observed that the term "free enthalpy" was used by O. Martin [**1941**]. (See Martin [**1941**], p. 82.) The name 'free enthalpy' is also mentioned by Lewis and Randall [**1924**], pp. 140–141, used by Truesdell and S. Bharatha [**1977**], Chapter 14, and recommended in Dickerson [**1969**], p. 191. Concerning the representation of these quantities by symbols, Lewis and Randall, [**1924**], p. 141, noted that the quantity $A^{(fe)}$ is designated as A by writers in the United States and often as F by European writers. They recommended F for the free energy because the tables in common use are tabulated with respect to F. Lewis and Randall also observed that the quantity $G^{(fe)}$ is usually designated as G by European writers and as F in the United States.

[3]Planck [**1922**], p. 113, stated that the free energy should be called the "free energy for isothermal processes" and referred to $TS = \mathcal{H} - A^{(fe)}$ as the "latent energy" (gebundene energie). He, p. 114, used the relation $dA^{(fe)} < W$, which states that the change in the free energy is less than the work done on the system in an isothermal process, to characterize irreversible isothermal processes. Planck, p. 118, also called the quantity represented here by the symbol $G^{(fe)}$ the "thermodynamic potential at constant pressure". For a discussion of the free energy (under the name 'Helmholtz free energy'), the free enthalpy (under the name 'thermodynamic potential'), and a number of relations involving them, see Planck, pp. 113–124. See also Dickerson [**1969**], Chapters 4 and 5.

written in TIS notation in terms of equilibrium quantities as

$$(19.3) \qquad G_k^{(\text{fe})} = A_k^{(\text{fe})} + \delta_k \text{P}_k^\epsilon = \mathcal{H}_k^{(\text{en})\epsilon} - [k_B \beta_k]^{-1} S_k^{(\text{n})\epsilon}.$$

In standard thermodynamics, the free energy is usually considered to be a function of the temperature and the volume and written $A^{(\text{fe})}(T, V)$ and used to represent systems when they are described in terms of their temperature and volume. The free enthalpy is usually considered a function of the temperature and the pressure and written $G^{(\text{fe})}(T, \text{P})$ and used to represent systems when they are described in terms of their temperature and pressure. In TIS, the boundary conditions determine the independent variables in a similar way, but a second step is required for systems described by temperature and pressure as the independent variables. The TIS volume descriptor is associated with the system projection operator and the formula for the pressure is used with the volume descriptor formalism and the relation $d\text{P}_k = 0$, calculated in (17.38), to express the volume as an implicit function of the temperature and pressure as $\delta(\beta_k, \text{P}_k)$. This step was used in equation (18.35) in the last chapter in the calculation of the heat capacity at constant pressure. A transformation of this type of a thermodynamic quantity from one set of independent variables to another is called a *Legendre* transformation.

A number of difficult problems are encountered when trying to interpret the free energy and the free enthalpy as thermodynamic functions in the TIS framework. These problems will be addressed next.

19.2 The Free Energy and Free Enthalpy

The concepts of free energy and free energy are central to equilibrium thermodynamics and have played an important role in the theory of irreversible processes and the theory of phase boundaries.[4] From a macroscopic point of view, formalisms employing the free energy and free enthalpy have been successful in predicting the formation of phase boundaries in calculations based on macroscopic system parameters. They are also used as intermediate quantities in a number of other thermodynamic calculations and are occasionally treated as observables in standard thermodynamics.

Although the free enthalpy and free energy have been considered fundamental to the analysis of phase change, it should be noted that they are not the only quantities that have been used for this purpose. In the early twentieth century, an alternative particle-based approach to the theory of the phase boundary problem, in which forces on particles at a phase boundary were computed directly without using the free energy or the free enthalpy, was employed with some success by Jones [**1924**], Fowler [**1936**], and others. Other alternative approaches have been associated with the Clapeyron equation.

---

[4]For further information, see, for example, the extensive discussion in Dickerson [**1969**] or the discussion in Planck [**1922**].

The standard equilibrium thermodynamic formulas defining the free energy and free enthalpy were stated in (18.40) in terms of the thermodynamic concepts of the Hamiltonian energy, thermature, entropy, and enthalpy. Explicit formulas for them in TIS notation in terms of the equilibrium partition function and its derivatives with respect to thermature and volume were given in (18.41) and (18.42). The derivatives of the free energy and free enthalpy with respect to thermature and volume were computed in (18.43) and (18.44).

In spite of their success predicting phase boundaries, the free energy and free enthalpy have not been given an interpretation as analog functions based in the underlying particle model that would allow them to be generalized to the nonequilibrium case. One problem is connected with the presence of the entropy term in the definitions of both the free energy and free enthalpy. This problem stems from the fact established in Chapter 14 that the macroscopic entropy is not well defined at absolute zero, which means that the entropy and therefore the free energy and free enthalpy do not have determinate values. Another problem is that the entropy is not grounded in the individual particles. On the other hand, entropy differences are well defined in thermodynamics. This is consistent with the TIS result (9.16) that entropy changes are associated with individual particles. These facts imply that useable thermodynamic relations based on the free energy or the free enthalpy should be in the form of either differential or difference equations.[5]

In trying to generalize these potentials to the nonequilibrium case, the definitions (18.41) and (18.42) are first extended to the general case by describing the system by $F_k(Q_k, P_k, t)$ rather than $F_k^\epsilon(Q_k, P_k)$ and replacing $\ln Z_k^\epsilon$ with $\ln Z_k(t)$. The generalized free energy and free enthalpy are expressed in these terms as

(19.4a)
$$A_k^{(\mathrm{fe})}(t) = -\frac{1}{\beta_k(t)} \ln Z_k(t),$$

(19.4b)
$$G_k^{(\mathrm{fe})}(t) = \frac{1}{\beta_k(t)} \left[\delta_k(t) \frac{\partial}{\partial \delta_k} - 1 \right] \ln Z_k(t).$$

By (5.34), the time derivatives of these quantities are

(19.5a)
$$\frac{dA_k^{(\mathrm{fe})}(t)}{dt} = -\frac{\dot{\beta}_k(t)}{\beta_k(t)} A_k^{(\mathrm{fe})}(t),$$

(19.5b)
$$\frac{dG_k^{(\mathrm{fe})}(t)}{dt} = -\frac{\dot{\beta}_k(t)}{\beta_k(t)} A_k^{(\mathrm{fe})}(t) + \dot{\delta}_k(t) \mathrm{P}_k(t) + \delta_k(t) \dot{\mathrm{P}}_k(t).$$

---

[5]See Gibbs [**1875**] (Gibbs [**1928**], p. 86) for the differential form, and Partington [**1949**], pp. 182–188, for the difference form, of the free enthalpy at equilibrium. The attempts of Nernst, Planck, and others to define an absolute entropy are discussed in QTS.

In an isothermal process, $\dot{\beta}_k(t) = 0$, and it follows that $dA_k^{(fe)}(t)/dt = 0$ and $dG_k^{(fe)}(t)/dt = \dot{\delta}_k(t)P_k(t) + \delta_k(t)\dot{P}_k(t)$.

The major problem in attempting to interpret the definitions (19.4) as particle-based functions is the presence of the partition function. The partition function is a thermodynamic object, but it is not a thermodynamic observable with a an associated particle analog function. This problem also occurs in the time derivatives of the potentials because $A_k^{(fe)}(t)$ is present in these derivatives. Because $A_k^{(fe)}(t)$ is not an observable, the time derivatives of $A_k^{(fe)}(t)$ and $G_k^{(fe)}(t)$ are not observables either.

In a second attempt to generalize the free energy and free enthalpy, the definitions (19.2) and (19.3) will be generalized to the nonequilibrium case and then the time derivative will be computed. An immediate snag is the fact that the calculation of the $dS_k(t)/dt$ term can only be carried through by using Clausius' inequality. When this is done, the relation $\mathcal{Q}_{\partial k}(t) \leq [k_B\beta_k(t)]^{-1}(dS_k(t)/dt)$ is used to replace $[k_B\beta_k(t)]^{-1}(dS_k(t)/dt)$ with $\mathcal{Q}_{\partial k}(t)$. This results in inequalities for the total time derivatives of the free energy and free enthalpy rather than a definite value that would be expected if they had a mechanical definition.

To pursue these issues, let us begin with the rate of change of the k *local free energy* for an isothermal process $\dot{\beta}_k(t) = 0$. This is expressed in TIS terms in the approximate inequality based on Clausius' inequality

$$(19.6) \qquad \frac{dA_k^{(fe)}(q,t)}{dt} \lessapprox \frac{d\mathcal{H}_k(q,t)}{dt} - \mathcal{Q}_{\partial k}(q,t) = \mathcal{W}_{k,b}(q,t) + \mathcal{W}_{\partial k}(q,t).$$

For the isothermal rate of change of the k *local free enthalpy*, it is easy to show that the approximate inequality is

$$(19.7) \qquad \frac{dG_k^{(fe)}(q,t)}{dt} \lessapprox \frac{d\mathcal{H}_k(q,t)}{dt} - \mathcal{Q}_{\partial k}(q,t) - \mathcal{W}_{\partial k}(q,t) = \mathcal{W}_{k,b}(q,t).$$

There is equality in each of these equations if and only if the process it is describing is reversible. These approximate inequalities are therefore not helpful either in providing a particle-based analog for the free energy and free enthalpy.

Let us consider the volume and thermature derivatives of the free energy and free enthalpy next. The volume derivatives of the equilibrium free energy and free enthalpy, calculated in (18.43b) and (18.44b), are generalizable to the nonequilibrium case because both $P_k(t)$ and $\delta_k(t)(\partial P_k(t)/\partial \delta_k)$ can be expressed as phase averages of analog functions. However, this is not true for their thermature derivatives calculated in (18.43a) and (18.44a). The sticking point lies with the appearance of the term $S_k^{(n)\epsilon}$ in (18.43a) and, using (18.43a), it appears in (18.44a) as well. The root of the problem in all these cases is the fact that the quantity $\beta_k^{-1}(t)\ln Z_k(t)$ appears in the definitions of both $A_k^{(\text{fe})}$ and $G_k^{(\text{fe})}$.

Because neither the free energy nor the free enthalpy can be generalized to the nonequilibrium case, they cannot be included as part of the general theory. In spite of this problem, differences of the form $\Delta A_k^{(\text{fe})} = A_k^{(\text{fe})}(\beta_{k2},\delta_{k2}) - A_k^{(\text{fe})}(\beta_{k1},\delta_{k1})$ and $\Delta G_k^{(\text{fe})} = G_k^{(\text{fe})}(\beta_{k2},\delta_{k2}) - G_k^{(\text{fe})}(\beta_{k1},\delta_{k1})$, for two points on an equilibrium manifold, do not depend on the absolute definition of the entropy and can be expressed in terms of measurable heat flows when the system moves from one equilibrium point to the other. In the standard theory of phase transitions, a phase change is predicted between two states at representing densities when $\Delta A_k^{(\text{fe})} = 0$ or $\Delta G_k^{(\text{fe})} = 0$. Setting these differences to zero has proved useful in determining phase transition thresholds.

19.3 The Chemical Potential

In his analysis of thermodynamic phases, Gibbs needed a quantity to determine the proportion of a substance in each phase. He [1875] ([1928], pp. 62–85), introduced the chemical potential $\mu_k^{(\text{fe})}$ for this purpose. If it is assumed that a system k is open and can exchange particles with another system l, the formula for the differential of the equilibrium free enthalpy can be extended from $dG_k^{(\text{fe})} = dH_k^{(\text{en})\epsilon} - T_k S_k^{(\text{n})\epsilon}$ to

$$(19.8) \qquad dG_k^{(\text{fe})} = dH_k^{(\text{en})\epsilon} - T_k S_k^{(\text{n})\epsilon} + \mu_k^{(\text{fe})\epsilon} d\mathcal{N}_k.$$

In Gibbs' theory, equality of the chemical potentials is another condition, in addition to equality of pressures and temperatures, that must be satisfied for the formation of a phase boundary.

The chemical potential was introduced to deal with open systems by Gibbs [1902], Chapter XV, who generalized his original concept of an ensemble of systems, which he then called a *petit ensemble,* to that of a *grand ensemble* of systems. The petit ensemble is a collection of representations of a closed system consisting of the same particles. The members of the ensemble differ only in the initial mechanical phase point (Q_k, P_k) assigned to the system. The grand ensemble is a collection of petit ensembles that differ in the number of particles in the system. Gibbs also generalized the notion of a canonical ensemble to that of a *grand canonical ensemble.* The grand canonical ensemble is in equilibrium not only with respect to the exchange of momentum, angular momentum, and energy, by its component systems through their boundaries, but also with respect to the exchange of particles. By this device, Gibbs extended the equilibrium thermodynamics of closed systems to the equilibrium thermodynamics of open systems.

Gibbs' definition of the chemical potential treated the mole number of each system as an independent variable and extended the definition of the system energy to include it.[6] In this way, the chemical potential as expressed in (19.8) represents the contribution of the number of particles to the total system energy.

In the case of a heterogeneous substance composed of a number of components, each component consists of a number of identical particles of a particular type. Let n_α be the number of moles of component α in a substance. The chemical potential of component α is then defined at equilibrium in terms of the system free enthalpy by the symbolic equation $\mu_\alpha^{(\text{fe})\epsilon} = (dG_k^{(\text{fe})\epsilon}/dn_\alpha)$. While it is straightforward to include heterogeneous systems in these calculations, the system will be assumed to be homogeneous for simplicity and the chemical potential will be defined using the system particle number rather than the mole number as the variable.

The k *chemical potential* per particle is defined in these terms as[7]

$$(19.9) \qquad \mu_k^{(\text{fe})\epsilon} = \frac{dG_k^{(\text{fe})\epsilon}}{d\mathcal{N}_k}.$$

In a system of $\mathcal{N}_k$ particles in a single phase, as Bent [**1965**], pp. 276–278, showed using standard results on the homogeneity of systems satisfying the maximum entropy condition, the chemical potential is the free enthalpy per particle:[8]

$$(19.10) \qquad \mu_k^{(\text{fe})\epsilon} = \frac{G_k^{(\text{fe})\epsilon}}{\mathcal{N}_k}.$$

This implies that the chemical potential is an intensive quantity in standard thermodynamics.

The derivative with respect to particle number differs from the derivatives previously introduced in that the number of particles in a system is discrete. The change of one particle from one phase to another involves a discontinuous change of the total kinetic and potential energy in each phase. Although Gibbs [**1902**], p. 204, footnote, argued that the chemical potential can be obtained as in (19.9) as the derivative of the mole number, it will be treated here as a difference equation because of the discontinuous nature of the transition. Let $\delta_i^{(p)}(\beta_k)$ be the specific volume of particle i in phase $p \in \{1,2\}$ at thermature β_k. A difference equation can be written for the change at equilibrium in $G_{k(p)}^{(\text{fe})\epsilon}$ when particle i moves into phase p from some

---

[6]See Gibbs [**1902**], p. 200, for the statistical mechanical and thermodynamic definitions of the free energy including the chemical potential coefficients. See equation (554), p. 204, for an expression of the chemical potential as a derivative of what Gibbs called the generic free energy with respect to particle number.

[7]For the use of this definition in the standard setting, see, for example, Bent [**1965**], p. 276, or Dickerson [**1969**], pp. 249–254. See also Bailyn [**1994**], Chapter 6.

[8]See also the demonstration of this in Bailyn [**1994**], p. 216, and the discussion in Lewis and Randall [**1924**], pp. 148–149.

other phase. The change in the free enthalpy of phase p with the gain of particle i takes the form

$$(19.11a) \qquad \mu_{i(p)}^{(\text{fe})\epsilon} = \Delta_i^{(p)} G_{k(p)}^{(\text{fe})\epsilon},$$

where the change in the free enthalpy is given by

$$(19.11b) \quad \Delta_i^{(p)} G_{k(p)}^{(\text{fe})\epsilon} = G_k^{(\text{fe})\epsilon}(\beta_k, \delta_k^{(p)} + \delta_i^{(p)}(\beta_k), \mathcal{N}_p + 1) - G_k^{(\text{fe})\epsilon}(\beta_k, \delta_k^{(p)}, \mathcal{N}_p)$$

and $\mathcal{N}_p \leq \mathcal{N}_k$ is the number of particles in phase p. The loss of a particle in phase p is accounted for in (19.11) by using $\mathcal{N}_p - 1$ in place of $\mathcal{N}_p + 1$ and $-\delta_i^{(p)}(\beta_k)$ in place of $+\delta_i^{(p)}(\beta_k)$. To avoid irrelevant details, let us work with a large number of particles, so that $\mathcal{N}_p \gg 1$ and the change in the free enthalpy due to the loss of particle i in phase p is $-\Delta_i^{(p)} G_{k(p)}^{(\text{fe})\epsilon}$, which is the negative of the gain when the particle enters phase p.

Assume that the pressure and temperature are held fixed. Making use of the fact that the pressure is constant and using the equilibrium definition (18.40) of the free enthalpy gives for the movement of particle i into phase p the relation

$$(19.12) \qquad \Delta_i^{(p)} G_{k(p)}^{(\text{fe})\epsilon} = H_{i(p)}^{\epsilon} + \delta_i^{(p)}(\beta_k) \mathrm{P}_{k(p)}^{\epsilon} - \mathcal{Q}_{\partial i(p)}^{\epsilon},$$

where the change in entropy times the thermature for particle i is expressed as a "heating function":

$$(19.13) \qquad \mathcal{Q}_{\partial i(p)}^{\epsilon} = [k_B \beta_k]^{-1} \Delta_i^{(p)} S_k^{(n)\epsilon}(\beta_k, \delta_k^{(p)}, \mathcal{N}_p).$$

With these results, the *i-particle equilibrium chemical potential* and the *k-equilibrium chemical potential* in phase p can be written in the form

$$(19.14a) \qquad \mu_{i(p)}^{(\text{fe})\epsilon} = H_{i(p)}^{\epsilon} + \delta_i^{(p)}(\beta_k) \mathrm{P}_{k(p)}^{\epsilon} - \mathcal{Q}_{\partial i(p)}^{\epsilon},$$

$$(19.14b) \qquad \mu_{k(p)}^{(\text{fe})\epsilon} = \frac{1}{\mathcal{N}_k} \sum_{i \in k} \mu_{i(p)}^{(\text{fe})\epsilon}.$$

Let us next compute the difference in the chemical potential for a transition from phase 1 to phase 2 of a substance. First, set the pressures of the phases equal, that is, $\mathrm{P}_{k(1)}^{\epsilon} = \mathrm{P}_{k(2)}^{\epsilon} = \mathrm{P}_k^{\epsilon}$, and, using (19.14), write the difference in the chemical potentials for the two phases in the form of an average one-particle exchange between the phases as

$$(19.15a) \qquad \Delta_{21} \mu_k^{(\text{fe})\epsilon} = \mu_{k(2)}^{(\text{fe})\epsilon} - \mu_{k(1)}^{(\text{fe})\epsilon},$$

$(19.15b)$

$$\Delta_{21} \mu_k^{(\text{fe})\epsilon} = \frac{1}{\mathcal{N}_k} \{ H_{k(2)}^{\epsilon} - H_{k(1)}^{\epsilon} + \mathrm{P}_k^{\epsilon} [\delta_k^{(2)}(\beta_k) - \delta_k^{(1)}(\beta_k)] - [\mathcal{Q}_{\partial k(2)}^{\epsilon} - \mathcal{Q}_{\partial k(1)}^{\epsilon}] \}.$$

A phase boundary will form when the chemical potentials are equal for two suitable densities of a substance that have the same pressure at a given temperature. This fact is used to rearrange the result (19.15) to obtain a relation valid when there is more than one phase

(19.16)
$$\Delta_{21}\mathcal{Q}_k^\epsilon = \mathcal{Q}_{\partial k(2)}^\epsilon - \mathcal{Q}_{\partial k(1)}^\epsilon = H_{k(2)}^\epsilon - H_{k(1)}^\epsilon + \mathrm{P}_k^\epsilon[\delta_{k(2)}(\beta_k) - \delta_{k(1)}(\beta_k)]$$
$$= H_{k(2)}^{(\mathrm{en})\epsilon} - H_{k(1)}^{(\mathrm{en})\epsilon} = \Delta_{21}\mathcal{H}_k^{(\mathrm{en})}.$$

The result (19.16) shows that the difference in the heating in the two phases due to a transition between phases is the same as the difference in the enthalpies of the two phases when the chemical potentials are equal. From the standpoint of the free enthalpy difference (19.12), a phase boundary is predicted to occur when the change in the free enthalpy in moving a particle from phase 2 into phase 1 equals the change in the free enthalpy in moving a particle phase 1 into phase 2. In EIS, these results will be connected with the Clapeyron equation mentioned in the last chapter.

From the TIS perspective, the fact that the chemical potential is defined in terms of the derivative or difference of the free enthalpy with respect to the transfer of particles between phases, means that it suffers from problems similar to those of the free enthalpy. While the system thermodynamic analogs, for example, do depend on the number of particles in the system, they do not depend explicitly on the system particle number as a parameter. This means that the difference formulas above must be used to make sense of the "derivative" with respect to particle number. Moreover, while heat flow is measurable and can be represented theoretically, there is no "heat function" or "heat function analog" defined in terms of the entropy in the Interacting Systems approach. It follows that only differences between the chemical potentials of two phases, but not the individual chemical potentials themselves, can be used in determining the formation of a phase boundary.

Finally, the chemical potential is represented in (19.8) as a contribution to the total system energy that is proportional to the number of particles. Because the classical version of the TIS theory allows only closed systems, the theory of phase change must be analyzed within a single system k and the phase boundary between phases 1 and 2 is represented as a boundary internal to that system. While it is true that the addition of a particle to a phase changes the total energy of that phase, the microscopic k system Hamiltonian does not have a separate component reflecting this fact. In the absence of electromagnetic fields, all of the energy components in the Hamiltonian are expressed in terms of the kinetic energy or by means of the interparticle potential functions. In other words, the k system microscopic equilibrium Hamiltonian $\mathcal{H}_k^\epsilon$ in TIS does not allow the inclusion of terms of the form $\mu_{k(1)}^{(\mathrm{fe})}\mathcal{N}_k^{(1)}$ and $\mu_{k(2)}^{(\mathrm{fe})}\mathcal{N}_k^{(2)}$ that represent in the standard theory the total contribution of the chemical potential for the two phases to the system energy. This means that Gibbs' theory of phase change based on the chemical potential cannot be generalized to nonequilibrium thermodynamics and that the chemical potential does not have a mechanical significance.

For these reasons, the chemical potential cannot be a fundamental part of the TIS theory and can only be used as an aid to computation in the equilibrium case. The problem is again due to the presence of a standard entropy term in the definition of the free enthalpy. The explanation of the significance of the vanishing of the difference in the free enthalpies of two phases in the calculation of phase change—in spite of the fact that the free enthalpy is not a particle based function—requires a deeper inquiry and will be taken up again as part of the theory of phases in EIS.

19.4 The Joule-Thomson Experiment

The Joule-Thomson experiment illustrates the application of some of the TIS formalism to an experimental situation. The Joule-Thomson experiment was originally proposed by Thomson, discussed in Thomson and Joule [1854], and given a theoretical and experimental treatment in Joule and Thomson [1862]. The Joule-Thomson experiment is illustrated in Figure 19.1.

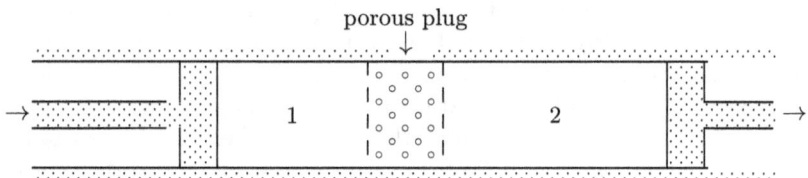

FIGURE 19.1.   The Joule-Thomson Experiment

The experimental arrangement consists of a cylinder with a piston at each end and a porous plug in the middle. The portion of the cylinder to the left of the plug is designated as chamber 1 and the portion to the right is chamber 2. Initially, a volume δ_{k1} of gas at pressure P_{k1} is introduced into chamber 1. The first piston is used to keep chamber 1 at the constant pressure P_{k1} as the gas passes through the porous plug and the second piston keeps the pressure in chamber 2 at P_{k2}, where $P_{k2} < P_{k1}$. The experiment is designed so that the process is adiabatic and the net heat entering or leaving the cylinder vanishes.

The TIS description of the Joule-Thomson experiment is based on the following assumptions:

1. The process is adiabatic:

(19.17a) $$\mathcal{Q}_{\partial k}(t) = 0;$$

2. The working done by nonboundary forces is neglected:

(19.17b) $$W_{k,b}(t) = 0;$$

3. The working at the boundary is approximated by the equilibrium relation:

(19.17c) $$W_{\partial k}(t) \approx -\mathrm{P}_k(t)\dot{\delta}_k(t) = -[\mathrm{P}_{k1}(t)\dot{\delta}_{k1}(t) + \mathrm{P}_{k2}(t)\dot{\delta}_{k2}(t)];$$

4. The pressures are constant:

(19.17d) $$\frac{d\mathrm{P}_{k1}(t)}{dt} = \frac{d\mathrm{P}_{k2}(t)}{dt} = 0.$$

5. External forces, other than surface forces, are neglected, so that, $\Phi_{k,x}(Q_k, t) \approx 0$. The relation $\mathcal{H}_k(t) \approx \mathcal{E}_k(t)$ that follows from this is used to replace $\mathcal{H}_k$ with $\mathcal{E}_k$.

Under these assumptions, axiom (3.9b) can be expressed in the form

(19.18) $$\frac{d\mathcal{E}_k(t)}{dt} = W_{\partial k}(t) \approx -\mathrm{P}_k(t)\dot{\delta}_k(t).$$

With this approximation, the relation

(19.19) $$\frac{d\mathcal{H}_k^{(\mathrm{en})}(t)}{dt} \approx \frac{d\mathcal{E}_k(t)}{dt} + \mathrm{P}_k(t)\dot{\delta}_k(t) \approx 0$$

follows. This justifies the description of the Joule-Thomson experiment as a constant enthalpy experiment.

In order to make progress, some specific aspects of the experimental situation will be used to approximate the relations needed. It is important to note that the use of variational methods in which the macroscopic temperature and volume are varied have limited value in TIS thermodynamics because the time partial derivative operator does not commute with the volume and thermature partial derivative operators. As an illustration, suppose that the variation of the Hamiltonian energy $\mathcal{H}_k(t)$ is interpreted as

(19.20) $$\delta\mathcal{H}_k(t) = \frac{\partial \mathcal{H}_k(t)}{\partial \beta_k}\delta\beta_k + \frac{\partial \mathcal{H}_k(t)}{\partial \delta_k}\delta\delta_k.$$

It follows immediately that

(19.21) $$\delta\mathcal{H}_k(t) = K_{k,V}(t)\delta\beta_k - [\mathrm{P}_k(t) - \Lambda_{k,V}(t)]\delta\delta_k.$$

If the variation of the work done on the system is interpreted as

(19.22) $$\delta W_k = -\mathrm{P}_k(t)\delta\delta_k,$$

and the variation of the heat in the system is interpreted as

$$(19.23) \qquad \delta \mathcal{Q}_k = K_{k,V}(t) \delta \beta_k + \Lambda_{k,V}(t) \delta \delta_k,$$

in accord with the calorimetry equation, it follows that

$$(19.24) \qquad \delta \mathcal{H}_k(t) = \delta \mathcal{Q}_k(t) + \delta W_k(t).$$

In spite of the appearance that this is a generalization of the equilibrium result, a comparison of (19.24) with the time partial derivative of the system energy (7.61b), and the variations of work (19.22) and heat (19.23) with definitions of the working (7.53), (7.56), and heating (7.58), shows that the macroscopic equation (19.24) is not interpretable in terms of the TIS spacetime version of the theory. These variational methods may be useful in some approximation schemes but they are only valid in special circumstances.

The definition (7.81c) of the enthalpy and the relation (11.9) are used next to obtain the volume partial derivative of the enthalpy in the form

$$(19.25) \qquad \frac{\partial \mathcal{H}_k^{(\mathrm{en})}(t)}{\partial \delta_k} = \Lambda_{k,V}(t) + \delta_k(t) \frac{\partial \mathrm{P}_k(t)}{\partial \delta_k}.$$

The chambers are held at fixed pressures. For this reason, in parallel with the equilibrium equation (18.35), the spacetime version of the *heat capacity at constant pressure,* which is defined by

$$(19.26) \qquad K_{k,\mathrm{P}}(t) = \left. \frac{\partial \mathcal{H}_k^{(\mathrm{en})}(t)}{\partial \beta_k} \right|_{\mathrm{P}_k},$$

will be used. The experiment is constructed so that the volumes change significantly but slowly, the pressures are fixed, and the temperature change is small and depends on the change in volume. These facts justify using the next approximation, which expresses the time derivative of the system enthalpy in terms of the volume derivative of the enthalpy multiplied by the time derivative of the volume as

$$(19.27) \qquad \frac{d\mathcal{H}_k^{(\mathrm{en})}(t)}{dt} \approx \frac{d\mathcal{H}_k^{(\mathrm{en})}(t)}{d\delta_k} \frac{d\delta_k(t)}{dt},$$

where the macroscopic total volume derivative of the enthalpy is expressed in terms of the changes in both the volume and the thermature caused by the movement of the gas through the porous plug. The relation (19.19) and the fact that $d\delta_k(t)/dt \neq 0$ is used to show that

$$(19.28) \qquad \frac{d\mathcal{H}_k^{(\mathrm{en})}(t)}{d\delta_k} \approx 0.$$

The experiment is performed slowly so that differences in the local momentum, angular momentum, and energy, will have time to even out. According to the definition (13.49d), the difference between the equilibrium

Hamiltonian and the ordinary Hamiltonian is entirely in the special form of the equilibrium external potential function. This special form insures that the total time derivative of equilibrium Hamiltonian will vanish in accord with the relation (13.49e). Because external forces have been neglected, the equilibrium Hamiltonian $\mathcal{H}_k^\epsilon$ can be used as an approximation of the general Hamiltonian $\mathcal{H}_k(t)$.

The total macroscopic volume derivative is computed using the equilibrium Hamiltonian with the relation (19.28) to obtain the result

$$(19.29) \qquad \frac{d\mathcal{H}_k^{(\mathrm{en})\epsilon}}{d\delta_k} = \frac{\partial \mathcal{H}_k^{(\mathrm{en})\epsilon}}{\partial \delta_k} + \frac{\partial \mathcal{H}_k^{(\mathrm{en})\epsilon}}{\partial \beta_k} \frac{\partial \beta_k(\delta_k)}{\partial \delta_k} \approx 0.$$

The equilibrium relation (17.37a) is used to replace $\Lambda_{k,V}^\epsilon$ with $-\beta_k(\partial \mathrm{P}_k^\epsilon/\partial \beta_k)$ and then (19.25) and (19.26) are used to give the TIS version of the k *Joule-Thomson coefficient* in the form

$$(19.30) \quad C_k^{(\mathrm{JT})}(t) = \left.\frac{d\beta_k(\delta_k(t))}{d\delta_k}\right|_{\mathcal{H}_k^{(\mathrm{en})}} \approx \frac{1}{K_{k,\mathrm{P}}^\epsilon}\left[\beta_k(\delta_k(t))\frac{\partial \mathrm{P}_k^\epsilon}{\partial \beta_k} - \delta_k(t)\frac{\partial \mathrm{P}_k^\epsilon}{\partial \delta_k}\right].$$

This expresses the rate of change of the thermature with respect to the change in volume as the gas moves through the plug from chamber 1 to chamber 2. The *Joule-Thomson temperature inversion,* which is an inversion in the direction of the thermature change with respect to the volume, is predicted when a system path in the equilibrium manifold passes through the point, if it exists, at which the relation $\beta_k(\delta_k(t))\partial \mathrm{P}_k^\epsilon/\partial \beta_k = \delta_k(t)\partial \mathrm{P}_k^\epsilon/\partial \delta_k$ holds.

The standard Joule-Thomson coefficient, $(\partial T/\partial P)_{H_k^{(\mathrm{en})}}$, is written in TIS notation as $C_k^{(\mathrm{JT})\mathrm{std}} = -[k_B\beta_k^2]^{-1}(\partial \beta_k/\partial \mathrm{P}_k)|_{\mathcal{H}_k^{(\mathrm{en})}}$.[9] The TIS computation for the standard equilibrium Joule-Thomson coefficient when the enthalpy is held constant first uses $d\mathcal{H}_k^{(\mathrm{en})\epsilon} = 0$ to establish the relation

$$(19.31) \qquad 0 = \frac{d\mathcal{H}_k^{(\mathrm{en})\epsilon}}{d\mathrm{P}_k} = \left.\frac{\partial \mathcal{H}_k^{(\mathrm{en})\epsilon}}{\partial \beta_k}\right|_{\mathrm{P}_k}\frac{d\beta_k}{d\mathrm{P}_k} + \frac{\partial \mathcal{H}_k^{(\mathrm{en})\epsilon}}{\partial \delta_k}\left.\frac{d\delta_k}{d\mathrm{P}_k}\right|_{\beta_k}.$$

To compute the last term, in which the thermature is fixed, the relation

$$(19.32) \qquad 0 = d\beta_k = \frac{\partial \beta_k}{\partial \mathrm{P}_k}d\mathrm{P}_k + \frac{\partial \beta_k}{\partial \delta_k}d\delta_k$$

---

[9]For its standard derivation using the free enthalpy, see Planck [**1922**], p. 131, or Dickerson [**1969**], pp. 177–178. The standard result for the Joule-Thomson coefficient is expressed in the usual notation as

$$\mu_k^{(\mathrm{JT})\mathrm{std}} = \frac{1}{C_{k,\mathrm{P}}}\left[T\frac{dV}{dT} - V\right].$$

is used to obtain

$$(19.33) \qquad \left. \frac{d\delta_k}{d\mathrm{P}_k} \right|_{\beta_k} = - \left. \frac{\partial \delta_k}{\partial \beta_k} \right|_{\mathrm{P}_k} \frac{d\beta_k}{d\mathrm{P}_k}.$$

By (19.33) and (19.31) with (19.26) it follows that

$$(19.34) \qquad C_k^{(\mathrm{JT})\mathrm{std}} = - \frac{1}{k_B \beta_k^2} \left. \frac{\partial \beta_k}{\partial \mathrm{P}_k} \right|_{\mathcal{H}_k^{(\mathrm{en})}} = \frac{1}{k_B \beta_k^2 K_{k,\mathrm{P}}^{\epsilon}} \left[\frac{\partial \mathcal{H}_k^{(\mathrm{en})\epsilon}}{\partial \delta_k} \left. \frac{d\delta_k}{d\mathrm{P}_k} \right|_{\beta_k} \right].$$

The result

$$(19.35) \qquad \frac{\partial H_k^{(\mathrm{en})\epsilon}}{\partial \delta_k} = \delta_k \frac{\partial \mathrm{P}_k}{\partial \delta_k} - \beta_k \frac{\partial \mathrm{P}_k}{\partial \beta_k}$$

is used with (19.33) and (19.34) to obtain the TIS version of the standard result for the *Joule-Thomson coefficient* in the form

$$(19.36) \qquad \begin{aligned} C_k^{(\mathrm{JT})\mathrm{std}} &= \frac{1}{k_B \beta_k^2 K_{k,\mathrm{P}}^{\epsilon}} \left[\delta_k(\beta_k) + \beta_k \frac{d\delta_k(\beta_k)}{d\beta_k} \right] \\ &= -\frac{1}{C_{k,\mathrm{P}}} \left[\delta_k(\beta_k) + \beta_k \frac{d\delta_k(\beta_k)}{d\beta_k} \right]. \end{aligned}$$

By (12.33), this result is equivalent to the standard formula. The inversion point, if one exists, occurs when $\delta_k(\beta_k) = -\beta_k(d\delta_k(\beta_k)/d\beta_k)$.

For purposes of comparison, the formula obtained by Planck [**1922**], p. 130–133, using van der Waals' equation to compute the Joule-Thomson coefficient for a mole of a substance is

$$(19.37) \qquad \begin{aligned} C_k^{(\mathrm{JT/vW})\mathrm{std}} &= \frac{\Delta T_k}{\Delta P_k} = \left[\frac{2a(\delta_k - b)^2 - RT_k b \delta_k^2}{RT_k \delta_k^3 - 2a(\delta_k - b)^2} \right] \frac{\delta_k}{C_{k,\mathrm{P}}} \\ &\approx \frac{1}{C_{k,\mathrm{P}}} \left(\frac{2a}{RT_k} - b \right), \end{aligned}$$

where a represents the effects of the attraction of particles for each other, b represents their effective volume, and the approximation is valid if a and b are sufficiently small. Planck observed that the approximate equation agrees quite well with experiment. The *temperature of inversion* based on this calculation is

$$(19.38) \qquad T_k = \frac{2a}{Rb}.$$

19.5 Thermodynamic Commutation Relations

The relation (18.32), which established that Z_k^ϵ is an equilibrium state function, is equivalent to $\left[\frac{\partial}{\partial \delta_k}, \frac{\partial}{\partial \beta_k}\right] \ln Z_k^\epsilon = 0$. The commutator of two derivation operators is a measure of the failure of Euler's criterion for successive differentiation with respect to two variables. From the standpoint of differential geometry, it represents the degree of failure of parallel transport around a small loop. For our purposes, it is a measure of the dependence of two operators on each other. While independent operator mappings commute and can be performed in any order, dependent operator mappings do not commute and the order in which the operations are performed matters. In this section, the commutators for each pair of operators taken from the set $\{\partial/\partial t, \partial/\partial \beta_k, \partial/\partial \delta_k\}$ will be examined.

Thermature and Volume Derivative Commutator.

Let us begin with the commutator of $\partial/\partial \beta_k$ and $\partial/\partial \delta_k$. The commutator of $\partial/\partial \beta_k$ and $\partial/\partial \delta_k$ is computed using their definitions (12.41) and (10.48). When acting on phase functions $G_k(Q_k, P_k, t)$, but not the system distribution, the operators $\partial/\partial t$, $\partial/\partial \delta_k$, and $\partial/\partial \beta_k$ are ordinary derivatives with respect to the independent variables or parameters t, δ_k, β_k. This means that the commutation relation $[\partial/\partial \beta_k, \partial/\partial \delta_k]G_k(Q_k, P_k, t) = 0$ is valid.

The thermature derivative operator was defined by its action on system distributions and on phase functions in Section 12.5. It was shown there that the thermature derivative of the constant function $\langle 1 \rangle_{kt}$ is zero as required. After (12.45), it was mentioned that $\partial/\partial \beta_k$ commutes with the volume derivative operator $d/d\Omega_k$ when they act on phase functions that are not system distributions because both are ordinary derivatives in that setting. Because the equilibrium system distribution depends explicitly on β_k, the action of the thermature partial derivative on F_k^ϵ was defined in Chapter 12 as the ordinary derivative with respect to β_k as a parameter. In the equilibrium case, it follows that the thermature and volume phase derivatives commute. For consistency, this commutativity is extended to the general system distribution and entropy also. Using the definitions (10.20) and (12.38) with this convention implies that

(19.39)
$$\frac{\partial}{\partial \beta_k} \frac{dS_k}{d\Omega_k} = \frac{1}{3\delta_k(t)} \frac{\partial}{\partial \beta_k} \left[Q_k \cdot \frac{\partial}{\partial Q_k} - P_k \cdot \frac{\partial}{\partial P_k}\right] S_k(Q_k, P_k, t)$$
$$= \frac{1}{3\delta_k(t)} \left[Q_k \cdot \frac{\partial}{\partial Q_k} - P_k \cdot \frac{\partial}{\partial P_k}\right] \frac{\partial}{\partial \beta_k} S_k(Q_k, P_k, t)$$
$$= \frac{d}{d\Omega_k} \frac{\partial S_k}{\partial \beta_k} = k_B \frac{dH_k(Q_k, P_k, t)}{d\Omega_k} = -k_B \mathsf{P}_k(Q_k, P_k, t).$$

To facilitate calculations, the partial volume derivative of $G_k(t)$ is expressed in the form

(19.40)
$$\frac{\partial G_k(t)}{\partial \delta_k} = \left\langle \frac{dG_k(Q_k, P_k, t)}{d\Omega_k} \right\rangle_{kt} - \frac{1}{k_B} \left\langle G_k(Q_k, P_k, t) \frac{dS_k}{d\Omega_k} \right\rangle_{kt}$$
$$- \frac{\partial \ln Z_k(t)}{\partial \delta_k} G_k(t).$$

The partial thermature derivative of this is

(19.41)
$$\frac{\partial^2 G_k(t)}{\partial \beta_k \partial \delta_k} = - \left\langle \frac{dG_k}{d\Omega_k} H_k \right\rangle_{kt} - \frac{1}{k_B} \left\langle G_k \left(\frac{\partial}{\partial \beta_k} \frac{dS_k}{d\Omega_k} \right) \right\rangle_{kt} + \frac{1}{k_B} \left\langle G_k H_k \frac{dS_k}{d\Omega_k} \right\rangle_{kt}$$
$$- \frac{\partial \ln Z_k(t)}{\partial \beta_k} \frac{\partial G_k(t)}{\partial \delta_k} - \frac{\partial^2 \ln Z_k(t)}{\partial \beta_k \partial \delta_k} G_k(t) - \frac{\partial \ln Z_k(t)}{\partial \delta_k} \frac{\partial G_k(t)}{\partial \beta_k}.$$

Similarly, the partial thermature derivative of $G_k(t)$ is written in the form

(19.42) $\quad \dfrac{\partial G_k(t)}{\partial \beta_k} = - \langle G_k \mathcal{H}_k \rangle_{kt} - \dfrac{\partial \ln Z_k(t)}{\partial \beta_k} G_k(t) = - \langle G_k \mathcal{H}_k \rangle_{kt} + H_k(t) G_k(t)$

and the partial volume derivative of this is computed to obtain

(19.43)
$$\frac{\partial^2 G_k(t)}{\partial \delta_k \partial \beta_k} = - \left\langle \frac{dG_k}{d\Omega_k} H_k \right\rangle_{kt} - \left\langle G_k \frac{dH_k}{d\Omega_k} \right\rangle_{kt} + \frac{1}{k_B} \left\langle G_k H_k \frac{dS_k}{d\Omega_k} \right\rangle_{kt}$$
$$- \frac{\partial \ln Z_k(t)}{\partial \beta_k} \frac{\partial G_k(t)}{\partial \delta_k} - \frac{\partial^2 \ln Z_k(t)}{\partial \delta_k \partial \beta_k} G_k(t) - \frac{\partial \ln Z_k(t)}{\partial \delta_k} \frac{\partial G_k(t)}{\partial \beta_k}.$$

Using (19.39) in (19.41) and then setting $G_k(Q_k, P_k, t) = 1$ and $G_k(t) = 1$ in both (19.41) and (19.43) expresses the second derivatives of $\ln Z_k(t)$ as phase averages of other quantities. The definitions (11.11) and (11.12) are used with (11.9) in the calculation below to obtain

(19.44)
$$\frac{\partial^2 \ln Z_k(t)}{\partial \beta_k \partial \delta_k} = \frac{\partial \beta_k(t)}{\partial \beta_k} P_k(t) + \beta_k(t) \frac{\partial P_k(t)}{\partial \beta_k}$$
$$= P_k(t) - \Lambda_{k,V}(t) = -\frac{\partial H_k(t)}{\partial \delta_k} = \frac{\partial^2 \ln Z_k(t)}{\partial \delta_k \partial \beta_k}$$

in accord with (12.46), (12.47), and (12.48). The relation (19.39) is used in (19.41) again and (19.44) is used with the difference between (19.41) and (19.43) to show that the volume and thermature operators commute and are therefore independent:

(19.45)
$$\left[\frac{\partial}{\partial \beta_k}, \frac{\partial}{\partial \delta_k} \right] G_k(t) = 0.$$

It is necessary that the commutator of the thermature and volume derivatives respect the constancy of the phase average $\langle 1 \rangle_{kt} = 1$. The requirement associated with this is obtained by setting $G_k(Q_k, P_k, t) = 1$ and $G_k(t) = 1$ in (19.45) with the result

$$(19.46) \qquad 0 = \left[\frac{\partial}{\partial \beta_k}, \frac{\partial}{\partial \delta_k} \right] \langle 1 \rangle_{kt} = -\frac{1}{k_B} \left\langle \left[\frac{d}{d\beta_k}, \frac{d}{d\delta_k} \right] S_k(Q_k, P_k, t) \right\rangle_{kt}.$$

This relation is valid by (19.39), so the formalism is consistent in this respect. As a verification, it can be shown explicitly that the commutator in (19.45) is valid at equilibrium as well. In addition, the formulas (10.61) and (12.51) can also be used to verify that the thermature and volume derivatives are (trivially) independent in the exact state case. This means that

$$(19.47) \qquad\qquad \left[\frac{\partial}{\partial \beta_k}, \frac{\partial}{\partial \delta_k} \right] G_k^\delta(t) = 0,$$

whenever the volume partial derivative of $G_k^\delta(t)$ is defined.

Time and Volume Derivative Commutator.

The next concern is the relation between the spacetime description of thermodynamic processes and their description as paths on the equilibrium manifold. Let us consider initiating a process for a system in a particular initial equilibrium state by modifying its macroscopic boundary conditions. At the end of the process, the boundary conditions are set in such a way to move it to a particular final equilibrium state. Compare this to a process on a quasistatic path that begins and ends at the same pair of equilibrium points as the first process but which remains on the system equilibrium manifold during the whole process. Since the entropy is a state function when the system is at equilibrium, it takes the same value at a given state on the equilibrium manifold no matter what path was used to reach that point. This means that the initial and final entropies will be the same for the two processes whether path the system took is quasistatic or not.

To investigate the incompatibilities in the spacetime and equilibrium descriptions, the commutation relations for the macroscopic operators $\partial/\partial t$ with $\partial/\partial \delta_k$ and $\partial/\partial t$ with $\partial/\partial \beta_k$ are computed. The partial time derivative of $G_k(t)$ obtained from (5.43) is

$$(19.48) \qquad \frac{\partial G_k(t)}{\partial t} = \left\langle \frac{dG_k(Q_k, P_k, t)}{dt} \right\rangle_{kt} + \left\langle \dot{\chi}_{\Omega_k(t)}(Q_k, P_k) G_k(Q_k, P_k, t) \right\rangle_{kt}.$$

The relations (19.48), (10.48), (10.52), and integration by parts are used next to compute the volume partial derivative of $\partial G_k(t)/\partial t$. This gives

(19.49)
$$\frac{\partial^2 G_k(t)}{\partial \delta_k \partial t} = \left\langle \frac{d^2 G_k}{d\Omega_k dt} \right\rangle_{kt} - \frac{1}{k_B} \left\langle \left[\frac{dG_k}{dt} + \dot{\chi}_{\Omega_k(t)} G_k \right] \frac{dS_k}{d\Omega_k} \right\rangle_{kt}$$
$$+ \left\langle \dot{\chi}_{\Omega_k(t)} \frac{dG_k}{d\Omega_k} \right\rangle_{kt} - \frac{\partial \ln Z_k(t)}{\partial \delta_k} \frac{\partial G_k(t)}{\partial t}.$$

The volume partial derivative of $G_k(t)$ is given in (19.40). Its time partial derivative is

(19.50)
$$\frac{\partial^2 G_k(t)}{\partial t \partial \delta_k} = \left\langle \frac{d^2 G_k}{dt d\Omega_k} \right\rangle_{kt} + \left\langle \dot{\chi}_{\Omega_k(t)} \frac{dG_k}{d\Omega_k} \right\rangle_{kt} - \frac{1}{k_B} \left[\left\langle \frac{dG_k}{dt} \frac{dS_k}{d\Omega_k} \right\rangle_{kt} \right.$$
$$\left. + \left\langle G_k \frac{d}{dt} \frac{dS_k}{d\Omega_k} \right\rangle_{kt} + \left\langle \dot{\chi}_{\Omega_k(t)} G_k \frac{dS_k}{d\Omega_k} \right\rangle_{kt} \right]$$
$$- \frac{\partial^2 \ln Z_k(t)}{\partial t \partial \delta_k} G_k(t) - \frac{\partial \ln Z_k(t)}{\partial \delta_k} \frac{\partial G_k(t)}{\partial t}.$$

Setting $G_k(Q_k, P_k, t) = 1$, $G_k(t) = 1$ in (19.50) and using (10.53) gives

(19.51)
$$\frac{\partial^2 \ln Z_k(t)}{\partial t \partial \delta_k} = -\frac{1}{k_B} \left\langle \frac{d}{dt} \frac{dS_k}{d\Omega_k} \right\rangle_{kt}.$$

The total phase time derivative of the volume phase derivative operator is the operator

(19.52)
$$\frac{d}{dt} \frac{d}{d\Omega_k} = \frac{d}{d\Omega_k} \frac{\partial}{\partial t} - \frac{\dot{\delta}_k(t)}{\delta_k(t)} \frac{d}{d\Omega_k} + \frac{1}{3\delta_k(t)} \left[V_k \cdot \frac{\partial}{\partial Q_k} + V_k^\mu Q_k^\nu \frac{\partial^2 \Phi_k(Q_k, t)}{\partial Q_k^\mu \partial Q_k^\nu} \right.$$
$$+ \frac{\partial \Phi_k(Q_k, t)}{\partial Q_k} \cdot \frac{\partial}{\partial P_k} + \frac{\partial \Phi_k(Q_k, t)}{\partial Q_{k,\mu}} P_k^\nu \frac{\partial^2}{\partial P_k^\mu \partial P_k^\nu}$$
$$\left. - \left\{ V_k^\mu P_k^\nu + Q_k^\mu \frac{\partial \Phi_k(Q_k, t)}{\partial Q_{k,\nu}} \right\} \frac{\partial^2}{\partial Q_k^\mu \partial P_k^\nu} \right].$$

The calculation of $d/d\Omega_k(d/dt)$ is similar. With these quantities, the time-volume phase commutator $[d/dt, d/d\Omega_k]$ is

(19.53)
$$\left[\frac{d}{dt}, \frac{d}{d\Omega_k} \right] = -\frac{\dot{\delta}_k(t)}{\delta_k(t)} \frac{d}{d\Omega_k} + \frac{1}{3\delta_k(t)} \left[2V_k \cdot \frac{\partial}{\partial Q_k} + \frac{\partial \Phi_k(Q_k, t)}{\partial Q_k} \cdot \frac{\partial}{\partial P_k} \right.$$
$$\left. + Q_k^\mu \frac{\partial^2 \Phi_k(Q_k, t)}{\partial Q_k^\mu \partial Q_k^\nu} \frac{\partial}{\partial P_{k,\nu}} \right].$$

Using (12.12a) and (5.34) directly yields
(19.54)
$$\left[\frac{\partial}{\partial t}, \frac{\partial}{\partial \delta_k}\right] \ln Z_k(t) = -\frac{1}{k_B} \left\langle \frac{d}{dt} \frac{dS_k}{d\Omega_k} \right\rangle_{kt} = \frac{\partial \beta_k(t)}{\partial t} P_k(t) + \beta_k(t) \frac{\partial P_k(t)}{\partial t}.$$

The definition (19.48) is used to compute the time derivative of phase averages and (19.40) is used to compute the partial volume derivatives of phase averages. With these, the calculations of the time and volume derivative commutators give us

$$(19.55) \quad \left[\frac{\partial}{\partial t}, \frac{\partial}{\partial \delta_k}\right] G_k(t) = \left\langle \left[\frac{d}{dt}, \frac{d}{d\Omega_k}\right] G_k(Q_k, P_k, t) \right\rangle_{kt}$$
$$- \frac{1}{k_B} \left\langle [G_k(Q_k, P_k, t) - G_k(t)] \frac{d^2 S_k(Q_k, P_k, t)}{dt d\Omega_k} \right\rangle_{kt}.$$

The relation $\left[\frac{\partial}{\partial t}, \frac{\partial}{\partial \delta_k}\right] \langle 1 \rangle_{kt} = 0$, needed for consistency, follows immediately.

The equilibrium case is simpler and can be computed directly. The total time derivative of the volume derivative of the entropy is
(19.56)
$$\frac{d}{dt} \frac{dS_k^\epsilon}{d\Omega_k} = -k_B \beta_k \frac{dP_k^\epsilon(Q_k, P_k, t)}{dt} = \frac{k_B \beta_k}{3\delta_k} V_k^\mu \left[3 \frac{\partial \Phi_k^\epsilon(Q_k)}{\partial Q_k^\mu} + Q_k^\nu \frac{\partial^2 \Phi_k^\epsilon(Q_k)}{\partial Q_k^\mu \partial Q_k^\nu}\right].$$

In the equilibrium rest frame of a system at equilibrium, odd powers of the velocity average to zero. This means that

$$(19.57) \qquad \left[\frac{\partial}{\partial t}, \frac{\partial}{\partial \delta_k}\right] Z_k^\epsilon = 0,$$

which shows that these operators are independent when acting on the local norm of the equilibrium state. Using this with the time independent analog $G_k(Q_k, P_k)$ and using (19.53) and (19.56) in (19.55) yields the result

$$(19.58) \quad \left[\frac{\partial}{\partial t}, \frac{\partial}{\partial \delta_k}\right] G_k^\epsilon = \left\langle 2V_k \cdot \frac{\partial G_k(Q_k, P_k)}{\partial Q_k} + \frac{\partial \Phi_k(Q_k, t)}{\partial Q_k} \cdot \frac{\partial G_k(Q_k, P_k)}{\partial P_k} \right.$$
$$\left. + Q_k^\mu \frac{\partial^2 \Phi_k(Q_k, t)}{\partial Q_k^\mu \partial Q_k^\nu} \frac{\partial G_k(Q_k, P_k)}{\partial P_{k,\nu}} \right\rangle_{k\epsilon}$$
$$- \frac{\beta_k}{3\delta_k} \left\langle [G_k(Q_k, P_k) - G_k^\epsilon] V_k^\mu \left[3 \frac{\partial \Phi_k^\epsilon(Q_k)}{\partial Q_k^\mu} + Q_k^\nu \frac{\partial^2 \Phi_k^\epsilon(Q_k)}{\partial Q_k^\mu \partial Q_k^\nu}\right] \right\rangle_{k\epsilon}.$$

For the commutation of the time and volume derivative operators on an exact state phase average, the relation

(19.59)
$$\frac{\partial G_k^\delta(t)}{\partial \delta_k} = -\int dQ_k \int dP_k \, \frac{d\chi_{\Omega_k(t)}(Q_k, P_k)}{d\Omega_k} F_k^\delta(Q_k, P_k, t) G_k(Q_k, P_k, t)$$
$$= -\left\langle \frac{d\chi_{\Omega_k(t)}(Q_k, P_k)}{d\Omega_k} G_k(Q_k, P_k, t) \right\rangle_{k\delta t},$$

obtained from (10.61), is used first to obtain

(19.60)
$$\frac{\partial}{\partial t}\frac{\partial G_k^\delta(t)}{\partial \delta_k} = -\frac{\dot{\delta}_k(t)}{\delta_k(t)}\frac{\partial G_k^\delta(t)}{\partial \delta_k} - \left\langle \left(\frac{d}{dt}\frac{d\chi_{\Omega_k(t)}}{d\Omega_k}\right) G_k + \frac{d\chi_{\Omega_k(t)}}{d\Omega_k}\frac{dG_k}{dt} \right\rangle_{k\delta t}.$$

Next, the result

(19.61)
$$\frac{\partial G_k^\delta(t)}{\partial t} = \left\langle \dot{\chi}_{\Omega_k(t)}(Q_k, P_k)G_k + \chi_{\Omega_k(t)}(Q_k, P_k)\frac{dG_k}{dt} \right\rangle_{k\delta t}$$

leads to

(19.62)
$$\frac{\partial}{\partial \delta_k}\frac{\partial G_k^\delta(t)}{\partial t} = -\left\langle \frac{d\dot{\chi}_{\Omega_k(t)}}{d\Omega_k}G_k + \frac{d\chi_{\Omega_k(t)}}{d\Omega_k}\frac{dG_k}{\partial t} \right\rangle_{k\delta t}.$$

Subtracting (19.62) from (19.60) yields

(19.63)
$$\left[\frac{\partial}{\partial t}, \frac{\partial}{\partial \delta_k}\right] G_k^\delta(t) = -\frac{\dot{\delta}_k(t)}{\delta_k(t)}\frac{\partial G_k^\delta(t)}{\partial \delta_k} + \left\langle G_k\left[\frac{d}{dt}, \frac{d}{d\Omega_k}\right]\chi_{\Omega_k(t)}(Q_k, P_k) \right\rangle_{k\delta t}.$$

Setting $G_k(Q_k, P_k, t) = 1$ and $G_k^\delta(t) = 1$ in (19.63) yields the consistency relation

(19.64)
$$\frac{d\langle 1\rangle_{k\delta t}}{dt} = \left\langle \left[\frac{d}{dt}, \frac{d}{d\Omega_k}\right]\chi_{\Omega_k(t)}(Q_k, P_k) \right\rangle_{k\delta t} = 0.$$

The formulas (19.59) and (19.63) are of limited usefulness because the singularity of the F_k^δ state in conjunction with the singularity of the time and volume derivative commutator acting on $\chi_{\Omega_k(t)}$ will usually make the phase average undefined. This is a consequence of the mismatch between the focus of the exact state on the precise particle trajectories and the focus in the boundary formalism on handling the particles reaching the boundary in terms of a stochastic interaction. If no particle trajectories reach the boundary, then $\partial G_k^\delta(t)/\partial \delta_k = 0$ and the phase average of the term containing the time and volume derivative commutator in (19.63) vanishes. As a consequence, the time and volume derivative commutator of $G_k^\delta(t)$ vanishes when no particles reach the boundary as is required for a closed system described by an exact state.

Time and Thermature Derivative Commutator.

For the time and thermature derivative commutator, the commutator of the logarithm of the partition function is first computed using (4.22c) with the result
(19.65)
$$\frac{\partial^2 \ln Z_k(t)}{\partial t \partial \beta_k} = -\frac{\partial \mathcal{H}_k(t)}{\partial t} = -\left\langle \frac{\partial \Phi_{k,x}(Q_k, t)}{\partial t} \right\rangle_{kt} - \left\langle \dot{\chi}_{\Omega_k(t)} \mathcal{H}_k(Q_k, P_k, t) \right\rangle_{kt}.$$

Using this with the relation (12.45) stipulating the value of $\partial \dot{\chi}_{\Omega_k(t)}/\partial \beta_k$ and computing the commutator gives
(19.66)
$$\left[\frac{\partial}{\partial t}, \frac{\partial}{\partial \beta_k} \right] G_k(t) = -\left\langle \frac{\partial \Phi_{k,x}}{\partial t} G_k \right\rangle_{kt} - \left\langle \dot{\chi}_{\Omega_k(t)} \mathcal{H}_k G_k \right\rangle_{kt} + \frac{\partial \mathcal{H}_k(t)}{\partial t} G_k(t).$$

Setting $G_k(Q_k, P_k, t) = G_k(t) = 1$ in (19.66) shows immediately that the commutator relation $\left[\frac{\partial}{\partial t}, \frac{\partial}{\partial \beta_k} \right] \langle 1 \rangle_{kt} = 0$ holds as required. This is a restatement of the justification for the definition (12.45) of the action of the thermature derivative on $\dot{\chi}_{\Omega_k(t)}(Q_k, P_k)$.

The time and thermature derivative commutator shows that the time and thermature variables are intertwined because of changing external potentials and the surface flux. For an isolated system $(k = t)$, (4.22c) can be used with (2.21c) and then (5.31) to show that

(19.67)
$$\left[\frac{\partial}{\partial t}, \frac{\partial}{\partial \beta_t} \right] G_t(t) = 0.$$

It is also easy to show that

(19.68)
$$\left[\frac{\partial}{\partial t}, \frac{\partial}{\partial \beta_k} \right] G_k^\epsilon = 0.$$

Finally, for the exact state phase averages, it follows trivially from (12.51) that

(19.70)
$$\left[\frac{\partial}{\partial t}, \frac{\partial}{\partial \beta_k} \right] G_k^\delta(t) = 0.$$

Using G_k^ϵ in the general result (19.45) for the thermature and volume partial derivative commutator shows that the thermature and volume variables are independent of each other for the equilibrium case. This demonstrates the consistency of the TIS definitions of the thermature and volume derivatives with equilibrium theory.

Several other operations used in the previous work do not commute with one of the operators considered in this chapter. It has been pointed out above that the reason the time derivative of the entropy for an isolated system, for example, is different from the time derivative of the coarse grained entropy of the system is because the procedure of coarse-graining the system distribution does not commute with the time derivative operator. The coarse-graining procedure is not part of the TIS theory, so the fact that it does not commute with the time derivative operator does not pose a problem.

Processes that take place in time will in general modify the thermature or volume. The fact that the commutators of the time partial derivative operator with the volume and thermature partial derivative operators do not vanish in general indicates that the equilibrium formalism and the associated dynamic assumptions of quasistatic processes and local thermodynamic equilibrium have limited value as a guide to the behavior of actual systems. This is another way of stating that real processes cannot be limited to equilibrium space or to a manifold in that space.

These facts show that the spacetime and equilibrium manifold descriptions are not compatible and cannot be used simultaneously to describe a system. The need for both of these incompatible approaches for a full characterization of thermodynamics is similar in some respects to that envisioned by Bohr [1930] when he stated that a concept similar to complementarity in atomic physics would be needed in reconciling the statistical mechanical and thermodynamic descriptions of a system.

Part III

The Maxwell-Boltzmann Approach

...it is to be feared that we shall have to be taught thermodynamics for several generations before we can expect beginners to receive as axiomatic the theory of entropy.

James Clerk Maxwell, *Review of Tait's Thermodynamics*, 1878

CHAPTER 20

The Theory of Maxwell and Boltzmann

One of the most important developments in the theory of the relation between thermodynamics and statistical mechanics was the evolution equation for a collection of particles that Boltzmann developed in 1872 out of transport equations published previously by Maxwell. In Part III, the theory developed by Maxwell and Boltzmann is presented in the TIS framework.

20.1 Introduction

The advent of kinetic theory in the middle of the nineteenth century was a powerful stimulus to thinking about how particles might explain macroscopic phenomena. The prospect of explaining the thermodynamic relations between macroscopic bodies in terms of particle behavior was of special interest. However, the number and complexity of the particle interactions impeded this step. Reducing this multiplicity of relations between the particles to a workable form required powerful techniques. Drawing on recently introduced ideas for representing mathematically the distribution of a quantity within a population, researchers began to introduce simple specialized particle distributions to represent these particle collections. The most significant early use of these distributions was by Clausius [1857]. Among other things, Clausius was concerned with how collisions between particles would affect rates of diffusion. To estimate this, he introduced the notion of a mean free path calculation and presented a formula for computing the rate of collisions between particles in a simple distribution.

An even more important advance was made by Maxwell [1860], [1866], who introduced more sophisticated statistical methods. To track the changes in a system of particles in time, Maxwell introduced the idea of a general statistical distribution of particles and extended Clausius' work on the rate of collisions of particles to calculate how these more general statistical distributions describing the particles will change in time. As another innovation, Maxwell systematically used particle analogs of physical quantities with the particle distribution to calculate the transport of these physical quantities as the mean values of currents of these analog functions.

Maxwell's work was soon generalized by Boltzmann [1872] into a dynamic theory. He presented an evolution equation for the particle distribution that is now called the *Boltzmann equation*. Rather than using a statistical distribution that includes all the particles, Boltzmann focused on the evolution of a representative one-particle probability distribution density. In creating the Boltzmann equation and its associated formalism, Boltzmann [1872] and [1882] relied heavily on Maxwell's work. Because of this, Maxwell claimed priority.[1] However, Boltzmann's [1872] work went beyond Maxwell's [1866] statistical description of a system of particles and the transport of physical quantities. He significantly reconceptualized the problem as a dynamic theory of the evolution of a representative probability distribution density that describes the system. This reconceptualization laid the groundwork for the path taken by many future investigations of statistical mechanics.

The Boltzmann equation has been used as the basis of many studies, including ones by Boltzmann himself, of measurable physical quantities such as diffusion, viscosity, and heat flow. The Boltzmann equation is nonlinear, so

---

[1]See Maxwell [1873b] ([1890], pp. 365–366) and Maxwell [1874] (Maxwell [1890], pp. 427–428).

various assumptions and approximation methods are required when using it so that explicit solutions can be found that will yield numbers to compare with experiment. The early approximation methods introduced by Boltzmann and others were later replaced with the more sophisticated methods introduced at the beginning of the twentieth century in the studies of David Enskog and Stanley Chapman.[2]

There are a number of problems associated with Boltzmann's theory. Among them are:

(i) there are six major interpretations of the one-particle state function $f(q, p, t)$;

(ii) there is a mathematical proof by Boltzmann that his entropy analog, $H(t)$, never decreases (TIS sign convention) followed later by verbal statements and plausibility arguments that sometimes it does decrease—but only rarely;

(iii) there are statements connecting the mechanical concept of a collision with the statistical concept of a correlation without linking them properly in the theory;

(iv) there is confusion between the statistical and stochastic aspects of Boltzmann's equation;

(v) there is the question of the necessity of taking the "thermodynamic limit" in the interpretation of the Boltzmann state.

These problems are usually not clearly addressed or understood in discussions of Boltzmann's equation. Adding to the problem, new terminology has sometimes been invented, or prior terminology extended to new situations, without physical or mathematical justification.

In Boltzmann's [1872] theory, particles are described by the one-particle probability distribution density $f(q, p, t)$. To describe collisions between particles, he drew on the work of Clausius and Maxwell who showed how particle collisions can modify the particle distribution. As pointed out by Brush [1976], Section 14.4, initial uses of probability in kinetic theory by Clausius and Maxwell were equivalent to the assumption of total ignorance or complete randomness at all times for the variables characterizing individual molecules. Boltzmann made use of an assumption, which came to be called the *molecular chaos* or *molecular disorder* assumption, that the rate of collision at location q of two particles in mechanical states (q, p_i) and (q, p_j) is proportional to $f(q, p_i, t)f(q, p_j, t)$.

---

[2]Carlo Cercignani has studied the Boltzmann equation for many years in Cercignani [1975], [1998], C. Cercignani, V. I. Gerasimenko, and D. Ya. Petrina [1997], and other works. See also the references cited in these works. For more details on the work of Enskog and Chapman, including a discussion of the assumptions on which Boltzmann's equation is based, see Chapman and Cowling [1939]. The "Introduction" and "Historical Summary" chapters contain useful historical information and references to original papers. See also R. Dugas [1959].

The calculation of the collision term was undertaken using methods similar to those of the standard approach to studying hydrodynamics. To define the hydrodynamic behavior of a system of particles, a subvolume of the system volume is considered that is small by macroscopic standards but is still large enough to contain many particles. External forces acting on this volume are assumed constant throughout the volume. Forces due to particles in the volume acting on particles outside the subvolume are summed into a net force and assumed to be acting from the center of the volume. This procedure leads to a macroscopic differential equation for the hydrodynamic behavior of the system under consideration.

Maxwell considered the collisions between particles i and j, with momenta p_i and p_j, in a small volume around the point q during a short time dt. The collision is represented in the rest frame of particle j. A plane containing particle j at its origin is constructed perpendicular to the velocity vector of particle i in this frame. Points in this plane are located using a polar coordinate system by a distance b from the origin, called the *impact parameter*, and an angle ϕ measured from a chosen axis in the plane. The size of an element of area in this plane is $bdbd\phi$.

The probability that the representative particle state in the infinitesimal neighborhood $d^3qd^3p_i$ of the mechanical state (q, p_i) is $f(q, p_i, t)d^3qd^3p_i$. Maxwell [1866] combined these elements into a formula for the number of collisions by particles with momenta p_i and p_j during the interval dt around the time t, which he called the *particle collision number* and was called the *Stosszahlansatz* by others. It is written in the form

$$(20.1) \qquad |g| \, f(q, p_i, t) f(q, p_j, t) \, d^3p_i d^3p_j \, b \, db \, d\phi dt,$$

where the vector $g^\mu = g^\mu_{ij} = m_i^{-1}p_i^\mu - m_j^{-1}p_j^\mu$ is the relative velocity vector of the colliding particles.

The collision term has been interpreted variously as the "number of collisions at q during the time interval dt", the "average number of collisions at location q during dt", or the "most probable number of collisions at q during dt", etc. The justification for adopting one or another of these (incompatible) interpretations is usually not given. Because $f(q, p, t)$ does not depend on the impact parameters (b, ϕ), this particle collision number calculation is based on the assumption that all values of the impact parameters (b, ϕ) have equal probability. To support this, Boltzmann assumed that the particles are "molecularly disordered, which means that they have a completely random, distribution" (Boltzmann [1896a], pp. 115–116) at each point in time. This assumption was also employed to justify the use of the product $f(q, p_i, t)f(q, p_j, t)$ in the particle collision number formula (20.1)

In addition to the assumptions listed above that are required to set up the basic theory, many authors have introduced other assumptions to facilitate calculations. Assumptions, for example, concerning the fluctuations of various quantities and their presumed rates of decay or "relaxation times" are often used in discussions of Boltzmann's equation. These are not supported by experimental evidence and are usually not well defined mathematically.[3] It is no wonder that there are some paradoxes associated with this subject. Much of the confusion can be traced back to Boltzmann himself, so his ideas will be analyzed in detail.

An important aspect of Boltzmann's work was the introduction of the function $\ln f(q,p,t)$ as a phase analog for the entropy. In the discussion below, the TIS version of Boltzmann's entropy phase analog will be defined as $-k_B \ln f(q,p,t)$. Boltzmann designated the phase average of his analog by $H(t)$ and proposed it as the thermodynamic entropy. He proved in his H theorem that $H(t)$ is nondecreasing as a function of time (TIS sign convention), which offered a compelling analog to irreversibility in thermodynamics. In addition, the Maxwell-Boltzmann exponential distribution is a stationary state of the Boltzmann equation and is therefore an analog of the thermodynamic equilibrium state.

Having articulated his H theorem, and associated it with the second law of thermodynamics, Boltzmann soon faced serious criticism. The conflict between the irreversibility embodied in the H theorem and the reversibility of the underlying mechanics was pointed out early by Thomson [1873] and Loschmidt [1876], part I.[4] A number of other criticisms were leveled by Loschmidt, primarily in reference to the barometer formula.[5] Many of Loschmidt's criticisms were off the mark, but a few struck home. In particular, his time reversal criticism, which will be discussed in Section 22.3, was important. Loschmidt [1876], part IV, pp. 222–224, also objected to Boltzmann's use of a version of the thermodynamic limit, in which the number of particles goes to infinity, in deriving his equations. Loschmidt, p. 223, felt that the equation should be valid for "zwei, drei Molecüle als für eben so viele Trillonen".[6] Finally, he stated, pp. 223–225, that allowing particle velocities to range from $-\infty$ to $+\infty$ would lead to errors in the case of a few particles.

Boltzmann [1878a] (Boltzmann [1909], Vol. II, pp. 284–285) responded to some of Loschmidt's objections. The objections to the barometer formula, which predicts that the temperature of a gas will be constant with height while the density decreases, were dealt with easily. He, p. 284, also agreed with Loschmidt's point that his work is only an approximate description of gases occurring in nature and said that this is true of all such theories. In particular, he stated that the proof that the equilibrium distribution is unique depends on his assumptions of a gas of massive particles for which

---

[3]See the complaints in Truesdell [1969], Chapter 9, on this.

[4]Henry Hollinger and Michael Zenzen [1985], p. 226, assert that these issues of irreversibility were discussed by Maxwell, Thomson, and Tait, as early as 1867. These issues are discussed in more detail in EIS, Part II.

[5]See the discussion of the barometer formula in Partington [1949], pp. 275–279.

[6]"... two or three molecules just as for so many trillions".

the duration in time of particle interactions is vanishingly small. On the other hand, Boltzmann dismissed Loschmidt's objection to his use of particle velocities in the range $-\infty$ to $+\infty$. He conceded that this is an error for the case of only a few particles, but maintained that for the gases found in nature, in which certainly the smallest spatial domains contain an enormous number of molecules, it is unthinkable that an significant error would be made by supposing that the greatest possible velocity of a molecule is very much greater than the mean velocity.

Continuing his response to Loschmidt, Boltzmann [1878a] stated that his proof of the equipartition of energy does not apply to gases with a finite number of particles or gases that contain particles of unequal masses. Boltzmann went on to maintain that there are some very special initial conditions for ideal gases of particles with mass for which the equipartition theorem does not apply. Boltzmann, pp. 286–288, finally defended his theory against Loschmidt's objection that the factored product $f(q,p,t)f(q,p',t)$ cannot be used to represent the number of collisions because the particles are not independent.

Other criticisms of Boltzmann's ideas came late in the nineteenth century. The most important is the conflict between the passage to the equilibrium state (in modern terminology: the uniform asymptotic stability in the large) of solutions of Boltzmann's equation and the implications of the Poincaré recurrence theorem, which states that an arbitrary mechanical state will eventually recur in an isolated system located in a fixed volume. This conflict was pointed out by Zermelo [1896a], [1896b]. Questions were also raised concerning the limitation on the validity of Boltzmann's equation due to the requirement that only binary collisions be considered. The analysis of the *Stosszahlansatz* was prominent in these discussions. The concepts of "molecular disorder" and "molecular chaos" that underlie the Stosszahlansatz were also discussed at length.[7]

In response to these criticisms, Boltzmann [1882] stated that his entropy analog, $H(t)$, could sometimes decrease as would be required by reversibility. He maintained that the initial state of the system determines what will actually occur. He went on to use the same mixing example as Maxwell and Thomson to illustrate the probabilistic nature of the second law.[8] Thus, both Maxwell and Boltzmann saw the second law as a "statistical law" which meant that it could be occasionally violated or violated in special circumstances.

---

[7]See Brush [1976], Chapter 14, for a review of these issues and a large set of references to the early discussions. For reference to more recent work, see Cercignani [1975], Kuhn [1978], and Truesdell and Muncaster [1980].

[8]See Brush [1976], p. 606. On mixing and entropy, see Boltzmann [1878b]. See also the discussion of mixing in Chapter 8 of this book and the references cited there. In further attempts to meet objections to his H theorem and justify the Stosszahlansatz, Boltzmann considerably muddies his conception of the probability issues.

After some time, the objections to Boltzmann's theory were almost forgotten. They were relegated to short discussions of the "paradoxes" of thermodynamics and the standard justifications, mostly based on Boltzmann's responses to his critics, were trotted out to soothe the doubters. Nevertheless, logical problems remained and these objections were never completely forgotten.[9]

While Boltzmann defended his theory at length, and conceded its limitations, his attempts to defend it from these criticisms made the conceptual situation worse. His responses to Loschmidt are not adequate when one goes beyond the form in which Loschmidt expressed them. Moreover, as indicated below and discussed in detail in EIS, Boltzmann's later views on the H theorem cannot be maintained. On a more fundamental level, Boltzmann's use of the molecular disorder assumption and the conclusions he drew from it, do not make sense when collisions are examined either from the point of view of particle mechanics or in the context of general $\mathcal{N}_k$-particle distributions adopted in Part I. These points will be considered in detail when specific aspects of Boltzmann's theory are taken up below.

From the TIS perspective, Boltzmann's equation is viewed as the result of an approximation procedure that reduces the evolution equation of the $\mathcal{N}_k$-particle Hamiltonian dynamics to a simpler one-particle stochastic evolution equation. In Part I of this book, the "independent systems approximation" was introduced in which the system distribution for an isolated system was replaced by the product of individual subsystem distributions and the exact interaction between particles in one subsystem and those in another at a shared system boundary was replaced by a stochastic operator called the "wall operator". In the TIS treatment of Boltzmann's equation, Boltzmann's one-particle theory is equivalent to extending the independence approximation to the individual particles themselves.

In TIS terms, moving from a Hamiltonian description of the underlying particles to Boltzmann's one-particle description means setting $\phi_{ij}(q_i - q_j) = 0$, for $i, j \in k$. These interactions are replaced by the Maxwell-Boltzmann stochastic operator, introduced below, in which the interparticle interactions are represented as collisions. It is shown below that the collision term is a stochastic average, in which the exact individual particle interactions are replaced with an averaged result. This has the consequence, from the TIS standpoint, of treating every particle as if it were an independent system. As such, the representation of the work of Maxwell and Boltzmann fits smoothly within the TIS framework and formalism.

---

[9] As an example of these logical issues, Jagdish Mehra [**1972**], pp. 22-23, reported that Dirac, who studied thermodynamics with R. H. Fowler, a leading expert at the time, preferred the approach of Gibbs to thermodynamics and did not appreciate the Boltzmann equation because "the collision term was 'not explained very well'."

In the development of the TIS version of Boltzmann's equation, the assumption that the system of particles is isolated results in Boltzmann's equation in its usual form. If the system is not isolated, a version of Boltzmann's equation with a boundary is obtained. The version of Boltzmann's equation presented here and the TIS interpretation of it meet Loschmidt's objections and those of other critics.

Treating each particle as an independent system is common to many calculations. In the quantum theory of the solid state, the Hartree and Hartree-Fock approximations are two important examples of this type (Kittel [**1963**]). Questions concerning the mathematical or physical validity of such approximation schemes are seldom explored, however.

There is an extensive literature on Boltzmann's equation. The work on Boltzmann's equation in this book is based primarily on the treatments of Maxwell [**1866**], Boltzmann [**1872**], Boltzmann [**1896a**], Cercignani [**1975**], Dorfman and van Beijeren [**1977**], Truesdell [**1969**] and Truesdell and Muncaster [**1980**]. For extensive historical references and a very useful summary of modern work, see Truesdell and Muncaster [**1980**].

20.2 The Boltzmann Equation

The Boltzmann equation will be developed now as an approximation to the evolution equation (5.16). An explicit approximation scheme will be used to obtain Boltzmann's equation from the interacting systems evolution equation in ways that are consistent with Boltzmann's approach. To do this, a procedure similar to that of Chapter 6 will be used. As part of this approximation procedure, the particle-particle interaction potential terms will be removed from the evolution equation and replaced by the operator. As in the general theory, an average external potential energy function is used to account for the influence of particles in other systems. After these steps, the $\mathcal{N}_k$-particle system distribution F_k can be replaced by a product of one-particle distributions and used in the approximate equation.

The introduction of an individual distribution for each particle in the TIS version of Boltzmann's equation represents a significant change from Boltzmann's original theory. Boltzmann's non-linear equation is reformulated as a large set of coupled linear equations. Because each particle is represented by an independent distribution, the TIS entropy analog is a particle sum function in the Boltzmann context. Each single particle component in this sum is essentially equivalent to Boltzmann's analog $\ln f(q, p, t)$. The particle sum analog functions in the general TIS theory are all valid in Boltzmann's theory. This overall approach to Boltzmann's equation as a set of coupled linear equations has the advantages of being more natural, realistic, accurate, and mathematically tractable than the original.

Consider a set $\{ f_i(q_i, p_i, t) \}$ of one-particle probability distribution density functions for $i \in k$ that are solutions of the evolution equation introduced below. These functions will be required to be well defined for $t \in [0, \alpha)$, where $\alpha > 0$, and to have finite norms. It will also be assumed that the thermodynamic functions that are phase averaged over members of the set $\{ f_i \}$, and Boltzmann's $H(t)$ function in particular, are mathematically well defined on these states. These mathematical assumptions will be analyzed in Part IV. The set $\{ f_i(q_i, p_i, t) \}$ of one-particle distributions replaces in the TIS theory the one-particle state $f(q, p, t)$ Boltzmann used.[10] The molecular chaos assumption is expressed formally in TIS terms, using the k *Boltzmann system distribution* $F_{[k]}(Q_k, P_k, t)$, as

$$(20.2) \qquad F_{[k]}(Q_k, P_k, t) = \prod_{i \in k} f_i(q_i, p_i, t),$$

where, by analogy to $[t]$, $[k] = \{i \mid i \in k\}$ is the set of particles in the system k viewed as separate subsystems. The Boltzmann system distribution $F_{[k]}(Q_k, P_k, t)$ approximates the system distribution $F_k(Q_k, P_k, t)$ in much the same way that $F_{[t]}(Q_t, P_t, t)$ approximates $F_t(Q_t, P_t, t)$.

Let us begin by specializing some of the formalism in Part I to the interacting system case. Using the assumptions (4.2) on $\phi_{ij}(q_i - q_j)$ stated in Chapter 4, the bounds on the external potential are

$$(20.3) \qquad -[\mathcal{N}_t - \mathcal{N}_k] \phi_0 \le \Phi_{i,x}(q, t) \le [\mathcal{N}_t - \mathcal{N}_k] \|\phi\|.$$

This is obtained by using the ϕ_0 as the lower bound and the semi-norm as the upper bound for the particle interaction potentials contributed by the $\mathcal{N}_t - \mathcal{N}_k$ external particles.

The Boltzmann Hamiltonian is a particle sum function. The *i-particle Boltzmann phase Hamiltonian* and k *Boltzmann phase Hamiltonian* depend on the time due to the external potentials. They are defined as

$$(20.4a) \qquad H_i^{(B)}(q_i, p_i, t) = K_i(p_i) + \Phi_{i,x}(q_i, t),$$

$$(20.4b) \qquad \mathcal{H}_k^{(B)}(Q_k, P_k, t) = \sum_{i \in k} H_i^{(B)}(q_i, p_i, t).$$

The *i-particle Boltzmann phase energy* and k *Boltzmann phase energy* are defined by

$$(20.5a) \qquad E_i^{(B)}(q_i, p_i, t) = K_i(p_i) + \tfrac{1}{2}\Phi_{i,x}(q_i, t),$$

$$(20.5b) \qquad \mathcal{E}_k^{(B)}(Q_k, P_k, t) = \sum_{i \in k} E_i^{(B)}(q_i, p_i, t).$$

---

[10]M. Chen [**1989**], p. 1329, has independently retained the separate particle identities in his version of the Boltzmann collision term. He has also independently made use of the work of Cercignani [**1975**] and Darrozès and Guiraud [**1966**] on the boundary conditions as well. He, p. 1330, defined a certain set of solutions to a linearized version of his entropy balance equation obtained from Boltzmann's equation as the "thermodynamic solution" of Boltzmann's equation.

For $i \in k$, the information in Chapter 4 is used to show easily that the kinetic energy is bounded by calculating the amount of energy that can be transferred to the k system from other systems and minimizing the internal potential energy. This bound takes the form

$$(20.6) \qquad 0 \leq K_i(p_i) \leq \mathcal{K}_k(P_k) \leq \mathcal{K}_k^u = \mathcal{C}_{k,\mathcal{H}} - \phi_0 \mathcal{N}_k(\mathcal{N}_k - 1).$$

From this is obtained a momentum bound for each $i \in k$

$$(20.7) \qquad |p_i| \leq p_k^u = \sup_{i \in k} \left[2 m_i \mathcal{K}_k^u\right]^{\frac{1}{2}}.$$

Lower and upper bounds of the Boltzmann phase Hamiltonian,

$$(20.8) \qquad H_k^{(B)l} \leq H_i^{(B)}(q_i, p_i, t) \leq H_k^{(B)u},$$

are also easily obtained, where

$$(20.9a) \qquad H_k^{(B)l} \geq -\mathcal{N}_k(\mathcal{N}_k - 1)\phi_0,$$

$$(20.9b) \qquad H_k^{(B)u} \leq \mathcal{C}_{k,\mathcal{H}} + H_k^{(B)l}.$$

The Boltzmann *i-particle phase volume set* and k *phase volume set* are

$$(20.10a)$$
$$\omega_i(t) = \{\, (q,p) \in d_k(t) \times \mathbf{R}^3 \mid H_k^{(B)l} \leq H_i^{(B)}(q,p,t) \leq H_k^{(B)u} \,\},$$
$$(20.10b)$$
$$\omega_k(t) = \cup_{i \in k} \omega_i(t).$$

If the particles are identical, it follows that $\omega_i(t) = \omega_j(t)$ for $i, j \in k$ so that

$$(20.11) \qquad \omega_k(t) = \omega_i(t) \qquad \text{for any } i \in k.$$

Note that $\omega_i(t)$ and $\omega_k(t)$ are precompact.

The total phase time derivative operator acting on phase functions is

$$(20.12) \qquad \frac{d}{dt} = \frac{\partial}{\partial t} + V_{2k}^{(B)} \cdot \nabla_{2k},$$

where the 3-dimensional bivelocity and bigradient are

$$(20.13a) \qquad v_{2i}^{(B)} = \left(v_i, -\frac{\partial \Phi_{i,x}(q_i, t)}{\partial q_i} \right),$$

$$(20.13b) \qquad \nabla_{2i} = \left(\frac{\partial}{\partial q_i}, \frac{\partial}{\partial p_i} \right),$$

and the $3\mathcal{N}_k$-dimensional versions are

$$(20.14a) \qquad V_{2k}^{(B)} = \times_{i \in k} v_{2i}^{(B)},$$

$$(20.14b) \qquad \nabla_{2k} = \times_{i \in k} \nabla_{2i}.$$

20.3 The Boltzmann Equation Approximation

In this section, the factorization of the system distribution implicit in the work of Maxwell and extended by Boltzmann will be presented in the context of an approximation procedure. The approximation of the evolution equation (5.16) in the form (6.15) will be modified by an approximation procedure specialized to the Boltzmann setting. The result will be an equation similar to that of Boltzmann [**1872**], p. 133.

The TIS version of the theory of Boltzmann's equation follows the same pattern as the general theory. An isolated system called the total system is chosen and subsystems of that system are defined. The evolution equation (5.16), with $k = t$, for the total system is to be approximated to obtain evolution equations for the subsystems. The first step is the marginalization of the total system evolution equation. This led in Chapter 6 to the equation (6.13) as the equation to approximate in moving from the total system to the subsystem representation. The approximation of (6.13) then led to an evolution equation of the form (6.15) for each subsystem $k \in [t]$. The Boltzmann equation approximation scheme begins with equation (6.15) for the system k and approximates it in accord with the TIS assumptions underlying the Boltzmann equation. The result is the TIS version of the Boltzmann evolution equation viewed as an approximation of the k system evolution equation.

The particle interaction potential appears in the quantity dF_k/dt, defined in equation (6.15) in the internal force terms $(\partial \Phi_{k,n}(Q_k)/\partial Q_k) \cdot (\partial F_k^t/\partial P_k)$ and in the external force terms $(\partial \Phi_{k,x}(Q_k)/\partial Q_k) \cdot (\partial F_k^t/\partial P_k)$. The first step in the program for modifying (6.15) is the replacement of the internal force term by a stochastic operator that will approximate the effects of collisions in modifying the distribution function. At the same time, the Hamiltonian $\mathcal{H}_k$ is changed to the corresponding Boltzmann Hamiltonian, $\mathcal{H}_k^{(B)}$. It is also assumed that the modification of the i-particle distribution density f_i due to collisions between particles i and j, for any $j \in k - i$, is independent of the modifications due to all other collisions. The total modification in f_i at a given time is the sum over $j \in k - i$ of each ij collision term.

To accomplish these steps, the one and two particle marginalizations of the evolution equation (6.15) are obtained next. The i-particle and ij-particle marginalizations of F_k are

(20.15a)
$$f_i^k(q_i, p_i, t) = \left[\prod_{s \in k-i} \int d^3 q_s \int d^3 p_s \right] F_k(Q_k, P_k, t),$$

and

(20.15b)
$$f_{ij}^k(q_i, p_i, q_j, p_j, t) - \left[\prod_{s \in k-i-j} \int d^3 q_s \int d^3 p_s \right] F_k(Q_k, P_k, t).$$

To obtain a one particle evolution equation, both sides of equation (6.15) are integrated over all variables in (Q_k, P_k) except q_i and p_i. This yields the marginalization of the k system evolution equation in the form

$$(20.16) \quad \frac{\partial f_i^k(q_i, p_i, t)}{\partial t} + v_i \cdot \frac{\partial f_i^k(q_i, p_i, t)}{\partial q_i} - \frac{\partial \Phi_{i,x}(q_i, t)}{\partial q_i} \cdot \frac{\partial f_i^k(q_i, p_i, t)}{\partial p_i}$$

$$= \sum_{j \in k-i} \left[\int d^3q_j \int d^3p_j \frac{\partial \phi_{ij}(q_i - q_j)}{\partial q_i} \cdot \frac{\partial f_{ij}^k(q_i, p_i, q_j, p_j, t)}{\partial p_i} \right],$$

where, as in (6.13), an integration by parts of q_j and p_j over all phase space for $j \in k-i$ shows that only the terms on the right hand side that are derivatives of q_i or p_i do not vanish. The sum of terms on the right hand side represents the interactions between particles $j \in k - i$ and particle $i \in k$. In the case of an isolated system ($k = t$), the external potential is set to $\Phi_{i,x}(q_i, t) = 0$ for each $i \in k$.

Following the pattern of Chapter 6, the following approximations steps will be used in (20.16):

1. Replace the one-particle marginalization $f_i^k(q_i, p_i, t)$ by $f_i(q_i, p_i, t)$ and replace the two-particle marginalization $f_{ij}^k(q_i, p_i, q_j, p_j, t)$ by the product $f_i(q_i, p_i, t) f_j(q_j, p_j, t)$;
2. Choose the initial condition $f_i(q_i, p_i, 0) = f_i^k(q_i, p_i, 0)$;
3. Replace the term on the right-hand side of (20.16), which represents the interactions between particles in k, by the operator $C[f] f_i(q_i, p_i, t)$.
4. Replace the k system distribution $F_k(Q_k, P_k, t)$ by $F_{[k]}(Q_k, P_k, t) = \prod_{i \in k} f_i(q_i, p_i, t)$

This sequence of steps is called the *Boltzmann equation approximation procedure*. It should be emphasized that there is no evidence that either Maxwell or Boltzmann approached their equations as an approximation in this way and Boltzmann himself (Boltzmann [**1872**] in Brush [**1965**], Vol. 1, p. 96) initially felt that his approach was exact.[11]

Using the above approximation procedure with equation (20.16) and (20.12) for the definition of the total time derivative operator in the Boltzmann case yields the *Boltzmann evolution equation*

$$(20.17) \qquad \frac{df_i(q_i, p_i, t)}{dt} = C[f] f_i(q_i, p_i, t),$$

which is usually referred to as the *Boltzmann equation*. The operator will be defined in the next chapter.

---

[11]The replacement of $f_{ij}^k(q_i, p_i, q_j, p_j, t)$ by $f_i(q_i, p_i, t) f_j(q_j, p_j, t)$ is equivalent to the molecular chaos assumption. There have been many "derivations" of Boltzmann's equation over the years. A recent approximation scheme, based on a different perspective than is used here, is given in Cercignani, Gerasimenko, and Petrina [**1997**], Chapter IV.

The TIS version of the Boltzmann formalism follows the pattern of the general theory. The particle distribution density $f_i(q_i, p_i, t)$ is not normalized and is required to meet the conditions

$$(20.18a) \qquad f_i(q_i, p_i, t) \geq 0,$$

$$(20.18b) \qquad Z_i(t) = \int d^3 q_i \int d^3 p_i \, f_i(q_i, p_i, t) \chi_{\omega_i(t)}(q_i, p_i) < \infty,$$

where $Z_i(t)$ is the local norm of $f_i(q_i, p_i, t)$ and is called the *i-particle partition function*. In keeping with the definition of the Boltzmann k system distribution density $F_{[k]}(Q_k, P_k, t)$ as the product of individual particle densities, the *k partition function* is expressed in terms of $Z_i(t)$ by

$$(20.19) \qquad Z_{[k]}(t) = \prod_{i \in k} Z_i(t).$$

It follows from (20.18b) that

$$(20.20) \qquad Z_{[k]}(t) < \infty.$$

20.4 The Evolution of Boltzmann Phase Averages

The *i-particle phase average* of a quantity $G_i(q_i, p_i, t)$ in system k is computed as in the general theory using a bracket notation in the form

$$(20.21)$$
$$G_i(t) = \frac{1}{Z_i(t)} \int d^3 q_i \int d^3 p_i \, f_i(q_i, p_i, t) \chi_{\omega_i(t)}(q_i, p_i) G_i(q_i, p_i, t)$$
$$= \langle G_i(q_i, p_i, t) \rangle_{it} .$$

The *k system phase average* is

$$(20.22) \qquad G_k(t) = \sum_{i \in k} G_i(t).$$

The time partial derivative of $G_i(t)$ is given by

$$(20.23)$$
$$\frac{\partial G_i(t)}{\partial t} = \frac{1}{Z_i(t)} \int d^3 q_i \int d^3 p_i \left[\chi_{\omega_i(t)}(q_i, p_i) \frac{\partial f_i(q_i, p_i, t) G_i(q_i, p_i, t)}{\partial t} \right.$$
$$\left. + \frac{\partial \chi_{\omega_i(t)}(q_i, p_i)}{\partial t} f_i(q_i, p_i, t) G_i(q_i, p_i, t) \right] - \frac{\partial \ln Z_i(t)}{\partial t} G_i(t).$$

By (20.17) and (20.12), the partial time derivative of $f_i(q_i, p_i, t)$ is

$$(20.24) \qquad \frac{\partial f_i(q_i, p_i, t)}{\partial t} = -v_{2i}^{(B)} \cdot \nabla_{2i} f_i(q_i, p_i, t) + C[f] f_i(q_i, p_i, t).$$

Using this in (20.23) and integrating by parts yields

(20.25)
$$\frac{\partial G_i(t)}{\partial t} = \frac{1}{Z_i(t)} \int d^3q_i \int d^3p_i \left\{ f_i(q_i, p_i, t) \left[\chi_{\omega_i(t)}(q_i, p_i) \frac{dG_i(q_i, p_i, t)}{dt} \right. \right.$$
$$\left. \left. + \frac{d\chi_{\omega_i(t)}}{dt} G_i(q_i, p_i, t) \right] + \chi_{\omega_i(t)} G_i(q_i, p_i, t) C[f] f_i(q_i, p_i, t) \right\} - \frac{\partial \ln Z_i(t)}{\partial t} G_i(t).$$

Setting $G_i(q_i, p_i, t) = 1$ and $G_i(t) = 1$ in (20.25) and rearranging gives the result

(20.26)
$$\frac{\partial Z_i(t)}{\partial t} = \int d^3q_i \int d^3p_i \left[f_i(q_i, p_i, t) \frac{d\chi_{\omega_i(t)}}{dt} + \chi_{\omega_i(t)} C[f] f_i(q_i, p_i, t) \right].$$

Anticipating the result (21.10) that the phase average of $C[f]f_i(q_i, p_i, t)$ yields 0, the $C[f]f_i(q_i, p_i, t)$ term is removed from (20.26). This leads to the final result

(20.27)
$$\frac{\partial Z_i(t)}{\partial t} = \int d^3q_i \int d^3p_i \, f_i(q_i, p_i, t) \frac{d\chi_{\omega_i(t)}(q_i, p_i)}{dt} = \left\langle \dot{\chi}_{\omega_i(t)}(q_i, p_i) \right\rangle_{it} = 0.$$

This result shows that the i-particle local norm $Z_i(t)$ is constant in time, which is consistent with and required by the fact that the k system is closed. Using this with (20.25) gives the result

$$\frac{\partial G_i(t)}{\partial t} = \left\langle \frac{dG_i(q_i, p_i, t)}{dt} \right\rangle_{it} + \left\langle \dot{\chi}_{\omega_i(t)}(q_i, p_i) G_i(q_i, p_i, t) \right\rangle_{it}$$

(20.28a)

$$+ \frac{1}{Z_i(t)} \int d^3q_i \int d^3p_i \, \chi_{\omega_i(t)}(q_i, p_i) G_i(q_i, p_i, t) C[f] f_i(q_i, p_i, t),$$

(20.28b)
$$\frac{\partial G_k(t)}{\partial t} = \sum_{i \in k} \frac{\partial G_i(t)}{\partial t}.$$

The local phase averages are defined similarly by

(20.29a) $$G_i(q, t) = \left\langle \delta(q - q_i) G_i(q_i, p_i, t) \right\rangle_{it},$$

(20.29b) $$G_k(q, t) = \sum_{i \in k} G_i(q, t).$$

The partial time derivative of a local phase average is defined in terms of its action on the thermodynamic quantity $G_i(q, t)$ by

(20.30) $$\frac{dG_i(q, t)}{dt} = \frac{\partial G_i(q, t)}{\partial t} + \frac{\partial G_i^\mu(q, t)}{\partial q^\mu}.$$

The total phase time derivative d/dt in the Boltzmann setting was defined in equation (20.12). Using this in the computation of the partial time derivative of (20.29a) gives the result

(20.31)
$$\frac{\partial G_i(q,t)}{\partial t} = \frac{1}{Z_i(t)} \int d^3 q_i \int d^3 p_i \, \delta(q-q_i) \left[\frac{\partial \chi_{\omega_i(t)}(q_i, p_i)}{\partial t} f_i(q_i, p_i, t) G_i(q_i, p_i, t) \right.$$
$$\left. + \chi_{\omega_i(t)}(q_i, p_i) \left\{ \frac{\partial f_i(q_i, p_i, t)}{\partial t} G_i(q_i, p_i, t) + f_i(q_i, p_i, t) \frac{\partial G_i(q_i, p_i, t)}{\partial t} \right\} \right].$$

Using (20.24) in (20.31), integrating by parts, and using the relation $\partial \delta(q - q_i)/\partial q = -\partial \delta(q - q_i)/\partial q_i$, it follows that the total time derivative of $G_i(q,t)$ is then given by

(20.32)
$$\frac{dG_i(q,t)}{dt} = \frac{1}{Z_i(t)} \int d^3 q_i \int d^3 p_i \, \delta(q - q_i) \left[\frac{d\chi_{\omega_i(t)}(q_i, p_i)}{dt} f_i(q_i, p_i, t) G_i(q_i, p_i, t) \right.$$
$$\left. + \chi_{\omega_i(t)}(q_i, p_i) \left\{ f_i(q_i, p_i, t) \frac{dG_i(q_i, p_i, t)}{dt} + C[f] f_i(q_i, p_i, t) G_i(q_i, p_i, t) \right\} \right]$$
$$= \left\langle \delta(q - q_i) \frac{dG_i(q_i, p_i, t)}{dt} \right\rangle_{it} + \left\langle \dot{\chi}_{\omega_i(t)}(q_i, p_i) G_i(q_i, p_i, t) \right\rangle_{it}$$
$$+ \frac{1}{Z_i(t)} \int d^3 q_i \int d^3 p_i \, \delta(q - q_i) \chi_{\omega_i(t)}(q_i, p_i) C[f] f_i(q_i, p_i, t) G_i(q_i, p_i, t).$$

This result can be separated into pieces referring to the interactions of particles with the wall and collisions with each other. For any thermodynamic quantity $G_i(q_i, p_i, t)$, let us define the *i-particle wall term* by

(20.33)
$$\left[\frac{dG_i(q,t)}{dt} \right]_w = \frac{1}{Z_i(t)} \int d^3 q_i \int d^3 p_i \, \delta(q - q_i) f_i(q_i, p_i, t)$$
$$\times \frac{d\chi_{\omega_i(t)}(q_i, p_i)}{dt} G_i(q_i, p_i, t)$$

and the *i-particle collision term* by
(20.34)
$$\left[\frac{dG_i(q,t)}{dt} \right]_c = \frac{1}{Z_i(t)} \int d^3 q_i \int d^3 p_i \, \delta(q - q_i) C[f] f_i(q_i, p_i, t) G_i(q_i, p_i, t).$$

With this notation, the total time derivative of $G_i(q,t)$ is written
(20.35)
$$\frac{dG_i(q,t)}{dt} = \left\langle \delta(q - q_i) \frac{dG_i(q_i, p_i, t)}{dt} \right\rangle_{it} + \left[\frac{dG_i(q,t)}{dt} \right]_w + \left[\frac{dG_i(q,t)}{dt} \right]_c.$$

There is a similar decomposition for $G_i(t)$.

CHAPTER 21

The Collision Operator

The collision operator $C[f_j]f_i(q_i, p_i, t)$ has been introduced as a stochastic operator that acts on Boltzmann particle states and represents the rate at which the Boltzmann state $f_i(q_i, p_i, t)$ is changed by a collision between particle i and a particle j in the state $f_j(q_j, p_j, t)$. In this chapter, the collision term is expressed as an integral operator and shown to be well defined in the specific settings in which it will be used.

21.1 The Collision Term

The Boltzmann operator $C[f]f_i(q_i, p_i, t)$ is a replacement for the term

$$\sum_{j \in k-i} \left[\int d^3 q_j \int d^3 p_j \, \frac{\partial \phi_{ij}(q_i - q_j)}{\partial q_i} \cdot \frac{\partial f_{ij}^k(q_i, p_i, q_j, p_j, t)}{\partial p_i} \right]$$

obtained in equation (20.16) of the TIS Boltzmann equation approximation procedure. The collision term that Boltzmann chose in place of this element of the Hamiltonian formalism is the subject of this chapter.

Let us begin with a review of the collision process. Consider a collision between particles i and j with masses m_i, m_j. Assume that the asymptotic incoming velocities are v_i and v_j and that the asymptotic outgoing velocities are v_i^* and v_j^*. The *asymptotic relative incoming velocity* and *asymptotic relative outgoing velocity* are then defined by

(21.1a) $$g = g_{ij} = v_i - v_j,$$
(21.1b) $$g^* = g_{ij}^* = v_i^* - v_j^*.$$

Using the conservation of energy and momentum in a collision collision it is easy to show that there exists a unit vector $\hat{l}$ such that[1]

(21.2a) $$v_i^* = v_i - \frac{2m_j(g \cdot \hat{l})\hat{l}}{m_i + m_j},$$

(21.2b) $$v_j^* = v_j + \frac{2m_i(g \cdot \hat{l})\hat{l}}{m_i + m_j}.$$

Furthermore, it follows immediately that

(21.3a) $$g^* - g = -2(g \cdot \hat{l})\hat{l},$$
(21.3b) $$|g^*| = |g|,$$
(21.3c) $$g^* \cdot \hat{l} = -g \cdot \hat{l},$$
(21.3d) $$\frac{\partial(v_i^*, v_j^*)}{\partial(v_i, v_j)} = -1,$$

where (21.3d) is the Jacobian of the transformation from (v_i, v_j) to (v_i^*, v_j^*). Observe that $\hat{l}$ lies in the plane determined by g^* and g. The direction of $\hat{l}$ is determined in (21.3a) up to a choice of sign.

In the usual fashion, the angle between g and g^* is designated by $\pi - 2\theta$ and θ is called the *scattering angle*. The angle θ ranges between 0 (head-on collision and $\pi/2$ (no scattering). The direction of $\hat{l}$ is specified by requiring

---

[1] See Cerciguani [1975], pp. 68–69, or Jeans [1925], for a derivation of formulas equivalent to (21.2).

that $g \cdot \hat{l} \geq 0$. An important relation connects the scattering angle and the angle between g and $\hat{l}$. Squaring both sides of (21.3a) and using (21.3b) gives

(21.4) $\quad 4(g \cdot \hat{l})^2 = |g - g^*|^2 = 2[g^2 - g \cdot g^*] = 2g^2[1 - \cos(\pi - 2\theta)] = 4g^2 \cos^2 \theta.$

Let σ be the angle between g and $\hat{l}$. It follows from (21.4) that there is a relation between σ and the scattering angle that takes the form

(21.5) $$\cos^2 \sigma = \frac{(g \cdot \hat{l})^2}{g^2} = \cos^2 \theta.$$

The choice $g \cdot \hat{l} > 0$ for the direction of $\hat{l}$ implies that $\sigma = \theta$, so the angle between g and $\hat{l}$ is the scattering angle and

(21.6) $$g \cdot \hat{l} = |g| \cos \theta.$$

Finally, it follows that $|g|$ is bounded by a number $g_0 > 0$ because $g = (m_i p_j - m_j p_i)/m_i m_j$ and the momenta are bounded.

Let us consider now a velocity frame in which particle i is at rest at the origin before the interaction. Let g define the z-axis of this frame and choose two mutually orthogonal unit vectors $\hat{c}_1$ and $\hat{c}_2$ such that both are orthogonal to g and that the ordered set of vectors $(\hat{c}_1, \hat{c}_2, g)$ are the axes of a right-handed coordinate system. The directions of the unit vectors $\hat{c}_1$ and $\hat{c}_2$ will be specified in more detail below.

Particle i is at the origin of the plane defined by $(\hat{c}_1, \hat{c}_2)$, called the *impact plane,* so the impact parameters (b, ϕ) in this plane define a vector in polar coordinates. The $(\hat{c}_1, \hat{c}_2)$ coordinate axes will define the axes of the vector (b, ϕ) in a polar integration over the plane. Unless the collision is head-on, the relative velocity vector g originating at the j particle will not intersect the $(\hat{c}_1, \hat{c}_2)$ plane at the origin.

According to Boltzmann's reasoning, the change in $f_i(q, p, t)$ at time t due to collisions comes from two sources. The particle i with momentum p' may acquire the momentum p at location q due to a collision with a particle j or, if it has momentum p at location q, it may lose the momentum p and acquire the momentum p^* due to a collision with a particle j. The Stosszahlansatz is the assumption that the probability flux density for a collision between a particle with phase coordinates (q_i, p_i) and one with phase coordinates (q_j, p_j) is

(21.7a) $\qquad\qquad b\, db d\phi |g_{ij}| f_i(q_i, p_i, t) f_j(q_j, p_j).$

In this case the momenta (p_i, p_j) will be converted into (p_i^*, p_j^*) and the particle i will be removed from the state (q_i, p_i). Similarly, the probability flux density that particle i will leave the state (q_i, p_i^*) and enter the state with the phase coordinates (q_i, p_i) is

(21.7b) $\qquad\qquad b\, db d\phi |g_{ij}| f_i(q_i, p_i^*, t) f_j(q_j, p_j^*, t).$

To obtain the total probability for the changes of the i-particle momentum p_i due to collisions, the probability (21.7a) that particle i will lose the momentum p_i is subtracted from the probability (21.7b) that particle i will acquire the momentum p_i and an integration over the impact plane (i.e., over b and ϕ) is performed followed by an integration over (q_j, p_j). This yields the ij-*Boltzmann collision term*

$$(21.8) \quad C[f_j]f_i(q_i, p_i, t) = \int d^3 q_j \int d^3 p_j \int_0^\infty db\, b \int_0^{2\pi} d\phi |g_{ij}|$$
$$\times \left[f_i(q_i, p_i^*, t) f_j(q_j, p_j^*, t) - f_i(q_i, p_i, t) f_j(q_j, p_j, t) \right].$$

Note that, by (21.2), the members of the set of momenta $\{ p_i^* \}$ for $i \in k$ are functions of the members of the set $\{ p_i \}$ and the masses. The i-*Boltzmann collision term* is defined in terms of these quantities as

$$(21.9) \qquad C[f]f_i(q_i, p_i, t) = \sum_{j \in k-i} C[f_j]f_i(q_i, p_i, t).$$

In the collision term calculation, it is assumed that each location (b, ϕ) in the collision plane is equally likely to represent the collision and these points are averaged over by the b and ϕ integrations. Both this assumption and the assumption that $f_{ij} = f_i f_j$ is always valid are clearly not compatible with Hamiltonian mechanics in general, so Boltzmann's equation is at best an approximation.

When the collision term is expressed as it was in (21.8) and (21.9), Boltzmann's equation is expressed as a set of N_k coupled equations and, for each $i, j \in k$, the ij term is linear in f_i and f_j. Using (21.3), it is easy to show that $|g_{ij}| d^3 p_i d^3 p_j = |g_{ij}^*| d^3 p_i^* d^3 p_j^*$. This implies that the two terms on the right in (21.8) cancel when both sides are integrated over (q_i, p_i) so that

$$(21.10) \qquad \int d^3 q_i \int d^3 p_i\, C[f]f_i(q_i, p_i, t) = 0$$

as required. It is also easy to show in a similar fashion that the total 2-particle momentum, angular momentum, and kinetic energy, are preserved by $C[f]$, that is,

$$(21.11a) \quad \int d^3 p_i\, p_i^\mu C[f_j]f_i(q_i, p_i, t) + \int d^3 p_j\, p_j^\mu C[f_i]f_j(q_j, p_j, t) = 0,$$

(21.11b)
$$\int d^3 p_i\, j_i^\mu(q_i, p_i) C[f_j]f_i(q_i, p_i, t) + \int d^3 p_j\, j_j^\mu(q_j, p_j) C[f_i]f_j(q_j, p_j, t) = 0.$$

(21.11c)
$$\int d^3 p_i\, K_i(p_i) C[f_j]f_i(q_i, p_i, t) + \int d^3 p_j\, K_j(p_j) C[f_i]f_j(q_j, p_j, t) = 0,$$

Cercignani [1998], pp. 267–270, has given a brief proof that the most general collision invariant takes the form $C = a + b \cdot p + cp^2$, for constants a and c and a constant vector b.

The requirements that the members of the set $\{\, f_i(q_i, p_i, t)\,\}$ of k system one-particle distribution densities must meet in order to be solutions of Boltzmann's equation will be investigated in this chapter. In addition, it can be shown by the methods of Part I that the conditions (20.18) on $f_i(t)$ are compatible with the evolution equation (20.17) and that the factorization $F_{[k]}(Q_k, P_k, t) = \prod_{i \in k} f_i(q_i, p_i, t)$ is preserved by the evolution of the system in accord with the Boltzmann equation.

21.2 The Collision Kernel

By (21.3b) and (21.8), the action of $C[f_j]$ on f_i can be written as an integral operator in the form
(21.12)

$$C[f_j] f_i(q_i, p_i, t) = \int d^3 q_j \int d^3 p_i^\natural \int d^3 p_j^\natural \, K_{ij}(p_i^\natural, p_j^\natural; p_i) f_i(q_i, p_i^\natural, t) f_j(q_j, p_j^\natural, t)$$

where the *collision kernel*, K_{ij} is defined by

(21.13) $K_{ij}(p_i^\natural, p_j^\natural; p_i) = \displaystyle\int_0^\infty db\, b \int_0^{2\pi} d\phi \int d^3 p_j |g_{ij}|$

$$\times \left[\delta(p_i^\natural - p_i^*) \delta(p_j^\natural - p_j^*) - \delta(p_i^\natural - p_i) \delta(p_j^\natural - p_j) \right]$$

$$= |g_{ij}^\natural| \int_0^\infty db\, b \int_0^{2\pi} d\phi \int d^3 p_j \left[\delta(p_i^\natural - p_i^*) \delta(p_j^\natural - p_j^*) - \delta(p_i^\natural - p_i) \delta(p_j^\natural - p_j) \right].$$

The facts that $|g^*| = |g|$ and $|g| \delta(p_i^\natural - p_i) \delta(p_j^\natural - p_j) = |g^\natural| \delta(p_i^\natural - p_i) \delta(p_j^\natural - p_j)$ and a similar relation between $|g^*|$ and $|g^\natural|$ were used to obtain the last line in (21.13). The reversal of integrations used in obtaining (21.12) and (21.13) from (21.8) is justified by a version of Fubini's theorem.[2]

The vector p_r, $r \in \{i, j\}$, can be decomposed into a vector $p_{r,l}^\mu$ parallel to $\hat{l}$ and a vector $p_{r,o}^\mu$ orthogonal to $\hat{l}$:

(21.14a) $p_{r,l}^\mu = (p_r \cdot \hat{l}) \hat{l}^\mu$,

(21.14b) $p_{r,o}^\mu = p_r^\mu - p_{r,l}^\mu$.

---

[2]If $f_i(q_i, p_i, t) \in L^1(\omega_i(t))$ for $i \in k$, theorem 40.4 in Treves [1967], pp. 416–417, can be adapted to prove (21.13) if the integrals in (21.8) and (21.13) both exist.

In terms of this decomposition, the results

(21.15a) $$p_i^{*\mu} = p_i^{\mu} + \frac{2m_i p_{j,l}^{\mu}}{m_i + m_j} - \frac{2m_j p_{i,l}^{\mu}}{m_i + m_j},$$

(21.15b) $$p_j^{*\mu} = p_j^{\mu} - \frac{2m_i p_{j,l}^{\mu}}{m_i + m_j} + \frac{2m_j p_{i,l}^{\mu}}{m_i + m_j},$$

are obtained from (21.1) and (21.2). In addition, these definitions imply that the relation $d^3 p_j = dp_{j,l} d^2 p_{j,o}$, where $p_{r,l} = |p_{r,l}^{\mu}|$ for $r = i$ or $r = j$, is also valid.

The integration of the delta measures in (21.13) over p_j is done most easily in a coordinate system $(\hat{o}_1, \hat{o}_2, \hat{l})$, where $\hat{o}_1$ and $\hat{o}_2$ are unit vectors orthogonal to $\hat{l}$ and to each other. It is not hard to verify that

(21.16) $$\int d^3 p_j \, \delta(p_i^{\natural} - p_i^*)\delta(p_j^{\natural} - p_j^*)$$

$$= \delta\left(\frac{m_i - m_j}{m_i + m_j} p_{i,l}^{\natural} + \frac{2m_i p_{j,l}^{\natural}}{m_i + m_j} - p_{i,l}\right) \delta(p_{i,o}^{\natural} - p_{i,o}).$$

With (21.16), the p_j integration can be performed in (21.13) to obtain

(21.17) $$K_{ij}(p_i^{\natural}, p_j^{\natural}; p_i) = |g_{ij}^{\natural}| \int_0^{\infty} db \, b \int_0^{2\pi} d\phi \, \delta(p_{i,o}^{\natural} - p_{i,o})$$

$$\times \left[\delta\left(\frac{m_i - m_j}{m_i + m_j} p_{i,l}^{\natural} + \frac{2m_i p_{j,l}^{\natural}}{m_i + m_j} - p_{i,l}\right) - \delta(p_{i,l}^{\natural} - p_{i,l}) \right].$$

A useful definition for the next step is the *reduced mass*

(21.18) $$\mu = \mu_{ij} = \frac{m_i m_j}{m_i + m_j}.$$

The representation

(21.19) $$\delta(p_i^{\natural} - p_i) = \frac{1}{[2\pi]^3} \int d^3 \lambda \, e^{-i\lambda \cdot (p_i^{\natural} - p_i)}$$

of the δ measure, where the λ integration is over all of $\mathbf{R}^3$, is used with (21.17) in conjunction with the $(\hat{o}_1, \hat{o}_2, \hat{l})$ coordinate system to show that $K_{ij}(p_i^{\natural}, p_j^{\natural}; p_i)$ can be written as

(21.20)

$$K_{ij}(p_i^{\natural}, p_j^{\natural}; p_i) = \frac{1}{[2\pi]^3} |g_{ij}^{\natural}| \int_0^{\infty} db \, b \int_0^{2\pi} d\phi \int d^3 \lambda \, e^{-i\lambda \cdot (p_i^{\natural} - p_i)}$$

$$\times \left[e^{2i\mu(\lambda \cdot l)(g^{\natural} \cdot l)} - 1 \right].$$

Fubini's theorem (again in the context of the integrals in (21.12)) is used to exchange the order of the λ and the b, ϕ integrations. This gives the result

$$(21.21) \qquad K_{ij}(p_i^\natural, p_j^\natural; p_i) = \frac{1}{[2\pi]^3} \int d^3\lambda \, e^{-i\lambda \cdot (p_i^\natural - p_i)} K_{ij}(\lambda, g^\natural),$$

where

$$(21.22) \qquad K_{ij}(\lambda, g^\natural) = |g^\natural| \int_0^\infty db \, b \int_0^{2\pi} d\phi \left[e^{2i\mu(\lambda \cdot \hat{i})(g^\natural \cdot \hat{i})} - 1 \right]$$

Using the expression

$$(21.23) \qquad g^\natural = \frac{m_i p_j^\natural - m_j p_i^\natural}{m_i m_j}$$

for $g^\natural$ in terms of the momenta $p_i^\natural$ and $p_j^\natural$, it is easy to show

$$(21.24) \qquad \frac{1}{m_i} \left(m_i \frac{\partial}{\partial p_i^\natural} + m_j \frac{\partial}{\partial p_j^\natural} \right) g^\natural = 0.$$

Let us define the operator $\mathcal{O}(p_i^\natural, p_j^\natural)$ by

$$(21.25) \qquad \mathcal{O}(p_i^\natural, p_j^\natural) = \left[1 - \frac{1}{m_i^2} \left(m_i \frac{\partial}{\partial p_i^\natural} + m_j \frac{\partial}{\partial p_j^\natural} \right)^2 \right]^3.$$

It follows easily by (21.22) and (21.24) that

$$(21.26) \quad \mathcal{O}(p_i^\natural, p_j^\natural) \left\{ e^{-i\lambda \cdot (p_i^\natural - p_i)} K_{ij}(\lambda, g^\natural) \right\} = [1 + \lambda^2]^3 e^{-i\lambda \cdot (p_i^\natural - p_i)} K_{ij}(\lambda, g^\natural).$$

Let us also define

$$(21.27) \qquad K_{ij}^{(b)}(\lambda, g^\natural) = \frac{K_{ij}(\lambda, g^\natural)}{[1 + \lambda^2]^3}.$$

Then, if the kernel $K_{ij}^{(b)}(p_i^\natural, p_j^\natural; p_i)$ is defined by

$$(21.28) \qquad K_{ij}^{(b)}(p_i^\natural, p_j^\natural; p_i) = \frac{1}{[2\pi]^3} \int d^3\lambda \, e^{-i\lambda \cdot (p_i^\natural - p_i)} K_{ij}^{(b)}(\lambda, g^\natural),$$

it follows that

$$(21.29) \qquad K_{ij}(p_i^\natural, p_j^\natural; p_i) = \mathcal{O}(p_i^\natural, p_j^\natural) K_{ij}^{(b)}(p_i^\natural, p_j^\natural; p_i).$$

Using (21.29) in (21.12) gives the result

$$(21.30) \quad C[f_j] f_i(q_i, p_i, t) = \int d^3 q_j \int d^3 p_i^\natural \int d^3 p_j^\natural \left[\mathcal{O}(p_i^\natural, p_j^\natural) K_{ij}^{(b)}(p_i^\natural, p_j^\natural; p_i) \right]$$

$$\times f_i(q_i, p_i^\natural, t) f_j(q_j, p_j^\natural, t).$$

21.3 The Scattering Angle

The net effect of the particle-particle interaction is expressed in Boltz-mann's theory in the scattering angle θ. In order to establish bounds on $K_{ij}(\lambda, g)$ and the integrals used in the last section, an asymptotic formula is needed for the scattering angle as the impact parameter, b, goes to infinity. In this section, only the particles i and j will be used and the i, j subscripts will be suppressed. They will be reintroduced later when needed.

Let $r = |q_i - q_j|$ be the distance between particles i and j and let $\phi(r) = \phi_{ij}(r)$ be the interparticle potential energy. The scattering angle for a collision between particles i and j can be expressed by the integral[3]

$$(21.31) \qquad \theta(g, b) = -b \int_{\infty}^{r_0(g,b)} \frac{dr}{r^2} \, [h(r, g, b)]^{-\frac{1}{2}}$$

where

$$(21.32) \qquad h(r, g, b) = 1 - \frac{2\phi(r)}{\mu g^2} - \frac{b^2}{r^2}$$

and

$$(21.33) \qquad r_0(g, b) = \sup_{r \in \mathbf{R}_+} \{\, r \mid h(r, g, b) = 0 \,\}.$$

The conditions on $\phi(r)$, expressed in (4.2), imply that h is a jointly continuous function mapping $(r, g, b) \in \mathbf{R}^3$ into $\mathbf{R}$. In addition, the conditions (4.2) can be used to show that the following inequalities are valid:

$$(21.34) \qquad -\infty \le h(r, g, b) \le 1 + \frac{2\phi_0}{\mu g^2}.$$

Furthermore, the continuity of h and the facts that $\lim_{r \to \infty} h(r, g, b) = 1$ and $\lim_{r \to 0} h(r, g, b) = -\infty$ imply that an $r_0 > 0$ satisfying (21.33) always exists. Finally, setting $r = r_0$ and $h = 0$ in (21.32) and then using (4.2b) gives

$$(21.35) \qquad r_0(g, b) = b \left[1 - \frac{2\phi(r_0)}{\mu g^2} \right]^{-\frac{1}{2}}$$

$$\ge b \left[1 + \frac{2\phi_0}{\mu g^2} \right]^{-\frac{1}{2}}.$$

This calculation shows with the help of (4.2d) that $r_0(g, b) \to b$ as $b \to \infty$, so $\phi(r_0) \to 0$ as $b \to \infty$ and $\lim_{b \to \infty} \theta(g, b) = \pi/2$. These facts imply that the scattering vanishes as the impact parameter increases without limit—as expected on physical grounds.

There are two cases that need to be considered for the behavior of $h' = \partial h / \partial r$ near $r = r_0$. These are illustrated in Figure 21.1. Let us begin with

---

[3]For further information on this calculation, see Goldstein [**1950**], Chapter III.

Case (a)   Case (b)

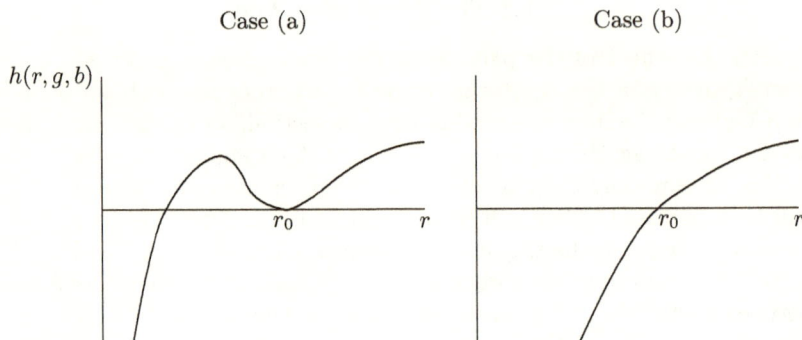

FIGURE 21.1.   Critical points of $h(r, g, b)$

case (a) in which $h(r_0, g, b) = h'(r_0, g, b) = 0$. If r_0 is viewed as an implicit function of g and b, it is possible that r_0 is discontinuous when case (a) holds. In order that this situation not lead to discontinuities or divergences in integrations involving $\theta(g, b)$, Sard's theorem (Sternberg [**1964**], p. 47) is invoked. Sard's theorem states that the set of critical values of h, i.e. the points at which

(21.36) $$\{ (r, g, b) \mid h(r, g, b) = h'(r, g, b) = 0 \},$$

is a set of measure zero in $\mathbf{R}^3$ if $h \in C^k(\mathbf{R}^3)$ and $k \geq 3$. The choice of $k \geq 3$ in (4.2a) insures that the set of critical values of h is a set of measure zero in $\mathbf{R}^3$.

Case (a) will be excluded from consideration. In the integration over the impact parameter b below, the asymptotic behavior of $\theta(g, b)$ as b becomes large is important. It will be shown that $\cos \theta(g, b) \sim O(1/|g|^2 b^3)$ as $b \to \infty$.

The transformation $y = r/b$ with $y_0(g, b) = r_0(g, b)/b$ will be used next to rewrite the scattering angle calculation in the form

(21.37) $$\theta(g, b) = - \int_{\infty}^{y_0(g,b)} \frac{dy}{y^2} [h(y, g, b)]^{-\frac{1}{2}}$$

where

(21.38) $$h(y, g, b) = \frac{1}{y^2} \left[y^2 - \frac{2y^2 \phi(by)}{\mu g^2} - 1 \right].$$

Next, the condition (4.2d) is used to show that the potential $\phi(r)$ satisfies the asymptotic relation $-A/r^3 < \phi(r) < A/r^3$ for all $r > r_1$, where $r_1 < \infty$ and $A > 0$ is a constant. Because $r_0 \to \infty$ as $b \to \infty$, there exists a value b_1 such that $r_0(g, b_1) > r_1$. Thus, for $b > b_1$, the asymptotic representation of

$\phi(r)$ can be used. In the asymptotic region for $\phi(r)$, it follows that

$$(21.39) \qquad \left[y^2 - 1 - \frac{2A}{\mu g^2 b^3 y}\right] \leq y^2 h(y, g, b) \leq \left[y^2 - 1 + \frac{2A}{\mu g^2 b^3 y}\right].$$

In order to simplify calculations, the facts that $y \geq y_0$ and $y_0 \sim 1$ for $b > b_1$ (with a correction of order $O(b^{-3})$) will be used to refine this approximation. The relation (21.35) can be used to show in this case that there is an $\epsilon \ll 1$ such that $1 - \epsilon \leq y_0 \leq 1 + \epsilon$. The quantity b_1 can be chosen large enough so that $|2/y| < 3$ for $b > b_1$ as well. Using this, an asymptotic limit of the function $h(r, g, b)$ can be defined by

$$(21.40) \qquad h^{ext(a)}(y, g, b, A) = y^2 - y_1^2(g, b, A),$$

where

$$(21.41) \qquad y_1(g, b, A) = \sqrt{1 - \frac{3A}{\mu g^2 b^3}}.$$

It is also assumed that $3A/\mu g^2 b^3 \ll 1$, which means that y_1 will be real. It then follows that $y_1(g, b, A) \leq y_0(g, b) \leq y_1(g, b, -A)$. Using (21.41), it is easy to show in this case that $h^{ext(a)}$ and h satisfy the inequalities

$$(21.42) \quad [h^{ext(a)}(y, g, b, A)]^{-\frac{1}{2}} \leq y^{-1}[h(y, g, b)]^{-\frac{1}{2}} \leq [h^{ext(a)}(y, g, b, -A)]^{-\frac{1}{2}}.$$

The representation of $\theta(g, b)$ given in (21.37) and the approximation $h^{ext(a)}(y, g, b, A)$ defined in (21.40) are used to define the asymptotic scattering angle $\theta(g, b, A)$ by

$$(21.43) \qquad \theta(g, b, A) = -\int_{\infty}^{y_1(g,b,A)} \frac{dy}{y} \, [h^a(y, g, b, A)]^{-\frac{1}{2}}$$

$$= -\int_{\infty}^{y_1(g,b,A)} \frac{dy}{y} \, [y^2 - y_1^2(g, b, A)]^{-\frac{1}{2}}.$$

For $b > b_1$, it is easy to show by the above inequalities and the properties of the integral in (21.43) that

$$(21.44) \qquad \theta(g, b, A) \leq \theta(g, b) \leq \theta(g, b, -A)$$

and that $\theta(g, b, \pm A)$ converges to $\theta(g, b)$ as $b \to \infty$. Moreover, $\theta(g, b, \pm A)$ can be integrated. For $\theta(g, b, -A)$ this yields the result

$$(21.45) \qquad \theta(g, b, -A) = \frac{\pi}{2y_1(g, b, -A)}.$$

This in turn can be closely approximated by

$$(21.46) \qquad \theta(g, b, -A) \approx \frac{\pi}{2}\left[1 - \frac{3A}{2\mu g^2 b^3}\right].$$

It follows immediately that
(21.47)
$$\cos\theta(g,b) \sim \cos\theta(g,b,-A) = -\sin\left[-\frac{3\pi A}{4\mu g^2 b^3}\right] \sim \frac{3\pi A}{4\mu g^2 b^3}, \qquad \text{for } b > b_1.$$

21.4 A Bound on the Collision Kernel

The next step is to show that $K_{ij}^{(b)}(p_i^\natural, p_j^\natural; p_i)$ is mathematically well defined. In (21.28), $K_{ij}^{(b)}(p_i^\natural, p_j^\natural; p_i)$ was defined as the Fourier transform of $K_{ij}^{(b)}(\lambda, g^\natural)$ with respect to $p_i^\natural - p_i$. To meet the requirements of the Fourier integral theorem, it must be shown that $K_{ij}^{(b)}(\lambda, g^\natural)$ is absolutely integrable in λ over the range $-\infty \leq \lambda^\mu \leq \infty$ and that this function and its derivative with respect to λ are both piecewise continuous in every finite interval in the same range.

An examination of the definition (21.27) of $K_{ij}^{(b)}(\lambda, g^\natural)$ shows readily that it is continuous in every finite interval. The derivative of $K_{ij}^{(b)}(\lambda, g^\natural)$ with respect to the vector λ is also clearly continuous in every finite interval. The next step is to show that it is absolutely integrable in λ over all $\mathbf{R}^3$.

The analysis will be performed in the coordinate system $(\hat{c}_1, \hat{c}_2, g)$. The direction of the unit vector $\hat{c}_2$ is defined by the vector cross product $g \times \lambda$ if $g \cdot \lambda \geq 0$ and by $-g \times \lambda$ otherwise. The unit vector $\hat{c}_1$ is then directed along $\hat{c}_2 \times g$. The vector λ is expressed in this coordinate system by

(21.48)
$$\lambda = |\lambda|(\sin\xi, 0, \cos\xi).$$

The unit vector $\hat{l}$ lies in the plane defined by $(g, g^\natural)$. In a coordinate system with g as the z-axis, the angle ϕ locates the direction of the component of $\hat{l}$ in the plane orthogonal to g, which is the impact plane. This has the consequence that $\hat{l}$ is written in this coordinate system in terms of the scattering angle θ and the angle ϕ on the impact plane as

(21.49)
$$\hat{l} = (\sin\theta\cos\phi, \sin\theta\sin\phi, \cos\theta).$$

It follows immediately from (21.3c), (21.6), and (21.49) that

(21.50a) $$g^\natural \cdot \hat{l} = -|g^\natural|\cos\theta,$$
(21.50b) $$\lambda \cdot \hat{l} = |\lambda|(\sin\xi\sin\theta\cos\phi + \cos\xi\cos\theta).$$

It is not hard to show using the relation between $g^\natural$ and g that $[K_{ij}(\lambda, g^\natural)]^* = K_{ij}(\lambda, g)$, where $*$ represents complex conjugation.

The inner integral in (21.22) is

(21.51) $$I_1(\lambda, g^\natural, b) = |g^\natural| \int_0^{2\pi} d\phi \left[e^{2i\mu(\lambda\cdot\hat{l})(g^\natural\cdot\hat{l})} - 1\right].$$

To demonstrate the absolute integrability of the Fourier transform $K_{ij}^{(b)}(\lambda, g^{\natural})$ of $K^{(b)}(p_i^{\natural}, p_j^{\natural}; p_i)$, the quantity $I_1(\lambda, g^{\natural}, b)/[1 + \lambda^2]^3$ will be integrated over b and then λ. In order to handle the convergence issues, the integrations over λ and b in this integration of $K_{ij}^{(b)}(\lambda, g^{\natural})$ will be divided as follows:

$$(21.52) \quad \int d^3\lambda \left[\theta(|\lambda| - b_1) \left\{ \int_0^{|\lambda|} db\,b + \int_{|\lambda|}^{\infty} db\,b \right\} + \theta(b_1 - |\lambda|) \int_0^{\infty} db\,b \right]$$
$$\times \frac{I_1(\lambda, g^{\natural}, b)}{[1 + \lambda^2]^3}.$$

The value b_1 has been chosen as above so that the scattering angle $\theta(g, b)$ is in the asymptotic region for $b > b_1$ and the approximation stated in (21.47) is valid.

The b integration under the integral over λ from 0 to b_1 will be computed first. This b integration is divided into two pieces. The first piece is from 0 to b_1. Using the fact that $|I_1(\lambda, b, g^{\natural})| \leq 4\pi|g^{\natural}|$, it follows that

$$(21.53) \quad K_{ij}^{(1,0)}(\lambda, g^{\natural}, b_1) = \int_0^{b_1} db\,b|I_1(\lambda, g^{\natural}, b)| \leq 2\pi|g^{\natural}|b_1^2.$$

The integration over λ is then bounded by

$$(21.54) \quad K_{ij}^{(b)(1,0)} = \frac{1}{[2\pi]^3} \int d^3\lambda \frac{K_{ij}^{(1,0)}(\lambda, g^{\natural}, b_1)}{[1 + \lambda^2]^3} \leq \tfrac{1}{16}|g^{\natural}|b_1^2.$$

where the inequality

$$(21.55) \quad \int d^3\lambda \frac{\theta(b_1 - |\lambda|)}{[1 + |\lambda|^2]^3} < 4\pi \int_0^{\infty} d|\lambda| \frac{|\lambda|^2}{[1 + |\lambda|^2]^3} = \frac{\pi^2}{4}$$

was used to obtain the result.

The only term that includes the angle ϕ in the integrand of $I_1(\lambda, g^{\natural}, b)$ is $\lambda \cdot \hat{l}$ defined in (21.50b). Expanding the exponential in the integrand of $I_1(\lambda, b, g^{\natural})$ gives

$$(21.56) \quad I_1(\lambda, g^{\natural}, b) = |g^{\natural}| \sum_{p=1}^{\infty} \frac{1}{p!} [2i\mu|\lambda||g^{\natural}| \cos\theta(g^{\natural}, b)$$
$$\times \int_0^{2\pi} d\phi \{\cos\theta(g^{\natural}, b)\cos\xi + \sin\theta(g^{\natural}, b)\cos\phi\sin\xi\}]^p$$
$$= |g^{\natural}| \sum_{p=1}^{\infty} \frac{1}{p!} [2i\mu|\lambda||g^{\natural}| \cos\theta(g^{\natural}, b)]^p$$
$$\times \sum_{j=0}^{p} \binom{p}{j} [\cos\theta(g^{\natural}, b)\cos\xi]^{p-j} [\sin\theta(g^{\natural}, b)\sin\xi]^j \int_0^{2\pi} d\phi\,[\cos\phi]^j.$$

The terms with j odd in this expansion are all zero. It follows that $2p - j \geq 2$ for each appearance of the quantity $\cos^{2p-j} \theta(g,b)$ in the resulting expansion after integrating over ϕ.

The inequality

$$(21.57) \qquad |\cos\theta(g,b)\cos\xi + \sin\theta(g,b)\cos\phi\sin\xi| = \frac{|\lambda \cdot \hat{l}|}{|\lambda|} \leq 1,$$

is obtained from (21.50b). This result implies, for $p \geq 1$, the relation

$$(21.58) \qquad \left| \int_0^{2\pi} d\phi \, (\cos\theta\cos\xi + \sin\theta\cos\phi\sin\xi)^p \right| \leq 2\pi.$$

This leads to an upper bound of the form

$$(21.59) \qquad |I_1(\lambda, g^\natural, b)| \leq 2\pi|g^\natural| \sum_{p\geq 1} \frac{1}{p!} [2\mu|\lambda||g^\natural|]^p \, |\cos\theta(g,b)|^p.$$

For the bound on the integration over b from b_1 to ∞, the bound on the integral of $|\cos\theta(g,b)|^p$ over this range is obtained first. Using the asymptotic expression (21.47) for $\cos\theta(g^\natural, b)$, it follows for $p \geq 1$ that

$$(21.60) \qquad \int_{b_1}^{\infty} db \, b \, |2\cos\theta(g,b)|^p \leq \left(\frac{3\pi A}{2\mu|g|^2} \right)^p \frac{b_1^{-3p+2}}{3p-2}$$
$$\leq b_1^2 [C(g,b_1)]^p,$$

where $C(g,b_1) = [3\pi A/2\mu|g|^2 b_1^3] < 1$. The expansion (21.59) is used with (21.60) to integrate $I_1(\lambda, g^\natural, b)$ over b from b_1 to ∞. Summing the series that results from this yields

$$(21.61) \quad K_{ij}^{(1,1)}(\lambda, g^\natural, b_1) = \int_{b_1}^{\infty} db \, b \, |I_1(\lambda, g^\natural, b)| \leq 2\pi b_1^2 |g^\natural| I_2(|\lambda|, g^\natural, b_1)$$

where $I_2(|\lambda|, g^\natural, b_1)$ is defined by

$$(21.62) \qquad I_2(|\lambda|, g^\natural, b_1) = e^{\mu|\lambda||g^\natural| C(g^\natural, b_1)} - 1.$$

Next, the fact that $I_2(|\lambda|, g^\natural, b_1)$ is an increasing function of $|\lambda|$ is employed along with (21.55) to show
(21.63)

$$K_{ij}^{(b)(1,1)} = \frac{1}{[2\pi]^3} \int d^3\lambda \, \theta(b_1 - |\lambda|) \frac{K_{ij}^{(1,1)}(\lambda, g^\natural, b_1)}{[1 + |\lambda|^2]^3} \leq \tfrac{1}{16} b_1^2 |g^\natural| I_2(b_1, g^\natural, b_1).$$

The relation $|I_1(\lambda, g^\natural, b)| < 4\pi|g^\natural|$ is used with the remaining integral to be computed in (21.52) to obtain

$$(21.64) \quad K_{ij}^{(b)(2,0)} = \frac{1}{[2\pi]^3} \int d^3\lambda \, \frac{\theta(|\lambda| - b_1)}{[1 + |\lambda|^2]^3} \int_0^{|\lambda|} db \, b |I_1(\lambda, g^\natural, b)|$$

$$\leq \frac{1}{\pi}|g^\natural| \int_{b_1}^\infty d|\lambda| \, \frac{|\lambda|^4}{[1 + |\lambda|^2]^3} \leq \frac{1}{\pi}|g^\natural| \int_0^\infty d|\lambda| \, \frac{|\lambda|^4}{[1 + |\lambda|^2]^3} = \frac{3}{16}|g^\natural|.$$

The fourth integral uses the result (21.61) with $|\lambda|$ in place of b_1. This yields

(21.65)

$$K_{ij}^{(b)(2,1)} = \frac{1}{[2\pi]^3} \int d^3\lambda \, \frac{\theta(|\lambda| - b_1)}{[1 + |\lambda|^2]^3} K_{ij}^{(1)}(\lambda, g^\natural, |\lambda|)$$

$$\leq \frac{1}{\pi}|g^\natural| \int_{b_1}^\infty d|\lambda| \, \frac{|\lambda|^4}{[1 + |\lambda|^2]^3} I_2(|\lambda|, g^\natural, |\lambda|).$$

Using the fact that $I_2(|\lambda|, g^\natural, |\lambda|)$ is a decreasing function of $|\lambda|$ for $|\lambda| \geq b_1$ it follows that

$$(21.66) \quad K_{ij}^{(b)(2,1)} \leq \frac{1}{\pi}|g^\natural| I_2(b_1, g^\natural, b_1) \int_{b_1}^\infty d|\lambda| \, \frac{|\lambda|^4}{[1 + |\lambda|^2]^3}$$

$$\leq \frac{3}{16}|g^\natural| I_2(b_1, g^\natural, b_1).$$

Putting these pieces together yields the bound

$$(21.67) \quad \frac{1}{[2\pi]^3} \int d^3\lambda \, |K_{ij}^{(b)}(\lambda, g^\natural)| \leq K_{ij}^{(b)},$$

where

(21.68)

$$K_{ij}^{(b)} = K_{ij}^{(b)(1,0)} + K_{ij}^{(b)(1,1)} + K_{ij}^{(b)(2,0)} + K_{ij}^{(b)(2,1)}$$

$$= \frac{1}{16}|g^\natural|[b_1^2 + 3][1 + I_2(b_1, g^\natural, b_1)].$$

This establishes the absolute integrability of $K_{ij}^{(b)}(\lambda, g^\natural)$ required for the existence of its Fourier transform.

It follows immediately from (21.67) that $K_{ij}^{(b)}(p_i^\natural, p_j^\natural; p_i)$ is bounded

$$(21.69) \quad |K_{ij}^{(b)}(p_i^\natural, p_j^\natural; p_i)| \leq \frac{1}{[2\pi]^3} \int d^3\lambda \, |K_{ij}^{(b)}(\lambda, g^\natural)| \leq K_{ij}^{(b)}.$$

It is not hard to show that $K_{ij}^b(p_i^\natural, p_j^\natural; p_i)$ is also continuous. This and the relation (21.29) show that $K_{ij}(p_i^\natural, p_j^\natural; p_i)$ is a *tempered distribution*.[4]

---

[4]See Treves [**1967**], p. 272.

The circumstances under which the collision kernel, $C[f_j]f_i(q_i, p_i, t)$, is well-defined can now be determined. Integrating (21.30) by parts gives the result

(21.70)

$$C[f_j]f_i(q_i, p_i, t) = \int d^3q_j \int d^3p_i^\natural \int d^3p_j^\natural \, K_{ij}^b(p_i^\natural, p_j^\natural; p_i)$$

$$\times \, [\mathcal{O}(p_i^\natural, p_j^\natural) f_i(q_i, p_i^\natural, t) f_j(q_j, p_j^\natural, t)].$$

If $(\partial^\gamma f_i(q_i, p_i^\natural, t)/(\partial p_i^{\natural\mu})^\gamma) \in L^2(\omega_i(t))$ for each i and μ, and for $\gamma \le 6$, then the definition (21.25) of $\mathcal{O}(p_i^\natural, p_j^\natural)$ and the expression (21.70) imply that the collision term $C[f_j]f_i(q_i, p_i, t)$ is well defined.

This completes the demonstration that the operator is well defined in the TIS formalism without the need for arbitrary cutoffs.

21.5 Special Cases

In order to make a connection with the literature on Boltzmann's equation, a set of special cases for the collision terms are computed. This calculation is a concrete realization of the abstract calculation performed in Section 21.4. The special cases are based on a purely repulsive central force law with the potential

(21.71)

$$\phi_{ij}(r) = \frac{a_{ij}(s)}{r^s}$$

for $s \ge 4$ and $a_{ij}(s) \le 0$. This potential obviously satisfies (4.2). The i and j indices will be suppressed for most of the rest of this section. Using (21.71) in (21.32) and a procedure initiated by Maxwell [1866], an application of the transformations

(21.72a)

$$u = \frac{b}{r},$$

(21.72b)

$$b^\star = b \left[\frac{\mu g^2}{2a(s)} \right]^{\frac{1}{s}}.$$

successively to (21.32) and (21.31) gives the scattering angle in terms of $b^\star$ as[5]

(21.73)

$$\theta(b^\star, g) = \int_0^{u_0} du \left[1 - u^2 - \left(\frac{u}{b^\star} \right)^s \right]^{-\frac{1}{2}}$$

where u_0 is defined by

(21.74)

$$u_0 = \inf \left\{ u \mid 1 - u^2 - \left(\frac{u}{b^\star} \right)^s = 0 \right\}.$$

---

[5]See the version in Jeans [1925], pp. 213–219, or Goldstein [1950], p.73.

Next, let $y = u/u_0$ and use (21.74) to obtain the formula

$$(21.75) \qquad \theta(b^\star, g) = u_0 \int_0^1 dy \, [1 - y^2]^{-\frac{1}{2}} \left[1 + \left(\frac{u_0}{b^\star} \right)^s \frac{1 - y^s}{1 - y^2} \right]^{-\frac{1}{2}}.$$

It is not hard to show that

$$(21.76) \qquad 1 \le \frac{1 - y^s}{1 - y^2} \le \frac{s}{2}.$$

This implies

$$(21.77) \qquad \left(\frac{u_0}{b^\star} \right)^s \le \left(\frac{u_0}{b^\star} \right)^s \frac{1 - y^s}{1 - y^2} \le \frac{s}{2} \left(\frac{u_0}{b^\star} \right)^s.$$

Suppose now that b_1 is chosen large enough so that for $b \ge b_1$ the relation

$$(21.78) \qquad \frac{s}{2} \left(\frac{u_0}{b^\star} \right)^2 \ll 1$$

is valid. In this case, the relation

$$(21.79) \qquad \left[1 + \left(\frac{u_0}{b^\star} \right)^s \frac{1 - y^s}{1 - y^2} \right]^{-\frac{1}{2}} \approx 1 - \frac{s}{4} \left(\frac{u_0}{b^\star} \right)^s$$

is used to show

$$(21.80) \qquad \theta(b^\star, g) \approx \frac{\pi}{2} \left[1 - \frac{s}{4} \left(\frac{u_0}{b^\star} \right)^s \right].$$

From this it follows for $b \ge b_1$ that

$$(21.81) \qquad |\cos \theta(b^\star, g)| \approx \frac{s\pi}{8} \left(\frac{u_0}{b^\star} \right)^s.$$

Let us define next

$$(21.82) \qquad x = \frac{s - 4}{s}$$

and note that the above assumptions imply $0 \le x \le 1$. The definition (21.52) can be used with (21.51) to write $K_{ij}(\lambda, g^\natural)$ in the form

$$(21.83) \qquad K_{ij}^{(x)}(\lambda, g^\natural) = \left[\frac{2a(s)}{\mu[g^\natural]^2} \right]^{\frac{1 - x}{2}} \int_0^\infty db^\star \, b^\star I_1^{(x)}(\lambda, g^\natural, b^\star),$$

where $\mu = \mu_{ij}$ is the reduced mass defined in (21.18) and $I_1^{(x)}(\lambda, g^\natural, b_1)$ is defined as in (21.51) with the scattering angle defined by (21.75). The sequence of calculations in Section 21.4 and using the integration scheme (21.52) will be repeated for the current case. The result is

$$(21.84) \qquad |K_{ij}^{(b)(x)}(p_i^\natural, p_j^\natural; p_i)| \le \frac{1}{[2\pi]^3} \int d^3\lambda \, |K_{ij}^{(b)(x)}(\lambda, g^\natural)| \le K_{ij}^{(b)(x)}.$$

where

(21.85) $$K_{ij}^{(b)(x)} = \tfrac{1}{16}|g^\natural|[b_1^2 + 3][1 + I_2^{(x)}(b_1, g^\natural, b_1)]$$

and

(21.86) $$I_2^{(x)}(\lambda, g^\natural, b_1) = e^{\frac{s}{4}\pi\mu|\lambda||g^\natural|\left(\frac{2a(s)}{\mu[g^\natural]^2}\right)\left(\frac{u_0}{b_1}\right)^s} - 1.$$

This shows that the collection of special cases reviewed here are well defined as well.

The boundary case $x = 0$ corresponds to an inverse fifth power repulsive central force law. This was the case considered in detail by Maxwell. The other boundary case is $x = 1$ which corresponds to a hard sphere gas with an inverse infinite power repulsive central force law. The hard sphere case will quickly be recomputed by a different method. Consider a collision between two hard spheres i and j with effective radii σ_i and σ_j, respectively. The distance between the centers of the particles at impact is

(21.87) $$\sigma_{ij} = \sigma_i + \sigma_j.$$

By the law of reflection, the relation between the impact parameter and the scattering angle θ is

(21.88) $$b^\star = \sigma_{ij}\sin\theta.$$

Using the vector $\hat{l}$ exhibited in (21.4) and the expressions for $\lambda \cdot \hat{l}$ and $g \cdot \hat{l}$ given in (21.50), it is easy to show that

(21.89) $$\left|K_{ij}^{(hs)}(\lambda, g)\right| \le 2\pi\sigma_{ij}$$

for all λ and g.

The results (21.81) and (21.84) are used with (21.26) and lead to the conclusion that a bound similar to (21.69) holds for each of these cases. Requiring that members of the set $\{f_i\}$ belong to an appropriate function space, similar to the requirements stated after (21.70), leads to the conclusion that the collision kernel is bounded for these special cases as expected.

CHAPTER 22

Boltzmann's Entropy and Ergodic Theory

The Boltzmann equilibrium state and the Boltzmann entropy analog, defined as a generalization of the analog $\ln f(q, p, t)$ that Boltzmann introduced in his H theorem, are discussed. Various criticisms of the H theorem and ergodic theory are also examined from the TIS perspective.

22.1 Limiting States

Two important limiting states in any version of thermodynamics are the asymptotic equilibrium state and the exact state. The Maxwell-Boltzmann version of these states will be introduced now. The first asymptotic equilibrium state introduced was Maxwell's [1860], [1866], equilibrium function. This function depended only on the particle kinetic energy and took the form of a Gaussian velocity distribution. Boltzmann [1875], [1878a] extended Maxwell's velocity distribution by working out a steady distribution for particles in a static field of force in connection with the barometer formula. The equilibrium distribution was subsequently extended to its current form that is valid for any field of force that allows a steady distribution.[1] As noted before, Boltzmann [1896b], pp. 220–221, ultimately recognized that the Maxwell distribution is only approximate if the number of particles in the system is finite.[2]

The TIS use of the equilibrium state differs from that of Maxwell and Boltzmann mainly due to the TIS use of unnormalized system distributions and the TIS limitation of expectation calculations for the k system to the system phase set $\Omega_k(t)$ in which the particle coordinates and momenta are both bounded. While Boltzmann stated in his response to Loschmidt's criticisms that allowing the momenta to range from $-\infty$ to ∞ does not introduce a "significant error" in the calculation of mean values when the number of molecules is sufficiently large, this is only true with regard to the calculation of the mean values of quantities at a particular time. While this idealization does offer some minor simplifications in the calculation of certain averages, as was noted in Chapter 9, it is nevertheless unrealistic, unnecessary, and does not allow keeping proper account of conserved quantities. Thus, while having little practical import, it is a blemish on Boltzmann's theory.

The version of thermodynamics and statistical mechanics developed in the Theory of Interacting Systems applies to any finite number of particles in a system in a bounded coordinate volume with a bounded total momentum. With these points in mind, let us follow the path set out in Chapter 13 and define the *i-particle asymptotic equilibrium distribution density* and *i-particle equilibrium partition function* for a system in the k equilibrium rest frame by

$$(22.1a) \qquad f_i^\epsilon(q_i, p_i) = e^{-\beta_k H_i^{(B)\epsilon}(q_i, p_i)},$$

$$(22.1b) \qquad Z_i^\epsilon = \int d^3 q_i \int d^3 p_i \, \chi_{\omega_i(t)}(q_i, p_i) e^{-\beta_k H_i^{(B)\epsilon}(q_i, p_i)}.$$

----

[1] See the two cases allowed for a reversible equilibrium distribution that were obtained in Chapter 13. In modern terms, any function that is a steady state of the Hamiltonian evolution equation in a system is a function of the invariants of the Hamiltonian.

[2] See also Boltzmann [1896a], p. 448.

Recall that the equilibrium rest frame of the k system is the frame in which
the walls are at rest (the system rest frame) and the macroscopic center of
momentum conditions $\mathcal{P}_k(t) = 0$ and $\mathcal{J}_k(t) = 0$ are valid (the particle rest
frame). While setting up the system rest frame is a macroscopic matter, the
requirement that the system is in the particle rest frame depends on the particle state. Another difference between this approach and the standard version
of Boltzmann's equation is the inclusion of the i-particle projection operator $\chi_{\omega_i(t)}(q_i, p_i)$ in (22.1b). Note also the difference between the definition
(22.1a) and the standard definition of the normalized Boltzmann i-particle
equilibrium probability distribution density by

$$(22.2) \qquad f_i^{(n)\epsilon}(q_i, p_i) = [Z_i^\epsilon]^{-1} e^{-\beta_k H_i^{(B)\epsilon}(q_i, p_i)}.$$

The equilibrium i-particle Boltzmann Hamiltonian, $H_i^{(B)\epsilon}(q_i, p_i, t)$ satisfies the condition

$$(22.3) \qquad \frac{dH_i^{(B)\epsilon}(q_i, p_i, t)}{dt} = \frac{\partial \Phi_{i,x}^\epsilon(q_i, t)}{\partial t} = 0,$$

so it will be written $H_i^{(B)\epsilon}(q_i, p_i)$ from now on. The relation (22.3) is just
the condition that the external potentials be constant in time, but otherwise
arbitrary. In parallel with the relation of system distributions in TIS, the k
Boltzmann equilibrium system distribution and the k *Boltzmann equilibrium
partition function* are

$$(22.4a) \qquad F_{[k]}^\epsilon(Q_k, P_k) = \prod_{i \in k} f_i^\epsilon(q_i, p_i)$$

with

$$(22.4b) \qquad Z_k^\epsilon = \prod_{i \in k} Z_i^\epsilon.$$

This is also referred to as the k *Boltzmann equilibrium state*.

The second limiting state is not based on the Maxwell-Boltzmann collision approximation for particle interactions. It is the single particle exact
state based on Hamiltonian mechanics, which is the single particle version of
the Hamiltonian mechanics of a subsystem introduced in Chapter 4. The k
system Hamiltonian $\mathcal{H}_k(Q_k, P_k, t)$ is the same as that introduced in the general theory in Chapter 4 and its solutions are the system specific trajectories
introduced there. This means that the i-particle Hamiltonian equations of
motion for $i \in k$ are

$$(22.5) \qquad \left(\frac{dq_i}{dt}, \frac{dp_i}{dt} \right) = \left(\frac{\partial \mathcal{H}_k(Q_k, p_i, t)}{\partial p_i}, -\frac{\partial \mathcal{H}_k(Q_k, p_i, t)}{\partial q_i} \right).$$

The i-particle system specific trajectory $(q_i^{(s)}(t), p_i^{(s)}(t))$ is a piecewise continuous solution of these equations and the relation $(q_i^{(s)}(t), p_i^{(s)}(t)) \in \omega_i(t)$

is valid for $t \in [0, \alpha)$ and $i \in k$. The *i-particle exact state* and the *k system exact state* are then defined as in the general case by

$$(22.6a) \qquad f_i^\delta(q_i, p_i, t) = \delta(q_i - q_i^{(s)}(t))\delta(p_i - p_i^{(s)}(t)),$$

$$(22.6b) \qquad F_{[k]}^\delta(Q_k, P_k, t) = \prod_{i \in k} f_i^\delta(q_i, p_i, t),$$

where $F_{[k]}^\delta(Q_k, P_k, t) = F_k^\delta(Q_k, P_k, t)$.

22.2 The Entropy Density and Entropy Current Density

For a system in the equilibrium state, Boltzmann [**1871a**] identified the temperature with the average kinetic energy of a particle, i.e.,

$$(22.7) \qquad T = \tfrac{2}{3}K_i^\epsilon = [k_B \beta_k]^{-1}.$$

This was mentioned briefly in Chapter 12 and rejected as a general definition of the temperature.

To get at the relationships involved in thermodynamic changes, Boltzmann used a variational approach. Before examining Boltzmann's equilibrium results, it is important to recall the demonstration in Chapter 19 that variational methods based on modifying the macroscopic parameters describing the system are limited to special cases in TIS thermodynamics because the partial derivatives with respect to thermature and volume do not commute with the partial derivative with respect to time.

In Boltzmann's variational approach to calculating thermodynamic quantities, he varied the Hamiltonian energy with respect to the thermature and other macroscopic parameters.[3] To use this method, the variation of the system distribution with respect to the thermature must be calculable. Boltzmann made his calculation with the equilibrium state, which fills the requirement for a system distribution that depends on macroscopic parameters. Boltzmann also assumed that the potential energy is constant in time but depends on a set of macroscopic constraint parameters $\{a_m\}$ and is written $\Phi_k^\epsilon(Q_k; a_m)$.[4] For simplicity, the system projection operator will be omitted from the calculations below and the integrations performed over all of phase space.

This calculation was made by Boltzmann prior to the introduction of the Boltzmann equation and is based on standard Hamiltonian mechanics. For the equilibrium phase average of the Hamiltonian, he used (TIS notation)

$$(22.8) \qquad \mathcal{H}_k^\epsilon = \frac{3\mathcal{N}_k}{2\beta_k} + \langle \Phi_k^\epsilon(Q_k; a_m) \rangle_{k\epsilon}.$$

---

[3]See Bailyn [**1994**], Chapter 10, for a discussion of Boltzmann's previous work and a discussion of the variational calculation that appeared in Boltzmann [**1871b**] and [**1868**].

[4]The use of constraint parameters such as these in the Hamiltonian was rejected in Chapter 4 as inappropriate for a general thermodynamics.

The parameters to be varied are β_k and the members of the set $\{a_m\}$. The variation of $\mathcal{H}_k^\epsilon$ is then

(22.9)
$$\delta\mathcal{H}_k^\epsilon = -\left(\frac{3\mathcal{N}_k}{2\beta_k^2}\right)\delta\beta_k + \delta\langle\Phi_k^\epsilon(Q_k;a_m)\rangle_{k\epsilon}.$$

The work done on the system by varying the members of the set $\{a_m\}$ is written

(22.10)
$$\delta\mathcal{W}_k = \langle\delta\Phi_k^\epsilon(Q_k;a_m)\rangle_{kt} = \sum_m\left\langle\frac{\partial\Phi_k^\epsilon(Q_k;a_m)}{\partial a_m}\right\rangle_{k\epsilon}\delta a_m.$$

The heat, $\delta\mathcal{Q}_k^\epsilon$, added to the system is the difference between the variation of the Hamiltonian δH_k^ϵ and the variation of the work done on the system. This implies that
(22.11)
$$\delta\mathcal{Q}_k^\epsilon = \delta\mathcal{H}_k^\epsilon - \delta\mathcal{W}_k = -\frac{3\mathcal{N}_k}{2\beta_k^2}\delta\beta_k + \delta\langle\Phi_k^\epsilon(Q_k;a_m)\rangle_{k\epsilon} - \langle\delta\Phi_k^\epsilon(Q_k;a_m)\rangle_{k\epsilon}.$$

For the entropy, Boltzmann used the standard equilibrium entropy expressed as

(22.12)
$$S_k^{(\mathrm{B1})} = S_k^{(\mathrm{n})\epsilon} = k_B[\beta_k\mathcal{H}_k^\epsilon + \ln Z_k^\epsilon].$$

The entropy analog corresponding to this entropy is

(22.13)
$$S_k^{(\mathrm{n})\epsilon}(Q_k,P_k) = k_B[\beta_k\mathcal{H}_k^\epsilon(Q_k,P_k) + \ln Z_k^\epsilon].$$

The variation of the macroscopic entropy presented in (22.12) is

(22.14)
$$\delta S_k^{(\mathrm{n})\epsilon} = k_B[\mathcal{H}_k^\epsilon\delta\beta_k + \beta_k\delta\mathcal{H}_k^\epsilon + \delta\ln Z_k^\epsilon].$$

It is easy to see by the definition of Z_k^ϵ that the variation of $\ln Z_k^\epsilon$ is

(22.15)
$$\delta\ln Z^\epsilon = -\langle\mathcal{H}_k^\epsilon(Q_k,P_k)\rangle_{k\epsilon}\delta\beta_k - \beta_k\langle\delta\Phi_k(Q_k;a_m)\rangle_{k\epsilon}.$$

By (22.14), (22.15), (22.9), and (22.11), it follows that

(22.16)
$$\delta S_k^{(\mathrm{n})\epsilon} = k_B\beta_k\left[-\left(\frac{3\mathcal{N}_k}{2\beta_k^2}\right)\delta\beta_k + \delta\langle\Phi_k^\epsilon(Q_k;a_m)\rangle_{k\epsilon} - \langle\delta\Phi_k(Q_k;a_m)\rangle_{k\epsilon}\right]$$
$$= k_B\beta_k\delta\mathcal{Q}_k^\epsilon.$$

For an ideal monatomic gas, Boltzmann showed that this result is in accord with the familiar thermodynamic results.

The next important step Boltzmann took was the development in Boltzmann [1872] of his nonequilibrium theory of the Boltzmann equation and a different analog for the entropy. Modifying the formula of Boltzmann [1872] slightly, the *i-particle Boltzmann phase entropy* is written in TIS notation as

(22.17)
$$S_i^{(\mathrm{B2})}(q_i,p_i,t) = -k_B\ln f_i(q_i,p_i,t).$$

This is different from Boltzmann's original definition, which was based on the normalized probability distribution density $f^{(n)}(q, p, t)$ rather than the set of unnormalized particle distribution densities $\{ f_i(q_i, p_i, t) \}$ used in TIS.

Since each particle is treated as an independent system in the Boltzmann approximation, the entropy can be defined locally and the entropies of individual particles can be added. The entropy in Boltzmann's theory is therefore a particle sum function, which makes it an extensive quantity. Taking the local phase average of $S_i^{(B2)}(q_i, p_i, t)$ over $f_i(q_i, p_i, t)$ using the definition (20.21) of the phase average, the *i-particle Boltzmann local entropy density* and the *k Boltzmann local entropy density* are obtained in the form[5]

(22.18a) $S_i^{(B2)}(q, t) = \left\langle \delta(q - q_i) S_i^{(B2)}(q_i, p_i, t) \right\rangle_{it}$,

(22.18b) $S_k^{(B2)}(q, t) = \sum_{i \in k} S_i^{(B2)}(q, t).$

Similarly, the *i-particle Boltzmann local entropy current density* and the *k Boltzmann local entropy current density* are defined by

(22.19a)

$$S_i^{(B2)\mu}(q, t) = \left\langle \delta(q - q_i) v_i^\mu S_i^{(B2)}(q_i, p_i, t) \right\rangle_{it},$$

(22.19b)

$$S_k^{(B2)\mu}(q, t) = \sum_{i \in k} S_i^{(B2)\mu}(q, t).$$

The equilibrium version of Boltzmann's second entropy is obtained by using f_i^ϵ in (22.17) and phase averaging to obtain the result

(22.20) $$S_k^{(B2)\epsilon} = k_B \beta_k \mathcal{H}_k^{(B)\epsilon}.$$

As noted above, these results for the equilibrium thermodynamic entropy differ from Boltzmann's, which is written as $S_k^{(B2)(n)\epsilon} = k_B \beta_k [\mathcal{H}_k^{(B)\epsilon} + \ln Z_k^\epsilon]$ in TIS notation.

By (20.17) and the definition (22.17) of $S_i^{(B2)}(q_i, p_i, t)$, it follows that

(22.21) $$f_i(q_i, p_i, t) \frac{dS_i^{(B2)}(q_i, p_i, t)}{dt} = -k_B C[f] f_i(q_i, p_i, t).$$

---

[5] Gibbs [**1902**], pp. 76, 86, spoke of the density of entropy or "specific entropy" in the equilibrium case. Clausius [**1878**], p. 214, probably understood his statement of Clausius' inequality in the form $dQ \geq TdS$ to be a locally defined relation that is integrated over the system or system boundary to obtain the global result, but he provided no statements concerning this or how to interpret this integration. Duhem [**1891**], pp. 34–37, used a more precise locally defined quantity.

By (21.10) this term yields 0 in the integral (22.24). It follows that

$$(22.22) \qquad \left\langle \delta(q - q_i) \frac{dS_i^{(B2)}(q_i, p_i, t)}{dt} \right\rangle_{ki} = 0.$$

The total time derivative of $S_i^{(B2)}(q, t)$ is then expressed using (20.35) as the sum of the rate of change at the wall and the rate of change due to collisions in the form

$$(22.23) \qquad \frac{dS_i^{(B2)}(q, t)}{dt} = \left[\frac{dS_i^{(B2)}(q, t)}{dt} \right]_w + \left[\frac{dS_i^{(B2)}(q, t)}{dt} \right]_c,$$

By suitably modifying the results of Chapter 6 to reflect the independent particles assumption made for the Maxwell-Boltzmann theory, it can be shown that

$$(22.24) \qquad \left[\frac{dS_i^{(B2)}(q, t)}{dt} \right]_w \geq k_B \beta_k(q, t) \mathcal{Q}_{\partial i}^*(q, t),$$

where $\beta_k(q, t)$ is the local inverse absolute temperature and $\mathcal{Q}_{\partial i}^*(q, t)$ is the local i-particle surface heating density of system k. There is equality in (22.24) and if and only if the system is in the equilibrium state.

The rate of change of the entropy due to collisions is the subject of the H theorem. The proof of Boltzmann [**1872**] that $S_k^{(B2)}(q, t)$ is non-decreasing will be modified slightly. Let us first sum (22.22) over $i \in k$. Then, by (21.8) and (21.9) and including a factor of $\frac{1}{2}$ to remove the double counting of the particles, it follows immediately that

(22.25)
$$\left[\frac{dS_k^{(B2)}(q, t)}{dt} \right]_c = -\tfrac{1}{2} k_B \sum_{i \in k} \sum_{j \in k - i} \int d^3 q_i \int d^3 p_i \int d^3 q_j \int d^3 p_j \int_0^\infty db\, b$$

$$\times \delta(q - q_i) \int_0^{2\pi} d\phi\, |g_{ij}| \left[f_i(q_i, p_i^*, t) f_j(q_j, p_j^*, t) - f_i(q_i, p_i, t) f_j(q_j, p_j, t) \right]$$

$$\times \ln \left[f_i(q_i, p_i, t) f_j(q_j, p_j, t) \right].$$

Similarly, renaming (p_i^*, p_j^*) to (p_i°, p_j°) and (p_i, p_j) to (p_i^*, p_j^*) gives

(22.26)
$$\left[\frac{dS_k^{(B2)}(q, t)}{dt} \right]_c = -\tfrac{1}{2} k_B \sum_{i \in k} \sum_{j \in k - i} \int d^3 q_i \int d^3 p_i^* \int d^3 q_j \int d^3 p_j^* \int_0^\infty db\, b$$

$$\times \int_0^{2\pi} d\phi\, \delta(q - q_i)\, |g_{ij}| \left[f_i(q_i, p_i^\circ, t) f_j(q_j, p_j^\circ, t) - f_i(q_i, p_i^*, t) f_j(q_j, p_j^*, t) \right]$$

$$\times \ln \left[f_i(q_i, p_i^*, t) f_j(q_j, p_j^*, t) \right].$$

The relation (21.3c) is used next with the fact that one set of limits of integration must be reversed to replace $d^3p_i^*d^3p_j^*$ by $d^3p_id^3p_j$ in (22.26). Note that replacing (p_i^*, p_j^*) by (p_i°, p_j°) and then (p_i, p_j) by (p_i^*, p_j^*) in (21.2) and using the relation $g^* \cdot \hat{l} = -g \cdot \hat{l}$ obtained from (21.3c), yields $p_i = p_i^\circ$ and $p_j = p_j^\circ$. These are used in (22.26), then (22.25) and (22.26) are added and the result is divided by 2. Boltzmann's [1872], p. 116, observation that $(y-x)\ln(x/y) \le 0$ for x and y real and non-negative, then leads to the H *theorem*:

$$(22.27) \qquad \left[\frac{dS_k^{(B2)}(q,t)}{dt}\right]_c \ge 0,$$

with equality if and only if the system is in the equilibrium state. Summing (22.24) over $i \in k$ and adding the result to (22.27) yields

$$(22.28) \qquad \frac{dS_k^{(B2)}(q,t)}{dt} \ge k_B\beta_k(q,t)\mathcal{Q}_{\partial k}^*(q,t)$$

for each $t \in [0, \alpha)$ and $q \in d_k(t)$. Inequality (22.28) is a version of Axiom 6 of the thermodynamics presented in Chapter 3 of Part I. For the isolated system t, the wall term vanishes: $[dS_t^{(B2)}(q,t)/dt]_w = 0$. This gives the usual isolated system result $dS_t^{(B2)}(q,t)/dt = [dS_t^{(B2)}(q,t)/dt]_c \ge 0$ for Boltzmann thermodynamics.[6]

Boltzmann presented another definition of the entropy in his 1872 paper that played only a minor role. He gave another more general and more important definition of the entropy in 1877. Because the 1877 version was significant in Planck's work on blackbody radiation and the development of early quantum theory, the discussion of both of these versions of the entropy will be postponed until QTS.

22.3 Challenges to the H Theorem

When Boltzmann first proposed the H theorem, he [1872], p. 117, felt that he had "rigorously proved that, whatever may be the initial distribution of kinetic energy, in the course of a very long time it must necessarily approach the one found by Maxwell."[7] Loschmidt argued four years later that for every state of increasing entropy there must be a state or reversed velocity for which the entropy is decreasing. As Klein [1973], pp. 71–74, has recounted it, Boltzmann found Loschmidt's argument "very seductive" and concluded that

---

[6]One may also arrange particles in groups such that they interact with particles in the same group and not with particles outside the group as was done in Chapter 6. This procedure was used by Eu [1979] to prove a generalized H theorem for these groups in the same way that Boltzmann proved his H theorem.

[7]After reviewing the Ehrenfest's account of Boltzmann's H theorem, Kaç [1959], p. 82, stated: "Although this argument is almost entirely lacking in mathematical rigor, it has a strong aura of plausibility." This "aura of plausibility" has taken many in.

entropy decreasing processes could occur. Boltzmann [**1877a**] then claimed that H could decrease when certain initial or other conditions obtained. He repeated this position, with embellishments, throughout the rest of his life.

In the face of the criticisms of the H Theorem mentioned in Chapter 8, Boltzmann argued that the H theorem is a "statistical theorem". By this he meant that the passage of a system to equilibrium was not certain, but highly probable: "One cannot therefore prove that, whatever may be the positions and velocities of the spheres at the beginning, the distribution must become uniform after a long time; rather one can only prove that infinitely more initial states will lead to a uniform one after a definite length of time than to a non-uniform one."[8] No such proof was given, however. The idea that Boltzmann is trying to express is that even though the H theorem predicts (in Gibbs terminology) that the ensemble of systems as a whole will evolve toward equilibrium, individual systems are not bound by this and a few systems will violate this prediction.[9] It is argued at length in EIS, Chapter 2, that Maxwell and Boltzmann have muddled their ideas here and are vacillating between different conceptions of a system distribution in the course of a single argument.

Tolman [**1938**], pp. 153–154, maintained that the reversibility argument is only an apparent difficulty because situations in which Boltzmann's equation is used refer to observational situations in which there is a range of states involved. Situations such as this are described by Boltzmann's state $f(q, p, t)$. Other observational situations require the use of exact mechanics. Because these are different observational situations, Tolman maintained that the two approaches can never come into conflict. However, the assumption of classical physics is that the particle system is always in one or another exact state and that this fact does not depend on the "observational situation". Although Tolman's solution to the reversibility problem shares some aspects of the TIS solution, it is not an adequate answer to the problem because it does not address the root of the conflict.

Kuhn [**1978**], p. 58, observed that Loschmidt's reversibility argument for an isolated system makes use of the reversal of the exact particle trajectories and may not actually apply to the state of the system and its evolution as Boltzmann understood it. This is indeed the case. Loschmidt's argument assumes that the entropy depends on the mechanical state (Q_k, P_k), that is, on the set of exact states $\{f_i^\delta\}$, and a reversible mechanical evolution equation. To facilitate comparisons, the equation of motion for the set of exact states

---

[8]See Boltzmann [**1877b**] (Brush [**1965**], Vol. 2, p. 192). See also Boltzmann [**1896a**], pp. 441–443.

[9]Pascual Jordan [**1933**], p. 17, characterized the idea of a statistical theorem as (TIS notation): "if an arbitrary initial state out of the manifold of all possible states is selected, it is overwhelmingly improbable that a state not satisfying $dH_k/dt \geq 0$ will be chosen." Jordan also maintained, following Schrödinger, that the H theorem is not asymmetric with respect to time.

of the system $\{f_i^\delta\}$ is needed. For simplicity, the investigation will be limited to the case of an isolated system. Set $\Phi_{i,x}(q_i, t) = 0$ and $k = t$ in (20.16); set also $\langle (d\chi_{\omega_i(t)}(q_i, p_i)/dt)G_k(q_i, p_i, t)\rangle_{it} = 0$; then use f_i^δ in place of f_i^k and $f_{ij}^\delta = f_i^\delta f_j^\delta$ in place of f_{ij}^k and integrate the right side over (q_j, p_j) for $j \in k-i$. The resulting exact state evolution equation is
(22.29)
$$\frac{\partial f_i^\delta(q_i, p_i, t)}{\partial t} + v_i \cdot \frac{\partial f_i^\delta(q_i, p_i, t)}{\partial q_i} - \sum_{j \in k-i} \frac{\partial \phi_{ij}(q_i - q_j)}{\partial q_i} \cdot \frac{\partial f_i^\delta(q_i, p_i, t)}{\partial p_i} = 0.$$

Notice that in the exact case, the particle interactions are represented by the interaction potentials instead of a stochastic collision term as in Boltzmann's equation.

Boltzmann [1896a], pp. 55–62, stated his own view of the effect of time reversal. He considered the particle collisions when time is moving forward to be governed by probability considerations because the molecular distribution is governed by the molecular disorder assumption. In the time reversed case, the distribution does not show molecular disorder because the collision must follow the precise reverse path. The Boltzmann equation (20.17) reflects Boltzmann's perspective for collisions moving forward in time. It is not reversible in time because the collision term does not change sign under a velocity reversal.

To illustrate the difference between the exact state equation and Boltzmann's equation, suppose a system described by $f_i(q, p, 0)$ at time $t = 0$ evolves into state $f_i(q, p, t_1)$ at time t_1. At t_1, suppose next that the velocities are precisely reversed and the system evolves from state $f_i(q, p, t_1)$ at t_1 to state $f_i(q, p, 2t_1)$ at time $2t_1$. It cannot be concluded using Boltzmann's equation that $f_i(q, p, 0) = f_i(q, p, 2t_1)$ except in the trivial equilibrium case, where $f_i = f_i^\epsilon$ at each point in time. This point was made in a slightly different form by Watson [1876], who noted that the system H function, $S_i^{(B2)}(t)$, would satisfy the conditions $S_i^{(B2)}(0) \le S_i^{(B)}(t_1) \le S_i^{(B2)}(2t_1)$ with equality if and only if the system is in an equilibrium state. The exact state equation (22.29), on the other hand, is reversible in time, so $f_i^\delta(q, p, 2t_1) = f_i^\delta(q, p, 0)$. Loschmidt's argument that the entropy will decrease in the reversed velocities case, however, does not follow from this fact because the entropy is not defined for a system represented by an f_i^δ state.[10]

There is a significant equivocation in the way in which Boltzmann refers to a system state in his claim that there are exceptional initial states for which $S_i^{(B2)}(t)$ decreases. He seems to vacillate, without realizing it, between a conception of the state in terms of the particle mechanical state (q_i, p_i) and the $f_i(q_i, p_i, t)$ state, which is a solution of the Boltzmann equation.

---

[10]See also Watson [1894].

Suppose that the notion of state Boltzmann was using in this case is the particle mechanical state (q_i, p_i). Then, according to Boltzmann's argument, the exceptional set of these mechanical states, for which the entropy $S_i^{(B2)}(t)$ will decrease in time, has Lebesgue measure zero in phase space. In TIS terminology, each member of a set of individual mechanical states is represented by an of exact state f_i^δ. However, because the entropy cannot be defined for singular states like f_i^δ, the corresponding entropy, $S_B^\delta(t)$, is undefined. This, coupled with the fact that Boltzmann's equation does not yield exact states as solutions, means that Boltzmann cannot use a set of mechanical states as the basis for his argument regarding the H theorem.

On the other hand, suppose Boltzmann is referring to the $f_i(q_i, p_i, t)$ distribution as the state. It will be shown in (24.23b) that there are no solutions $f_i(q_i, p_i, t)$ of Boltzmann's equation, such that the local norm of f_i is zero or the support of f_i is a set of Lebesgue measure zero in phase space. It follows from this argument that $S_i^{(B2)}(t)$ cannot decrease and therefore that Zermelo's [**1896b**] criticisms of Boltzmann's views are valid, namely, that Boltzmann's [**1896b**] reasoning on the "H curve" is not mathematically sound. It follows also from this discussion that the view of the H theorem as "statistical theorem", which implies that there is a certain set of system mechanical states (Q_k, P_k) for which it is not true, is a misunderstanding of the nature of the H theorem and an improper application of Boltzmann's formalism.[11]

Loschmidt was using the (Q_k, P_k) or (q_i, p_i) mechanical state in his reversibility criticism of Boltzmann. This requires an evolution equation like (22.29) with the $\{f_i^\delta\}$ states. Boltzmann's H theorem is based on the Boltzmann evolution equation (20.17) and the $\{f_i\}$ states. This means that Loschmidt and Boltzmann were working at cross-purposes in their discussion.

Physically, if the Boltzmann equation (20.17) is our choice to describe the system in the forward velocities case, it must also be our choice to describe the system in the reversed velocities case. Otherwise, we must consistently use the exact state and its evolution equation (22.29) to describe both cases. To use the Boltzmann equation to describe the forward velocities case and what amounts to the exact state evolution equation to describe the reversed velocities case, as Boltzmann and his contemporaries did, leads to paradox.

Finally, there are various computer studies have been cited recently as proving that the H theorem and Boltzmann equation are correct. An examination of the papers discussing the programs involved shows that these claims are wide of the mark.[12]

---

[11] See EIS for more details.

[12] Prigogine's [**1980**], pp. 407–410, statement that computer studies "completely verify Boltzmann's predictions" concerning the fluctuations in H and the behavior of H when the velocities are reversed is wrong. An examination of how the programs were set up (see Wood [**1973**]) shows that the studies Prigogine refers to are a molecular dynamics calculation by

22.4 Ergodic Theory

In the late nineteenth century, there was an effort to establish the law of the equipartition of energy on a mechanical footing. Maxwell [**1879b**], p. 713, criticized the usual assumptions of short-range forces and binary collisions in the attempts to demonstrate equipartition as a consequence of the mechanical laws.[13] He went on to say: "The only assumption which is necessary for the direct proof is that the system, if left to itself in its actual state of motion, will, sooner or later, pass through every phase which is consistent with the equation of energy." The idea is that special assumptions beyond those of Hamiltonian mechanics are not required to show that equipartition will prevail. It is simply necessary to show that the moving phase point $(Q_k(t), P_k(t))$ will spend the vast majority of its time in regions of phase space within a neighborhood of states that are indistinguishable macroscopically from true equipartition.

Maxwell's idea was an extension of earlier ideas that he and Boltzmann had explored in their parallel attempts to provide a justification for using phase averages to represent thermodynamic quantities. These ideas came to be called *ergodic theory*.[14]

From a general perspective, the conceptual tension experienced by workers contemplating the application of probability theory and statistics to dynamical problems is similar to the questions surrounding any application of statistics to a situation. To illustrate the statistical aspects of the ergodic problem, let us express it in the notation of Part I of this book. Consider an isolated system, t, which is following the trajectory $(Q_t(t), P_t(t))$, with the initial condition (Q_t^0, P_t^0). Consider also a thermodynamic quantity G_t with phase analog $G_t(Q_t, P_t)$. The exact value of this thermodynamic quantity for the system t at time t is $G_t(Q_t(t), P_t(t))$. In terms of the TIS formalism, this is the $F_t^\delta(Q_t, P_t, t)$ phase average of G_t:

$$(22.30) \qquad G_t^\delta(t) = G_t(Q_t(t), P_t(t)) = \langle G_t(Q_t, P_t) \rangle_{t\delta t}.$$

Contrast this with the description of the same system by the probability density description $F_t(Q_t, P_t, t)$. The phase average of $G_t(Q_t, P_t)$ with this

---

Adler and Wainwright and an exact state initial condition calculation (TIS terminology) by Orban and Bellemans. Neither of these is based on the Boltzmann equation as the evolution equation for the "molecules". Nor is it clear how the value for H is computed because no extended distributions are defined.

[13] He referred here explicitly to the usual assumptions as stated by Watson [**1876**], for example.

[14] See the discussion of some of these issues in relation to ergodic theory in Truesdell [**1960**], pp. 21–30. For a useful discussion of ergodic theory, its history and purposes, criticism of the early ideas, and remarks on quasiergodic theory, see Tolman [**1938**], pp. 65–70. For a discussion of the work of von Neumann and Birkhoff on ergodic theory, see also Kurth [**1960**], pp. 66–76.

probability density is

(22.31) $G_t(t) = \langle G_t(Q_t, P_t) \rangle_{kt}$.

In general, it is the case that $G_t(t) \neq G_t^\delta(t)$. However, both of these quantities claim to express a single thermodynamic concept. Reconciling these competing viewpoints was the primary purpose of ergodic theory.

Let us consider (22.31) first. Boltzmann [**1871b**] suggested that the isolated system probability density, $F_t(Q_t, P_t, t)$, is proportional to the time the system spends in the volume element $dQ_t dP_t$ centered at (Q_t, P_t). This in essence states that the relative probability for finding the system in a state (Q_t, P_t), as expressed by the system distribution $F_t(Q_t, P_t, t)$, is actually a relative frequency. If this is true, the time average of the exact thermodynamic quantity $G_t^\delta(t)$ over a sufficiently long time interval will give the same result as taking the phase average of $G_t(Q_t, P_t, t)$ with the state $F_t(Q_t, P_t, t)$. The time average of the exact thermodynamic quantity $G_t^\delta(t)$ over the time interval α is defined, in parallel with the definition of the system distribution average defined in (5.59), by

(22.32) $\overline{G_t^\delta}(\alpha) = \frac{1}{\alpha} \int_0^\alpha dt\, G_t^\delta(t).$

As to the choice of a particle distribution to use for the phase average in (22.31), Maxwell [**1879b**] elaborated Boltzmann's idea further and suggested that the phase averaging be done with a stationary state. His suggestion is equivalent to phase averaging $G_t(Q_t, P_t)$ with the microcanonical state $F_t^\gamma(Q_t, P_t, t)$.[15]

The question of representing the stationary state of an isolated system by the F_t^γ state and the justification of replacing time averages with phase averages has been addressed in somewhat different ways by Boltzmann [**1896a**], pp. 310–312, Gibbs [**1902**], Chapter XII, P. and T. Ehrenfest [**1912**], pp. 21–26, and then pursued in the form of ergodic theory by Birkhoff [**1931**], Birkhoff and Koopman [**1932**], Hopf [**1932**], and von Neumann [**1932a**], [**1932b**], followed later by many others.

The time-independent F_t^γ phase average of the time-independent function $G_t(Q_t, P_t)$ is

(22.33) $G_t^\gamma = \langle G_t(Q_t, P_t) \rangle_{t\gamma}$.

The program of ergodic theory, then, is to show that the computations (22.32) and (22.33) give the same result. That is, the problem was taken to be that of demonstrating, for the F_t^δ phase average of each $G_t \in \mathcal{O}_t$, that the following relation

(22.34) $\lim_{\alpha \to \infty} \overline{G_t^\delta}(\alpha) = G_t^\gamma$

---

[15]See, however, the caveats in Maxwell [**1879b**] (Maxwell [**1890**], p. 714).

is true. Note that the time average in this calculation does not depend on the initial point (Q_t^0, P_t^0).[16]

Since $G_t(Q_t, P_t)$ is an arbitrary continuous function of (Q_t, P_t), the definition of time averaging displayed in (5.59) and the uniqueness properties of Radon measures are used[17] to rephrase the ergodic problem as being that of establishing

$$(22.35) \qquad \lim_{\alpha \to \infty} \overline{F_t^\delta}(Q_t, P_t; \alpha) = F_t^\gamma(Q_t, P_t) \qquad \text{a.e.}$$

for almost any F_t^δ state. The idea behind the attempt to justify phase averaging by ergodic methods is that as the point representing the system moves in phase space, its path is supposed to fill the bounded portion of phase space that is accessible to the system. When this is the case, the value of the time average of a phase analog function will approach the microcanonical average to any desired degree of accuracy. This is the point of equation (22.34).

Let us now turn to an evaluation of the ergodic program. The feeling of Maxwell and Boltzmann that ergodic theory can contribute to the justification and illumination of the relation between statistical mechanics and thermodynamics is wide of the mark. First, a counterexample to the assumption of what is now called "metric transitivity" or "metric indecomposability" was given in a simple example by Rayleigh [1892], p. 555. A theorem of Birkhoff [1931] established definitively the statement of ergodic theory as pertaining to the existence of the time average limit only on a metrically indecomposable invariant subset of phase space.[18] More recently, it has been shown that the claims of ergodic theory are not true for the case of classical C^∞ Hamiltonian mechanics. The work of Markus and Meyer [1974], pp. 49–50, for example, indicates that almost no C^∞ Hamiltonian system in classical mechanics is ergodic.[19] Therefore, the ergodic program generally fails in the C^∞ Hamiltonian setting.

The most important criticism of ergodic theory is that it is irrelevant to the justification of the use of the phase averaging procedure in obtaining thermodynamics from statistical mechanics.[20] First, observe that time averages of thermodynamic quantities are not usually important—except for the relatively brief duration of a measurement process.[21] Moreover, the duration of a measurement is far too short for the system to visit more than a very small neighborhood within a very large region of phase space typically open

---

[16]Formulations of the ergodic program can be found in many places. The statement presented here is in accord, for example, with that of Arnold and Avez [1968], pp. 15–17.

[17]See Treves [1967], p. 221.

[18]See Kurth [1960], pp. 74–75 for further details.

[19]See also Arnold and Avez [1968], p. 18.

[20]See Farquhar [1972], for example, for a demonstration that ergodicity does not imply irreversibility and vice versa.

[21]See also Jaynes [1967], pp. 92–95, on this.

to it. In addition, none of the quantities displayed in the axioms of Chapter 3 involve time averaging in their definition. Nor are time averages required for the definitions of equilibrium quantities in Chapter 18.

The time variation of a local thermodynamic quantity is often what we are interested in. This variation obviously disappears in the α time averaging of the ergodic approach and its does not appear at all in equilibrium theory. Furthermore, the original conception on which this approach rests, namely that the probability distribution density for a system at a point (Q_t, P_t) reflects the relative time spent by the system in the neighborhood of that point, is also not true in terms of how the system distribution is defined and employed. This point will be discussed further in Chapter 23.

The justification of the use of phase averaging must come, as has been emphasized before, out of an analysis of the use and justification of statistics in physical theory. The phase average of a quantity is a mean value, which makes it an unbiased estimator for the value the quantity will take in the actual system. While the choice of an estimator has physical significance, the justification of the use of statistical quantities to represent the system must be given within the framework of statistical theory and is not a matter of physics.

Interpretations of the Boltzmann State

Boltzmann's equation can be used as the foundation for a version of nonequilibrium thermodynamics. While this is straightforward using the formalism above, there remain some issues concerning the interpretation of the set of one-particle probability distribution densities that define the state of the system. There are also questions connected with the interpretation of the statistical aspects of Boltzmann's theory. These issues of interpretation will be sorted out in this chapter.

23.1 Boltzmann Thermodynamics

The Boltzmann theory leads to a version of thermodynamics just as the Gibbs based theory developed in Part I did. For an isolated system t, $\Phi_{i,x}(q_i, t) = 0$ for $i \in t$ and the Boltzmann theory is a local theory. With the exception of the entropy axiom, which is replaced in Boltzmann's theory by the H theorem, the axioms of the thermodynamics based on Boltzmann's equation are given by an appropriate modification of the local version of the axioms stated in Chapter 3. As an example, the pressure tensor in the Boltzmann thermodynamics of an isolated system is the momentum current density tensor: $P_t^{\mu\nu}(q, t) = \mathcal{P}_t^{\mu\nu}(q, t)$. The change in the form of the entropy in moving from the general theory to Boltzmann's theory was discussed in the last chapter. As stated in there, the entropy axiom for an isolated system is then written as $dS_t^{(B2)}(q, t)/dt \geq 0$, with equality if and only if the system is in an equilibrium state. For a system that is not isolated, the quasilocal axioms of Part I apply and the entropy axiom is given by equation (22.29).

23.2 Interpretations of the Boltzmann System State

Let us turn now to an analysis of the concept of state used by Maxwell, Boltzmann and others in connection with statistical mechanics and the Boltzmann equation. The initial research of Krönig and then Clausius on kinetic theory made use of statistical assumptions (uniform behavior or complete randomness) without introducing a general distribution for the system. Nor did they consider a dynamical relationship between the arrangement of particles at one time and at another.

Maxwell discussed on several occasions the need to introduce statistical methods into kinetic theory and statistical mechanics.[1] He [1860] was the first to introduce a distribution into his calculations as an explicit mathematical object and his [1866] treatment of transport equations introduced the dynamical element that was missing before.

Boltzmann made the concept of a particle distribution the foundation of his work on particle theory and thermodynamics. However, as discussed in the last chapter, Boltzmann [1868] used two meanings for the distribution function, namely, that the distribution function determines "the fraction of any suitably long time interval during which the velocity of any particular molecule had values within the prescribed limits" or "the fraction of the total number of molecules in the gas which had velocities within the prescribed limits at any given moment."[2] The difference in the views of Maxwell and Boltzmann was pointed out by Boltzmann [1882], p. 300: "inasmuch as Boltzmann measures the probability of a condition by the time during

---

[1]See Maxwell [1872], p. 309, [1873b] ([1890], pp. 373–374), [1875] ([1890], pp. 427–428).

[2]See Klein [1973], p. 62, on this.

which a system possesses this condition on the average, whereas Maxwell considers innumerable similarly constructed systems with all possible initial conditions."[3] Maxwell's ensemble view was clearly the one adopted by Gibbs [1902] later. Boltzmann went on to say that the "ratio of the number of systems which are in that condition to the total number of systems determines the probability in question." This is a restatement of two perspectives on the system distribution which appeared first in Boltzmann [1868]. These points of view concerning the system distribution are not equivalent, however. This is a point that Boltzmann [1882] either misses or does not comment on.

Yet another notion of the system state was given by Boltzmann [1897] (Brush [1965], Vol. 2, pp. 243–244). Here he said "I have always measured the probability of a state, independently of its temporal duration, by the 'extension' (as Zermelo calls it) of its corresponding region." Boltzmann is referring to the definition in Zermelo [1896a] (Brush [1965], Vol. 2, pp. 231–232), obtained from Poincaré, that the "probability of occurrence of a certain property of the molecular states, for example for a definite value of the function S, can be measured only by the 'extension' γ of the 'region' g of all possible states which have this property, divided of course by the total extension Γ of the region G containing all possible states." The regions referred to here are sets of phase points and the assumption that the probability of an event is proportional to its Lebesgue measure on the set of phase points.

In reflecting on the definition of the system distribution around 1896, Boltzmann [1896a], p. 61, stated that he interpreted $f(q, p, t)d^3qd^3p$ as the number of molecules in d^3qd^3p rather than the probability of finding a molecule in d^3qd^3p because the former conception is "more perspicuous" and lends itself to permutations. He also stated, p. 62, that the distribution law for molecular velocities is not precisely correct as long as the number of molecules is not mathematically infinite. However, as will be discussed after the various interpretations of the Boltzmann state have been introduced, the statistical interpretation of the state is not compatible with the probabilistic interpretation, so this interpretation is incompatible with the probabilistic basis of his [1872] theory of the Boltzmann equation.

23.3 Formal Expressions for the Boltzmann System State

There are a number of interpretations of the Boltzmann state which reflect differences in the viewpoint on what it represents. The issue of interpretation was introduced above in the discussion of the conflict between Boltzmann and Loschmidt. Other issues are connected with the mathematical techniques that Boltzmann and those after him employed to obtain solutions of the Boltzmann equation and to draw conclusions about these solutions. To support the analysis of the different interpretations, it is necessary to understand their connections with each other and with the underlying particles.

---

[3]See also Boltzmann [1882], pp. 305, 309, and Maxwell [1890], p. 721.

To facilitate calculations, Boltzmann used a combination of continuous and discrete techniques in his study of the Boltzmann evolution equation, the particle state that is its solution, and the changes in this state over time. One of his standard techniques was to use a discrete analysis of the particle state to obtain a formula that he then maximized to obtain results that are then deemed valid at equilibrium. Taking the limit of these results as the cell size diminishes, he drew conclusions concerning continuous distributions. He applied these conclusions in particular to continuous probability distribution densities that are solutions of the Boltzmann equation.

The differences and incompatibilities of these two approaches to defining system states and drawing conclusions about them will be presented briefly here. To do this, it is necessary to supply some formalism for translating a continuous distribution to a discrete distribution so that both can be talked about from a common standpoint. This formalism will be used to point out some of the problems in Boltzmann's conceptions. A more detailed analysis of the issues raised by moving between a continuous and a discrete representation of a system state is pursued in EIS for the classical case and QTS for the quantum case.

Consider a system k with a fixed volume set d_k. Partition the set $\omega_k(t)$ in the k phase space into a collection of small, open phase volume sets. The set π_k is the index set of the partition. The partition consists of small phase domains $\Delta_k^a = \Delta_k^a(q_a, p_a)$ labeled by $a \in \pi_k$, where (q_a, p_a) represents a point in $\Delta_k^a(q_a, p_a)$. As before, let $\overline{\Delta}_k^a(q_a, p_a)$ be the closure of $\Delta_k^a(q_a, p_a)$. This partition satisfies the condition

$$(23.1) \qquad \cup_{a \in \pi_k} \overline{\Delta}_k^a(q_a, p_a) = \bar{\omega}_k.$$

The point $(q_a, p_a) \in \Delta_k^a(q_a, p_a)$ is called the *address* of $\Delta_k^a(q_a, p_a)$ for each $a \in \pi_k$.

The set $\{\Delta_k^a(q_a, p_a)\}$ will be used as the set of cells for a "coarse-grained" distribution or for combinatorial analyses such as those Boltzmann used. The Lebesgue measure of $\Delta_k^a(q_a, p_a)$ is $\mu(\Delta_k^a(q_a, p_a))$, where the measure μ is defined in (2.10). In terms of this formalism, the different system states that have been proposed and their descriptions are listed in Table 23.1. In this table, the normalized versions of the states have been used to correspond to those that are standard in the literature.

A mathematical expression for these coarse-grained states will be given now in terms of the TIS formalism for the Boltzmann equation. Because the states $f_i^{(n)}$ are normalized, the partition function satisfies $Z_i(t) = Z_i = 1$. The states f_i^δ and $f_i^{(n)}$ have been defined above. In accord with the state

Table 23.1: Interpretations of the Boltzmann State

1. $f_i^\delta(q_i, p_i, t)$ exact i-particle state description at time t.

2. $f_i^{(n)}(q_i, p_i, t)$ probability density of finding particle i in state (q_i, p_i) at time t.

3. $f_i^{(n)e}(\Delta_k^a(q_a, p_a), t)$ exact density of particle i in cell $\Delta_k^a(q_a, p_a)$ at time t.

4. $f_i^{(n)c}(\Delta_k^a(q_a, p_a), t)$ coarse-grained (average) probability density of finding particle i in cell $\Delta_k^a(q_a, p_a)$ at time t.

5. $\bar{f}_i^{(n)}(\Delta_k^a(q_a, p_a), \alpha)$ relative time particle i spends in cell $\Delta_k^a(q_a, p_a)$ during the time interval $[0, \alpha)$.

6. $f_i^{\max}(q_i, p_i, t)$ most probable exact i-particle state (q_i, p_i) at time t (the *mode* of the distribution).

descriptions in Table 23.1, let us make the following additional definitions

(23.2)
$$f_i^{(n)e}(\Delta_k^a, t) = \frac{1}{\mu(\Delta_k^a(q_a, p_a))} \int d^3 q_i \int d^3 p_i \, f_i^\delta(q_i, p_i, t) \chi_{\Delta_k^a(q_a, p_a)}(q_i, p_i),$$

(23.3)
$$f_i^{(n)c}(\Delta_k^a, t) = \frac{1}{\mu(\Delta_k^a(q_a, p_a))} \int d^3 q_i \int d^3 p_i \, f_i^{(n)}(q_i, p_i, t) \chi_{\Delta_k^a(q_a, p_a)}(q_i, p_i),$$

(23.4)
$$\bar{f}_i^{(n)}(\Delta_k^a(q_a, p_a), \alpha) = \frac{1}{\alpha} \int_0^\alpha dt \, f_i^{(n)e}(\Delta_k^a(q_a, p_a), t).$$

The exact count of particle i in cell $\Delta_k^a(q_a, p_a)$ at time t is

(23.5) $N_i^{e,a} = N_i^e(\Delta_k^a(q_a, p_a), t) = \mu(\Delta_k^a(q_a, p_a)) f_i^{(n)e}(\Delta_k^a(q_a, p_a), t).$

Similarly, the coarse count of particle i in cell $\Delta_k^a(q_a, p_a)$ at time t is

(23.6) $N_i^{c,a} = N_i^c(\Delta_k^a(q_a, p_a), t) = \mu(\Delta_k^a(q_a, p_a)) f_i^{(n)c}(\Delta_k^a(q_a, p_a), t).$

From these definitions, it follows that

(23.7) $N_i^e(\omega_k) = \sum_{a \in \pi_k} N_i^e(\Delta_k^a(q_a, p_a), t) = 1$

because

$$(23.8) \qquad N_i^e(\Delta_k^a(q_a, p_a), t) = \begin{cases} 1, & \text{if } (q_i^{(s)}(t), p_i^{(s)}(t)) \in \Delta_k^a(q_a, p_a) \\ 0, & \text{otherwise,} \end{cases}$$

where $(q_i^{(s)}(t), p_i^{(s)}(t))$ is the i-particle system specific trajectory. It also follows that the relation
$$(23.9)$$
$$N_i^c(\omega_k) = \sum_{a \in \pi_k} N_i^c(\Delta_k^a(q_a, p_a), t) = \sum_{a \in \pi_k} \mu(\Delta_k^a(q_a, p_a)) f_i^{(n)c}(\Delta_k^a(q_a, p_a), t) = 1$$

is valid. This follows from the facts that $f_i^{(n)}(q_i, p_i, t)$ is normalized in the phase volume set ω_k and

$$(23.10) \qquad \sum_{a \in \pi_k} \chi_{\Delta_k^a(q_a, p_a)}(q_i, p_i) = \chi_{\omega_k}(q_i, p_i).$$

The coarse count of particles in k is

$$(23.11) \qquad N_k^c(\omega_k) = \sum_{i \in k} N_i^c(\omega_k) = \mathcal{N}_k.$$

It also follows immediately from the above definitions that

$$(23.12) \qquad \sum_{a \in \pi_k} \mu(\Delta_k^a(q_a, p_a)) \bar{f}_i^{(n)}(\Delta_k^a(q_a, p_a), \alpha) = 1.$$

Boltzmann used the number of particles in a cell as the basis for his combinatorial calculations. The exact number of particles located in the cell $\Delta_k^a(q_a, p_a)$ at time t is

$$(23.13)$$
$$\begin{aligned} N_k^{e,a} &= N_k^e(\Delta_k^a(q_a, p_a), t) = \mu(\Delta_k^a(q_a, p_a)) f_k^{(n)e}(\Delta_k^a(q_a, p_a), t) \\ &= \sum_{i \in k} N_i^e(\Delta_k^a(q_a, p_a), t), \end{aligned}$$

where

$$(23.14) \qquad f_k^{(n)e}(\Delta_k^a(q_a, p_a), t) = \sum_{i \in k} f_i^{(n)e}(\Delta_k^a(q_a, p_a), t).$$

The total number of particles in all the cells is given by the sum of the quantity $N_k^e(\Delta_k^a(q_a, p_a))$ over $a \in \pi_k$ as

$$(23.15) \qquad N_k^c(\omega_k(t)) = \sum_{a \in \pi_k} N_i^e(\Delta_k^a(q_a, p_a), t) = \mathcal{N}_k.$$

The average of a particle analog quantity $G_i(q_i, p_i, t)$ can be computed using the $N_i^e(\Delta_k^a(q_a, p_a))$ distribution as

(23.16a) $$G_i^e(t) = \sum_{a \in \pi_k} G_i(q_a, p_a, t) N_i^e(\Delta_k^a(q_a, p_a), t),$$

(23.16b) $$G_k^e(t) = \sum_{i \in k} G_i^e(t).$$

Boltzmann [1868], p. 50, also used the time-averaged state $\bar{f}_i^{(n)}$ interchangeably with $f_i^{(n)e}$ in some cases. As discussed above, $\bar{f}_i^{(n)}$ is usually associated with ergodic assumptions. However, it cannot serve as the time dependent state describing an evolving system as required by Boltzmann's equation. It will not be pursued further here.

Boltzmann's [1877a] combinatorial approach was based on the quantity called N_k^e here. He calculated the combinatorial probability that a given mechanical state (Q_k, P_k) will occur by dividing the k phase space into cells of equal measure and assuming that the probability a given particle will occupy a given cell is equal to the probability that it will occupy any other cell—subject to the constraints of a fixed particle number and a fixed total energy. This state will be called the *uniform discrete state* here (TIS terminology). Boltzmann showed that this is the most probable discrete state, which means that it is the *mode* of the coarse-grained distribution. Boltzmann claimed that the uniform discrete state is the equivalent of the Maxwell distribution in the limit of vanishing cell size.

Summarizing the work he did in Boltzmann [1877b], Boltzmann [1896a], pp. 55–58, presented the idea that $H(t)$, which is the entropy at a given time for a particle in a state $f^{(n)}(q, p, t)$, represents the probability that particle i can be correctly described by distribution. Speaking loosely, and using TIS notation, this is the idea that $S_i^{(B2)}(t)$ represents the probability of finding particle i in the state $f_i(q_i, p_i, t)$. Using this interpretation, Boltzmann maintained that the H-theorem shows that systems move from less probable states to more probable states. Because the most probable discrete state is the analog of equilibrium, this is equivalent to the statement that the particle state will move to equilibrium.

There are a number of serious problems associated with the notion of a state that Boltzmann was using and his view of the 'probability of a distribution' in reference to a collection of particles. Leaving this aside for a future discussion, let us focus first on the mathematical problems associated with moving between continuous and discrete distributions.

This first problem with this point of view is that Boltzmann's claim that the uniform discrete state is the Maxwell-Boltzmann equilibrium state in the limit of vanishing cell size is incorrect. In showing this, Boltzmann's ideas will be described in TIS notation for conceptual consistency. Boltzmann [1872], pp. 94, 95, 133, and [**1896a**], p. 61, required that the cell $\Delta_k^a(q_a, p_a)$, used in the definition of N_k^e in (23.13), is small by macroscopic standards but large enough to contain many molecules. To use the calculus, he needs the limit of N_k^e as $\mu(\Delta_k^a(q_a, p_a)) \to 0$. However, as Poincaré [**1906**], pp. 371–373, pointed out, this limit is a singular state. In the notation used here, the following replacement is made for $N_i^{e,a}$, the i-component of $N_k^{e,a}$, in the limit of vanishing domain size

$$(23.17a) \quad N_i^{e,a} = \mu(\Delta_k^a(q_a, p_a))f_i^{(n)e}(\Delta_k^a(q_a, p_a), t) \to d^3 q_i d^3 p_i\, f_i^{\delta}(q_i, p_i, t).$$

This step is used to convert the sum over $a \in \pi_k$ into a Riemannian integral as Boltzmann did. On the other hand, because the Boltzmann equation does not support a singular solution of the form $f_i^{\delta}(q_i, p_i, t)$, Boltzmann must use the state $f_i^{(n)c}$ instead of $N_i^{e,a}$. The limit of $f_i^{(n)c}$ as $\mu(\Delta_k^a(q_a, p_a)) \to 0$ is

$$(23.17b) \quad N_i^{c,a} = \mu(\Delta_k^a(q_a, p_a))f_i^{(n)c}(\Delta_k^a(q_a, p_a), t) \to d^3 q_i d^3 p_i\, f_i^{(n)}(q_i, p_i, t).$$

Because the H theorem is based on the state $f^{(n)}(q, p, t)$ and not on the state $f^{\delta}(q, p, t)$, it follows from that Boltzmann's dynamical claim based on the H theorem that systems evolve from less probable states to more probable states cannot be supported using the distribution $N_i^{e,a}$. On the other hand, the coarse-grained distribution $N_i^{c,a}$ is approximate and not an exact reflection of the particle i mechanical state (q_i, p_i) or its distribution at any time. As time goes by, it is possible for the true mechanical state of particle i to deviate arbitrarily far from the coarse-grained mechanical state assigned to it. The same is true for a distribution of states representing the probability of finding particle i in a given mechanical state. It follows that nothing of dynamical significance has been shown by this form of coarse graining and the associated combinatorial calculation.

Proceeding to the conceptual issues, Kuhn [**1978**], p. 59, has observed that Boltzmann introduced a significant shift in his use of the "state" concept by employing the combinatorial approach with its attendant assumptions. This is a different shift than the one between $f(q, p, t)$ and (q, p) discussed in connection with Loschmidt's criticisms in the last chapter. In the current case, Boltzmann vacillated between his original (molecular) conception and a combinatorial (molar) conception as he tried to defend his H theorem against the challenge of reversibility and the later challenge of recurrence.

The main criticism of Boltzmann's ideas is that the dynamical theory that Boltzmann has proposed for systems evolving from less probable to more probable distributions is based on his combinatorial work, which is not compatible with Hamiltonian dynamics. In TIS, this is expressed in the statement that the procedure of assigning particles to cells does not commute with the time derivative operator. From the standpoint of the state concept 6 in Table 23.1, for example, it is not hard to show that the mode of a distribution does not evolve into the mode of a new distribution except in the static equilibrium case. For these and other reasons connected with the *ad hoc* character of the maximum probability approach, the state concept 6 in Table 23.1 will not be considered further.

On the basis of these considerations, it may be concluded that the mechanical ideas on which Boltzmann [1872] established his evolution equation and his use of the calculus as he understood it implies that $f_i(q_i, p_i, t)$ is essentially the only state conception on which Boltzmann's dynamical theory can be erected in anything like the form in which he presented it.

23.4 Boltzmann's Assumptions

After Maxwell introduced statistical distributions into physics, Boltzmann proceeded to exploit them, along with associated ideas of statistics and probability, in his search for the proper relation of statistical mechanics to thermodynamics and to help obtain solutions to evolution equations. These ideas have not always been used within the framework needed to justify their employment, however. Some aspects of this issue that are germane to the discussion of Boltzmann's assumptions will be introduced now.

Boltzmann made use of both the *molecular chaos* assumption and the *Stosszahlansatz* concerning the one-particle probability distribution density in setting up his theory. The molecular chaos assumption, as Boltzmann [1896a], pp. 41-42, used it, is a statement concerning the randomness of the locations and velocities of the particles in a gas. Boltzmann felt that this assumption was necessary "to prove the theorems of gas theory." Kaç [1959], pp. 112–113, correctly equated the molecular chaos assumption with the assumption that the joint N_k particle distribution can always be factored.[4] The Stosszahlansatz is based on this assumption in that terms of the form $f_i(q, p_i, t)f_j(q, p_j, t)$ rather than $f_{ij}(q, p_i, p_j, t)$ are used in the definition (21.8) of the collision term.

In reference to the Stosszahlansatz, statements have been made to the effect that "collisions create correlations"[5] or that "collisions destroy correlations" and at least one author claims that both are true. It is also sometimes claimed that the Maxwell-Boltzmann approximation applies only to dilute gases because the particles are then far enough apart for them to be uncorrelated.

Leaving aside the fact that these uses of the term correlation are not related to the statistical concept with the same name, what is usually meant by a 'correlation' between two particles is being able to predict the path of

---

[4]For more information on molecular chaos or molecular disorder, see Brush [1976], pp. 616–626.

[5]Burbury [1894] claimed that the Stosszahlansatz cannot be valid for both the "forward" and the "reversed" velocity cases on the grounds that the positions and momenta of two molecules after a collision are not independent but "correlated". See also P. and T. Ehrenfest [1912], footnote 65, pp. 85–86, and Brush [1976], pp. 620–624 on this.

particle j, $q_j(t)$, by observing the path of particle i, $q_i(t)$. However, when the conditions of the implicit function theorem are met, the coordinate trajectory $q_i(t)$ can be inverted and written as $t = t(q_i)$. In this case, the locations of all other particles are 'correlated' with the location of particle i through their mutual time dependence.

The issue of correlations in the context of the Stosszahlansatz is really the question of whether the system distribution function can be factored or not. However, this does not depend on whether the particles have collided or not or how far apart they are. In the general theory, the exact state, F_k^δ, and the equilibrium state, F_k^ϵ, are always factorable. Moreover, an initial state $F_k(0)$ can always be chosen as a factorable state. The factorability of subsequent states, $F_k(t)$, for $t > 0$, depends on the evolution equation chosen to describe the system. For the evolution equation (5.16), the state will not in general remain factorable. For the Boltzmann equation, on the other hand, an initially factorable state will remain so.

Factorization was used as part of the TIS approximation procedure leading to Boltzmann's equation in Chapter 20. However, the TIS approach is based on the Independence Approximation, as an epistemological requirement on the representation of a collection of systems carved out of a larger system, rather than the molecular chaos assumption which is an assumption about the physics of the particles.

According to the results of Part I, neither collisions nor the mixing of particles affect the entropy of an isolated system described by Hamiltonian mechanics. That is consistent with the fact that, if a system is isolated and no stochastic approximations are used in the Hamiltonian mechanics, the change in the system entropy is zero regardless of whether the particles are colliding or not. The Boltzmann stochastic collision approximation, on the other hand, results in the gradual erasing of the initial state of the system and leads to a unique equilibrium state. It is therefore true that the state of a system with Boltzmann's equation as its evolution equation, whether an isolated system or a subsystem of an isolated system with fixed walls at a fixed uniform temperature, will eventually become an equilibrium state.

Boltzmann's theory is a statistical theory but not in the sense in which he and Maxwell understood this to mean. It is statistical in that it employs probability distribution densities and uses statistical methods. It is not statistical in the sense that the formulas for the change in entropy are occasionally violated in the way that Maxwell and Boltzmann envisioned. This is a subtle, but important point, which is connected both with the conception of entropy as a mechanical aspect of a material system and with the mathematical representation of it. Apart from the views of Maxwell and Boltzmann, it has been have argued in Part I above that statistical mechanics is irreducibly statistical with respect to formulas involving the entropy because the entropy is not defined for a system described by the exact state.

The stochastic element in the Boltzmann collision term is in the calculation of the collision probabilities. Each point, labeled by (b, ϕ) on the $(\hat{c}_1, \hat{c}_2)$ plane, is assumed to be an equally probable source for a collision It was mentioned above that this is clearly not generally in accord with the exact mechanical description of the system. The integration over the variables (p_j, b, ϕ) is also an averaging procedure. These facts confirm that Boltzmann's theory is not exact in the sense of the underlying mechanics. Thus there is no real conflict between Boltzmann's irreversible theory and the mechanical theory of an isolated system with reversibility and recurrence because the assumptions underlying the theories are different.

Regarding the statistical versus probability aspects of particle distributions, the view of system states in terms of probability has been chosen in TIS as the correct one. A more complete treatment of these issues and their implications will be presented in EIS. The discussion of the connection between entropy and probability will be resumed in the context of quantum mechanics in QTS.

23.5 The Value of Boltzmann's Equation

As has been pointed out above, irreversible equations describing viscosity, particle diffusion, and heat flow, cannot be derived from a theory based on mechanics that uses exact distributions and reversible mechanical equations. This is due to the fact that there are some F_k^δ states for which these irreversible equations do not hold. As a theory based on the "average" behavior of the system under collisions, Boltzmann's theory offers the possibility of capturing an important aspect of the usual behavior of systems without being true for any one F_k^δ state. Practical calculations with Boltzmann's theory, or any other theory connecting thermodynamics with statistical mechanics, requires special assumptions on the particle state and special approximations so that the calculation may proceed. The calculation of the viscosity, for example, in the theories of Enskog and Chapman depends on system states "near" equilibrium and explicitly makes use of this fact. Boltzmann's equation does have the advantage of being mathematically tractable and physically perspicuous, which makes it a good candidate for helping us understand the general behavior of systems.

As emphasized by Truesdell [1969], the value of Boltzmann's equation lies in the way it helps us to understand and model the system it is describing—not in faithfully reflecting all aspects of reality. This point of view is in keeping with Boltzmann's favorite motto from Goethe's Faust, part 2,

Alles Vergängliche
Ist Nur ein Gleichnis

Part IV

Mathematical Foundations

La théorie cinétique des gaz laisse encore subsister bien des points embarrassants pour ceux qui sont accoutumés à la riguer mathématique; bien des résultats, insuffisamment précisés, se présentent sous une forme paradoxale et semblent engendrer des contradictions quis ne sont d'ailleurs qu'apparents.

Henri Poincaré, *Sur la Théorie Cinétique des Gaz*, 1906

CHAPTER 24

Basic Assumptions

The mathematical analysis of the major assumptions underlying the thermodynamics of the Theory of Interacting Systems is the subject of Part IV. The aim is to provide a mathematical foundation for the theory that can be made rigorous.

24.1 Mathematical Orientation

The classical version of the Theory of Interacting Systems is completed by an examination of its mathematical foundations. The four major concerns are showing that:

1. the equations of the theory are mathematically well defined;
2. the thermodynamic functions that belong to the set of observables are at least once continuously differentiable in their arguments;
3. solutions exist for the TIS evolution equation (5.16) and for the TIS version of Boltzmann's equation (20.17); and
4. under the proper boundary conditions, the equilibrium state is uniformly asymptotic in the large.

The main tools used to accomplish these goals are based on the compactness of the closures of $\Omega_k(t)$ and $\omega_k(t)$, the choice of appropriate function spaces, various embedding and fixed point theorems, Lyapunov methods, and Jensen's inequality.

The approach to mechanics from the standpoint of Hilbert space was pioneered by Koopman [**1931**]. His work was followed and extended into the domain of ergodic theory by Birkhoff [**1931**], Birkhoff and Koopman [**1932**], and von Neumann [**1932a**], [**1932b**]. In Chapter 5, the system state was defined in relation to the requirements of thermodynamic observables by means of a topological space duality. This duality will be made concrete here by the choice of an appropriate function space. It will be shown that the mathematical structures of TIS are all well defined for system distributions that belong to an appropriate Sobolev Hilbert space.[1] Other aspects of the theory, such as those concerned with the F_k^δ states, are shown to be well defined in the wider space of tempered distributions.

The microscopic and macroscopic components of the theory are connected expectation value or phase averaging calculations. The formalism for calculating expectation values is based on the relative probability space $(\Omega_k(t), \Sigma_k(t), \boldsymbol{P}_k(t))$ introduced in Chapter 5. The space of events at time t is $\Omega_k(t)$, which is the set of allowed system states. The object $\Sigma_k(t)$ is the σ field of subsets of $\Omega_k(t)$, which includes the empty set $\emptyset$ and $\Omega_k(t)$. The measure $\boldsymbol{P}_k(t)$ is a non-negative, time-dependent, bounded measure on this field. This means that $\boldsymbol{P}_k(t)$ is the time-dependent map $\boldsymbol{P}_k(t) : \Sigma_k(t) \to \mathbf{R}$ that maps a phase set $A \in \Sigma_k(t)$ into the real numbers. The values taken on by this map fall in the range $0 \leq \boldsymbol{P}_k(A, t) \leq \boldsymbol{P}_k(\Omega_k(t), t) < \infty$. The system distribution density $F_k(Q_k, P_k, t)$, which is the Radon-Nikodym derivative of

---

[1]Sobolev Hilbert spaces, their boundary spaces and embedding theorems, and Lyapunov functionals have been independently employed by Chen [**1989**] to show that the equilibrium state is asymptotically stable for his partially linearized version of Boltzmann's equation. While his results are obtained in a manner similar in some respects to the work presented here, they apply to an evolution equation that has a foundation very different from that underlying the TIS version of Boltzmann's equation.

$\mathcal{P}_k(t)$ with respect to the Lebesgue measure on the system phase volume set $\Omega_k(t)$, is a major focus of the TIS formalism.

The partial time derivative of the expectation functions of physical quantities introduces singular quantities at the boundary due to the presence of the system projection operator in these phase averaging calculations. These singular quantities are not defined within the volume set $d_k(t)$ belonging to system k and are not consequences of the evolution equation for the underlying particle system. They are consequences of introducing boundaries and the associated TIS macroscopic boundary formalism. The analysis of the mathematical aspects of the theory at the boundary is therefore an essential concern in Part IV.

For the TIS theory to be well defined mathematically, it is necessary that the integrals defining thermodynamic quantities be well defined and that solutions of the evolution equation exist. The question of the existence of solutions of the system evolution equation (5.16), which is often called the "Liouville equation" in standard statistical mechanics, has often been discussed.[2] For the Boltzmann equation, there have also been a number of proofs of the existence of solutions based on particular approximations and various assumptions on the initial conditions and boundary conditions.[3] None of the existence proofs for solutions of the Liouville equation or Boltzmann's equation meets the needs of the work here, however.

The rest of this chapter is concerned with showing that the properties of the elements of the function space selected are sufficient to insure that the thermodynamics developed in Parts I–III of this book is well defined. By this is meant, for $k = t$, $k = [t]$ or $k \in [t]$, that (1) the primitive local thermodynamic densities are continuous and at least once differentiable in t and q for $t \in [0, \alpha)$ and $q \in d_k(t)$; and (2) the global quantities are continuous and at least once differentiable in t. In this chapter, the solutions F_k and F_t of the evolution equation (5.16) are assumed to exist. The existence of these solutions will be taken up in the next chapter.

24.2 Bounds on Phase Functions

Some of the assumptions stated in Part I on the particle-particle interaction potential, $\phi_{ij}(q_i - q_j)$, will be reviewed first. The distance of closest approach of two particles, r_c, was defined in equation (4.28). The *distance of*

---

[2]For a treatment of many of the issues discussed in this part of the book in the context of probability densities with continuous derivatives, see Kurth [**1960**], Chapter V.

[3]For further information on these, see the review of, and extensive references to, the existence proofs for the Boltzmann equation given in Truesdell and Muncaster [**1980**]. The assumptions required to demonstrate the existence of some of the integrals considered in this book have been discussed by Truesdell and Muncaster [**1980**] as well.

greatest separation is now defined as

(24.1) $$r_g = \sup_{t \in [0,\alpha)} \sup_{i,j \in k} \{\, |q_i - q_j| \mid q_i, q_j \in d_t(t) \,\}.$$

This is used next to define the set $\mathbf{R}_{cg}$ by

(24.2) $$\mathbf{R}_{cg} = \{\, r \mid r_c \le r \le r_g \,\}.$$

This means that the value of $|q_i - q_j|$ always falls within the bounded set $\mathbf{R}_{cg}$.

To avoid repeating assumptions, let us assume for the remainder of the book that $|q_i - q_j| \in \mathbf{R}_{cg}$ for $i, j \in t$. At various points the following set of bounds, which are collected here for convenience, were computed:

(24.3a) $$|p_i| \le p_k^u,$$

(24.3b) $$0 \le H_i(p_i) \le \mathcal{H}_k(P_k) \le \mathcal{K}_k^u,$$

(24.3c) $$\mathcal{H}_k^l \le \mathcal{H}_k(Q_k, P_k, t) \le \mathcal{H}_k^u,$$

(24.3d) $$-[\mathcal{N}_t - \mathcal{N}_k]\phi_0 \le \Phi_i(q, t) \le [\mathcal{N}_t - \mathcal{N}_k] \, \|\phi\| \,,$$

(24.3e) $$\mathcal{H}_B^l \le H_{Bi}(q_i, p_i, t) \le \mathcal{H}_B^u.$$

For those system phase vectors of the general theory for which $(Q_k, P_k) \in \Omega_k(t)$, the various forms of the system Hamiltonian energy, as well as the formulas for the momentum and angular momentum, are bounded for any system k, including $k = t$ and $k = [t]$. If the i-particle mechanical state in the Boltzmann case satisfies $(q_i, p_i) \in \omega_i(t)$ for $i \in k$, the components of the theory are bounded for any particle in the system. Some other important bounds that follow from the previous assumptions are:

(24.4a) $$|q_i| \le q_k^u = \sup_{i \in k} \sup_{t \in [0,\alpha)} \sup_{q_i \in d_k(t)} |q_i| \,,$$

(24.4b) $$|v_i| \le v_k^u = \sup_{i \in k} \frac{p_k^u}{m_i}.$$

Furthermore, it is assumed that the volume and boundaries have been chosen so that

(24.5) $$v_{\partial k}^{(\alpha)} = \sup_{t \in [0,\alpha)} \sup_{q \in \partial d_k(t)} |v_{\partial k}(q, t)| < \infty.$$

Because the boundary velocity is chosen to fit particular macroscopic circumstances rather than computed from the microscopic particle mechanics, it is assumed to be bounded for all time so that $v_{\partial k}^{(\alpha)} < v_{\partial k}^{\max} < \infty$.

In Chapter 4, the general constitutive properties required of all the interaction potentials of the theory were examined. Some of the properties of the particle interaction potentials are fundamental to the structure of the general theory in the sense that if $\phi_{ij} = 0$ for each $i, j \in k$, the resulting theory is a local ideal gas theory; if $\phi_{ij}(r) = 0$ for $r > r_0$ and $\phi_{ij}(r) = \infty$ for $0 \le r \le r_0$, it is a local hard-sphere theory; if $\phi_{ij}(r)$ has a microscopic range, that is, if

$\phi_{ij}(r) = 0$ for $r > r_{\text{max}}$, it is a quasilocal theory; if $\phi_{ij}(r)$ has an arbitrarily long range, it is a global theory. In the development of the formalism, it was assumed that the particle interaction potentials are independently additive. This assumption is not strictly correct and can be replaced at the expense of invalidating a number of decompositions of the system potential into its components. At the system level, the system potential $\Phi_k(Q_k, t)$ and the rest of the theory are not invalidated by the failure of the additivity of particle potentials, however.

For potentials bounded in the semi-norm (4.2h), the continuity of the system potential $\Phi_k(Q_k, t)$ and its derivatives can be used with the mean value theorem in Banach spaces (Dieudonné [1960], p. 155) to show immediately that $\Phi_k(Q_k, t)$ satisfies a Lipschitz condition:

$$(24.6) \qquad \left| \Phi_k(Q_k^\natural, t) - \Phi_k(Q_k, t) \right| \leq \|\phi\| \left\| Q_k^\natural - Q_k \right\|.$$

24.3 Continuity and Differentiability of Phase Analogs

It was assumed in (4.2) that the members of the set $\{\phi_{ij}\}$, $i, j \in t$ of interparticle interaction potentials are continuously differentiable on $\Omega_t(t)$ to order m, where $m \geq 3$. This fact can be used to show that $\Phi_i(Q_k, t) \in C^m(\Omega_k(t))$ for $i \in k$. It is also easy to show that $\partial^n \Phi_i(q_i, t)/\partial q_i^n$ is continuous for $n \leq m$. Excluding the entropy and temperature from consideration for now, these facts imply that all the other primitive thermodynamic observable analogs are continuous and differentiable at least once.

Assume that $G_i(Q_k, p_i, t) \in C^m(\Omega_k(t) \times \theta_\alpha)$ is valid and that $G_i(Q_k, p_i, t)$ is bounded for $\theta_\alpha = \{t \mid t \in [0, \alpha)\}$ and $i \in k$. By (5.16) and (5.38), this implies

$$(24.7a) \qquad \frac{\partial G_i(q, t)}{\partial t} = \left\langle \delta(q - q_i) \frac{\partial G_i(Q_k, p_i, t)}{\partial t} \right\rangle_{kt}$$
$$+ \frac{1}{Z_k(t)} \int dQ_k \int dP_k \, \delta(q - q_i) G_i \left[-\chi_{\Omega_k(t)} V_{tk} \cdot \nabla_{2k} F_k + \frac{\partial \chi_{\Omega_k(t)}}{\partial t} F_k \right].$$

The time derivative of the G_k current density $G_k^\mu(q, t)$ is similar with an additional factor v_i^μ before $\partial G_i/\partial t$ the phase average term and before G_i in the explicit integrand. Next, for the G_k current density, it follows from a calculation of $\partial G_i(q, t)/\partial q^\mu$, which parallels the calculation of $\partial G_i(q, t)/\partial t$ in (5.38) and uses (5.40) followed by an integration by parts, that

$$(24.7b) \qquad \frac{\partial G_i^\mu(q, t)}{\partial q^\mu} = \left\langle \delta(q - q_i) v_i^\mu \frac{\partial G_i(Q_k, p_i, t)}{\partial q_i^\mu} \right\rangle_{kt}$$
$$+ \frac{1}{Z_k(t)} \int dQ_k \int dP_k \, \delta(q - q_i) v_i^\mu G_i(Q_k, p_i, t) \left[\frac{\partial F_k}{\partial q_i^\mu} \chi_{\Omega_k(t)} + F_k \frac{\partial \chi_{\Omega_k(t)}}{\partial q_i^\mu} \right].$$

There is a similar definition of $\partial G_i(q,t)/\partial q^\mu$ that is obtained by setting $v_i^\mu = 1$ in the right hand side of (24.7b). With the exception of the singular local surface flux density term containing $\partial \chi_{\Omega_k(t)}/\partial q^\mu$, the local derivatives of the thermodynamic density $G_i(q,t)$ and its current $G_i^\mu(q,t)$ with respect to t and q are bounded for $t \in [0, \alpha)$ and $q \in d_k(t)$ if the conditions

$$(24.8a) \qquad \int dQ_k \int dP_k \chi_{\Omega_k(t)}(Q_k, P_k) \left| \delta(q - q_i) V_{2k} \cdot \nabla_{2k} F_k \right| < \infty,$$

and

$$(24.8b)$$
$$\int dQ_k \int dP_k \, \chi_{\Omega_k(t)}(Q_k, P_k) \left| \delta(q - q_i) \dot{\chi}_{\Omega_k(t)}(Q_k, P_k) F_k(t) \right| < \infty,$$

are both satisfied. If this is true, then $dG_i(q,t)/dt$ is bounded as well. For the global case, the time derivative of $G_k(t)$ is bounded if it is true that

$$(24.9a) \qquad \int dQ_k \int dP_k \, \chi_{\Omega_k(t)}(Q_k, P_k) |V_{2k} \cdot \nabla_{2k} F_k(Q_k, P_k, t)| < \infty$$

and

$$(24.9b) \qquad \int dQ_k \int dP_k \, \chi_{\Omega_k(t)}(Q_k, P_k) |\dot{\chi}_{\Omega_k(t)}(Q_k, P_k) F_k(t)| < \infty.$$

Let us next look at some specific bounds that will be needed. By the definition (4.2h) of the particle interaction potential semi-norm and the definitions of the components of the Hamiltonian in Section 4.4, along with the results of this chapter, it follows that

$$(24.10) \qquad \left\{ \Phi_{i,n}(Q_k), \frac{\partial \Phi_{i,n}(Q_k)}{\partial q_i}, \frac{\partial^2 \Phi_{i,n}(Q_k)}{\partial q_i^2} \right\} \leq [\mathcal{N}_k - 1] \, \|\phi\| .$$

Using this with the i-particle versions of (7.71) and (7.72) and the definition (4.2h) of $\|\phi\|$, it is easy to show

$$(24.11) \qquad \left\{ \left| \frac{\partial \Phi_i(q_i, t)}{\partial t} \right|, \left| \frac{\partial \Phi_i(q_i, t)}{\partial q_i} \right| \right\} \leq \{v_k^u, 1\} \, \|\phi\|$$

and

$$(24.12) \quad \left\{ \left| \frac{\partial^2 \Phi_i(q_i, t)}{\partial t^2} \right|, \left| \frac{\partial^2 \Phi_i(q_i, t)}{\partial t \partial q_i} \right|, \left| \frac{\partial^2 \Phi_i(q_i, t)}{\partial q_i^2} \right| \right\} \leq \{(v_k^u)^2, v_k^u, 1\} \, \|\phi\| .$$

By (24.1) and (24.11), it also follows that

$$(24.13) \qquad \left| (q_i^\mu - q_j^\mu) \frac{\phi_{ij}(q_i - q_j)}{\partial q_{i,\nu}} \right| \leq r_g \, \|\phi\| .$$

As a physical quantity, $G_i(Q_k, p_i, t) \in C^m(\Omega_k(t))$ is bounded for each $t \in [0, \alpha)$ on the set $\Omega_k(t)$ by a quantity $G_i^b(t) < \infty$. This bound is expressed in the form

(24.15)
$$|G_i(q, t)| \leq G_i^b(t)\nu_i(q, t).$$

For the associated current density of G_i (q, t), a similar calculation shows

(24.16)
$$|G_i^\mu(q, t)| \leq v_k^u G_i^b(t)\nu_i(q, t).$$

24.4 Function Space Considerations

In Chapter 5 a duality was set up between the set of observable analogs and the allowable system distributions. This duality will be used now as a guide to defining a function space for the system distribution that will insure that the members of the set of observables, including the entropy and temperature, are well defined. This duality will also be used to show that all the integrals presented in Part I are well defined.

Let us first add some formalism that will be useful in what follows. Consider the L^2 spaces $L^2(\Omega_k(t))$ and $L^2(\omega_i(t))$ defined on the sets $\Omega_k(t)$ and $\omega_i(t)$, respectively. By $G \in L^2$ is meant any member of an equivalence class of L^2 functions that differ from each other only on a set of Lebesgue measure zero. For $A, B \in L^2(\Omega_k(t))$ and $a, b \in L^2(\omega_i(t))$, the k system inner products are defined by

(24.17a)
$$(A, B)_{L^2} = \int dQ_k \int dP_k \, \chi_{\Omega_k(t)}(Q_k, P_k)A(Q_k, P_k)B(Q_k, P_k),$$

(24.17b)
$$(a, b)_{L^2} = \int d^3q_i \int d^3p_i \, \chi_{\omega_i(t)}(q_i, p_i)a(q_i, p_i)b(q_i, p_i).$$

The local L^1 and L^2 norms defined next in terms of this inner product are limited to $\Omega_k(t)$ or $\omega_i(t)$ and are parallel to the local norm defined in Chapter 5. The local L^2 norms associated with the inner products (24.17) are

(24.18a)
$$\|A\|_{L^2} = (A, A)_{L^2}^{\frac{1}{2}},$$

(24.18b)
$$\|a\|_{L^2} = (a, a)_{L^2}^{\frac{1}{2}}.$$

In particular, the local norms of two special functions are important. These are the local L^2 norms of the constant unit functions $1_k \in L^2(\Omega_k(t))$ and $1_i \in L^2(\omega_i(t))$:

(24.19a)
$$\|1_k\|_{L^2} = [\mu(\Omega_k(t))]^{\frac{1}{2}},$$

(24.19b)
$$\|1_i\|_{L^2} = [\mu(\omega_i(t))]^{\frac{1}{2}},$$

where the Lebesgue measure μ is defined in (2.10). Under the previous assumptions on the Lebesgue measures of $\Omega_k(t)$ and $\omega_i(t)$, these norms satisfy the relations

(24.20a) $$0 < \|1_k\|_{L^2} < \infty,$$
(24.20b) $$0 < \|1_i\|_{L^2} < \infty.$$

The local $L^1(\Omega_k(t))$ and $L^1(\omega_i(t))$ norms for $G_k(Q_k, P_k)$ and $g_i(q, p)$ are

(24.21a) $$\|G_k\|_{L^1} = (|G_k|, 1_k)_{L^2},$$
(24.21b) $$\|g_i\|_{L^1} = (|g_i|, 1_i)_{L^2}.$$

The Schwartz inequality in L^2 is used next to obtain

(24.22a) $$\|G_k\|_{L^1} \leq \|G_k\|_{L^2} \|1_k\|_{L^2},$$
(24.22b) $$\|g_i\|_{L^1} \leq \|g_i\|_{L^2} \|1_i\|_{L^2}.$$

These results show that the local L^1 norms are bounded it the local L^2 norms are bounded.

To simplify notation, the maps introduced in Chapter 5 will be used. The function $\mathfrak{F}_k(t)$ will represent the phase map $\mathfrak{F}_k(t) : \Omega_k(t) \rightarrow \mathbf{R}$ defined by $\mathfrak{F}_k(t) : (Q_k, P_k) \rightarrow F_k(Q_k, P_k, t)$ and $\mathfrak{f}_i(t)$ will represent the map $\mathfrak{f}_i(t) : \omega_i \rightarrow \mathbf{R}$ defined by $\mathfrak{f}_i(t) : (q_i, p_i) \rightarrow \mathfrak{f}_i(q_i, p_i, t)$. In this notation, the normalizability requirements on $F_k(Q_k, P_k, t)$ and $\mathfrak{f}_i(q_i, p_i, t)$ are expressed as $0 < \|\mathfrak{F}_k(t)\|_{L^1} = Z_k < \infty$ and $0 < \|\mathfrak{f}_i(t)\|_{L^1} = Z_i < \infty$. Using these with (24.20) and (24.22), it follows that

(24.23a) $$\|\mathfrak{F}_k(t)\|_{L^2} > 0,$$
(24.23b) $$\|\mathfrak{f}_i(t)\|_{L^2} > 0.$$

This shows that none of the states of a system, described in terms of these L^2 distributions, can have zero norms. Moreover, the support of these functions cannot have Lebesgue measure zero in $\Omega_k(t)$ or $\omega_i(t)$, respectively, either.

The k system phase entropy, introduced in Chapter 9, is

(24.24) $$S_k(Q_k, P_k, t) = -k_B \ln F_k(Q_k, P_k, t).$$

The results (14.32) and (3.16a) provided the lower and upper bounds on the entropy:

(24.25) $$-\infty = S_k^\delta(t) \leq S_k(t) \leq S_k^\epsilon < \infty.$$

The goal is to show that $S_k(t)$ is well defined for $F_k(Q_k, P_k, t) \in L^2(\Omega_k(t))$. The facts that $x \ln x$ has a minimum $-e^{-1}$ at $x = e^{-1}$ and is bounded for $x < \infty$ are used with the fact that $F_k(Q_k, P_k, t) \geq 0$ to show first
(24.26)
$$\chi_{\Omega_k(t)}(Q_k, P_k) F_k(Q_k, P_k, t) |\ln F_k(Q_k, P_k, t)| \leq \chi_{\Omega_k(t)}(Q_k, P_k)(1 + [\mathfrak{F}_k(t)]^2).$$

Next, the definition of the local L^2 norm shows that the phase average of the system distribution density $F_k(Q_k, P_k, t)$ is the square of the local norm of F_k

$$(24.27) \qquad \langle \mathfrak{F}_k(t) \rangle_{kt} = \|\mathfrak{F}_k(t)\|_{L^2}^2 \,.$$

From (24.26) and (24.27), it follows that

$$(24.28) \qquad |\mathfrak{S}_k(t)| \leq k_B \langle |\ln \mathfrak{F}_k(t)| \rangle_{kt} \leq k_B \left(\|1_k\|_{L^2}^2 + \|\mathfrak{F}_k(t)\|_{L^2}^2 \right),$$

where $\mathfrak{S}_k(t)$ is the map $\mathfrak{S}_k(t) : (Q_k, P_k) \to S_k(Q_k, P_k, t)$. Thus, if the $L^2(\Omega_k(t))$ norm $\|\mathfrak{F}_k(t)\|_{L^2}$ of $\mathfrak{F}_k(t)$ is bounded, the function $\mathfrak{S}_k(t)$ is bounded as well.

Consider next $\beta_k(t)$ and $\beta_k^{\mu\nu}(t)$. The definition (12.12a) of $\beta_k(t)$ and (12.12b) of $\beta_k(Q_k, P_k, t)$ shows that $\beta_k(t)$ is well defined and bounded if $P_k(t) > 0$ and the phase average of $dS_k(Q_k, P_k, t)/d\Omega_k$ is bounded. The latter bound follows from the bound on the elements of Q_k and P_k obtained from (24.4) and the bounds on integrals containing $\partial F_k(Q_k, P_k, t)/\partial Q_k$ and $\partial F_k(Q_k, P_k, t)/\partial P_k$ that follow in turn from (24.9) by setting $v_i^\mu = 1$ and $v_i^\nu = 0$ if $\nu \neq \mu$ and setting all other $v_j = 0$. This procedure is repeated for each μ and each $i \in k$.

To sum the progress to this point: the system states were assumed to be L^2, that is, that $F_k \in L^2(\Omega_k(t))$ and $f_i \in L^2(\omega_i(t))$. It was also assumed that the observable analogs in the set $\mathcal{O}_k^a$ defined in Chapter 5 and their derivatives to order m, where $m \geq 3$, are continuous. These assumptions are sufficient to conclude that global observables of the form $\langle G_k \rangle_{kt}$ for $G_k \in \mathcal{O}_k^a$ and their total time derivatives are well defined. The thermodynamic observables, entropy, temperature and pressure, are well defined as well.

The phase averages of the singular functions $\dot{\chi}_{\Omega_k(t)} G_k$, $\delta(q - q_i)G_i$, and $\delta(q - q_i)\dot{\chi}_{\Omega_k(t)} G_i$ are examined next. These quantities contain singular delta measures of 1, 3, and 4 dimensions, respectively. In the context of a phase integration, only the last form results in a singular thermodynamic function equivalent to a 1-dimensional delta measure.

To assist in computing bounds with these quantities, let us define the sets

(24.29a)
$$\partial_i \Omega_k(t) = \{ (Q_k, P_k) \in \Omega_k(t) \mid q_i \in \partial d_k(t) \} \,,$$

(24.29b)
$$\vdash_i \Omega_k(t) = \{ (Q_k, P_k) \in \Omega_k(t) \mid q_i = q \} \,,$$

(24.29c)
$$\partial_i \vdash_i \Omega_k(t) = \{ (Q_k, P_k) \in \Omega_k(t) \mid q_i = q, q \in \partial d_k(t) \} \,.$$

Let us also define, for the Boltzmann case, the sets

(24.30a) $\qquad \partial_i \omega_i(t) = \{ (q_i, p_i) \in \omega_i(t) \mid q_i \in \partial d_k(t) \} ,$

(24.30b)
$$\vdash_i \omega_i(t) = \{ (q_i, p_i) \in \omega_i(t) \mid q_i = q \} ,$$

(24.30c)
$$\partial_i \vdash_i \omega_i(t) = \{ (q_i, p_i) \in \omega_i(t) \mid q_i = q, q \in \partial d_k(t) \} .$$

The abbreviations (1) $\partial_i L^2$ for the space $L^2(\partial_i \Omega_k(t))$, (2) $\vdash_i L^2$ for the space $L^2(\vdash_i \Omega_k(t))$ and (3) $\partial_i \vdash_i L^2$ for the space $L^2(\partial_i \vdash_i \Omega_i(t))$ will simplify the notation. Abbreviations are defined for the L^2 spaces in the Boltzmann case similarly. The following inner products are based on these spaces:

(24.31a)
$$(A, B)_{\partial_i L^2} = \int dQ_k \int dP_k \, \chi_{\Omega_k(t)} \frac{\dot{\chi}_{d_k(t)}(q_i, v_i)}{\chi_{d_k(t)}(q_i)} A(Q_k, P_k) B(Q_k, P_k),$$

(24.31b)
$$(A, B)_{\vdash_i L^2} = \int dQ_k \int dP_k \, \chi_{\Omega_k(t)}(Q_k, P_k) \delta(q - q_i) A(Q_k, P_k) B(Q_k, P_k),$$

(24.31c)
$$(A, B)_{\partial_i \vdash_i L^2} = \int dQ_k \int dP_k \chi_{\Omega_k(t)} \delta(q - q_i) \frac{\dot{\chi}_{d_k(t)}(q_i, v_i)}{\chi_{d_k(t)}(q_i)} A(Q_k, P_k) B(Q_k, P_k).$$

Again, there are similar formulas for the Boltzmann quantities $(a, b)_{\partial_i L^2}$, $(a, b)_{\vdash_i L^2}$ and $(a, b)_{\partial_i \vdash_i L^2}$. The quantities in (24.31) are summed over $i \in k$ to define next

(24.32a)
$$(A, B)_{\partial_k L^2} = \sum_{i \in k} (A, B)_{\partial_i L^2},$$

(24.32b)
$$(A, B)_{\vdash_k L^2} = \sum_{i \in k} (A, B)_{\vdash_i L^2},$$

(24.32c)
$$(A, B)_{\partial_k \vdash_k L^2} = \sum_{i \in k} (A, B)_{\partial_i \vdash_i L^2}.$$

As before, there are similar functions defined for the Boltzmann case.

This notation will be applied to the total time derivative of the entropy. Recall equation (9.15) for the total time derivative of the entropy:

(24.33)
$$\frac{dS_k(t)}{dt} = \left\langle \dot{\chi}_{\Omega_k(t)}(Q_k, P_k) S_k(Q_k, P_k, t) \right\rangle_{kt} .$$

Using (24.31), this is written as

(24.34)
$$\left| \frac{dS_k(t)}{dt} \right| \leq \sum_{i \in k} |(\mathfrak{F}_k(t), \mathfrak{S}_k(t))_{\partial_i L^2}| .$$

By (24.26), it can be put in the form

$$(24.35) \qquad \left|\frac{dS_k(t)}{dt}\right| \le k_B \sum_{i \in k} \left[\|1_k\|_{\partial_i L^2} + \|\mathfrak{F}_k(t)\|^2_{\partial_i L^2}\right].$$

For the Boltzmann case, the local entropy was defined in (22.17) and (22.18) by

$$(24.36) \qquad S_i^{(B2)}(q,t) = -k_B \langle \delta(q - q_i) \ln f_i(q_i, p_i, t)\rangle_{it}.$$

Proceeding as before, this has the bound

$$(24.37) \qquad \left|S_i^{(B2)}(q,t)\right| \le k_B[\|1_i\|_{\vdash_i L^2} + \|f_i\|^2_{\vdash_i L^2}].$$

The time derivative of the global Boltzmann entropy also has the bound

$$(24.38) \qquad \left|\frac{dS_i^{(B2)}(t)}{dt}\right| \le k_B[\|1_i\|_{\partial_i L^2} + \|f_i\|^2_{\partial_i L^2}].$$

Because the velocities in the system are bounded, the result (22.19) is used to obtain similar results for the Boltzmann entropy current density vector.

The function space requirements on the functions of the theory are based on the following assumptions: The fact that V_{2k} is bounded in $\Omega_k(t)$ and the relation (24.22) can be used to show that the quantity defined in (24.9a) is well defined if $\nabla_{2k}\mathfrak{F}_k(t) \in \partial_i L^2$. A similar result is obtained for (24.9b) if $V_{2k} \cdot \nabla_{2k}\mathfrak{F}_k(t) \in L^2$. For (24.8), $\mathfrak{F}_k(t) \in \vdash_i L^2$ and $\mathfrak{F}_k(t) \in \partial_i \vdash_i L^2$ are required. For the other quantities defined in this chapter, it is required that $\mathfrak{F}_k(t) \in L^2$ be true as well. Similarly, in the Boltzmann case, it is required that the following relations hold: $\nabla_i f_i(t) \in \partial_i \vdash_i L^2$, $f_i(t) \in \partial_i L^2$, and $f_i(t) \in L^2$. These requirements will be taken account of in the next chapter.

24.5 Sobolev Spaces

Sobolev spaces will be used extensively in the next chapter, so their important properties are collected here. Let us begin with some notational considerations and definitions. Let $\mathbf{N}$ be the set of non-negative integers and let $\mathbf{N}^{6\mathcal{N}_k}$ be the integer space defined as the $6\mathcal{N}_k$-dimensional direct product of $\mathbf{N}$. Consider a vector γ in this space. The components of $\gamma \in \mathbf{N}^{6\mathcal{N}_k}$ are written as $\gamma_{i,x}^\mu$, where $1 \le \mu \le 3$, $i \in k$ and $x \in \{1, 2\}$. The length of γ is

$$(24.39a) \qquad |\gamma_i| = \sum_\mu \sum_{x \in \{1,2\}} \gamma_{i,x}^\mu,$$

$$(24.39b) \qquad |\gamma| = \sum_{i \in k} |\gamma_i|.$$

The vector γ is used to label the components of the gradient operator ∇_{2k}:

$$(24.40) \qquad \nabla_k^{\gamma_{i,1}^\mu} = \left(\frac{\partial}{\partial q_i^\mu}\right)^{\gamma_{i,1}^\mu}, \qquad \nabla_k^{\gamma_{i,2}^\mu} = \left(\frac{\partial}{\partial p_i^\mu}\right)^{\gamma_{i,2}^\mu}.$$

With this notation, the real Sobolev space $H^s = H^s(\Omega_k(t))$, which is sometimes written $W^{s,2}(\Omega_k(t))$, is defined for $s \in \mathbf{N}$ as the space of real valued functions with support in $\Omega_k(t)$ such that if $G \in H^s(\Omega_k(t))$ then $\nabla_k^\gamma G \in L^2(\Omega_k(t))$ for $|\gamma| \le s$. Under this definition, $H^0(\Omega_k(t))$ is $L^2(\Omega_k(t))$. For $G, M \in H^s(\Omega_k(t))$, the inner product and norm on H^s are defined using the formalism defined in (24.17a) and (24.18a) as

$$(24.41a) \qquad\qquad (G, M)_s = \sum_{|\gamma| \le s} (\nabla_k^\gamma G, \nabla_k^\gamma M)_{L^2},$$

$$(24.41b) \qquad\qquad \|G\|_s = (G, G)_s^{\frac{1}{2}}.$$

It is not hard to show that $H^s(\Omega_k(t))$, equipped with the inner product (24.41a), is a real Hilbert space.[4]

Observe next that the natural injection of $H^s(\Omega_k(t))$ into $H^m(\Omega_k(t))$ for $m \le s$ is continuous, that is, if $\iota_{sm} : H^s \to H^m$, then

$$(24.42) \qquad\qquad \|\iota_{sm} G\|_m \le \|G\|_s.$$

The injection ι_{sm} will not be indicated explicitly when it is used below.

One important result of the theory of Sobolev spaces needed for the work here is the collection of embedding theorems. Assume that $\partial d_k(t)$ is *piecewise uniformly C^∞ regular* (Adams [1975], p. 67). In the case dealt with here, this means that $\partial d_k(t)$ is a 2-dimensional compact piecewise smooth surface for which $d_k(t)$ lies on one side of $\partial d_k(t)$. It follows that $D_k(t)$ is uniformly C^∞ regular and that $\partial D_k(t)$ is a $(3\mathcal{N}_k - 1)$-dimensional compact piecewise smooth surface. Since $\partial d_k(t)$ is piecewise smooth and compact, it follows that (10.2) is satisfied:

$$(24.43) \qquad\qquad \alpha_k(t) = \int d^2 q \, \chi_{\partial d_k(t)}(q) < \infty.$$

There is also a finite constant M_P defined for $(Q_k, P_k) \in \Omega_k(t)$ by

$$(24.44)$$
$$M_P = \sup_{Q_k \in D_k(t)} \int dP_k \, \theta(\mathcal{P}_k^u - \mathcal{P}_k)\theta(\mathcal{P}_k - \mathcal{P}_k^l)\theta(\mathcal{J}_k^u - \mathcal{J}_k)\theta(\mathcal{J}_k - \mathcal{J}_k^l)$$
$$\times \, \theta(\mathcal{H}_k^u - \mathcal{H}_k)\theta(\mathcal{H}_k - \mathcal{H}_k^l).$$

---

[4]For further information on Sobolev spaces, see Treves [1967], Chapter 31, and Adams [1975].

An inequality is obtained by setting $G_k(Q_k, P_k, t) = 1$ in (5.29b) and using the definition (5.20) of $\dot{\chi}_{d_k(t)}(q_i, v_i)$ with the definition (5.26) of $\dot{\chi}_{\Omega_k(t)}(Q_k, P_k)$ to show

$$(24.45) \quad \int dQ_k \int dP_k |\dot{\chi}_{\Omega_k(t)}(Q_k, P_k)| \leq [v_k^u + v_{\partial k}^{(\max)}] M_P \alpha_k(t) [\delta_k(t)]^{N_k - 1}.$$

The particular embedding theorems needed here for each $i, j \in k$ with $i \neq j$ are[5]

$$(24.46a) \qquad H^s(\Omega_k(t)) \subset L^2(\partial_i \Omega_k(t)), \qquad s \geq 1,$$

$$(24.46b) \qquad H^s(\Omega_k(t)) \subset L^2(\vdash_i \Omega_k(t)), \qquad s \geq 2,$$

$$(24.46c) \qquad H^s(\Omega_k(t)) \subset L^2(\partial_i \partial_j \Omega_k(t)), \qquad s \geq 2.$$

Relation (24.46a) means, for example, that for each $G \in H^s$, the restriction

$$(24.47) \qquad G|_{\partial_i \Omega_k(t)} \in L^2(\partial_i \Omega_k(t))$$

is valid. Choosing $s \geq 2$, it follows that the distributions $F_k(t)$ and $f_i(t)$ meet the conditions stated at the end of the last section.

It follows from (24.47) that there is a continuous injection $\iota_{s\partial i}$ of the Sobolev space $H^s(\Omega_k(t))$ into $L^2(\partial_i \Omega_k(t))$ and a constant, $c_{s\partial i}$, such that for $G \in H^s(\Omega_k(t))$ and $i \in k$

$$(24.48) \qquad \|\iota_{s\partial i} G\|_{\partial_i L^2} \leq c_{s\partial i} \|G\|_s \leq c_{s\partial} \|G\|_s .$$

The constant $c_{s\partial}$ is the supremum of $c_{s\partial i}$ over $i \in k$. Similarly, for $\vdash_i \Omega_k(t)$, there is an injection $\iota_{s\delta i}$ of $H^s(\Omega_k(t))$ into $L^2(\vdash_i \Omega_k(t))$ and a constant $c_{s\delta i}$ such that, for $G \in H^s(\Omega_k(t))$ and $i \in k$

$$(24.49) \qquad \|\iota_{s\delta i} G\|_{\vdash_i L^2} \leq c_{s\delta i} \|G\|_s \leq c_{s\delta} \|G\|_s ,$$

where $c_{s\delta}$ is the supremum of $c_{s\delta i}$ over $i \in k$. As before, the injections $\iota_{s\partial i}$ and $\iota_{s\delta i}$ will not be indicated explicitly when they are used below.

---

[5]See Carroll [**1969**], pp. 60–62, Adams [**1975**], pp. 113–115, or Browder [**1961**].

CHAPTER 25

The F_k and f_i States

The question of the existence and uniqueness of the F_k and f_i states as solutions of the system specific equations of motion is examined in this chapter using the contraction mapping theorem. Global solutions of the equations of motion are also discussed briefly.

25.1 The Local F_k State

The main objective of this chapter is to show the existence of solutions F_k and f_i of the equations of motion of the system. Although the states F_k and f_i, as solutions of their respective microscopic evolution equations, are not required to be limited to the phase spaces $\Omega_k(t)$ or $\omega_i(t)$ of system k, our interest in this chapter is in the state both inside $d_k(t)$ and on the boundary $\partial d_k(t)$. On the boundary, the stochastic boundary formalism keeps the particles confined to the system and allows fluxes of physical quantities to flow between systems. Showing that the modification of the state at the boundary by this formalism fits within the requirements of the TIS approach is part of the concern.

To begin, a space of continuous maps from the time interval $[0, \alpha)$ into the space $H^s(\Omega_k(t))$ is defined by

$$(25.1) \qquad X_{ks}^\alpha = C([0, \alpha), H^s(\Omega_k(t))).$$

Let $F_k \in X_{ks}^\alpha$ be the map $F_k : t \to \mathfrak{F}_k(t)$, where $\mathfrak{F}_k(t)$ is the image in $H^s(\Omega_k(t))$ induced by the map F_k. This space has a norm defined by

$$(25.2) \qquad \|F_k\|_{s,\alpha} = \sup_{0 \le t < \alpha} \|\mathfrak{F}_k(t)\|_s .$$

Let us next define the map $A_s : H^s \times H^s \to \mathbf{R}$ for $F_k \in X_{ks}^\alpha$, $G \in H^s(\Omega_k(t))$, where G is independent of the time, and $0 \le t < \alpha$ by[1]

$$(25.3)$$
$$A_s(\mathfrak{F}_k(t), G) = (\mathfrak{F}_k(0), G)_s$$
$$+ \int_0^t du \left[(-\chi_{\Omega_k(u)} V_{2k} \cdot \nabla_{2k} \mathfrak{F}_k(u), G)_0 + (\mathfrak{F}_k(u), G)_{\partial_k L^2} \right].$$

This map is based on the integral over time from 0 to t of $d(\mathfrak{F}_k(t), G)_s/dt$, which is the time derivative of the inner product of $G(Q_k, P_k)$ with the system distribution density $F_k(Q_k, P_k, t)$. It makes use of $\partial F_k(Q_k, P_k, t)/\partial t$, as defined in (5.32), and uses the definitions of the inner products given in (24.31) and (24.32).

The first term in the integrand of the inner product is well defined because, by the natural injections, $\nabla_{2k}^\gamma \mathfrak{F}_k(t)$ and G will both belong to H^0 (i.e., L^2) if $\mathfrak{F}_k(t), G \in H^s$ and $s \ge 1$ and $|\gamma| \le 1$. Let M_{2k}^α be a finite upper bound of V_{2k} for $(Q_k, P_k) \in \Omega_k(t)$ and $t \in [0, \alpha)$. By the global bounds (24.3), (24.4), and $v_{\partial k}^{\max}$, there is a constant M_{2k} such that $M_{2k}^\alpha < M_{2k}$ for all α. Then the Schwartz inequality in H^0 can be used followed by (24.41) and (24.42) to

---

[1] This method has been used by Lions and Magenes [**1968**], Carroll, and other authors. See Carroll [**1969**], Chapter 2, Sections 3 and 4, Chapter 3, Section 4, and the references given there.

show

(25.4) $$|(-V_{2k} \cdot \nabla_{2k} \mathfrak{F}_k(t), G)_0| \le M_{2k} \|\mathfrak{F}_k(t)\|_s \|G\|_s .$$

It also follows that

(25.5) $$|(\mathfrak{F}_k(0), G)_s| \le \|\mathfrak{F}_k(0)\|_s \|G\|_s .$$

For the second term in the integrand in (25.3), the embedding theorem (24.46a) is used with the injection (24.48) and (24.4b) is used with (24.5) to write a bound on the inner products as

(25.6) $$|(\mathfrak{F}_k(t), G)_{\partial_k L^2}| \le M_{2k} c_{s,\partial}^2 \|\mathfrak{F}_k(t)\|_s \|G\|_s .$$

Let us define next

(25.7) $$C_s = M_{2k}(1 + c_{s,\partial}^2)$$

and use it with (25.4), (25.5), (25.6), and (25.7), in (25.3) to obtain

(25.8) $$|A_s(\mathfrak{F}_k(t), G)| \le (1 + \alpha C_s) \|F_k\|_{s\alpha} \|G\|_s$$

for $F_k \in X_{ks}^\alpha$, $G \in H^s$.

By (25.3) and (25.8) it is clear that $A_s(\mathfrak{F}_k(t), \cdot)$ is a linear functional on $H^s(\Omega_k)$ for each $t \in [0, \alpha)$. Because H^s is a Hilbert space, a theorem of Frigyes Riesz (F. Riesz and B. Sz.-Nagy [**1953**]) states that for each $t \in [0, \alpha)$, there exists an element $\mathfrak{L}_k(t)$ in the dual of H^s (which is identified with an element of H^s by means of the canonical isometry) such that for each $G \in H^s$, the relation[2]

(25.9) $$(\mathfrak{L}_k(t), G)_s = A_s(\mathfrak{F}_k(t), G)$$

is valid. To show that there is a solution of the evolution equation (5.16), it is necessary to show that there is an $\mathfrak{F}_k(t)$ such that $\mathfrak{L}_k(t) = \mathfrak{F}_k(t)$.

Assume that $\mathfrak{F}_k(t)$ is a solution of (5.16), so that $\mathfrak{F}_k(t) = \mathfrak{L}_k(t)$ in (25.9), and that $G(Q_k, P_k) \ge 0$. This implies that $(\mathfrak{F}_k(t), G)_s \ge 0$ if and only if it is also true that $A_s(\mathfrak{F}_k(t), G) \ge 0$. As the initial condition, let us assume $\mathfrak{F}_k(0) \ge 0$, a.e., which implies that $A_s(\mathfrak{F}_k(0), G) \ge 0$, a.e. Computing next the derivative of $A_s(\mathfrak{F}_k(t), G)$ with respect to the time yields

(25.10)
$$\frac{dA_s(\mathfrak{F}_k(t), G)}{dt} = -(V_{2k} \cdot \nabla_{2k} \mathfrak{F}_k(t), G)_s + (\mathfrak{F}_k(t), G)_{\partial_k L}$$
$$= (\mathfrak{F}_k(t), V_{2k} \cdot \nabla_{2k} G)_s + (\mathfrak{F}_k(t), G)_{\partial_k L}.$$

Thus, if $\mathfrak{F}_k(t) = 0$ a.e. for $(Q_k, P_k) \in \text{supp } G$, then $dA_s(\mathfrak{F}_k(t), G)/dt = 0$ a.e. and if $dA_s(\mathfrak{F}_k(t), G)/dt = 0$ a.e. for all G, it follows that $\mathfrak{F}_k(t) = 0$. It follows also that the relation $A_s(\mathfrak{F}_k(t), G) = 0$ for all G implies $\mathfrak{F}_k(t) = 0$ a.e., which implies in turn that $dA_s(\mathfrak{F}_k(t), G)/dt = 0$. Together, these results mean

---

[2] Treves [**1967**], pp. 326, 330.

that the quantity $A_s(\mathfrak{F}_k(t), G)$ cannot become negative, that is, the relation $A_s(\mathfrak{F}_k(t), G) \geq 0$ is valid, if $A_s(\mathfrak{F}_k(0), G) \geq 0$. Because $G > 0$ is arbitrary and $\mathfrak{F}_k(0) \geq 0$, it may be concluded that $\mathfrak{F}_k(t) \geq 0$ for each $t \in [0, \alpha)$. It follows that condition (5.14a) is satisfied by $\mathfrak{F}_k(t)$.

In light of the relation (25.9), let us define a map $T_s : H^s \to H^s$ in terms of $\mathfrak{F}_k(t)$ and $\mathfrak{L}_k(t)$ by

$$(25.11) \qquad\qquad T_s \mathfrak{F}_k(t) = \mathfrak{L}_k(t).$$

With this definition it follows, for each $G \in H^s$, that

$$(25.12) \qquad\qquad (T_s \mathfrak{F}_k(t), G)_s = A_s(\mathfrak{F}_k(t), G).$$

By (25.3), (25.12) and the uniqueness of Radon measures, it also follows that (25.13)

$$T_s \mathfrak{F}_k(t) = \mathfrak{F}_k(0) + \int_0^t du \left[-\chi_{\Omega_k(u)} V_{2k} \cdot \nabla_{2k} + \frac{\partial \chi_{\Omega_k(u)}}{\partial t} \right] \mathfrak{F}_k(u) \qquad \text{a.e.,}$$

which, as indicated, is valid almost everywhere. Equation (25.13), shows that $T_s \mathfrak{F}_k(t)$ is a distribution for which $T_s \mathfrak{F}_k(0) = \mathfrak{F}_k(0)$ at time $t = 0$.

Consider next the continuity of the distribution $T_s \mathfrak{F}_k(t)$ as a function of t. Let t_0, $t_0 + t \in [0, \alpha)$ and $\mathfrak{F}_k(t) \in H^s$. Then, since $T_s \mathfrak{F}_k(t) \in H^s$ also, the result (25.12) is used with the bounds defined previously to write

(25.14)
$$\| T_s \mathfrak{F}_k(t_0 + t) - T_s \mathfrak{F}_k(t_0) \|_s^2 = A_s(\mathfrak{F}_k(t_0 + t), T_s \mathfrak{F}_k(t_0 + t) - T_s \mathfrak{F}_k(t_0))$$
$$- A_s(\mathfrak{F}_k(t_0), T_s \mathfrak{F}_k(t_0 + t) - T_s \mathfrak{F}_k(t_0))$$
$$= \int_{t_0}^{t_0 + t} du \left[(-\chi_{\Omega_k(u)} V_{2k} \cdot \nabla_{2k} \mathfrak{F}_k(u), T_s \mathfrak{F}_k(t_0 + t) - T_s \mathfrak{F}_k(t_0))_0 \right.$$
$$\left. + (\mathfrak{F}_k(u), T_s \mathfrak{F}_k(t_0 + t) - T_s \mathfrak{F}_k(t_0))_{\partial_k L^2} \right]$$
$$\leq t C_s \| F_k \|_{s\alpha} \| T_s \mathfrak{F}_k(t_0 + t) - T_s \mathfrak{F}_k(t_0) \|_0 .$$

The natural injection is used again with (25.14) to show

$$(25.15) \qquad\qquad \| T_s \mathfrak{F}_k(t_0 + t) - T_s \mathfrak{F}_k(t_0) \|_s \leq t C_s \| F_k \|_{s\alpha} .$$

It follows that $T_s \mathfrak{F}_k(t)$ is a continuous function of t for $t \in [0, \alpha)$ if $\| F_k \|_{s\alpha}$ is bounded. The map T_s is therefore a map of X_{ks}^α into X_{ks}^α.

Assume now that the local norm $\| \mathfrak{F}_k(t) \|_{L^1}$ of $\mathfrak{F}_k(t)$ is finite and satisfies the condition (5.14b) for $t \in [0, \alpha)$. The next step is to demonstrate that the local norm of $T_s \mathfrak{F}_k(t)$ is finite and satisfies (5.14b). To obtain the local norm of $T_s \mathfrak{F}_k(t)$, integrate (25.13) over (Q_k, P_k). Then use Fubini's theorem, integration by parts, the relation $\nabla_{2k} \cdot V_{2k} = 0$, the definition (5.3) of $Z_k(t)$, the relation (5.29), and then the expression of $Z_k(t)$ in terms of $\mathfrak{F}_k(t)$ to obtain

$$(25.16) \qquad\qquad \| T_s \mathfrak{F}_k(t) \|_{L^1} = \| \mathfrak{F}_k(0) \|_{L^1} = Z_k(0).$$

This result shows that the constancy of the local norm stated in (5.35) is a consequence of this calculation as well. It is also clear by (25.13) that $\operatorname{supp} \mathfrak{F}_k(t) \subset \Omega_k(t)$ implies $\operatorname{supp} T_s \mathfrak{F}_k(t) \subset \Omega_k(t)$ and the supports of $T_s \mathfrak{F}_k(t)$ and $\mathfrak{F}_k(t)$ will coincide when there is a solution of the equation $T_k \mathfrak{F}_k(t) = \mathfrak{F}_k(t)$.

To show that a solution exists, let $Z_{ks}^\alpha \subset X_{ks}^\alpha$ be the non-empty bounded set defined by

$$(25.17) \qquad Z_{ks}^\alpha = \{\, G_k \in X_{ks}^\alpha \mid \|\mathfrak{G}_k(t)\|_{L^1} = 1, \|\mathfrak{G}_k(t) - \mathfrak{F}_k(0)\|_s \leq 1 \,\}.$$

Let $\{\, G_{k,n} \,\}$ be a Cauchy sequence in Z_{ks}^α with limit G_k. Then $\{\mathfrak{G}_{k,n}(t)\}$ is a Cauchy sequence in L^2 with limit $\mathfrak{G}_k(t)$. Since the support of $\mathfrak{G}_k(t)$ is the complement of the largest open set in Ω_k on which it vanishes, it is a closed set. If the supports of $\{\mathfrak{G}_{k,n}(t)\}$ fall within $\Omega_k(t)$, then the support of $\mathfrak{G}_k(t)$ will fall within $\Omega_k(t)$ as well. By (24.21a), $\{\mathfrak{G}_{k,n}\}$ is also a Cauchy sequence in L^1 which implies $\|\mathfrak{G}_k(t)\|_{L^1} = \lim_{n\to\infty} \|\mathfrak{G}_{k,n}(t)\|_{L^1} = 1$. The properties of Cauchy sequences can also be used to show that $\|\mathfrak{G}_{k,n}(t) - \mathfrak{F}_k(0)\|_s \leq 1$ implies $\|\mathfrak{G}_k(t) - \mathfrak{F}_k(0)\|_s \leq 1$. These results mean that Z_{ks}^α is closed in X_{ks}^α.

Let us define next a constant η by

$$(25.18a) \qquad \eta = [\|\mathfrak{F}_k(0)\|_0 + 1]^{-1}$$

and use (24.23a) to show that

$$(25.18b) \qquad \eta < 1.$$

For $G \in Z_{ks}^\alpha$, the definition (25.18a) and the relation (25.18b) are used with the triangle inequality, $\|\mathfrak{G}_k(t)\|_0 - \|\mathfrak{F}_k(0)\|_0 \leq \|G_k(t) - \mathfrak{F}_k(0)\|_0 \leq \|\mathfrak{G}_k(t) - \mathfrak{F}_k(t)\|_s \leq 1$ to show

$$(25.19) \qquad \|\mathfrak{G}_k(t)\|_0 \leq \eta^{-1}.$$

Suppose $F_k \in Z_{ks}^\alpha$. It follows from (25.19) and the definition of Z_{ks}^α that $\|\mathfrak{F}_k(t)\|_s < \eta^{-1}$ and therefore $\|F_k\|_{s\alpha} < \eta^{-1}$. Next, if $t_0 = 0$ is set in (25.15) and the relation $T_s \mathfrak{F}_k(0) = \mathfrak{F}_k(0)$ is used, it is possible to show

$$(25.20) \qquad \|T_s \mathfrak{F}_k(t) - \mathfrak{F}_k(0)\|_s \leq t C_s \eta^{-1}.$$

Let us choose α so that

$$(25.21) \qquad \alpha < \eta [C_s]^{-1}$$

and then take the supremum of (25.20) over $t \in [0, \alpha)$ to show that T_s is a map of Z_{ks}^α into Z_{ks}^α.

Let us now choose $F_1, F_2 \in Z_{ks}^{\alpha}$ with $F_1(0) = F_2(0)$. By (25.12) and (25.3), it follows that

$$(25.22) \quad \|T_s\mathfrak{F}_1(t) - T_s\mathfrak{F}_2(t)\|_s^2 = A_s(\mathfrak{F}_1(t) - \mathfrak{F}_2(t), T_s\mathfrak{F}_1(t) - T_s\mathfrak{F}_2(t))$$

$$= \int_0^t du \left[(-\chi_{\Omega_k(u)}V_{2k} \cdot \nabla_{2k}(\mathfrak{F}_1(u) - \mathfrak{F}_2(u)), T_s\mathfrak{F}_1(t) - T_s\mathfrak{F}_2(t))_0 \right.$$

$$\left. + (\mathfrak{F}_1(u) - \mathfrak{F}_2(u), T_s\mathfrak{F}_1(t) - T_s\mathfrak{F}_2(u))_{\partial_k L^2}\right]$$

$$\leq tC_s\|F_1 - F_2\|_{s\alpha}\|T_s\mathfrak{F}_1(t) - T_s\mathfrak{F}_2(t)\|_s.$$

Take the supremum of both sides over $t \in [0, \alpha)$ and use (25.21) to obtain

$$(25.23) \qquad\qquad \|T_sF_1 - T_sF_2\|_{s\alpha} \leq \eta\|F_1 - F_2\|_{s\alpha}.$$

Since $\eta < 1$, it follows that T_s is a contraction map on the closed, bounded set $Z_{ks}^{\alpha} \subset X_{ks}^{\alpha}$. The contraction mapping theorem[3] implies then that there is a unique $F_k \in Z_{ks}^{\alpha}$ that satisfies $T_sF_k = F_k$. This F_k is the unique, non-negative, normalizable, local solution of (5.16) with $F_k(t) \in H^s(\Omega_k(t))$ for $s \geq 2$ and $t \in [0, \alpha)$.

For the isolated system case, there are no external forces and no moments of momentum cross the system boundary. This means that the quantity $(\mathfrak{F}_t(u), T_s\mathfrak{F}_t(t_0 + t) - T_s\mathfrak{F}_t(t_0))_{\partial_k L^2} = 0$ in (25.14) and the proof goes forward as before.

25.2 The Local f_i State

The procedure used in this section is very similar to that in the last section. In this case, the space Y_{ks}^{α} is defined as the direct product space of the maps $t \to H^s(\omega_i(t))$ for $i \in k$ by

$$(25.24) \qquad\qquad Y_{ks}^{\alpha} = \times_{i \in k} C([0, \alpha), H^s(\omega_i(t))).$$

Let $F_{[k]} \in Y_{ks}^{\alpha}$ be the column vector with components f_i such that for $i \in k$, $f_i \in C([0, \alpha), H^s(\omega_i(t)))$ is valid. The norm $\|f_i(t)\|_s$ in $H^s(\omega_i(t))$ induces the norm

$$(25.25) \qquad\qquad \|\mathfrak{F}_{[k]}(t)\|_s = \sum_{i \in k} \|f_i(t)\|_s;$$

and the norm in Y_{ks}^{α} is

$$(25.26) \qquad\qquad \|F_{[k]}\|_{s\alpha} = \sup_{0 \leq t < \alpha} \|\mathfrak{F}_{[k]}(t)\|_s.$$

As in (24.42), if $g \in H^s(\omega_i(t))$ and $m \leq s$, it follows that

$$(25.27) \qquad\qquad \|g\|_m \leq \|g\|_s.$$

___
[3]See R. H. Martin [1976], p. 114.

In addition, for $s > 1$, it is also true that

(25.28) $$H^s(\omega_i(t)) \subset L^2(\omega_i(t)),$$

which implies, as before, that

(25.29) $$\|g\|_{L^2} \leq c_{Bi}\|g\|_s.$$

In parallel with (25.6), for $f_i(t), g \in H^s(\omega_i(t))$ it can be shown that

(25.30) $$|(f_i(t), g)_{\partial_i L^2}| \leq c_{B\partial}\|f_i\|_s\|g\|_s.$$

The Boltzmann collision term was defined in (21.8) and (21.9). Both the total energy and the total momentum are conserved in a collision The bound (24.4a) is valid for the sum of any two particle momenta. It follows that $(q, p_i + p_j) \in \omega_i(t)$ and $(q, p_i^\natural + p_j^\natural) \in \omega_i(t)$ for the collision $(p_i, p_j) \rightarrow (p_i^\natural, p_j^\natural)$. This means that the relation

(25.31) $$\operatorname{supp} C[f]f_i \subset \omega_i(t)$$

is valid.

Let us define

(25.32)
$$L_{ij}(q, p, t; f) = \int d^3q_i \int d^3p_i \int d^3q_j \int d^3p_j\, \delta(q - q_i)K_{ij}^b(p_i, p_j; p)$$
$$\times \mathcal{O}(p_i, p_j)f_i(q_i, p_i, t)f_j(q_j, p_j, t)$$

where $K_{ij}^b(p_i, p_j, p)$ is defined in equation (21.28) and $\mathcal{O}(p_i, p_j)$ is defined in (21.25). By (21.70), the relation

(25.33) $$C[f]f_i(q, p, t) = \sum_{j \in k-i} L_{ij}(q, p, t; f)$$

follows. The objective is to show that $|\int d^3q \int d^3p\, L_{ij}(q, p, t; f)|$ is bounded for $t \in [0, \alpha)$. By (25.32), choosing $s \geq 6$, and using the bound $K_{ij}^{(b)}$ on $K_{ij}^b(p_i, p_j; p)$ stated in (21.69), it follows that

(25.34) $$\left|\int d^3q \int d^3p\, L_{ij}(q, p, t; f)\right| = \left|\int d^3q \int d^3p \int d^3q_i \int d^3p_i\right.$$
$$\times \left.\int d^3q_j \int d^3p_j\, \delta(q - q_i)K_{ij}^b(p_i, p_j; p)\mathcal{O}(p_i, p_j)f_i(q_i, p_i, t)f_j(q_j, p_j, t)\right|$$
$$\leq K_{ij}^{(b)}\|f_i(t)\|_6\|f_j(t)\|_6$$

Let us define the bound

(25.35) $$K_k^{(b)} = \sum_{i \in k}\sum_{j \in k-i} K_{ij}^{(b)} < \infty.$$

By (25.34), the bound on the inner product of the collision term with $g \in H^s(\omega_i(t))$ is then given by

(25.36)

$$|(C[f]\mathfrak{f}_i(t), g)_0| = \left| \sum_{j \in k-i} \int d^3q \int d^3p\, g(q,p) L_{ij}(q,p,t;f) \right|$$

$$\leq K_k^{(b)} \|g\|_0 \|\mathfrak{f}_i(t)\|_6 \|\mathfrak{F}_{[k]}(t)\|_6.$$

This shows that the quantity $(C[f]\mathfrak{f}_i(q_i, p_i, t), g)_0$ is bounded for any $g \in H^s(\omega_i(t))$ if $f_i \in H^s(\omega_i(t))$ and $s \geq 6$.

The relation

(25.37) $$f_i^1 f_j^1 - f_i^2 f_j^2 = (f_i^1 - f_i^2) f_j^1 + (f_j^1 - f_j^2) f_i^2$$

can be used in a calculation parallel to (25.36) with a summation over $i \in k$ to obtain

(25.38)

$$|(C[f^1]\mathfrak{F}_{[k]}^1(t) - C[f^2]\mathfrak{F}_{[k]}^2(t), g)_0| \leq K_k^{(b)} \|g\|_s (\|\mathfrak{F}_{[k]}^1(t)\|_s + \|\mathfrak{F}_{[k]}^2(t)\|_s)$$

$$\times \|\mathfrak{F}_{[k]}^1(t) - \mathfrak{F}_{[k]}^2(t)\|_s.$$

This proves that $(C[f]\mathfrak{F}_{[k]}(t), g)_0$ is locally Lipschitz in $\mathfrak{F}_{[k]}(t)$.

The proof of the previous section can now be repeated rapidly. Define the map $A_s : H^s(\omega_i(t)) \times H^s(\omega_i(t)) \to \mathbf{R}$ for $f \in Y_{ks}^\alpha$, $g \in H^s(\omega_i(t))$ in terms of the i-th component, f_i, of $F_{[k]}$ by

(25.39)
$$A_s(\mathfrak{f}_i(t), g) = (\mathfrak{f}_i(0), g)_s$$

$$+ \int_0^t du \left(\left[-\chi_{\omega_i(u)} v_{2i} \cdot \nabla_{2i} + \frac{\partial \chi_{\omega_i(u)}}{\partial t} + C[f] \right] \mathfrak{f}_i(u), g \right)_0.$$

Since $A_s(\mathfrak{f}_i(t), \cdot)$ is a linear functional on $H^s(\omega_i(t))$, the transformation $T_s : H^s \to H^s$ is defined as in (25.12) so that for any $g \in H^s(\omega_i(t))$ the relation

(25.40) $$(T_s\mathfrak{f}_i(t), g)_s = A_s(\mathfrak{f}_i(t), g)$$

is valid. This transformation can be written

(25.41) $$T_s\mathfrak{f}_i(t) = \mathfrak{f}_i(0) + \int_0^t du[-\chi_{\omega_i(u)} v_{2i} \cdot \nabla_{2i} + \dot{\chi}_{\omega_i(u)} + C[f]]\mathfrak{f}_i(u) \qquad \text{a.e.}$$

It is easy to show by the methods of the previous section that T_s is a map of Y_{ks}^α into Y_{ks}^α.

The proof that $\mathfrak{f}_i(t) \geq 0$ if $\mathfrak{f}_i(t)$ is a solution of Boltzmann's equation follows the same pattern used in the previous section for the $\mathfrak{F}_k(t)$ case with the same result. The only new addition is the collision term, which is the last term on the right in (25.41). It can be concluded immediately from the form of the collision term as stated in (21.8) and (21.9) that the collision term cannot reduce $f_i(t)$ below zero, since, if $f_i(q_i, p_i, t) = 0$, it follows that

$$(25.42) \qquad C[f]f_i(q_i, p_i, t) \geq 0.$$

It also follows that

$$(25.43) \qquad \|T_s f_i(t)\|_{L^1} = \|f_i(0)\|_{L^1} = Z_i(0).$$

Let m_{ki} be an upper bound of v_{2i} on $\omega_i(t)$ for $t \in [0, \alpha)$ and let

$$(25.44) \qquad m_k = \sup_{i \in k} m_{ki}.$$

Use the constant

$$(25.45) \qquad \eta_B = (\|\mathfrak{F}_{[k]}(0)\|_s + 1)^{-1} \leq 1$$

to replace the η of (25.18a) and, in light of (25.30), (25.36) and (25.44), use

$$(25.46) \qquad C_{sB} = m_k(1 + c_{B\partial}^2) + K_k^{(b)} \eta_B^{-2}$$

to replace the C_s of (25.7). With these replacements T_s can be shown to be a contraction map on the closed, bounded set

$$(25.47) \qquad Z_{sB}^\alpha = \{ g \in Y_{ks}^\alpha \mid \|g(t)\|_{L^1} = 1, \|g(t) - \mathfrak{F}_{[k]}(0)\|_s \leq 1 \},$$

if α is chosen so that $\alpha < \eta_B^2(C_{sB})^{-1}$. This shows that there exists a unique, non-negative, normalizable, local solution of Boltzmann's equation (20.17). For an isolated system, set $\dot{\chi}_{\omega_i(t)} = 0$ and the proof is repeated to obtain a unique, non-negative, normalizable, local solution of Boltzmann's equation in its usual form.

25.3 Global Solutions

The quantity $-\partial \Phi_k(Q_k, t)/\partial q_i$ represents the force on particle i at all points (q_i, p_i) for $(Q_k, P_k) \in \Omega_k(t)$. This amounts to the assumption of conservative forces.

To extend the solution beyond the interval $t \in [0, \alpha)$, observe first that the quantity C_s, defined in (25.7), is independent of the time and of (Q_k, P_k). Consider the norm $\|F_k(\alpha)\|$ of the state $\mathfrak{F}_k(t)$ at $t = \alpha$. Because $F_k \in Z_{ks}^\alpha$, the triangle inequality can be used to conclude that $\|F_k(\alpha)\| \leq \|\mathfrak{F}_k(0)\| + 1$. Next, let α_n be the end point of the n-th time interval, with $\alpha_0 = 0$ and $\alpha_1 = \alpha$ as chosen in (25.21). In parallel with (25.1), define $X_{ks}^{\alpha_n}$ by $X_{ks}^{\alpha_n} = C([\alpha_{n-1}, \alpha_n), H^s(\Omega_k))$. In parallel with (25.17) use the definition
$$(25.48)$$
$$Z_{ks}^{\alpha_n} = \{ G \in X_{ks}^{\alpha_n} \mid \text{supp}\, \mathfrak{G}(t) \in \Omega_k(t), \|\mathfrak{G}(t)\|_{L^1} = 1, \|\mathfrak{G}(t) - \mathfrak{F}_k(\alpha_{n-1})\| \leq 1 \}.$$

It follows by induction that $\|\mathfrak{G}(t)\| \leq \|\mathfrak{F}_k(0)\| + n$ for $t \in [\alpha_{n-1}, \alpha_n)$. Using this as in (25.19), let $\eta_n = [\|\mathfrak{F}_k(0)\| + n]^{-1}$ and replace (25.21) with the choice $\alpha_n < \eta_n[C_s]^{-1}$. By the definition of η_n, it is easy to show that

$$(25.49) \qquad \frac{\text{const.}}{n+1} < \alpha_n < \frac{\text{const.}}{n}$$

will satisfy this requirement for $n \geq 1$. Since the sum $\sum_{n=1}^{\infty} \alpha_n$ diverges, any desired time interval can be covered with a sequence of these α's. This shows that under the TIS assumptions, the solution $\mathfrak{F}_k(t)$ may be extended to arbitrarily large time intervals. Similar results hold for the solution set $\{f_i\}$ of the Boltzmann equation.

CHAPTER 26

Limiting States: F_k^δ and F_k^ϵ

A number of limiting states in the function spaces in which system and particle probability distributions are defined have been identified and discussed in previous chapters. The existence of the exact and equilibrium states is demonstrated here and the circumstances under which they are well defined is discussed. The exact states are also shown to be extreme measures in the sense of Choquet and the equilibrium states are shown to be uniformly asymptotic in the large when the boundary conditions are fixed.

26.1 Existence of the F_k^δ State

The formulas in this chapter are valid for $k = t$, $k = [t]$, and $k \in [t]$.

For each $t \in [0, \alpha)$, let $(Q_k^{(s)}(t), P_k^{(s)}(t))$ be a continuous system specific solution of Hamilton's equations

$$(26.1a) \qquad \dot{Q}_k = \frac{\partial \mathcal{H}_k(Q_k, P_k, t)}{\partial P_k},$$

$$(26.1b) \qquad \dot{P}_k = -\frac{\partial \mathcal{H}_k(Q_k, P_k, t)}{\partial Q_k}.$$

This trajectory is made definite by choosing the k system mechanical state (Q_k^0, P_k^0) as the initial condition and requiring $Q_k^{(s)}(0), P_k^{(s)}(0)) = (Q_k^0, P_k^0)$. The k system exact state F_k^δ has been defined in terms of this trajectory by

$$(26.2) \qquad F_k^\delta(Q_k, P_k, t) = \delta(Q_k - Q_k^{(s)}(t))\delta(P_k - P_k^{(s)}(t)).$$

The objectives in this section are to show (1) that a continuous system specific trajectory, $(Q_k^{(s)}(t), P_k^{(s)}(t))$, exists and is unique, with $(Q_k^{(s)}(t), P_k^{(s)}(t)) \in \Omega_k(t)$, for $t \in [0, \alpha)$, and (2) that the F_k^δ state is equivalent to this trajectory.

Given the k system trajectory, the F_k^δ state is constructed from it as shown in (26.2), so F_k^δ exists whenever the solution $(Q_k^{(s)}(t), P_k^{(s)}(t))$ exists. Assume now that the F_k^δ state is given. The i-particle component of the k system trajectory is obtained by computing the total time derivative in the following formula:

$$(26.3) \qquad \left(\dot{q}_i^{(s)\mu}(t), \dot{p}_i^{(s)\mu}(t)\right) = \frac{d}{dt}\langle\!\langle (q_i^\mu, p_i^\mu) \rangle\!\rangle_{k\delta t}.$$

Carrying out these computations in accord with (5.42), the relations $dq_i^\mu/dt = v_i^\mu$ and $dp_i^\mu/dt = -(\partial \mathcal{H}_k/\partial q_{i,\mu})$ imply that
$$(26.4)$$
$$\left(\dot{q}_i^{(s)\mu}(t), \dot{p}_i^{(s)\mu}(t)\right) = \left(v_i^{(s)\mu}(t), -\left\langle \frac{\partial \mathcal{H}_k(Q_k, P_k, t)}{\partial q_{i,\mu}} \right\rangle_{k\delta t}\right) + \left\langle \dot{\chi}_{\Omega_k(t)} v_i^\mu \right\rangle_{k\delta t}.$$

Because this calculation concerns the microscopic system evolution equation and the system boundary formalism does not apply to this equation, the relations $\chi_{\Omega_k(t)}(Q_k, P_k) = 1$ and $\dot{\chi}_{\Omega_k(t)}(Q_k, P_k) = 0$ are used in (26.4) to remove the system projection operator from the phase average calculations. The relation $v_i^\mu = \partial \mathcal{H}_k(Q_k, P_k, t)/\partial p_{i,\mu}$ is then used in (26.4) and the phase average is computed to show that the Hamilton equations (26.1) are obtained again:

$$(26.5) \quad \left(\dot{q}_i^{(s)\mu}(t), \dot{p}_i^{(s)\mu}(t)\right) = \left(\frac{\partial \mathcal{H}_k(Q_k^{(s)}, P_k^{(s)}, t)}{\partial p_{i,\mu}}, -\frac{\partial \mathcal{H}_k(Q_k^{(s)}, P_k^{(s)}, t)}{\partial q_{i,\mu}}\right).$$

This implies the equivalence of the representation of the system by the Hamiltonian equations and by the exact state.

To prove that the trajectory exists, a standard existence theorem in Banach spaces will be used to provide the result.[1] Consider the $6\mathcal{N}_k$-vector $\nabla_{2k}\mathcal{H}_k(Q_k, P_k, t)$. By the definition of $\mathcal{K}_k(P_k)$ given in Chapter 4, it follows that

(26.6)
$$\tfrac{1}{2}\left|P_k^\natural \cdot V_k^\natural - P_k \cdot V_k\right| = \tfrac{1}{2}\left|(P_k^\natural - P_k) \cdot V_k + P_k^\natural \cdot (V_k^\natural - V_k)\right|$$
$$= \left|(P_k^\natural - P_k) \cdot V_k\right| \le M_1 \left|P_k^\natural - P_k\right|,$$

where the relation $P_k^\natural \cdot (V_k^\natural - V_k) = V_k^\natural \cdot (P_k^\natural - P_k)$ and the bound $|V_k| \le M_1$ were used. For the internal potential $\Phi_{k,n}(Q_k, t)$, the Lipschitz condition (24.6) is used with (26.6) to show that the Hamiltonian $\mathcal{H}_k(Q_k, P_k, t)$ satisfies the Lipschitz condition

(26.7)
$$\left|\mathcal{H}_k(Q_k^\natural, P_k^\natural, t) - \mathcal{H}_k(Q, P, t)\right| \le L \left|(Q_k^\natural, P_k^\natural) - (Q_k, P_k)\right|$$

for some finite constant L. Since $\overline{\Omega_k}(t)$ is compact and $\mathcal{H}_k(Q_k, P_k, t)$ is continuous on $\overline{\Omega_k}(t)$, it is uniformly continuous, i.e., there is a bound M such that

(26.8)
$$|\mathcal{H}_k(Q_k, P_k, t)| \le M$$

for all $(Q_k, P_k) \in \Omega_k(t)$.

Let us next choose a set of constants $\mathcal{C}_t$ to describe the total system and its subsystems. At $t = 0$, choose $Q_k(0)$ such that $|q_i - q_j| > r_c$ for $i, j \in k$ and let $d_k(0)$ be an open, connected, bounded volume set such that $Q_k(0) \in D_k(0)$. Since the velocities are always bounded, there is a bounded volume set d_m such that $D_k(t) \subset D_m = \times_{i \in k} d_m$ and $Q_k(t) \in \overline{D_m}$ during the time interval $t \in [0, \alpha)$. In this time interval, the wall operator for each system will keep particles in a given sub-volume if they are in that sub-volume initially. There is also a bound on the momentum, angular momentum, and energy that can be obtained from other systems. These facts mean that the trajectory $(Q_k^{(s)}(t), P_k^{(s)}(t))$, if it exists, will fall within the $\Omega_k(t)$ manifold.

Let us define next a closed ball in $\mathbf{R}^{6\mathcal{N}_k}$ of radius $b > 0$ with center $(Q_k^{(s)}(0), P_k^{(s)}(0))$ by

(26.9)
$$B_b(Q_k^{(s)}(0), P_k^{(s)}(0)) = \left\{ (Q_k, P_k) \mid \left|(Q_k, P_k) - (Q_k^{(s)}(0), P_k^{(s)}(0))\right| \le b \right\}.$$

---

[1]See, for example, Abraham and Marsden [**1978**], pp. 62–65, Miller [**1971**], pp. 26–33 or Carroll [**1969**], p. 124. For a general discussion of similar existence proofs in the isolated system setting, see Kurth [**1960**].

For the value of α, choose

(26.10) $\alpha = b/M > 0.$

Choose next the radius of the ball b and the initial point $(Q_k^{(s)}(0), P_k^{(s)}(0))$ so that

(26.11) $B_b(Q_k^{(s)}(0), P_k^{(s)}(0)) \subset \Omega_k(t)$

for each $t \in [0, \alpha)$. Then, by the above, there is a unique $C^1(\Omega_k(t))$ trajectory, $(Q_k^{(s)}(t), P_k^{(s)}(t)) \in B_b(Q_k^{(s)}(0), P_k^{(s)}(0))$ for $t \in [0, \alpha)$ with the initial condition $(Q_k^{(s)}(0), P_k^{(s)}(0)) \in \Omega_k(0)$.[2]

The fact that the Lipschitz condition (26.7) for the system Hamiltonian depends only on $|(Q_k^\natural, P_k^\natural) - (Q_k, P_k)|$ and is independent of the time implies that this construction can be continued from a new initial point near the boundary of the ball. As long as the conditions on the existence of a trajectory are met, this continuation can be extended indefinitely into a global trajectory that is a solution of Hamilton's equations.

From the mathematical perspective, the closure $\overline{\Omega_k}(t)$ of the set $\Omega_k(t)$ is compact and Hausdorff. The thermodynamic analogs, which are also called the observable analogs of the theory, are continuous functions on $\Omega_k(t)$ into **R**. The set of measures on $\Omega_k(t)$, which are represented by the system distribution densities, are defined in a space dual to that of the observable analogs. The set of normed measures on $\Omega_k(t)$, which are represented by the phase averaging brackets, can be shown to be a compact convex set. The extreme points of this convex set are just the Dirac measures, that is, the F_k^δ states.[3] When the entropy of states F_k is used as the criterion, the F_k^δ states are the lower bound and the equilibrium states F_k^ϵ are the upper bound.

26.2 The F_k^ϵ State

Let us turn now to a consideration of the equilibrium case. In keeping with the results of Chapter 13, an equilibrium state of a system is possible only if (a) β_w, the wall thermature, is uniform spatially and constant in time; and (b) the walls are at rest in the system rest frame, so the volume set d_k and the boundary set ∂d_k are not time dependent and the wall velocity satisfies $v_{\partial k}(q, t) = 0$ for each $q \in d_k$ and each t. These assumptions require that the wall system, designated by l, is fixed in the reference state L_l and that an unbiased wall, $W_{kl}^{(u)}$ is being used. For the purposes of this section, it will be assumed that $F_k(t) \in H^s(\Omega_k)$.

The steady state of the k system, which is invariant at the walls under these conditions, is the unique asymptotic equilibrium state introduced

---

[2] See Abraham and Marsden [1978], p. 70.
[3] See Choquet [1969], Vol. II, p. 112, and Vol. I, p. 217.

in equation (13.49). The equilibrium state is written in terms of the wall thermature β_w in the form

$$(26.12) \qquad F_k^\epsilon(Q_k, P_k) = e^{-\beta_w \mathcal{H}_k^\epsilon(Q_k, P_k)}$$

and its local norm is

$$(26.13) \qquad Z_k^\epsilon = \int dQ_k \int dP_k \, \chi_{\Omega_k}(Q_k, P_k) e^{-\beta_w \mathcal{H}_k^\epsilon(Q_k, P_k)}.$$

The equilibrium Hamiltonian is defined in (13.49d).

The objective is to show that, under these boundary conditions, the state (26.12) is uniformly asymptotically stable in the large. That is, for any initial time $t_0 \geq 0$ and any initial system state $F_k(t_0)$, it follows that $F_k(t) \to F_k^\epsilon$ (a.e.) as $t \to \infty$. In the Boltzmann case, it is to be shown that any initial state $F_{[k]}(t_0)$ will evolve into $F_{[k]}^\epsilon$ as $t \to \infty$.

Let us focus our attention first on the F_k^ϵ state. It will be shown after this case is completed that similar results follow immediately for the Boltzmann case. Let

$$(26.14) \qquad X_{ks} = C(\mathbf{R}_+, H^s(\Omega_k))$$

be the space of continuous functions from $\mathbf{R}_+$, the non-negative real numbers, into $H^s(\Omega_k)$. This space will be used with the map $F_k : t \to \mathfrak{F}_k(t)$. For $F_k, F_k^\epsilon \in X_{ks}$, the χ^2 functional is defined in terms of the versions of F_k and F_k^ϵ by

(26.15)

$$\chi^2(F_k, F_k^\epsilon) = \int dQ_k \int dP_k \, \chi_{\Omega_k}(Q_k, P_k) \frac{[F_k(Q_k, P_k, t) - F_k^\epsilon(Q_k, P_k)]^2}{F_k^\epsilon(Q_k, P_k)}.$$

It is modeled on the χ^2 function for comparing two distributions in statistics.[4] It is obviously the case that

$$(26.16) \qquad \chi^2(F_k, F_k^\epsilon) \geq 0$$

with equality if and only if $F_k(t) = F_k^\epsilon$ (a.e.). Furthermore, equations (26.15) and (26.16) lead to the relation

$$(26.17) \qquad \left\langle \frac{\mathfrak{F}_k(t)}{\mathfrak{F}_k^\epsilon} \right\rangle_{kt} \geq 2 - \frac{Z_k^\epsilon}{Z_k(t)} \geq 0,$$

with equality if and only if $\mathfrak{F}_k(t) = \mathfrak{F}_k^\epsilon$ (a.e.).

Let us now define the distribution

$$(26.18) \qquad \mathfrak{Y}_k(t) = \mathfrak{F}_k(t) - \mathfrak{F}_k^\epsilon$$

---

[4]See Haas [**1936c**], pp. 426–427, for a similar formula. The relative entropy of two probability measures, defined by J. M. Jauch and J. G. Barron and expressed in TIS notation as $S_k(F_k(t), F_k^\epsilon)$, could also be used. See Skagerstam [**1974**] for a mathematical discussion of this entropy.

and observe that for $(Q_k, P_k) \in \Omega_k$, it follows that
(26.19)
$$\chi_{\Omega_k}(Q_k, P_k)e^{\beta_w \mathcal{H}_k^l} \le \chi_{\Omega_k}(Q_k, P_k)[F_k^\epsilon(Q_k, P_k)]^{-1} \le \chi_{\Omega_k}(Q_k, P_k)e^{\beta_w \mathcal{H}_k^u}.$$

Using (26.18) and (26.19) in (26.15) yields

(26.20) $$e^{\beta_w \mathcal{H}_k^l}\|\mathfrak{Y}_k(t)\|_0^2 \le \chi^2(F_k, F_k^\epsilon) \le e^{\beta_w \mathcal{H}_k^u}\|\mathfrak{Y}_k(t)\|_0^2.$$

Next, the relations $\mathfrak{Y}_{k1}(t) - \mathfrak{Y}_{k2}(t) = \mathfrak{F}_{k1}(t) - \mathfrak{F}_{k2}(t)$ and

(26.21) $$|\mathfrak{Y}_{k1}^2(t) - \mathfrak{Y}_{k2}^2(t)| \le |\mathfrak{Y}_{k1}(t) + \mathfrak{Y}_{k2}(t)||\mathfrak{Y}_{k1}(t) - \mathfrak{Y}_{k2}(t)|$$

are used with the Schwartz inequality to obtain

(26.22)
$$\begin{aligned} |\chi^2(F_{k1}, F_k^\epsilon) - \chi^2(F_{k2}, F_k^\epsilon)| \le{}& e^{\beta_w \mathcal{H}_k^u}\|\mathfrak{Y}_{k1}(t) + \mathfrak{Y}_{k2}(t)\|_0 \\ &\times \|\mathfrak{F}_{k1}(t) - \mathfrak{F}_{k2}(t)\|_0. \end{aligned}$$

Since $\mathfrak{Y}_{k1}$ and $\mathfrak{Y}_{k2}$ are both L^2, it follows that $\chi^2(\mathfrak{F}_k(t), F_k^\epsilon)$ is continuous and locally Lipschitz in $\mathfrak{F}_k(t)$.

It follows from (10.12a) that

(26.23a) $$\frac{\partial \chi_{\Omega_k}(Q_k, P_k)}{\partial t} = 0$$

whenever $v_{\partial k}(q, t) = 0$ and $\partial \Phi_k(Q_k, t)/\partial t = 0$. These are the current assumptions concerning the k system, so the result $\langle \dot\chi_{\Omega_k}\rangle_{kt} = 0$ in (5.29) implies that $\langle V_{2k} \cdot \nabla_{2k}\chi_{\Omega_k}(Q_k, P_k)\rangle_{kt} = 0$. Using integration by parts, these facts imply that

(26.23b)
$$\begin{aligned} \int dQ_k \int dP_k\, \chi_{\Omega_k}(Q_k, P_k)(V_{2k} \cdot \nabla_{2k}\mathfrak{F}_k(t)) &= \langle V_{2k} \cdot \nabla_{2k}\chi_{\Omega_k}(Q_k, P_k)\rangle_{kt} \\ &= \langle \dot\chi_{\Omega_k}(Q_k, P_k)\rangle_{kt} = 0. \end{aligned}$$

The total time derivative of $\chi_{\Omega_k}(Q_k, P_k)$ is expressed in this case by
(26.23c)
$$\dot\chi_{\Omega_k}(Q_k, P_k) = \chi_{\Omega_k}(Q_k, P_k)\sum_{i \in k}\frac{1}{\chi_{\omega_k}(q_i, p_i)}\int_{\partial k} d^2 r\, \delta(q_i - r)\hat{n}_{\partial k}(r) \cdot v_i.$$

It is clear in (26.23c) that $\dot\chi_{\Omega_k}(Q_k, P_k)$ does not depend on the time. This means that

(26.23d) $$\frac{\partial \dot\chi_{\Omega_k}(Q_k, P_k)}{\partial t} = 0.$$

Taking the time derivative of (26.15) and using $\partial \mathfrak{F}_k^\epsilon/\partial t = 0$ with (26.23) yields

(26.24)
$$\frac{d\chi^2(F_k, F_k^\epsilon)}{dt} = 2\left\langle \frac{1}{\mathfrak{F}_k^\epsilon} \frac{\partial \mathfrak{F}_k(t)}{\partial t} \right\rangle_{kt}$$
$$= -2\left\langle [\mathfrak{F}_k^\epsilon]^{-1} V_{2k} \cdot \nabla_{2k}\mathfrak{F}_k(t) \right\rangle_{kt}$$
$$= 2\left\langle \dot{\chi}_{\Omega_k}(Q_k, P_k)[\mathfrak{F}_k^\epsilon]^{-1}\mathfrak{F}_k(t) \right\rangle_{kt},$$

where integration by parts and $V_{2k} \cdot \nabla_{2k}\mathfrak{F}_k^\epsilon = 0$ was also used. Notice that the change in $\chi^2(F_k, F_k^\epsilon)$ takes place only at the boundary.

Jensen's inequality will be used to show that $d\chi^2(F_k(t), F_k^\epsilon)/dt$ is non-positive and is 0 if and only if $F_k = F_k^\epsilon$. A weight function very similar to that of (9.19) is defined next by

(26.25)
$$U_i(P_k^\natural, P_l^\natural) = \frac{W_{kl}^{(u)}|\hat{n}_{\partial k} \cdot u_i^\natural-|F_k^{\epsilon\natural}F_l^{\epsilon\natural}}{|\hat{n}_{\partial k} \cdot u_i^+|F_k^\epsilon F_l^\epsilon},$$

where $u_i^\pm$ is defined in (6.30) and $u_i = v_i$ because $v_{\partial k} = 0$. The weight function U_i is normalized, as in (9.20a), so that

(26.26)
$$\int dP_k \int dP_l \, U_i(P_k^\natural, P_l^\natural) = 1.$$

It is also true that

(26.27a)
$$\int dP_k^\natural \int dP_l^\natural \, U_i \frac{F_k^\natural F_l^\natural}{F_k^{\epsilon\natural}F_l^{\epsilon\natural}} = \frac{F_k F_l}{F_k^\epsilon F_l^\epsilon},$$

(26.27b)
$$\int dP_k^\natural \int dP_l^\natural \, U_i \frac{F_k^\natural}{F_k^{\epsilon\natural}} = \frac{F_k}{F_k^\epsilon}.$$

Using Jensen's inequality, a relation similar to (9.21) is obtained, where in this case $C(F_k)$ is now the convex function $C(F_k) = [F_k]^2$ with the result

(26.28)
$$C\left(\int dP_k^\natural \int dP_l^\natural \, U_i \frac{F_k^\natural}{F_k^{\epsilon\natural}}\right) \leq \int dP_k^\natural \int dP_l^\natural U_i C\left(\frac{F_k^\natural}{F_k^{\epsilon\natural}}\right)$$

Multiplying both sides of this relation by $|\hat{n}_{\partial k} \cdot u_i^+|F_k^\epsilon F_l^\epsilon$ and integrating both sides over (P_k, P_l), gives the result

(26.29)
$$\int dP_k \int dP_l \, |\hat{n}_{\partial k} \cdot u_i^+| \frac{[F_k]^2 F_l^\epsilon}{F_k^\epsilon} \leq \int dP_k^\natural \int dP_l^\natural \, |\hat{n}_{\partial k} \cdot u_i^\natural-| \frac{[F_k^\natural]^2 F_l^{\epsilon\natural}}{F_k^{\epsilon\natural}}.$$

Next, both sides are integrated over (Q_k, Q_l), the dummy integration variables are changed from $(P_k^\natural, P_l^\natural)$ to (P_k, P_l), and the results are rearranged. This gives as in Chapter 9

(26.30)
$$\left\langle \dot{\chi}_{d_k}(q_i, v_i)[\mathfrak{F}_k^\epsilon]^{-1}\mathfrak{F}_k(t) \right\rangle_{kt} \leq 0.$$

Since, by (26.24), $d\chi^2(F_k, F_k^\epsilon)/dt$ is twice the sum of (26.30) over $i \in k$, it follows that

$$(26.31) \qquad \frac{d\chi^2(F_k, F_k^\epsilon)}{dt} \leq 0,$$

with equality if and only if $\mathfrak{F}_k(t) = \mathfrak{F}_k^\epsilon$ (a.e.).

Next, by (26.20), $\|\mathfrak{Y}_k(t)\|_0 \to 0$ implies $\chi^2(F_k, F_k^\epsilon) \to 0$; and, similarly, $\|\mathfrak{Y}_k(t)\|_0 \to \infty$ implies $\chi^2(F_k, F_k^\epsilon) \to \infty$. These properties of $\chi^2(F_k, F_k^\epsilon)$ along with (26.31), make it suitable as a Lyapunov function for F_k.[5]

With the help of (26.23d), the time derivative of (26.24) gives the second time derivative of $\chi^2(F_k, F^\epsilon)$ in the form

$$(26.32) \qquad \frac{d^2\chi^2(F_k, F_k^\epsilon)}{dt^2} = 4 \left\langle \dot{\chi}_{\Omega_k}(Q_k, P_k) \frac{1}{\mathfrak{F}_k^\epsilon} \frac{\partial \mathfrak{F}_k(t)}{\partial t} \right\rangle_{kt}.$$

Using $\partial \mathfrak{F}_k(t)/\partial t = -V_{2k} \cdot \nabla_{2k} \mathfrak{F}_k(t)$ in (26.32) and the upper bound stated in (26.18) with the upper bound M_{2k} on V_{2k} used in (25.4), (26.32) can be rewritten in the inner product form:

$$(26.33) \qquad \left| \frac{d^2\chi^2(F_k, F_k^\epsilon)}{dt^2} \right| \leq 4M_{2k}e^{\beta_w \mathcal{H}_k^u} \sum_{i \in k} (\mathfrak{F}_k(t), \nabla_{2k} \mathfrak{F}_k(t))_{\partial_i L^2}.$$

Since $\mathfrak{F}_k(t) \in H^s(\Omega_k)$ and $s \geq 1$, the embedding theorem (24.46a) is used to show that $\nabla_{2k} \mathfrak{F}_k(t)$ belongs to $L^2(\partial_i \Omega_k)$. It follows that

$$(26.34) \qquad \left| \frac{d^2\chi^2(F_k, F_k^\epsilon)}{dt^2} \right| < \infty.$$

The inequalities (26.16), (26.31), and (26.34) imply that[6]

$$(26.35) \qquad \lim_{t \to \infty} \chi^2(F_k, F_k^\epsilon) = 0.$$

This, in turn, implies by (26.31) that

$$(26.36) \qquad \lim_{t \to \infty} \mathfrak{F}_k(t) = \mathfrak{F}_k^\epsilon \qquad \text{a.e.}$$

which proves the assertion.

For the case of Boltzmann's equation, let

$$(26.37) \qquad Y_{ks} = \times_{i \in k} C(\mathbf{R}_+, H^s(\omega_i))$$

and, for $F_{[k]}, F_{[k]}^\epsilon \in Y_{ks}$,

$$(26.38) \qquad \chi^2(F_{[k]}, F_{[k]}^\epsilon) = \sum_{i \in k} \int d^3 q_i \int d^3 p_i \frac{[f_i(q_i, p_i, t) - f_i^\epsilon(q_i, p_i)]^2}{f_i^\epsilon(q_i, p_i)}.$$

---

[5]See Miller [1971], p. 342, and Driver [1962], p. 415.
[6]See Miller [1971], pp. 345–346.

The proof above can then be repeated with the same result, namely, that

(26.39) $$\lim_{t\to\infty} \mathfrak{F}_{[k]}(t) = \mathfrak{F}_{[k]}^\epsilon \qquad \text{a.e.}$$

In Chapter 9, the reference distribution L_k and then the equilibrium distribution F_k^ϵ were obtained as a consequence of maximizing S_k over Ω_k for Ω_k fixed with fixed momenta and energy. In the light of the interpretation of S_k developed in Chapter 16, this is equivalent to minimizing the information we have about system k. In accord with the results of Chapter 14 on the extreme values of S_k, it follows that the F_k^δ states represent states for which we have maximal information and that F_k^ϵ represents the state for which we have minimal information about system k.

CHAPTER 27

Dénouement

The work on the general classical thermodynamics of interacting systems is complete. Some final remarks are added as a survey of this new landscape and its *raison d'etre*. A final epilogue completes the book.

27.1 Thermodynamics and the Theory of Interacting Systems

The TIS synthesis of the diverse elements of macroscopic mechanics into a unified formalism that is consistent with the classical mechanics of the underlying particle system is complete. The theory includes aspects of classical macroscopic mechanics that are usually associated with the Newtonian mechanics of macroscopic bodies, hydrodynamics, standard thermodynamics, and various specialized theories of the behavior of gases, fluids, and solids. The scope of the theory is determined by the natural set of functions and relations that follow from the structures introduced to coordinate the macroscopic and microscopic levels of description. This is general thermodynamics in its widest sense operating at the proper level of abstraction.

The primary difference between this work and previous treatments is the TIS requirement that all the formulas of macroscopic mechanics and thermodynamics are rooted in the microscopic particle level of description. This has led for the first time to an acceptable nonequilibrium thermodynamics that includes a particle interpretation of the formulas and relations associated with the derivatives of thermodynamic quantities. It has also led to an adequate treatment of the entropy and the second law of thermodynamics in systems that are interacting with each other. The interest in the conserved quantities exchanged between macroscopic systems was motivated by the connections between the macroscopic and microscopic descriptions and led to the formulation of the TIS boundary formalism as part of the description of this interaction.

The fundamental requirements that the theory be mathematically cogent and physically motivated have conspired to modify the resulting formalism from that inherited from standard equilibrium thermodynamics. In spite of the facts that equilibrium thermodynamics is over 200 years old, the notion of entropy is over 135 years old, and the nonequilibrium thermodynamics provided by Maxwell and Boltzmann is about 130 years old, it was found in this investigation that many of the fundamental principles and much of the wisdom accumulated during this period cannot be retained as part of a general theory. In some cases, quantities could not be provided with particle analogs and were excluded from the general theory. In other cases, common practices, such as taking the thermodynamic limit, employing quasistatic processes, assuming weakly interacting systems, coarse-graining the system distribution, etc., were rejected as illegitimate in TIS.

The entropy in particular has been a source of what might be called a body of folklore concerning the various forms that it can take, the relations between these forms and other thermodynamic quantities, and its role in the evolution of particle systems. Unfounded theorems and results, some of which were loosely based on phenomenological observations and others simply on assumptions and conjectures, have been proposed and employed over the years. Later authors have quoted these as established results and used them to create further structures and relations. Still others have provided particle models in an attempt to give a physical interpretation for the growth of entropy at the particle level in terms of increased mixing, disorder, complexity, or some other macroscopic concept. Some of these beliefs are very deeply held and their truth is thought by many to be unquestionable.

These deeply held beliefs are one reason that the works of the original authors and their critics have played such an important role in this book. To establish the legitimacy of the basic ideas on which TIS is based, it was necessary to examine and come to terms with each version of these ideas in previous work. Earlier versions of thermodynamics were expressed and interpreted within the TIS formalism so that the continuity of ideas and their development over time could be explored. Some of these early works, such as that on pressure by Daniel Bernoulli and Boltzmann's attempt to define a temperature analog, provided calculations that were used in the inquiry that resulted in the definitions of particular thermodynamic quantities in the general theory.

The overall guiding principles of the Theory of Interacting Systems are the requirements that (i) thermodynamics must be established in a context suitable for describing a collection of all the systems that can interact with each other, (ii) the resulting thermodynamics must be consistent with the underlying mechanics, and (iii) equilibrium thermodynamics must follow from the nonequilibrium formalism. Creating a formalism consistent with these requirements was the key to working out some crucial relations involving the entropy and the transmission of conserved quantities between systems.

Another important principle in TIS thermodynamics is that all forces and all exchanges of physical quantities are ultimately between particles and must act at the locations of particles. This means that only those physical quantities that can be viewed as properties of the individual particles can be given a local interpretation and represented as local thermodynamic functions. This principle was used to exclude the calculation of the pressure based on Clausius' virial equation on the grounds that the treatment of the pressure as a force at the boundary is not warranted. This is because many of the particles contributing to that aspect of the force across the boundary embodied in the virial terms Ξ_k, which are obtained from the potential energy in the TIS formalism, are not located at the boundary. This principle was also used to show that the Euler-Cauchy and Navier-Stokes equations are specific kinds of approximations for fluid motion and that using them is justified only in limited circumstances.

The principle that every observable thermodynamic quantity must be represented by a microscopic analog function is also of fundamental importance in providing a way of distinguishing acceptable quantities from those that are not. The rejection of the thermodynamic potentials, for example, such as the free energy, free enthalpy, chemical potential, and some of their thermodynamic derivatives, was based on the failure to find a suitable analog. This association of observables with analogs was also the basis for calculating the time, volume, and thermature, partial derivatives of quantities defined in the theory and for defining the properties of thermodynamic quantities under Galilean transformations.

Another important principle is that macroscopic information and parameters cannot be used in the evolution equation for the particle system. This principle was used to reject some claims that the entropy, temperature, volume, or other macroscopic quantities, can influence the direction of the evolution of a system of particles in spite of the fact that they do not play a role in the equations of motion for the system. Statements such as 'isolated systems not at equilibrium evolve in such a way that their entropy increases' were therefore rejected because they do not have general validity.

Two approximations were introduced as fundamental to the development of the theory. The first is the requirement that the members of the collection of subsystems discussed in the theory are represented as independent of each other in the sense of probability. This essential assumption was introduced for epistemological reasons. The second approximation is the quasilocality assumption. This approximation is not a required aspect of the theory, but leads to the most interesting thermodynamics.

A significant innovation of the theory was the introduction of the system projection operator as part of the macroscopic phase averaging procedure. A boundary formalism was introduced in conjunction with this to maintain the closure of the systems and to account for what happens to particles reaching the boundary. The system projection operator works with the boundary formalism so that the time derivative of the system projection operator is the boundary flux operator. When the partial time derivative of a macroscopic thermodynamic function is computed, the total time derivative of the system projection operator obtained as part of the calculation is the boundary flux operator. This formalism therefore automatically tracks the transmission of fluxes of physical quantities through a boundary and maintains the ledger on conserved quantities. Only with such a formalism can the conservation of specific quantities in thermodynamics be seamlessly demonstrated and the connection with mechanics maintained.

Several special system states were introduced as part of the theory. These are the exact state, the equilibrium or canonical state, the absolute zero state, the microcanonical state, and the Khinchin equilibrium state. The absolute zero state was shown to be the exact state that is the limit of an equilibrium state as the temperature decreases to zero. The microcanonical state is rarely used in TIS and the Khinchin equilibrium state is not included in the general theory.

The work of Maxwell and Boltzmann, leading to the theory of Boltzmann's equation, was shown to follow from the general theory when the independent systems approximation is extended to the level of the particles themselves and the interactions between particles within a system are replaced by a collision term. This led to a local thermodynamics in which there is a local entropy and a local entropy current. The collision term was shown to be a stochastic approximation that transforms a particle i into and out of the mechanical state (q_i, p_i) during the evolution of the system in time. It was also demonstrated, without the need for arbitrary cutoffs, that the kernel of the representation of the collision term in the form of an integral equation is a tempered distribution.

A number of elements of standard thermodynamics were not included in the general TIS theory and are not allowed. Among these elements are the use of the thermodynamic limit, coarse-grained system distributions, phenomenological relations, local thermodynamic equilibrium, special particle correlations and their assumed relative decay rates, weakly interacting systems, reciprocity principles, normal systems, monocyclic systems, molecular chaos, Stosszahlansatz, and many other procedures and assumptions that have sometimes been used.

In the classical TIS theory of the relation between the macroscopic and microscopic worlds of interacting systems, the concept of probability has an essential role. The epistemological requirement of independence for the descriptions of systems means that the only microscopic particle coordinates allowed in the representation of a given system are the coordinates of the particles that belong to that system. This implies that the external force experienced by one system due to another system is the phase average of the force over the coordinates of the particles in the other system. Moreover, the TIS requirement of closure in the classical version of the theory means that particles reaching the wall are reflected back into the system to which they belong. A stochastic boundary formalism was introduced to account for this effect of the boundary on the system description. These probabilistic elements can only be dispensed with at the level of a single isolated system. The implication is that a probabilistic description is required for the components of a world composed of macroscopic systems constructed out of microscopic particles.

In its most abstract sense, the Theory of Interacting Systems was characterized as a metatheory concerned with the relation of macroscopic and microscopic mechanics. The extension of the TIS formalism introduced above to the quantum and relativistic cases requires minimal changes on the macroscopic level. Although the microscopic mechanics obeyed by the particles is different in these different theories, as are some aspects of the phase averaging map connecting the two levels, the thermodynamic formalism itself is quite insensitive to the underlying mechanics.

27.2 The Mathematical Aspects of TIS

The most important mathematical innovations in TIS are the introduction of function space considerations into the representation of the system distribution, the representation of the boundary and its motion in relation to the particle dynamics, the use of operators for thermodynamic transformations and derivatives, and the amalgamation of all the possible system states within a single mathematical formalism. These innovations allowed the theory to be put on a solid mathematical footing.

Finding a mathematical framework in which all of the states that can be attributed to a system in statistical mechanics have a place has meant finding a function space in which the states F_k, F^ϵ, F_k^δ, F_k^0, and F_k^γ, can coexist. The importance of this unified formalism is that it exhibits the roles of each of these states clearly and allows a meaningful comparison between them. This unified formalism is crucial to the reconciliation of the conceptual problems surrounding the entropy and the resolution of the paradoxes of thermodynamics.

Although an existence proof was given for the exact state F_k^δ, the primary focus in the mathematical discussions of Part IV has been on the general state F_k and the equilibrium state F_k^ϵ, which are central to most of thermodynamics. These states are required to belong to the family of Sobolev spaces $H^s(\Omega_k(t))$ that have a rich Hilbert space structure. This structure provides powerful tools for representation and analysis as well as a large body of mathematical results. These analytical tools, including the employment of embedding theorems to represent system boundaries as an integral part of the formalism, were used to advantage in the proofs of existence for the F_k state and the $F_{[k]}$ state of the TIS version of Boltzmann's theory.

The availability of powerful mathematical tools also has a practical advantage in that they can also be used to help find solutions or approximate solutions to the system evolution equation. As one example, the L^2 space structure and the local Lipschitz properties of $\chi_{\Omega_k(t)}(Q_k, P_k)F_k(t)$, $\chi_{d_k}(q_i, v_i)f_i(t)$, and $C[f]f_i(t)$ can be helpful in the construction of iterated solutions of the evolution equations.

There is some latitude in the mathematical choices made here. While the members of the family of Sobolev spaces $H^s(\Omega_k(t))$ are based on the requirement that elements of the theory and their derivatives to a given order belong to L^2 spaces, there are more general Sobolev spaces, based on L^p spaces and designated by $W^{s,p}(\Omega_k(t))$ with $H^s(\Omega_k(t)) = W^{s,2}(\Omega_k(t))$, that would do as well. Results parallel to those of the L^2 version of the proofs can be demonstrated using the L^p representation theorem in place of the L^2 version in places such as (25.9), etc.

For the system evolution equations in the Sobolev spaces chosen here, it was found that $s = 2$ for the space $H^s(\Omega_k(t))$ is the minimum index for which all of the elements of the theory are well defined. On the other hand, if s is chosen so that $s > m + 3\mathcal{N}_k$, the elements F_k of $H^s(\Omega_k(t))$ are at least m times continuously differentiable functions in (Q_k, P_k).[1] This case yields the "classical" existence proofs for distributions in phase space. In the Boltzmann case, $s = 6$ is required as the minimum index. By the above formula with $\mathcal{N}_k = 1$, the states in $H^6(\omega_i(t))$ are twice continuously differentiable in (q_i, p_i).

It was shown in Chapter 25 that the local thermodynamic densities, $G_k(q, t)$ are continuous and once differentiable in q and t with a singular boundary flux term. It also follows that the global surface densities, $G_{\partial k}(t)$, are continuous and differentiable in t. These results confirm the assumptions made on $G_k(q, t)$ and $G_{\partial k}(t)$ in Chapters 2 and 3 of Part I. It is also true that by an appropriate choice of s, higher derivatives of thermodynamic densities can be shown to exist, if needed.

The choice of working with an isolated system, with fixed and finite values for the total particle number, total phase momentum, total phase angular momentum, and total phase energy, and treating interacting systems as subsystems of this total system, allows the formalism to keep track of conserved quantities and maintain a connection with the particle mechanics. This connection is not possible in the usual formulations of thermodynamics which use arbitrary boundary conditions in undefined environments, and employ infinite heat reservoirs, thermodynamic limits, coarse-grained distributions, and quasistatic processes. On the other hand, the TIS formalism does not have the means to maintain a temperature and pressure at the boundary of a system that are rigidly fixed no matter what the system is doing. This means that the assumption in Chapter 26 that the boundary temperature has the fixed value β_w can only be approximately true.

27.3 Epistemological Reflections

Several related epistemological themes have played a role in the work in this book. These were centered on two issues that lie at the heart of thermodynamics. The first is the division of the world into parts for study. This standard procedure is used to limit the scope of an inquiry to a particular physical system. In classical physics, this procedure was felt to be benign if it was considered at all.

In the classical version of TIS presented here, it was found that a particle system, representing a macroscopic object, could be treated as independent of other systems representing other macroscopic objects only if its microscopic description is disengaged from the microscopic descriptions of all the systems around it. The Independent Systems Approximation was introduced to serve this purpose. While this was done to serve the needs of the thermodynamic formalism, it is an aspect of the more general concern, first raised by Bohr in quantum mechanics, with the need for the subject/object distinction in physics. Although the work in this volume was stimulated by a

---

[1]See Treves [1967], p. 331.

conjecture made by Bohr, it proceeded for many years in areas independent of his thought. In the end, the significance of the subject/object distinction in classical physics emerged and the importance of Bohr's thought to physics as a whole became clearer. These ideas had implications both for an adequate conceptual perspective on the physics of macroscopic systems constructed out of microscopic particles and for the formalism required to give an adequate representation of physical systems. In that way, the work in this volume, along with the work on other epistemological issues in the subsequent volumes, contributed to a deeper understanding of Bohr's work and allowed the completion of Volume 1.

The second major epistemological issue in thermodynamics and statistical mechanics is concerned with the resolution of the paradoxes of thermodynamics. These paradoxes are associated with the conflict between the irreversibility of thermodynamics and the reversibility and recurrence properties of the microscopic particle mechanics. The solution to this problem was based on the recognition that the properties of a macroscopic description in thermodynamics depends on the choice of the microscopic representation of the underlying particle system. The range of possibilities for the choice of a state to describe a microscopic system, from F_k^δ through F_k to F_k^ϵ, with specialized states F_k^γ and F_k^0, has implications for the physical quantities and their relations that can be part of the corresponding thermodynamic functions obtained as phase averages from them.

The most significant consequence of the choice of microscopic representations is the status of the macroscopic thermodynamic entropy $S_k(t)$. It was found in Chapter 16 that if a thermodynamic process is reversible, the entropy must not change in time in either the forward or backward direction or it must be undefined. Because $dS_t(t)/dt = 0$ for the rate of change of the entropy in the isolated system t, the evolution of the t system is reversible. Of the collection of microscopic states that play a role in the TIS theory, only the general F_k state for a subsystem k of the $[t]$ system leads to an irreversible thermodynamics.

These points are supported by the mathematics associated with the various microscopic descriptions. The formalism that was developed and used to establish the lower bound on the rate of change in the entropy (Axiom 6, Chapter 3) applies only to functions belonging to $H^s(\Omega_k(t))$ or $H^s(\omega_i(t))$. Since $F_k^\delta(Q_k, P_k, t)$ does not belong to a Sobolev space H^s and the phase entropy analog $-k_B \ln F_k^\delta(Q_k, P_k, t)$ is too singular to give a finite value for its phase average $S_k^\delta(t)$, the distribution $F_k^\delta(t)$ cannot be used in place of $F_k(Q_k, P_k, t)$ in the proofs concerned with Clausius' inequality. It was established in this way that the theory cannot be used to predict that the entropy of a system described by the exact state will increase. It follows that the F_k and F_k^δ descriptions of the system cannot be brought into conflict with each other because both descriptions cannot be used simultaneously to describe a given system.

Another aspect of the TIS analysis is the interplay between part and whole. This issue was examined with regard to its implications for physics in RSO. A concern with both the part and the whole in the TIS theory presented here allowed the solution of certain problems that could not even be expressed adequately in standard thermodynamics. As an example, there are problems with the principles in the standard thermodynamics concerned with the entropy. As Partington [**1949**], p. 194, stated it, "The principle that there is a *net* increase in the entropy in all irreversible changes is so general that a *formal* proof which does not depend on the dissipation of energy, or some other very general assertion, can hardly be given." (Italics Partington's.) It was also shown in Chapter 16 that the choice of how to divide the whole into parts has an effect on the value of the entropy calculated for the divided system. This emphasizes, again, the unique epistemological status for entropy in the theory as a connecting link between the physics of the particle system and the macroscopic information we have about the system.

It was well known from the nineteenth century that relations involving the entropy could not be established mathematically as a consequence of the laws of mechanics. This was dealt with in TIS by showing that the resolution of the paradoxes of thermodynamics requires an epistemological solution. Understanding the conventional and arbitrary aspects of the description of microscopic particle systems by independent system distributions and the effect of this arbitrariness on the entropy that could be attributed to these systems is necessary for the resolution of the paradoxes.

In the TIS version of thermodynamics, it is not mechanics or statistical mechanics that is being modified in the resolution of these paradoxes but our understanding of how a collection of macroscopic interacting mechanical systems composed of microscopic particles can be described consistently from the perspectives of both the part and whole. The unexpected intrusion of probability into classical thermodynamics that followed from this, as an indispensable aspect of the independence of systems in the TIS formalism, is another fundamental break with the worldview inherited from materialist philosophy. This is one more step down the path initiated in the middle of the nineteenth century by Gustav Kirchhoff, who proclaimed in the introduction to his mechanics that it is the purpose of mechanics to describe the motions of the world as simply as possible. With this step, he set in motion a revolution that has come to see that the description of a physical system itself plays an essential role in physics. These issues will be studied further in EIS and QIS.

27.4 Epilogue

The erosion of mountains, the progression of sand dunes into forests, and the evolution of jungles into deserts, are parts of both a dynamic ecology and the reaction of a vast collection of particles to pushes and pulls. Each of these descriptions is correct. Reconciling them was the issue.

The reconciliation of thermodynamics and mechanics provided by the Theory of Interacting Systems removed the limitation of previous theories to equilibrium settings and animated the theory of the macroscopic world. This brought thermodynamics to life and moved it out of the frozen Victorian world in which it was born. It can operate now at the speed of the engines it describes.

The possibility of obtaining an answer to the question of whether to blame the butterflies in Brazil for gales in the Bahamas depends on whether the macroscopic approximations of microscopic mechanical behavior and the tools we have to manipulate this information are sufficiently precise. But we can turn back the rising tide of entropy that Rudolph Clausius proclaimed in 1865. Whether the physical description of the universe is expressed as a deterministic march of atoms, in which William Thomson foresaw the heat death of the universe, or as a colloquy of quantum waves, it is no longer necessary or correct to view the evolution of physical systems as being controlled by the need for entropy to increase. Understood in this way, entropy is more like beauty than we ever imagined.

CIS Bibliography

[1978]   R. Abraham and J. Marsden, *Foundations of Mechanics*, 2nd ed., The Benjamin Cummings Publ. Co., Inc., Reading Mass., 1978.

[1998]   M. Adams, Z. Dogic, S. L. Keller and S. Fraden, *Entropically driven microphase transitions in mixtures of colloidal rods and spheres*, Nature **393** (1998), 349–352.

[1975]   R. A. Adams, *Sobolev Spaces*, Academic Press, New York, 1975.

[1958]   S. Asakura and F. Oosawa, *Interaction between particles suspended in solutions of macromolecules*, Journal of Polymer Science **33** (1958), 183–192.

[1974]   V. I. Arnold, *Mathematical Methods of Classical Mechanics*, 1974, translated by K. Vogtmann and A. Weinstein, Springer Verlag, New York, 1978.

[1968]   V. I. Arnold and A. Avez, *Ergodic Problems of Classical Mechanics*, W. A. Benjamin, Inc., New York, 1968.

[1994]   M. Bailyn, *A Survey of Thermodynamics*, AIP Press, New York, 1994.

[1972]   O. C. de Beauregard and M. Tribus, *Information Theory and Thermodynamics*, Helvetia Physica Acta **47** (1972), 238–247; it is reprinted as pp. 173–182 in H. S. Leff and A. F. Rex [**1990**].

[1982]   C. H. Bennett, *The Thermodynamics of Computation—A Review*, Int. J. Theoretical Physics **21** (1982), 905–940.

[1965]   H. Bent, *The Second Law*, Oxford Univ. Press, New York, 1965.

[1955]   P. G. Bergmann and J. L. Lebowitz, *New Approach to Nonequilibrium Processes*, Physical Review **99** (1955), 578–587.

[1738]   D. Bernoulli, *On the Properties and Motions of Elastic Fluids, Especially Air*; this is an extract from Bernoulli's Hydrodynamica..., translated by J. P. Berryman and printed as pp. 57–65 in Brush [**1965**], Vol. 1.

[1992]   G. Bierhalter, *Von L. Boltzmann bis J. J. Thomson: die Versuche einer mechanicschen Grundlegung der Thermodynamik (1866–1890)*, Archive for History of Exact Sciences **44** (1992), 25–75.

[1931]   G. D. Birkhoff, *Proof of the Ergodic Theorem*, Proc. National Academy of Sciences **17** (1931), 656–660.

[1932]   G. D. Birkhoff and B. O. Koopman, *Recent Contributions to the Ergodic Theory*, Proc. National Academy of Sciences **18** (1932), 279–282.

[1803]   J. Black, *Lectures on the Elements of Chemistry*, compiled and edited posthumously by J. Robison, W. Creech, Edinburgh, 1803.

[1959]   J. M. Blatt, *An Alternative Approach to the Ergodic Problem*, Progress in Theoretical Physics **22** (1959), 745–756.

[1962]   J. de Boer and G. E. Uhlenbeck, eds., *Studies in Statistical Mechanics*, North-Holland Publishing Co., Amsterdam, 1962.

[1946]   N. N. Bogoliubov, *Problems of a Dynamical Theory in Statistical Physics*, an English translation of this paper appears in de Boer and Uhlenbeck [**1962**].

[1930]   N. Bohr, *Faraday Lecture: Chemistry and the Quantum Theory of Atomic Constitution*, J. Chemical Society **pt. I** (1932), 349–384; the lecture was presented in 1930.

[1866]   L. Boltzmann, *Über die mechanische Bedeutung des zweiten Hauptsatzes der Wärmetheorie*, Sitzungsberichte der Kaiserlichen Akademie der Wissenshaft, Mathematisch-Naturwissenshaftliche Klasse **53, II** (1866), 195–220; it is reprinted in Boltzmann [**1909**], Vol. I, pp. 9–33.

[1868]   _____, *Studien über das Gleichgewicht der lebendingen Kraft zwischen bewegten materiellen Punkte*, Sitzungsberichte der Kaiserlichen Akademie der Wissenshaft, Mathematisch-Naturwissenshaftliche Klasse **58, II** (1868), 517–560; it is reprinted in Boltzmann [**1909**], Vol. I, pp. 49–96.

[1871a]   _____, *Einige allgemeine Sätze über Wärmegleichgewicht*, Sitzungsberichte der Kaiserlichen Akademie der Wissenshaft, Mathematisch-Naturwissenshaftliche Klasse **63, II** (1871), 679–711; it is reprinted in Boltzmann [**1909**], Vol. I, pp. 259–287.

[1871b]   _____, *Analytischer Beweis des zweiten Hauptsatzes der mechanischen Wärmetheorie aus den Sätzen über das Gleichgewicht der lebindigen Kraft*, Sitzungsberichte der Kaiserlichen Akademie der Wissenshaft, Mathematisch-Naturwissenshaftliche Klasse **63, II** (1871), 712–732; it is reprinted in Boltzmann [**1909**], Vol. I, pp. 288–308.

[1872]   _____, *Weitere Studien über das Wärmegleichgewicht unter Gasmolekülen*, Sitzungsberichte der Kaiserlichen Akademie der Wissenshaft, Mathematisch-Naturwissenshaftliche Klasse **66, II** (1872), 275–370; reprinted in Boltzmann [**1909**], Vol. I, pp. 316–402. This is also printed in an English translation as "Further Studies on the Thermal Equilibrium of Gas Molecules" in Brush [**1965**], Vol. 2, pp. 88–175.

[1875]   _____, *Über das Wärmegleichgewicht von Gasen auf welche äussere Kräfte wirken*, Sitzungsberichte der Kaiserlichen Akademie der Wissenshaft, Mathematisch-Naturwissenshaftliche Klasse **72, II** (1875), 427–457; it is reprinted in Boltzmann [**1909**], Vol. II, pp. 1–30.

[1876]   _____, *Über die Aufstellung und Integration der Gleichungen welche die Molecularbewegung in Gases bestimmen*, Sitzungsberichte der Kaiserlichen Akademie der Wissenshaft, Mathematisch-Naturwissenshaftliche Klasse **74, II** (1876), 503–552; it is reprinted in Boltzmann [**1909**], Vol. II, pp. 55–102.

[1877a]   _____, *Bemerkungen über einige Probleme der mechanischen Wärmetheorie*, Sitzungsberichte der Kaiserlichen Akademie der Wissenshaft, Mathematisch-Naturwissenshaftliche Klasse **75, II** (1877), 62–100; it is reprinted in Boltzmann [**1909**], Vol. II, pp. 112–148.

[1877b]   _____, *Über die Beziehung zwischen dem zweiten Hauptsatze der mechanische Wärmetheorie und der Wahrscheinlichkeitsrechnung respektive den Sätzen über das Wärmegleichgewicht*, Sitzungsberichte der Kaiserlichen Akademie der Wissenshaft, Mathematisch-Naturwissenshaftliche Klasse **76, II** (1877), 373–435; it is reprinted in Boltzmann [**1909**], Vol. II, pp. 164–223.

[1878a]   _____, *Weitere Bemerkungen über einige Probleme der mechanische Wärmetheorie*, Sitzungsberichte der Kaiserlichen Akademie der Wissenshaft, Mathematisch-Naturwissenshaftliche Klasse **78, II** (1878), 7–46; it is reprinted in Boltzmann [**1909**], Vol. II, pp. 250–288.

[1878b]   _____, *Über die Beziehung der Diffusionsphanomene zum zweiten Hauptsatze der mechanischen Wärmetheorie*, Sitzungsberichte der Kaiserlichen Akademie der Wissenshaft, Mathematisch-Naturwissenshaftliche Klasse **78, II** (1878), 733–763; it is reprinted in Boltzmann [**1909**], Vol. II, pp. 289–317.

[1882]   _____, *On Boltzmann's Theorem on the average Distribution of Energy in a System of Material Points*, Philosophical Magazine [5] **14** (1882), 299–312; it is printed in the original German in Boltzmann [**1909**], Vol. II, pp. 582–595.

[1894]   _____, *On Maxwell's Method of deriving the Equations of Hydrodynamics from the Kinetic Theory of Gases*, p. 579 in the British Association for the Advancement of Science (1894); (1894 Report), John Murray, London, this is reprinted as pp. 506–507 in Boltzmann [**1909**].

[1895] ———, *On Certain Questions of the Theories of Gases*, Nature **51** (1895), 413–415; it is reprinted in Boltzmann [**1909**], Vol. III, pp. 543–544.

[1896a] ———, *Lectures on Gas Theory*, translated by S. G. Brush, Univ. of California Press, Berkeley, 1964; it is a translation of "Vorlesungen über Gastheorie," parts I and II, Leipzig, J. A. Barth, 1896, 1898.

[1896b] ———, *Entgegnung auf die wärmetheoretischen Betrachtungen des Hrn. E. Zermelo*, Annalen der Physik **57** (1896), 773–784; it is reprinted in Boltzmann [**1909**], Vol. III, pp. 567–578. It also appears in an English translation as "Reply to Zermelo's Remarks on the Theory of Heat" in Brush [**1965**], Vol. 2, pp. 218–228.

[1897] ———, *Zu Hrn. Zermelos Abhandlung "Über die mechanische Erklärung irreversibler Vorgänge"*, Annalen der Physik **60** (1897), 392–398; it is reprinted in Boltzmann [**1909**], Vol. III, pp. 579–586. An English translation appears as "On Zermelo's Paper 'On the Mechanical Explanation of Irreversible Processes' " in Brush [**1965**], Vol. 2, pp. 238–245.

[1909] ———, *Wissenschaftliche Abhandlungen*, voles. I, II, III, J. A. Barth, Leipzig, 1909.

[1921] M. Born, *Kritische Betrachtungen zur traditionellen Darstellung der Thermodynamik*, Physikalische Zeitschrift **22** (1921), 218–224, 249–254, 282–286.

[1949] M. Born and H. S. Green, *A General Kinetic Theory of Liquids*, Cambridge Univ. Press, Cambridge, 1949.

[1996] W. Breymann, T. Tél and J. Vollmer, *Entropy Production for Open Dynamical Systems*, Physical Review Letters **77** (1996), 2945–2948.

[1962] L. Brillouin, *Science and Information Theory*, Academic Press, New York, 1962.

[1961] F. E. Browder, *On the Spectral Theory of Elliptic Differential Operators I.*, Mathematische Annalen **142** (1961), 22–130.

[1930] C. Brunold, *L'Entropie: son role dans le développement historique de la thermodynamique*, Masson et Cie, Éditeurs, Paris, 1930.

[1961] S. G. Brush, *Functional Integrals and Statistical Physics*, Reviews of Modern Physics **33** (1961), 79–92.

[1962] ———, *Development of the Kinetic Theory of Gases. VI. Viscosity*, American Journal of Physics **30** (1962), 269–281.

[1965] ———, *Kinetic Theory 1: The Nature of Gases and Heat; 2: Irreversible Processes; 3: The Chapman-Enskog Solution of the Transport Equation for Moderately Dense Gases*, Pergamon Press, Oxford, 1965, 1966, 1972.

[1970] ———, *The Wave Theory of Heat: A Forgotten Stage in the Transition from the Caloric Theory to Thermodynamics*, British Journal for the History of Science **5** (1970), 145–167.

[1976] ———, *The Kind of Motion we call Heat, Book 2: A History of the Kinetic Theory of Gases in the 19th Century, Statistical Physics and Irreversible Processes*, North Holland Publ. Co, New York, 1976.

[1985] ———, *Statistical Physics and the Atomic Theory of Matter*, Princeton University Press, Princeton, N. J., 1985.

[1891] G. H. Bryan, *Report of a Committee consisting of Messrs. J. Larmor and G. H. Bryan on the present state of our knowledge of Thermodynamics, specially with regard to the Second Law*, British Association for the Advancement of Science (1891); Report of the 61st Meeting, John Murray, London, 1892.

[1894] S. H. Burbury, *Boltzmann's Minimum Theorem*, Nature **51** (1894), 78.

[1909] C. Carathéodory, *Untersuchungen über die Grundlagen der Thermodynamik*, Mathematische Annalen **67** (1909), 355–386; an English translation by J. Kestin appears in Kestin [**1976**], pp. 229–256.

[1919] ———, *Über den Wiederkehrsatz von Poincaré*, Sitzungsberichte der Deutschen Akademie der Wissenschaft, Physikalisch-mathematischen Klasse (1919), 733–763.

[1824]   N. -L. -S. Carnot, *Réflexions sur la puissance motrice du feu et sur les moyens propre à la developer*, A. Blanchard, Paris, 1824; translated into English by R. H. Thurston and titled "Reflections on the motive power of Heat accompanied by An Account of Carnot's Theory by Sir William Thomson (Lord Kelvin),", John Wiley & Sons, New York, 1897.

[1969]   R. Carroll, *Abstract Methods in Partial Differential Equations*, Harper and Row, New York, 1969.

[1975]   C. Cercignani, *Theory and Application of the Boltzmann Equation*, Elsevier, New York, 1975.

[1998]   ———, *Ludwig Boltzmann: The Man Who Trusted Atoms*, Oxford University Press, Oxford, 1998.

[1997]   C. Cercignani, V. I. Gerasimenko and D. Ya. Petrina, *Many-Particle Dynamics and Kinetic Equations*, Kluwer Academic Publishers, Dordrecht, 1997.

[1977]   G. J. Chaitlin, *Algorithmic Information Theory*, IBM Journal of Research and Development **21** (1977), 350–359.

[1939]   S. Chapman and T. G. Cowling, *The Mathematical Theory of Non-Uniform Gases*, Cambridge University Press, Cambridge, 1964.

[1989]   M. Chen, *On the thermodynamic solution of the Boltzmann equation and nonlinear irreversible thermodynamics*, Journal of Mathematical Physics **30** (1989), 1329–1337.

[1969]   G. Choquet, *Lectures on Analysis*, Vol. I, "Integration and Topological Vector Spaces," Vol. II, "Representation Theory" and Vol. III, "Infinite Dimensional Measures and Problem Solutions", W. A. Benjamin, Inc., New York, 1969.

[1834]   É. Clapeyron, *Mémoire sur la Puissance Motrice de la Chaleur*, Journal de l'Ecole Polytechnique **XIV** (1834), 153–190; a translation by E. Mendoza with the title "Memoir on the Motive Power of Fire" appears as pp. 73–105 in Mendoza [**1977**]. It is also reprinted in part as pp. 36–51 in Kestin [**1976**].

[1850]   R. Clausius, *Über die bewegende Kraft der Wärme und die Gesetz, welche sich daraus für de Wärmelehre selbst ableiten lassen*, Annalen der Physik und Chemie **155** (1850), 368–397; an English translation is "On the Moving Force of Heat and the Laws Regarding the Nature of Heat Itself which are Deducible Therefrom," *Philosophical Magazine [4]* **2** (1850) pp. 1–20, 102–119. It is also printed in another translation by W. F. Magie as pp. 107–152 in Mendoza [**1977**].

[1854]   ———, *Über eine veränderte Form des zweiten Hauptsatzes der mechanischen Wärmetheorie*, Annalen der Physik und Chemie **169** (1854), 481–506; this appears in English translation as "On a modified form of the second fundamental theorem in the mechanical theory of heat," *Philosophical Magazine [4]* **12** (1856), pp. 81–98.

[1856]   ———, *Ueber die Anwendung der mechanischen Wämetheorie auf die Dampfmaschine*, Annalen der Physik **97** (1856), 441–476, 513–559; it is printed in translation as "On the Application of the Mechanical Theory of Heat to the Steam Engine," *Philosophical Magazine [4]* **12** (1856), pp. 241–265, 338–354, 426–443.

[1857]   ———, *Ueber die Art der Bewegung, welche wir Wärme nennen*, Annalen der Physik **100** (1857), 353–380; it is printed in translation as "The Nature of Motion We call Heat," *Philosophical Magazine [4]* **14** (1857), pp. 108–127, and in S. G. Brush [**1965**], Vol. 1, pp. 111–134.

[1862]   ———, *Ueber die Anwendung des Satzes von der Aequivalenz der Verwandlungen auf die innere Arbeit*, Annalen der Physik und Chemie **192** (1862), 73–113; an English translation appears as "On the Application of the Theorem of the Equivalence of Transformations to the Internal Work of a Mass of Matter," *Philosophical Magazine [4]* **24** (1862), pp. 81–97, 201–213. It is also reprinted in J. Kestin [**1976**], pp. 133–161.

[1865]   _____, *Über vershiedene für die Anwendung bequeme Formen der Hauptgleichungen der mechanishen Wärmetheorie*, Annalen der Physik **125** (1865), 353–400; it is translated by R. B. Lindsay and printed as "On Different Forms of the Fundamental Equations of the Mechanical Theory of Heat and their Convenience for Application" in J. Kestin [**1976**], pp. 162–193.

[1866]   _____, *Ueber umkehrbare und nicht umkehrbare Vorgänge in ihre Beziehung auf die Wärmetheorie*, Zeitschrift für Mathematik und Physik **11** (1866), 445–462.

[1870]   _____, *Ueber einen auf die Wärme anwendbaren mechanischen Satz*, Sitzungsberichte der Niedderrheinischen Gesellschaft, Bonn (1870), 114–119; it appears in an English translation as "On a Mechanical Theorem Applicable to Heat," *Philosophical Magazine [4]* **40** (1870), pp. 122–127, and in S. G. Brush [**1965**], Vol. 1, pp. 172–178.

[1878]   _____, *Ueber die Beziehung der durch Diffusion geleisteten Arbeit zum zweiten Hauptsatze der mechanischen Wärmetheorie*, Ann. der Physik **4** (1878), 341–343; it appears in English translation as "On the Relation of the Work Performed by Diffusion to the Second Proposition of the Mechanical Theory of Heat," *Philosophical Magazine [5]* **6** (1878), pp. 237–238.

[1879]   _____, *Die Mechanische Wärmetheorie*, F. Vieweg u. Sohn, Braunschweig, 1879; it is translated by W. R. Browne and appears with the title "The Mechanical Theory of Heat," Macmillan & Co., London, 1879.

[1973]   E. G. D. Cohen and W. Thirring, eds., *The Boltzmann Equation, Theory and Applications*, Springer-Verlag, New York, 1973.

[1963]   B. Coleman and W. Noll, *The Thermodynamics of Elastic Materials with Heat Conduction and Viscosity*, Archive for Rational Mechanics and Analysis **13** (1963), 169–178.

[1890]   E. P. Culverwell, *Note on Boltzmann's Kinetic Theory of Gases and on Sir W. Thomson's Address to Section A British Association 1884*, Philosophical Magazine [5] **30** (1890), 95–99.

[1939]   D. van Dantzig, *On the Phenomenological Thermodynamics of Moving Matter*, Physica **6** (1939), 673–704.

[1992]   O. Darrigol, *From c-Numbers to q-Numbers*, University of California Press, Berkeley, California, 1992.

[1966]   J. Darrozès and J. -P. Guiraud, *Généralisation formelle du théorème H en présence de parois. Application*, Comptes rendus hebdomadaires des Seances de l'Académie des Sciences (Paris) **A 262** (1966), 1368–1371.

[1970]   E. E. Daub, *Entropy and Dissipation*, Historical Studies in the Physical Sciences **2** (1970), 321–354.

[1985]   K. G. Denbigh and J. S. Denbigh, *Entropy in relation to incomplete knowledge*, Cambridge Univ. Press, Cambridge, 1985.

[1969]   R. E. Dickerson, *Molecular Thermodynamics*, W. A. Benjamin Inc., New York, 1969.

[1960]   J. Dieudonné, *Foundations of Modern Analysis*, Academic Press, New York, 1960.

[1996]   A. D. Dinsmore, A. G. Yodh and D. J. Pine, *Entropic control of particle motion using passive surface microstructures*, Nature **383** (1996), 239–242.

[1998]   A. D. Dinsmore, D. T. Wong, P. Nelson and A. G. Yodh, *Hard Spheres in Vesicles: Curvature-Induced Forces and Particle-Induced Curvature*, Physical Review Letters **80** (1998), 409–412.

[1934]   F. G. Donnan, *Activities of Life and the Second Law of Thermodynamics*, Nature **133** (1934), 99.

[1934]   F. G. Donnan and E. A. Guggenheim, *Activities of Life and the Second Law of Thermodynamics*, Nature **133** (1934), 530, 869; **134** (1934), 255.

[1977]   J. R. Dorfman and H. van Beijeren, *The Kinetic Theory of Gases*, pp. 65–179 in B. J. Berne, ed., Statistical Mechanics, part B. Time Development Processes, Plenum Press, New York, 1977.

[1962]   R. Driver, *Existence and Stability of a Delay Differential System*, Archive for Rational Mechanics and Analysis **10** (1962), 401–426.

[1959]   R. Dugas, *La Théorie Physique au sens de Boltzmann et ses Prolongements Modernes*, E'ditions du Griffon, Neuchatel-Suisse, 1959.

[1891]   P. Duhem, *Sur la Continuité entre l'état Liquide et l'état Gazeux et sur la Théorie Général des Vapeurs*, Université de France; Travaux et Mèmoires des Facultés de Lille **1** (1891), 1–105.

[1903]   _____, *Recherches sur l'Hydrodynamique Premier Serie*, Gauthier-Villars, Paris, 1903; it is reprinted in a new edition as "En vente au service de Documentation et D'Information Technique de l'Aéronautique," Paris, 1961.

[1911]   _____, *Traité d'Energetique ou de Thermodynamique Générale*, in two volumes, Gauthier-Villars, Paris, 1911.

[1820]   P. L. Dulong and A. T. Petit, *Recherches sur la Mesure des températures et sur les lois de la communication de la chaleur*, Journal de l'Ecole Polytechnique **XI** (1820), 189–294.

[1998]   J. Earman and J. D. Norton, *Exorcist XIV: The Wrath of Maxwell's Demon. Part I. From Maxwell to Szilard*, Studies in History of Modern Physics **29** (1998), 435–471.

[1935]   A. Eddington, *The Nature of the Physical World*, Cambridge University Press, Cambridge, 1953.

[1912]   P. Ehrenfest and T. Ehrenfest, *The Conceptual Foundations of the Statistical Approach in Mechanics*, Cornell University Press, Ithaca, New York, 1959; it is translated by M. J. Moravcsik from *Begriffliche Grundlagen der Statistische Auffassung in der Mechanik*, "Encyclopädie der mathematischen Wissenschaften mit Einschluss ihrer Anwendung," Bd. 4, Teil 32, Teubner, Leipzig, 1912.

[1920]   P. Ehrenfest and V. Trkal, *Deduction of the dissociation-equilibrium from the theory of quanta and a calculation of the chemical constant based on this*, Proceedings of the Amsterdam Academy **23** (1920), 162–183; this paper also appears in German as, *Ableitung des Dissociationsgleichgewicht aus der Quantentheorie und darauf beruhende Berechnung der Chemischen Konstanten*, Annalen der Physik **65** (1921), 609–628; it also appears as pp. 414–439 in P. Ehrenfest, *Collected Scientific Papers*, edited by M. Klein, North-Holland Publishing Co., Amsterdam, 1959.

[1902]   A. Einstein, *Kinetische Theorie des Wärmegleichgewichtes und des zweiten Hauptsatzes der Thermodynamik*, Annalen der Physik **9** (1902), 417–433; this is reprinted as pp. 56–75 in Vol. 2 of Einstein [**1996**]. An English translation appears as pp. 30–47 in volume 2 of the English translation of these papers.

[1903]   _____, *Eine Theorie der Grundlagen der Thermodynamik*, Annalen der Physik **11** (1903), 170–187; this is reprinted as pp. 76–97, in Vol. 2 of Einstein [**1996**]. An English translation appears as pp. 48–67 in volume 2 of the English translation of these papers.

[1996]   _____, *The Collected Papers of Albert Einstein*, edited by M. J. Klein, et al., Vols. 1–9, Princeton University Press, Princeton, New Jersey, 1987-1996.

[1931]   P. S. Epstein, *Critical Appreciation of Gibbs Statistical Mechanics*, pp. 461–519 in Haas [**1936d**].

[1937]   _____, *Textbook of Thermodynamics*, John Wiley & Sons, Inc, New York, 1937.

[1979]   B. C. Eu, *Kinetic Theory of Dense Fluids. I*, Annals of Physics **118** (1979), 187–229.

[1996]   B. C. Eu and L. S. García-Colin, *Irreversible processes and temperature*, Physical Review **E 54** (1996), 2501–2512.

[1972]   I. E. Farquhar, *Ergodicity and Related Topics*, pp. 29–104 in J. Biel and J. Rae, eds., "Irreversibility in the Many-Body Problem", Plenum Press, New York, 1972.

[1936]   R. H. Fowler, *Statistical mechanics, the theory of the properties of matter in equilibrium*, 2nd ed., Cambridge University Press, Cambridge, 1936.

[1873]   J. W. Gibbs, *A method of geometrical representation of the thermodynamic properties of substances by means of surfaces*, Transactions of the Connecticut Academy **II** (1873), 382–404; it is reprinted in Gibbs [**1928**], Vol. I, pp. 33–54.

[1875]   ———, *On the Equilibrium of Heterogeneous Substances*, Transactions of the Connecticut Academy **III** (1875–77), 108–248; ibid. (1877–78), 343–524, It is reprinted in Gibbs [**1928**], Vol. I, pp. 55–353.

[1902]   ———, *Elementary Principles in Statistical Mechanics developed with especial reference to the rational foundation of thermodynamics*, Yale University Press, New Haven, Rhode Island, 1902; it is reprinted in Gibbs [**1928**], Vol. II, part I. It was also reprinted in 1981 by Ox Bow Press, Woodbridge, Connecticut.

[1928]   ———, *The Collected Works of J. Willard Gibbs*, Vols. I and II, Longmans Green & Co., New York, 1928.

[1950]   H. Goldstein, *Classical Mechanics*, Addison-Wesley Publishing Co., Reading Massachusetts, 1950.

[1961]   H. Grad, *The Many Faces of Entropy*, Communications on Pure and Applied Mathematics **XIV** (1961), 323–354.

[1947]   H. S. Green, *General Kinetic Theory of Liquids II. Equilibrium Properties*, Proceedings Royal Society of London **189** (1947), 103–117; it is reprinted as Chapter II in Born and Green [**1949**].

[1967]   M. Gurtin and W. Williams, *An Axiomatic Foundation for Continuum Thermodynamics*, Archive for Rational Mechanics and Analysis **26** (1967), 3–117.

[1955]   D. ter Haar, *The Foundations of Statistical Mechanics*, Reviews of Modern Physics **27** (1955), 289–338.

[1936a] A. Haas, *Gibbs and the Statistical Conception of Physics*, pp. 161–178 in Haas [**1936d**].

[1936b] ———, *The Chief Results of Gibbs' Statistical Mechanics*, pp. 179–296 in Haas [**1936d**].

[1936c] ———, *Special Commentary on Gibbs' Statistical Mechanics*, pp. 297–460 in Haas [**1936d**].

[1936d] ———, *A Commentary on the Scientific Writings of J. W. Gibbs*, Vol. 2, Theoretical Physics, Yale University Press, New Haven, 1936.

[1928]   R. V. L. Hartley, *Transmission of Information*, Bell System Technical Journal **7** (1928), 535–563.

[1882]   H. von Helmholtz, *Die Thermodynamik chemischer Vorgänge*, Sitzungsberichte der konigliche preussischen Akademie der Wissenschaften, Physikalisch-mathematische Klasse (Berlin) **I** (1882), 22–39; it is reprinted as, 958–978 in von Helmholtz [**1895**], Vol. 2.

[1886]   ———, *Ueber die physikalische Bedeutung des Princips der kleinsten Wirkung*, Journal der reine angewandte Mathematik **100** (1886), 137–166, 213–222; it is reprinted in von Helmholtz [**1895**], **3**, 203–248.

[1887]   ———, *Zur Geshichte des Princips der kleinsten Wirkung*, Sitzungsberichte der konigliche preussischen Akademie der Wissenschaften, Physikalisch-mathematische Klasse (Berlin) **II** (1887), 225–236; it is reprinted in von Helmholtz [**1895**], **3**, 249–263.

[1805]   ———, *Wissenshaftliche abhandlunqen*, in 3 volumes, 1882–1895, J. A. Barth, Leipzig, 1895.

[1958]   T. L. Hill, *Statistical Mechanics: Principles and Selected Applications*, Dover
          Publications, Inc., New York, 1987.

[1954]   J. O. Hirschfelder, C. F. Curtiss and R. B. Bird, *Molecular Theory of Gases and
          Liquids*, John Wiley & Sons, Inc., New York, 1954.

[1985]   H. Hollinger and M. Zenzen, *The Nature of Irreversibility: A Study of its Dy-
          namics and Physical Origin*, D. Reidel Publishing Co., Dordrecht, Netherlands,
          1985.

[1932]   E. Hopf, *On the Time Average Theorem in Dynamics*, Proc. National Academy
          of Sciences **18** (1932), 93–100.

[1963]   K. Huang, *Statistical Mechanics*, John Wiley & Sons, New York, 1963.

[1981]   K. Hutchinson, *Rankine, Atomic Vortices and the Entropy Function*, Archives
          Internationales D'Histoire des Sciences **31 # 106** (1981), 72–134.

[1950]   J. H. Irving and J. G. Kirkwood, *The Statistical Mechanical Theory of Transport
          Processes. IV. The Equations of Hydrodynamics*, Journal of Chemical Physics **18**
          (1950), 817–829; it is reprinted in Kirkwood [**1967**], pp. 52–75.

[1957]   M. Jammer, *Concepts of Force*, Harper and Brothers, New York, 1962.

[1972]   J. M. Jauch and J. G. Báron, *Entropy, Information and Szilard's Paradox*, Hel-
          vetia Physica Acta **45** (1972), 220–232; it is reprinted as pp. 160–172 in H. S. Leff
          and A. F. Rex [**1990**].

[1957]   E. T. Jaynes, *Information Theory and Statistical Mechanics. I, II*, Physical Re-
          view **106** (1957), 620–630; ibid. **108** (1957), 171–190.

[1965]   _____, *Gibbs vs Boltzmann Entropy*, American Journal of Physics **33** (1965),
          391–398.

[1967]   _____, *Foundations of Probability Theory and Statistical Mechanics*, pp. 77–101
          in M. Bungé, ed., *Delaware Seminar in the Foundations of Physics*, Springer
          Verlag, New York, 1967.

[1978]   _____, *Where Do We Stand on Maximum Entropy?*, in R. D. Levine and M.
          Tribus, eds., "The Maximum Entropy Formalism," pp. 15–118, The MIT Press,
          Cambridge, Mass., 1978.

[1925]   J. Jeans, *The Dynamical Theory of Gases*, 4th edition, Dover Publications Inc.,
          New York, 1954.

[1934]   _____, *Activities of Life and the Second Law of Thermodynamics*, Nature **133**
          (1934), 174, 612, 986.

[1924]   J. E. Jones (Lennard-Jones), *On the Determination of Molecular Fields, I, II*,
          Proceedings of the Royal Society (London) **A106** (1924), 441–462, 463–477.

[1933]   P. Jordan, *Statistische Mechanik auf quantentheoretischer grundlage*, Friedr. Vie-
          weg & Sohn, Braunschweig, 1933.

[1988]   D. Jou, J. Casas-Vázquez and G. Lebon, *Extended irreversible thermodynamics*,
          Reports on Progress in Physics **51** (1988), 1107–1179.

[1862]   J. P. Joule and W. Thomson, *On the Thermal Effects of Fluids in Motion–Part
          IV*, Philosophical Transactions of the Royal Society London **152 pt II** (1862),
          579–589.

[1959]   M. Kaç, *Some Probabilistic Aspects of the Boltzmann Equation*, pp. 379–400 in
          Cohen and Thirring [**1973**].

[1967]   A. Katz, *Principles of Statistical Mechanics: The Information Theory Approach*,
          W. H. Freeman and Company, San Francisco, 1967.

[1976]   J. Kestin, *The Second Law of Thermodynamics*, Dowden Hutchinson and Ross,
          Inc., Stroudsburg Pennsylvania, 1976.

[1943]   A. I. Khinchin, *Mathematical Foundations of Statistical Mechanics*, translated by
          G. Gamow, Dover Publications, Inc., New York, 1949.

[1951]   _____, *Quantum Statistics*, 1951, translation edited by I. Shapiro, The Graylock
          Press, Albany, New York, 1960.

[1946]   J. Kirkwood, *The Statistical Mechanical Theory of Transport Processes*, Journal of Chemical Physics **14** (1946), 180–201; *Errata*, ibid. **14** (1946), 347; it is reprinted in Kirkwood [**1967**], pp. 1–30.

[1949]   ———, *The statistical mechanical theory of irreversible processes*, Nuovo Cimento Suppl. **VI** (1949), 233–239.

[1967]   ———, *Selected Topics in Statistical Mechanics*, edited by R. W. Zwanzig, Gordon and Breach, New York, 1967.

[1963]   C. Kittel, *Quantum Theory of Solids*, John Wiley and Sons, New York, 1963.

[1963]   M. J. Klein, *Planck, Entropy, and Quanta, 1901–1906*, pp. 83–108 in D. E. Gershenson and D. A. Greenberg, eds., *The Natural Philosopher*, Vol. 1, Blaisdell Publ. Co., New York, 1963.

[1970]   ———, *Maxwell, His Demon and the Second Law of Thermodynamics*, American Scientist **58** (1970), 84–97.

[1973]   ———, *The Development of Boltzmann's Statistical Ideas*, Acta Physica Austriaca **Suppl. X** (1973), 53–106; it is reprinted in Cohen and Thirring [**1973**], pp. 53–106.

[1931]   B. O. Koopman, *Hamiltonian Systems and Hilbert Space*, Proceedings of the National Academy of Sciences **17** (1931), 315–318.

[1949]   H. A. Kramers, *Presidential Address*, International Conference on Statistical Mechanics at Florence, 1949, Nuovo Cimento Suppl. **VI** (1949), 157–159.

[1856]   A. K. Krönig, *Grundzüge einer Theorie der Gase*, Annalen der Physik **99** (1856), 315–322.

[1950]   N. S. Krylov, *Works on the Foundations of Statistical Physics*, 1950, Princeton University Press, Princeton, New Jersey, 1979; this is a translation by A. H. Migdal, Ya. G. Sinai and Yu. L. Zeeman of "Raboty po obosnovaniiu statisticheskoi fiziki," Moscow, 1950.

[1967]   A. Kugler, *Exact Expressions for the Energy Current and Stress Tensor Operators in Many-Particle Systems*, Zeitschrift für Physik **198** (1967), 236–241.

[1960]   T. Kuhn, *Engineering Precedent for the Work of Sadi Carnot*, Archives Internationales D'Histoire des Sciences #**50–51** (1960), 251–255.

[1978]   ———, *Black-body Theory and the Quantum Discontinuity 1894–1912*, At the Clarendon Press, Oxford, 1978.

[1960]   R. Kurth, *Axiomatics of Classical Statistical Mechanics*, Pergamon Press, Oxford, 1960.

[1979]   R. G. Laha and V. K. Rohatgi, *Probability Theory*, John Wiley & Sons, New York, 1979.

[1961]   R. Landauer, *Irreversibility and Heat Generation in the Computing Process*, IBM Journal of Research and Development **5** (1961), 183–191.

[1999]   J. Lebowitz, *Statistical Mechanics: A Selective Review of Two Central Issues*, printed as pp. 581–600 in B. Bederson, ed., *More Things in Heaven and Earth: A Celebration of Physics at the Millennium*, American Physical Society, Springer-Verlag, New York, 1999.

[1957]   J. Lebowitz and P. G. Bergmann, *Irreversible Gibbsian Ensembles*, Annals of Physics **1** (1957), 1–23.

[1957]   J. Lebowitz and H. L. Frisch, *Model of Nonequilibrium Ensemble: Knudsen Gas*, Physical Review **107** (1957), 917–923.

[1990]   H. S. Leff and A. F. Rex, eds., *Maxwell's Demon: Entropy, Information, Computing*, Princeton University Press, Princeton, 1990.

[1963]   S. Lefschetz, *Differential Equations: Geometric Theory*, 2nd ed., 1963, Dover Publications Inc., New York, 1977.

[1998]   H. N. W. Lekkerkerker and Al. Stroobants, *Ordering Entropy*, Nature **393** (1998), 305–306.

[1930] G. N. Lewis, *The Symmetry of Time in Physics*, Science **71** (1930), 569–577.

[1924] G. N. Lewis and M. Randall, *Thermodynamics*, 2nd edition, revised by K. S. Pitzer and L. Brewer (1st edition: 1924), McGraw-Hill Book Company, Inc, New York, 1961.

[1983] G. Lindblad, *Non-Equilibrium Entropy and Irreversibility*, Mathematical Physics Studies, Vol. 5, D. Reidel Publishing Co., Dordrecht, 1983.

[1968] J. L. Lions and E. Magenes, *Problèmes aux limites non homogènes et applications*, vol. 1, Dunod, Paris, 1968.

[1876] J. Loschmidt, *Über den Zustand des Wärmegleichgewicht eines Systems von Korpern mit Rucksicht auf die Schwerkraft, I*, Sitzungsberichte der Kaiserlichen Akademie der Wissenshaft, Mathematisch-Naturwissenshaftliche Klasse **73, II** (1876), 128—142; *IV*, ibid. **II 76** (1877), 209–225.

[1893] E. Mach, *The Science of Mechanics: A Critical and Historical Account of Its Development*, translated by T. McCormack, Open Court Publishing Co., LaSalle, Illinois, 1960.

[1896] ———, *Principles of the theory of heat: historically and critically elucidated*, first edition 1896; translated from the 2nd edition 1900, D. Reidel, Boston, 1986.

[1968] G. W. Mackey, *Induced Representations of Groups and Quantum Mechanics*, W. A. Benjamin, New York, 1968.

[1962] B. Mandelbrot, *The Role of Sufficiency and Estimation in Thermodynamics*, The Annals of Mathematical Statistics **33** (1962), 1021–1038.

[1964] ———, *On the Derivation of Statistical Thermodynamics from purely Phenomenological Principles*, Journal of Mathematical Physics **5** (1964), 164–171.

[1967] H. Margenau and J. Stamper, *Nonadditivity of Intermolecular Forces*, pp. 129–160 in P-O. Lowdin, ed., "Advances in Quantum Chemistry," Academic Press, New York, 1967.

[1974] L. Markus and K. R. Meyer, *Generic Hamiltonian Systems are Neither Integrable nor Ergodic*, "American Mathematical Society Memoirs #144" (1974), American Mathematical Society, Providence RI.

[1941] O. Martin, *Eine Ableitung des thermochemischen Gleichgewichtes*, Mitteilungen aus den forschungsanstalten Konzernestelle der Gutehoffnungshütte actienverein für bergbau und hüttenbetrieb, Nürnberg **9** (1941), 77–84.

[1959] P. Martin and J. Schwinger, *Theory of Many Particle Systems. I*, Physical Review **115** (1959), 1342–1373.

[1976] R. H. Martin, *Nonlinear Operators and Differential Equations in Banach Spaces*, John Wiley & Sons, New York, 1976.

[1869] F. Massieu, *Sur les fonctions caractéristique des divers fluides*, Comptes Rendus hebdomadaires des Séances de l'Académie des Sciences **LXIX** (1869), 858–862, 1057–1061.

[1957] D. Massignon, *Méchanique Statistique des Fluides: Fluctuations et Proprietes Locales*, Dunod, Paris, 1957.

[1860] J. C. Maxwell, *Illustrations of the Dynamical Theory of Gases*, Philosophical Magazine [4] **19** (1860), 19–32; ibid. [4] **20**, 21–37; it is reprinted in Maxwell [**1890**], Vol. 1, pp. 377–409 and in Brush [**1965**], Vol. 1, pp. 148–171.

[1866] ———, *On the Dynamical Theory of Gases*, first published in 1866, Philosophical Magazine [4] **35** (1867), 129–145, 185–217; it is reprinted in Maxwell [**1890**], Vol. 2, pp. 26–78 and in Brush [**1965**], Vol. 2, pp. 23–87.

[1870] ———, *Address to the Mathematical and Physical Sections of the British Association*, 1870; it is printed in *British Association for the Advancement of Science 40th Meeting*, pp. 1–9, John Murry, London, 1871, and reprinted in Maxwell [**1890**], Vol. 2, pp. 215–229.

[1872] ———, *Theory of Heat*, 2nd ed., Longmans Green and Co, London, 1872.

[1873a] ———, *On the Final State of a System of Molecules in Motion Subject to Forces of any Kind*, Nature **8** (1873), 537–538; it is reprinted in Maxwell [**1890**], Vol. 2, pp. 351–354.

[1873b] ———, *Molecules*, Nature **8** (1873), 437–441; it is reprinted in Maxwell [**1890**], Vol. 2, pp. 361–378.

[1874] ———, *Van der Waals on the Continuity of the Gaseous and Liquid States*, Nature **10** (1874), 477–480; it is reprinted in Maxwell [**1890**], Vol. 2, pp. 409–413.

[1875] ———, *On the Dynamical Evidence of the Molecular Constitution of Bodies*, 1875, Nature **11** (1874–75), 357–359, 374–377; it is reprinted in Maxwell [**1890**], Vol. 2, pp. 418–438.

[1877] ———, *Matter and Motion*, Dover Publications, Inc., New York, 1952.

[1878a] ———, *Diffusion*, Encyclopedia Brittanica, 9th ed., **7** (1878), 214; it is reprinted in Maxwell [**1890**], Vol. 2, pp. 625–646.

[1878b] ———, *Tait's Thermodynamics*, Nature **17** (1878), 257–259, 278–280; reprinted in Maxwell [**1890**], Vol. 2, pp. 660–671.

[1879a] ———, *Appendix to 'On Stresses in Rarified Gases arising from Inequalities of Temperature'*, Philosophical Transactions of the Royal Society, part I (1879); it is reprinted in Maxwell [**1890**], Vol. 2, pp. 703–712.

[1879b] ———, *On Boltzmann's Theorem on the average distribution of energy in a system of material points*, Proceedings Cambridge Philosophical Society **12** (1879), 547–570; it is reprinted in Maxwell [**1890**], Vol. 2, pp. 713–741.

[1890] ———, *The Scientific Papers of James Clerk Maxwell, Vols. 1 and 2*, Dover Publications Inc., New York, 1952.

[1940] J. E. Mayer and M. G. Mayer, *Statistical Mechanics*, John Wiley & Sons, Inc., New York, 1940.

[2002] P. McEvoy, *The Theory of Interacting Systems, Vol. 1, Niels Bohr: Reflections on Subject and Object*, (RSO), MicroAnalytix, San Francisco, 2002.

[—] ———, *The Theory of Interacting Systems, Vol. 3, Equilibrium Theory*, (EIS) to be published, MicroAnalytix, San Francisco.

[—] ———, *The Theory of Interacting Systems, Vol. 4, Quantum Theory*, (QIS) to be published, MicroAnalytix, San Francisco.

[—] ———, *The Theory of Interacting Systems, Vol. 5, Quantum Thermodynamics*, (QTS) to be published, MicroAnalytix, San Francisco.

[1935] D. McKie and N. H. V. Heathcote, *The Discovery of Specific and Latent Heats*, Edward Arnold & Co., London, 1935.

[1972] J. Mehra, *The Golden Age of Theoretical Physics*, pp. 17–59 in Abdus Salam and Eugene P. Wigner, *Aspects of Quantum Theory*, Cambridge University Press, Cambridge, 1972.

[1977] E. Mendoza, ed., *Reflections on the Motive Power of Fire, by Sadi Carnot, and other Papers on the Second Law of Thermodynamics by É. Clapeyron and R. Clausius*, Peter Smith, Glouchester, Mass, 1977.

[1971] R. K. Miller, *Nonlinear Volterra Integral Equations*, W. A. Benjamin, New York, 1971.

[1969] M. Morse and S. S. Cairns, *Critical Point Theory in Global Analysis and Differential Topology*, Academic Press, New York, 1969.

[1953] P. M. Morse and H. Feshbach, *Methods of Theoretical Physics, parts 1 and 2*, McGraw-Hill Book Company, Inc., New York, 1953.

[1990] W. Muschik, *Aspects of Non-equilibrium Thermodynamics*, World Scientific Publishing Co., Singapore, 1990.

[1906a] W. Nernst, *Ueber die Berechnung chemischer Gleichgewichte aus thermischen Messungen*, Nach. von der kgl. Gesellschaft der Wissenschaften zu Göttingen, Math. Phys. Kl. **Heft 1** (1906), 1–40.

[1906b]   _____, *Über die Beziehungen zwischen Wärmeentwicklung und maximaler Arbeit bei kondensierten Systemen*, Sitzungsberichte der kgl. preuss. Akademie der Wissenschaften (Berlin) **V** (1906), 933–940.

[1926]   _____, *New Heat Theorem*, Methuen & Co., Ltd., London, 1926; it is a translation of "Grundlagen des neuen Wärmesatzes" (first edition, 1917) by Guy Barr of the second German edition (1924).

[1927]   J. von Neumann, *Thermodynamik quantenmechanischer Gesamtheiten*, Nachrichten der Gesellshaft der Wissenshaften zu Göttingen, Mathematisch-Physikalishe Klasse, **1927, Heft 1**, 273–291.

[1932a]   _____, *Proof of the Quasi-ergodic Hypothesis*, Proc. National Academy of Sciences **18** (1932), 70–82.

[1932b]   _____, *Physical Applications of the Ergodic Hypothesis*, Proc. National Academy of Sciences **18** (1932), 263–266.

[1932c]   _____, *Mathematical Foundations of Quantum Mechanics*, Princeton University Press, Princeton, NJ, 1955.

[1955]   W. Noll, *Der Herleitung der Grundgleichungen der Thermodynamik der Kontinua aus der Statistichen Mechanik*, Archive for Rational Mechanics and Analysis **4** (1955), 627–646.

[1924]   H. Nyquist, *Certain Factors Affecting Telegraph Speed*, Bell System Technical Journal **3** (1924), 324–346.

[1994]   D. Oliver, *The Shaggy Stead of Physics: Mathematical Beauty in the Physical World*, Springer-Verlag, New York, 1994.

[1931]   L. Onsager, *Reciprocal Relations in Irreversible Processes. I*, Physical Review **37** (1931), 405–426; II, Ibid, 2265–2279.

[1949]   _____, *The effects of shape on the interaction of colloidal particles*, Annals of the New York Academy of Sciences **51** (1949), 627–659.

[1949]   J. R. Partington, *An Advanced Treatise on Physical Chemistry*, Vols. 1–5, Longmans, Green and Co, London, 1949.

[1933]   W. Pauli, *Die Allgemeinen Prinzipien der Wellenmechanik*, pp. 83–272 in H. Geiger and J. Scheel, eds., Handbüch der Physik, Vol. 24, Part I, second edition, J. Springer, Berlin, 1933.

[1979]   O. Penrose, *Foundations of Statistical Mechanics*, Reports on Progress in Physics **42** (1979), 1937–2006.

[1989]   R. Penrose, *The Emperor's New Mind*, Oxford University Press, Oxford, England, 1989.

[1887]   M. Planck, *Ueber das Princip der Vermeherung der Entropie*, Annalen der Physik und Chemie **30** (1887), 562–582.

[1904]   _____, *Über die mechanische Bedeutung der Temperatur und der Entropie*, this appears as pp. 113–122 in *Boltzmann-Festschrift*, J. A. Barth, Leipzig, 1904. It is also reprinted as pp. 79–88 in Planck [**1958**], Vol. II.

[1913]   _____, *Vorträge über die kinetische Theorie der Materie und der Elektrizität*, Wolfskehl Commission lectures at Göttingen, B. G. Teubner, Leibzig, 1914.

[1922]   _____, *Treatise on Thermodynamics*, Dover Publications, Inc., New York, 1945; it is a translation by A. Ogg from the seventh German edition (first edition: 1897) of , *Vorlesungen über Thermodynamik*, J. A. Barth, Leipzig, 1922.

[1932]   _____, *Introduction to Theoretical Physics*, Vol. V, "Theory of Heat," translated by H. L. Brose, Macmillan and Co., Ltd., London, 1932.

[1958]   _____, *Physikalische Abhandlungen und Vorträge*, in 3 volumes, Friedr. Vieweg & Sohn, Braunschweig, 1958.

[1890]   H. Poincaré, *Sur les équations de la Dynamique et le Problème des trois corps*, Acta Mathematica **13** (1890), 1–270.

[1892]   _____, *Thermodynamique*, Georges Carré, Paris, 1892.

[1906]   _____, *Réflexions sur la théorie cinétique des gaz*, Journal de Physique **45** (1906), 369–403; it is reprinted in "Oeuvres de Henri Poincaré," Gauthier-Villars, Paris, 1954, Vol. IX, pp. 587–619.

[1981]   T. M. Porter, *A statistical survey of gases: Maxwell's social physics*, Historical Studies in the Physical Sciences **12:1** (1981), 77–116.

[1791]   P. Prévost, *Mémoire dur L'Equilibre du Feu*, J. de Physique **38** (1791), 314–322; it is translated by D. B. Brace and printed in "The Laws of Radiation and Absorption, Memoires by Prévost, Stewart, Kirchhoff, Kirchhoff and Bunsen," pp. 1–13, American Book Co., New York, 1901.

[1962]   I. Prigogine, *Non-equilibrium Statistical Mechanics*, John Wiley & Sons, New York, 1962.

[1973]   _____, *The Statistical Interpretation of Non-Equilibrium Entropy*, pp. 401–450 in Cohen and Thirring [**1973**].

[1980]   _____, *From Being to Becoming: Time and Complexity in the Physical Sciences*, W. H. Freeman and Company, San Francisco, 1980.

[1973]   I. Prigogine, C. George, F. Henin and L. Rosenfeld, *A Unified Formulation of Dynamics and Thermodynamics*, Chemica Scripta **4** (1973), 5–32.

[1939]   H. Primakoff and T. Holstein, *Many-Body Interaction in Atomic and Nuclear Systems*, Physical Review **55**, 1218–1234.

[1854]   W. J. M. Rankine, *On the Geometrical Representation of the Expansive Action of Heat, and the Theory of Thermo-dynamic Engines*, Philosophical Transactions of the Roy. Soc. (London) **144** (1854), 115–175.

[1892]   J. W. Strutt baron Rayleigh, *Remarks on Maxwell's Investigations Respecting Boltzmann's Theorem*, Philosophical Magazine [5] **33** (1892), 356–359; reprinted as pp. 554–557 in "Scientific Papers," Vol. III, Cambridge University Press, Cambridge, 1902.

[1953]   F. Riesz and B. Sz.-Nagy, *Functional Analysis*, translated by L. Boron, Frederick Ungar Publ. Co, New York, 1953.

[1951]   J. Rothstein, *Information, Measurement, and Quantum Mechanics*, Science **114** (1951), 171–175.

[1969]   D. Ruelle, *Statistical Mechanics: Rigorous Results*, W. A. Benjamin, New York, 1969.

[1954]   J. A. Schouten, *Ricci-Calculus: An Introduction to Tensor Analysis and its Geometrical Applications*, second edition, Springer–Verlag, Berlin, 1954.

[1943]   E. Schrödinger, *What is Life?*, Cambridge University Press, Cambridge, 1945.

[1959]   J. Serrin, *Mathematical Principles of Classical Fluid Mechanics*, pp. 125–263 in S. Flügge, ed., Handbüch der Physik, Vol. VIII/1, Fluid Dynamics I, Springer Verlag, Berlin, 1959.

[1949]   C. E. Shannon and W. Weaver, *The Mathematical Theory of Communication*, University of Illinois Press, Urbana, Illinois, 1962.

[1990a]   S. Sieniutycz and P. Salamon, *Diversity of Nonequilibrium Theories and Extremum Principles*, pp. 1–38 in Sieniutycz and Salamon [**1990b**].

[1990b]   S. Sieniutycz and P. Salamon, eds., *Advances in Thermodynamics, Volume 3, Nonequilibrium Theory and Extremum Principles*, Taylor & Francis, New York, 1990.

[1974]   B. Skagerstam, *On the Mathematical Definition of Entropy*, Zeitschrift für Naturforschung **29a** (1974), 1239–1243.

[1913]   M. von Smoluchowski, *Gültigkeitsgenzen des zweiten Haupsatzes der Wärmetheorie*, pp. 87–121 in Planck [**1913**].

[1965]   J. Stecki and H. S. Taylor, *On the Areas of Equivalence of the Bogoliubov Theory and the Prigogine Theory of Irreversible Processes in Classical Gases*, Reviews of Modern Physics **37** (1965), 762–773.

[1964]   S. Sternberg, *Lectures on Differential Geometry*, Prentice-Hall, Englewood Cliffs,
          New Jersey, 1964.

[1849]   G. G. Stokes, *On the Theories of the Internal Friction of Fluids in Motion, and
          of the Equilibrium and Motion of Elastic Solids*, Cambridge Philosophical Society
          8 (1849), 287–319.

[1925a]   L. Szilard, *Über die Ausdehnung der phänomenologischen Thermodynamik aft
          die Schwankungserscheinungen*, Zeitschrift für Physik **32** (1925), 753–788; the
          original article is reprinted as pp. 34–69 and an English translation appears as pp.
          70–102 in Szilard [**1972**].

[1925b]   _____, *Über die Entropieverminderung in einem thermodynamischen System bei
          Eingriffen intelligenter Wesen*, accepted in 1925 as his Habilitationschrift as Pri-
          vatdozent at the University of Berlin, Zeitschrift für Physik **53** (1929), 840–856;
          an English translation appeared in Behavioral Science **9** (1964), pp. 301–310; the
          original article is reprinted as pp. 103–119 and the English translation as pp.
          120–129 in Szilard [**1972**].

[1972]   _____, *The Collected Works of Leo Szilard*, edited by B. T. Feld and G. W.
          Szilard, M. I. T. Press, Cambridge, Mass, 1972.

[1885]   P. G. Tait, *Properties of Matter*, Adam and Charles Black, Edinburgh, 1885.

[1966]   G. R. Talbot and A. J. Pacey, *Some Early Kinetic Theories of Gases: Herapath
          and his Predecessors*, British J. History of Science **3** (1966), 133–149.

[1972]   C. J. Thompson, *Mathematical Statistical Mechanics*, Princeton Univ. Press,
          Princeton, N. J., 1972.

[1849]   J. Thomson, *Theoretical considerations on the effects of pressure in lowering the
          freezing point of water*, Transactions of the Royal Society of Edinburgh **16** (1849),
          575–580.

[1848]   W. Thomson (Lord Kelvin), *On an absolute thermometric scale founded on Car-
          not's theory of the motive power of heat, and calculated from Regnault's obser-
          vations*, Philosophical Magazine [3] **33** (1848), 313–317; it is reprinted as pp.
          100–106 in Thomson [**1882**], Vol. I, and as pp. 52–58 in Kestin [**1976**].

[1849]   _____, *An account of Carnot's theory of the motive power of heat; with numerical
          results deduced from Regnault's experiments with steam*, Transactions of the Royal
          Society of Edinburgh **16 pp. 575–580.** (1849); it is reprinted as pp. 156–164 in
          Thomson [**1882**], Vol. I.

[1851]   _____, *On the Dynamical Theory of Heat with Numerical Results Deduced from
          Mr. Joule's Equivalent of a Thermal Unit and M. Regnault's Observations on
          Steam*, Philosophical Magazine [4] **4 pp. 8–21, 105–117, 168–176.** (1851); it
          is reprinted in Thomson [**1882**], Vol. I, pp. 174–210, and in J. Kestin [**1976**], pp.
          106–132.

[1852]   _____, *On a Universal Tendency in Nature to the Dissipation of Energy*, Philo-
          sophical Magazine [4] **4 pp. 304–306.** (1852); it is reprinted in Thomson [**1882**],
          Vol. I, pp. 511–514, and in Kestin [**1976**], pp. 194–197.

[1854]   _____, *On the Dynamical Theory of Heat, Part V Thermo-electric Currents*,
          Philosophical Transactions of the Royal Society of Edinburgh **21** (1854), 123–171;
          it is reprinted in Thomson [**1882**], Vol. I, pp. 232–291.

[1873]   _____, *The Kinetic Theory of the Dissipation of Energy*, Proceedings of the
          Royal Society at Edinburgh **VIII** (1873), 325–334; it is reprinted in Thomson
          [**1911**], Vol. V, pp. 11–20, and in Brush [**1965**], Vol., 2, pp. 176–187.

[1882]   _____, *The Mathematical and Physical Papers*, Vol. I of 6 volumes, Cambridge,
          At the University Press, 1882.

[1891]   _____, *On Some Test Cases for the Maxwell-Boltzmann Doctrine Regarding Dis-
          tribution of Energy*, Nature **44** (1891), 355–358.

[1911] ———, *The Mathematical and Physical Papers*, Vol. V of 6 volumes, edited by J. Larmor, Cambridge, At the University Press, 1911.

[1854] W. Thomson and J. P. Joule, *On the Thermal Effects of Fluids in Motion*, Philosophical Transactions of The Royal Society of London **144** (1854), 321–364.

[1961] L. Tisza, *The Thermodynamics of Phase Equilibrium*, Annals of Physics **13** (1961), 1–92.

[1963] L. Tisza and P. M. Quay, *Statistical Mechanics of Equilibrium*, Annals of Physics **25** (1963), 164–179.

[1983] M. Toda, R. Kubo and N. Saito, *Statistical Physics I; Equilibrium Statistical Mechanics*, Springer-Verlag, Berlin, 1983.

[1938] R. C. Tolman, *The Principles of Statistical Mechanics*, Dover Publications, Inc., New York, 1979.

[1967] F. Treves, *Topological Vector Spaces, Distributions and Kernels*, Academic Press, New York, 1967.

[1960] C. Truesdell, *Ergodic Theory in Classical Statistical Mechanics*, pp. 22–56 in P. Caldirola, ed., "Ergodic Theories, Course XIV of the Enrico Fermi School of Physics," Academic Press, New York, 1961.

[1969] ———, *Rational Thermodynamics*, McGraw–Hill, New York, 1969.

[1980] ———, *The Tragicomical History of Thermodynamics 1822–1854*, Springer-Verlag, New York, 1980.

[1986] ———, *What Did Gibbs and Carathéodory Leave Us About Thermodynamics*, pp. 101–124 in J. Serrin, ed., "New Perspectives in Thermodynamics," Berlin, Springer-Verlag, 1986.

[1977] C. Truesdell and S. Bharatha, *The concepts and logic of heat engines, rigorously constructed upon the foundation laid by S. Carnot and F. Reech*, Springer-Verlag, New York, 1977.

[1980] C. Truesdell and R. G. Muncaster, *Fundamentals of Maxwell's Kinetic Theory of a Simple Monatomic Gas Treated as a Branch of Rational Mechanics*, Academic Press, New York, 1980.

[1960] C. Truesdell and R. Toupin, *The Classical Field Theories*, pp. 226–790 in S. Flügge, ed., "Handbüch der Physik, Vol. III/1, Principles of Classical Mechanics and Field Theory," Springer Verlag, Berlin, 1960.

[1963] G. E. Uhlenbeck and G. W. Ford, *Lectures in Statistical Mechanics*, American Mathematical Society, Providence, Rhode Island, 1963.

[1927] H. D. Ursell, *The Evaluation of Gibbs Phase Integral for Imperfect Gases*, Proc. Cambridge Philosophical Society **23** (1927), 685–697.

[1881] J. L. van der Waals, *Die Continuität des Gasförmigen und Flüssigen Zustandes*, 1881; it is a translation by F. Roth of "Over de continuiteit van den gas vloeistof-toestand," 1873, with additional material added, J. A. Barth, Leipzig.

[1846] J. J. Waterston, *I. On the Physics of Media that are Composed of Free and Perfectly Elastic Molecules in a State of Motion*, (1846), Transactions of the Royal Society of London **183A** (1892), 1–80.

[1876] H. W. Watson, *A Treatise on the Kinetic Theory of Gases*, The Clarendon Press, Oxford, 1876.

[1894] ———, *Boltzmann's Minimum Theorem*, Nature **51** (1894), 105.

[1978] A. Wehrl, *General Properties of Entropy*, Reviews of Modern Physics **50** (1978), 221–260.

[1963] E. P. Wigner and M. M. Yanase, *Information Contents of Distributions*, Proceedings of the National Academy of Sciences (USA) **49** (1963), 910–918.

[1973] W. W. Wood, *A Review of Computer Studies in the Kinetic Theory of Fluids*, pp. 451–490 in Cohen and Thirring [**1973**].

[1966]   W. Yourgrau, A. van der Merwe and G. Raw, *Treatise on Irreversible and Statis-
         tical Thermophysics*, Dover Publications, Inc., New York, 1982.
[1896a] E. Zermelo, *Ueber einen Satz der Dynamik und die mechanische Wärmetheorie*,
         Annalen der Physik **57** (1896), 485–494; it is translated as "On a Theorem in
         Dynamics and the Mechanical Theory of Heat" in Brush [**1965**], Vol. 2, pp. 208–
         217.
[1896b] ———, *Ueber mechanische Erklärungen irreversibler Vorgänge. Eine Antwort
         auf Hrn. Boltzmann's 'Entgegnung'*, Annalen der Physik **59** (1896), 793–801; it is
         translated as "On the Mechanical Explanation of Irreversible Processes" in Brush
         [**1965**], Vol. 2, pp. 229–237.
[1971]   D. N. Zubarev, *Nonequilibrium Statistical Thermodynamics*, Consultants Bureau,
         New York, 1974; translated by P. J. Shepard and edited by P. Gray and P. G.
         Shepard from "Neravnovesnîa statisticheskaîa termodinamika," Moscow, 1971.
[1984]   W. H. Żurek, *Maxwell's Demon, Szilard's Engine and Quantum Measurements*,
         pp. 151–161 in G. T. Moore and M. O. Scully, eds., *Frontiers of Nonequilibrium
         Statistical Mechanics*, Plenum Press, New York, 1986; it is reprinted as pp. 249–
         259 in Leff and Rex [**1990**].

Index

Index

www.ingramcontent.com/pod-product-compliance
Lightning Source LLC
Chambersburg PA
CBHW021023210326
41598CB00016B/892